Series: Environmental and Agricultural Nanotechnology
Series Editor: Kamel A. Abd-Elsalam

Volume

Smart Nanomaterials for Environmental Remediation

Books in the Series

- Biopolymer Nanoparticles: A Novel Approach for Eco-conscious Environmental Applications
  Ayat F. Hashim, Rabab M.A. El-Sherif, Farah K. Ahmed and Sabu Thomas (eds.), 2026
- Green Nanotechnology for Stress-Resilient Crops and Sustainable Agroecosystems
  Jen-Tsung Chen (ed.), 2026
- Soil Nanotechnology
  Akhtar Hameed and Muhammad Zahid (eds.), 2026
- Smart Nanomaterials for Environmental Remediation
  Megha Barot and Azamal Husen (eds.), 2027
- NanoFarming: The Future of Sustainable Agriculture
  Sumera Javad and Shagufta Naz (eds.), 2027
- Algal Nanotechnology: Synthesis and Applications
  Kamel A. Abd-Elsalam and Mohamed A. Mosa (eds.), 2027

Series: Environmental and Agricultural Nanotechnology

Series Editor: Kamel A. Abd-Elsalam

Volume

Smart Nanomaterials for Environmental Remediation

Editors

Megha Barot

Research and Development Cell
Parul University, Vadodara, Gujarat, India
and
Department of Environmental Science
Parul Institute of Applied Sciences
Parul University, Vadodara, Gujarat, India

Azamal Husen

Wolaita Sodo University, Wolaita, Ethiopia
and
Department of Biotechnology
Graphic Era (Deemed to be University)
Dehradun, Uttarakhand, India

CRC Press
Taylor & Francis Group
Boca Raton London New York

CRC Press is an imprint of the
Taylor & Francis Group, an **informa** business
A SCIENCE PUBLISHERS BOOK

First edition published 2027
by CRC Press
2385 NW Executive Center Drive, Suite 320, Boca Raton FL 33431

and by CRC Press
4 Park Square, Milton Park, Abingdon, Oxon, OX14 4RN

CRC Press is an imprint of Taylor & Francis Group, LLC

Library of Congress Cataloging-in-Publication Data (applied for)

ISBN: 978-1-032-87766-2 (hbk)
ISBN: 978-1-032-90607-2 (pbk)
ISBN: 978-1-003-55880-4 (ebk)

DOI: 10.1201/9781003558804

Typeset in Times New Roman
by Prime Publishing Services

Preface to the Series

Welcome to the "Environmental and Agricultural Nanotechnology" book series, a specialty forum revealing the breakthrough promise of nanotechnology in two of the most vital areas of human endeavor: environment and agriculture. As Kamel A. Abd-Elsalam, Series Editor from the Plant Pathology Research Institute, Agricultural Research Center (ARC), I'm pleased to present this series of publications by leading figures in the field.

Nanotechnology is a promise technology, with creative solutions to the majority of the globe's most contentious problems. It presents innovative alternatives in environmental applications to prevent pollution, protect resources, and effectively address environmental effects. Consider new nanomaterials with previously unavailable pollutant absorption properties from air, water, and soil, or new sensors with specific environmental detection and cleansing. These are no longer concepts on paper anymore; they are becoming things and hard facts in a rush to redefine our way of environmental stewardship.

Parallel to the concerns, nanotechnology is also going to revolutionize agriculture, ushering in new avenues of food production and sustainable land management. This includes higher yields from crops through improved delivery of nutrients, optimization of nutrient use efficiency, enhanced fertility of the soil, and development of climate change-resilient agro-systems. The combination of the nanotechnological advances is the door to a future where farming is not only productive but considerably more sustainable.

This book series is to be a comprehensive treatise, taking the reader through the fundamentals of and concepts in nanoscience, nanomaterials, and their unique properties in relation to energy, environmental and agronomic applications. We cover the specifics of the application of nanotechnology in controlling environmental pollution, e.g., its worth in contaminant decontamination and environmental monitoring. Moreover, we detail how nanotechnology can be applied in sustainable agricultural use and resource management, such as the application of targeted agrochemical delivery as well as crop adaptation to climate change.

Through these volumes, we hope to impart a deep appreciation of recent advances and envision future breakthroughs. This volume is a must-read for researchers, academics, policymakers, and entrepreneurs interested in leveraging nanotechnology for a sustainable and green future. In our opinion, by bridging the gap between frontier nanoscience and practical use in the environment and agriculture, we can contribute importantly to global efforts for sustainable development and higher living standards.

Preface

In today's world, both terrestrial and marine ecosystems face unprecedented threats due to climate change and the widespread use of non-(bio)degradable, non-renewable, carbon-intensive, and environmentally disruptive materials. Urban populations are particularly affected, with a 27% higher risk of asthma directly linked to escalating air pollution. Over recent decades, governments across the globe have recognized the severity of these challenges, prompting policy reforms targeting industrialization and urban development. From a technological perspective, material science has been pivotal in steering global efforts toward sustainable and innovative solutions addressing climate change, agricultural productivity, polluted ecosystem restoration, food safety, and food security. This interdisciplinary field has continually attracted researchers from physics, chemistry, and life sciences. Simultaneously, nanotechnology has shown significant promise in tackling environmental adversities.

By integrating material science with nanotechnology, numerous high-performance products have emerged. These include carbon nanotube-based solar panels and black phosphorus nanosheets for sustainable energy; nano-formulated fertilizers and pesticides for enhanced agricultural yields; carbon-derived, photocatalytic, and biodegradable nanomaterials for environmental cleanup; and nano-biosensors and smart packaging systems to ensure food safety and security. The present book is dedicated to both foundational and practical aspects of environmental remediation, with a special focus on the role of smart nanomaterials. Nanomaterials and smart nanomaterials, while often related, differ in functionality and responsiveness. Nanomaterials are defined by structural elements smaller than 100 nanometers and are distinguished by their unique properties, such as superior strength, enhanced electrical conductivity, and heightened chemical reactivity. Examples include quantum dots, carbon nanotubes, and metal-based nanoparticles. Smart nanomaterials, in contrast, are engineered to respond dynamically to physical stimuli like pH, temperature, humidity, and magnetic fields. These materials can alter their characteristics in real time, lending them to applications such as targeted drug delivery, self-repairing surfaces, and adaptive sensing systems. Therefore, while all smart nanomaterials fall under the category of nanomaterials, not all nanomaterials qualify as "smart." Smart nanomaterials are revolutionizing the field of environmental remediation through their responsiveness and efficiency. These materials are capable of detecting, capturing, and neutralizing contaminants in air, water, and soil. For instance, photocatalytic nanomaterials like titanium dioxide can degrade organic pollutants under sunlight, while carbon-based variants such as graphene oxide enhance filtration processes. Nano-structured adsorbents are being deployed to remove heavy metals, and biodegradable nanomaterials naturally decompose after neutralizing toxins. Their adaptability to environmental conditions makes them superior alternatives to conventional remediation methods, paving the way for more sustainable pollution control practices. With ongoing innovations, smart nanomaterials hold the potential to dramatically improve global environmental health.

A twofold strategy is essential for effective environmental remediation, involving both the advancement of scientific innovation and the promotion of public awareness. Technological progress, including the development of smart nanomaterials and improvements in waste management systems, provides powerful tools for mitigating pollution. At the same time, enhancing environmental literacy among the public is crucial to encourage responsible behavior and shape effective policy reforms. The publication of timely and accessible literature, such as books and journal articles, plays a significant role in supporting both scientific progress and public engagement. Numerous such publications can be found globally, issued by a wide range of academic and scientific publishers. Disseminating updated fundamental and applied knowledge is critical to supporting technological progress and educating stakeholders. Readers of this book must grasp key areas such as material science, smart nanomaterials, technological breakthroughs, waste management practices, ecosystem restoration techniques, traditional remediation strategies, and policy frameworks promoting sustainability. By integrating academic research with actionable insights, such works contribute to a shared vision of sustainability, empowering individuals to make informed choices for pollution control and ecological recovery.

In this context, the current book, namely *Smart Nanomaterials for Environmental Remediation*, aims to deliver both fundamental principles and advanced applications of smart nanomaterials in environmental restoration. It covers synthesis techniques, material properties, and real-world commercial uses of bioactive nanomaterials while also delving into legal, health, and safety aspects. Moreover, it highlights recent developments in nanotechnology aimed at alleviating pollution across air, water, and soil systems, offering a unique perspective compared to existing literature in the field. We are grateful to all contributors for readily accepting our invitation, sharing their knowledge in specialized areas of research, and adjusting to suggestions for improving the quality of their contributions.

Megha Barot
Azamal Husen

Contents

CHAPTER-1

Introduction to Smart Nanomaterials

Noopur Khare,[1,2] *Pragati Khare,*[3] *Tanvi Jain,*[2] *Sachidanand Singh,*[4,*] *Megha Barot*[5,6] and *Azamal Husen*[7,8]

1. Introduction

Nanotechnology has revolutionized various scientific and industrial domains by enabling the manipulation of materials at the nanoscale, typically in the range of 1 to 100 nanometers. Among the numerous advancements in this field, smart nanomaterials have emerged as a transformative class of materials with remarkable adaptive and responsive properties. These materials possess the ability to change their physical, chemical, or biological characteristics in response to external stimuli such as temperature, pH, light, magnetic fields, or electrical signals (Khan et al. 2022). This unique feature makes them highly valuable for applications in medicine, environmental science, electronics, and energy storage. The concept of smart materials originates from the need for dynamic and multifunctional materials that can actively interact with their environment. Traditional materials, though functional, often lack the adaptability required for advanced applications (Singh et al. 2024a). Smart nanomaterials overcome these limitations by offering superior control over their behavior at the molecular and atomic levels. These materials can self-assemble, self-heal, and exhibit shape-memory properties, making them ideal for cutting-edge technologies such as drug delivery systems, biosensors, and self-cleaning surfaces (Behera and Behera 2022).

The development of smart nanomaterials is a result of the rapid evolution of nanoscience, materials engineering, and computational modeling. Initially, nanotechnology was focused on passive nanomaterials like nanoparticles and nanocomposites, which offered enhanced mechanical and electrical properties (Singh and Melnik 2022). However, with advancements in nanofabrication techniques, scientists began exploring active or intelligent nanomaterials that could interact dynamically with their environment. This shift marked a significant milestone in materials science,

[1] Bhai Gurdas Institute of Engineering and Technology, Sangrur, Punjab, India. https://orcid.org/0000-0002-2594-920X.
[2] Faculty of Biotechnology, Institute of Biosciences and Technology, Shri Ramswaroop Memorial University, Barabanki, Uttar Pradesh, India. https://orcid.org/0000-0002-9203-8561; https://orcid.org/0000-0002-1817-9577.
[3] BPS Educational Institution of Pharmacy, Etah, U.P., India.
[4] Department of Biotechnology, School of Energy and Technology, Pandit Deendayal Energy University, Gandhinagar, Gujarat, India.
[5] Research and Development Cell, Parul University, Vadodara, Gujarat, India. https://orcid.org/0000-0002-9680-0760.
[6] Department of Environmental Science, Parul Institute of Applied Sciences, Parul University, Gujarat, India.
[7] Wolaita Sodo University, P.O. Box:138, Wolaita, Ethiopia. https://orcid.org/0000-0002-9120-5540.
[8] Department of Biotechnology, Graphic Era (Deemed to be University), Dehradun 248002, Uttarakhand, India.
* Corresponding author: drsachinbioinfo@gmail.com; https://orcid.org/0000-0002-2567-4779

DOI: 10.1201/9781003558804-1

leading to the integration of nanomaterials with biotechnology, electronics, and environmental technologies (Chaikittisilp et al. 2022). Smart nanomaterials have gained substantial attention due to their multifunctionality and versatility. For instance, in the biomedical field, they are widely used in targeted drug delivery, where nanoparticles transport therapeutic agents directly to diseased cells, minimizing side effects. In environmental applications, they aid in water purification by selectively removing pollutants (Konstantopoulos et al. 2022). Additionally, in the electronics industry, these materials are being employed in flexible and wearable devices, enabling the next generation of smart electronics. Smart nanomaterials can be classified based on their composition and functionality. Stimuli-responsive nanomaterials are a class of advanced materials that react to specific environmental triggers. For instance, pH-sensitive hydrogels are used in targeted drug delivery systems, while thermoresponsive polymers find applications in biomedical implants due to their ability to change properties with temperature variations. Another innovative category is self-healing nanomaterials, which can autonomously repair damage, making them highly valuable in protective coatings, structural components, and biomedical applications. Additionally, shape-memory nanomaterials have the unique ability to return to their original shape after deformation, making them ideal for use in biomedical implants and smart textiles. Their unique optical, electrical, magnetic, and mechanical properties enable their integration into diverse applications, ensuring their relevance in modern scientific advancements (Singh et al. 2024b).

2. Classification of Smart Nanomaterials

Smart nanomaterials are an advanced class of nanomaterials that exhibit unique properties, enabling them to respond dynamically to external stimuli such as temperature, pH, light, magnetic fields, or electric signals. These materials can be classified based on functionality and composition, each category offering specialized applications in fields such as biomedicine, electronics, environmental sciences, and energy storage (Mancardi et al. 2023).

2.1 Classification Based on Functionality

2.1.1 Stimuli-Responsive Nanomaterials

Stimuli-responsive nanomaterials change their physical or chemical properties in response to external environmental conditions. These can be further categorized based on the type of stimulus they respond to as seen in (Figure 1). pH-responsive nanomaterials are engineered to undergo structural or chemical modifications when exposed to varying pH levels. This unique feature makes them particularly effective in drug delivery systems, where they can release therapeutic agents in response to the pH of specific environments, such as the acidic microenvironment of cancerous tissues or different regions of the gastrointestinal tract (Zhang et al. 2021). Thermoresponsive nanomaterials, such as poly(N-isopropylacrylamide) (PNIPAM), exhibit changes in solubility or phase at defined temperatures. These materials are widely utilized in controlled drug release systems and smart coatings, where temperature acts as a trigger for activation or release (Papadakis et al. 2023). Photo-responsive nanomaterials react to specific wavelengths of light—ultraviolet (UV), visible, or infrared (IR)—by altering their structure or functional behavior. These materials find extensive applications in smart windows, light-controlled drug delivery systems, and photodynamic therapy, particularly in the treatment of cancer (Yan et al. 2022). Magneto-responsive nanomaterials, such as iron oxide (Fe_3O_4) nanoparticles, respond to external magnetic fields. Their ability to be directed and controlled remotely makes them ideal for targeted drug delivery, biomedical imaging, and magnetic hyperthermia therapy used in cancer treatment (Soares et al. 2021). Electro-responsive nanomaterials, on the other hand, exhibit changes in their electrical properties when subjected to an applied voltage. These materials are increasingly being integrated into electronic sensors, smart textiles, and bioelectronic devices, enhancing the functionality and adaptability of modern biomedical and wearable technologies (Alayande et al. 2020).

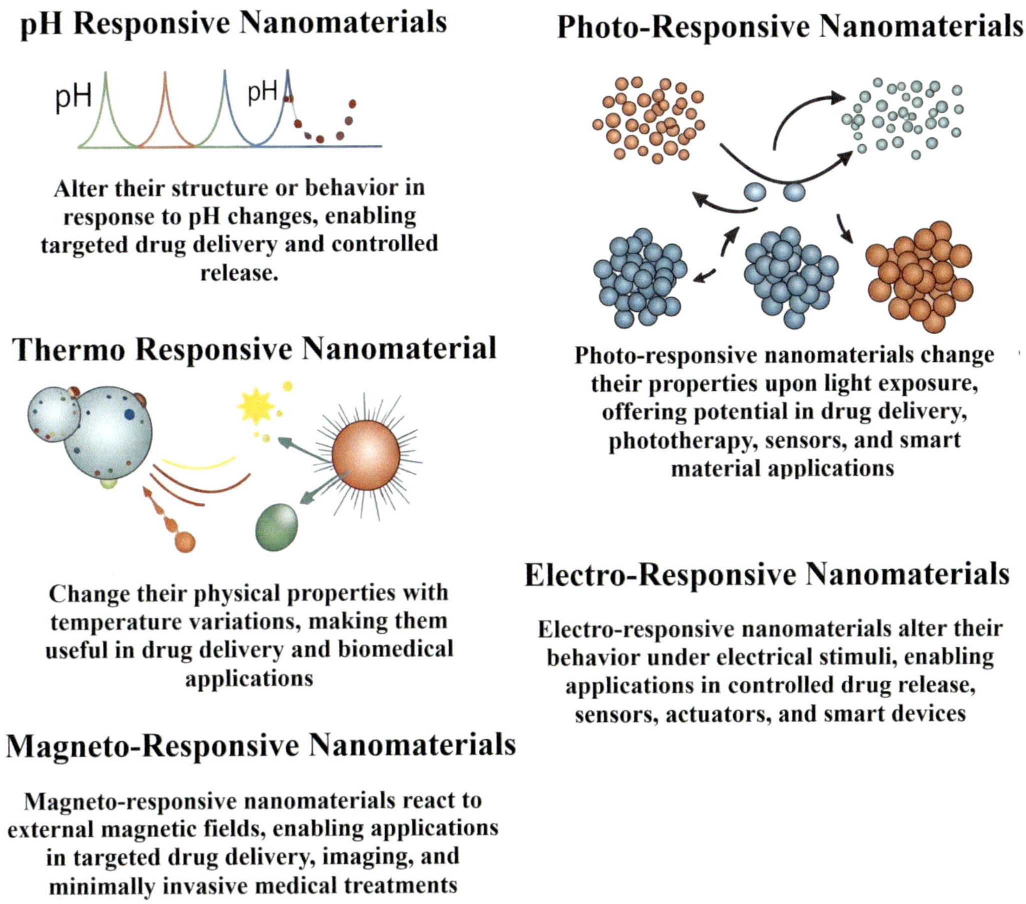

Figure 1: Stimuli-Responsive Nanomaterials.

2.1.2 Self-Healing Nanomaterials

Self-healing nanomaterials have the ability to repair structural damage autonomously. They incorporate nanocapsules or microcapsules filled with healing agents that release upon damage. These materials find applications in automotive coatings, biomedical implants, and aerospace engineering, where longevity and durability are critical (Zhai et al. 2020).

2.1.3 Shape-Memory Nanomaterials

Shape-memory nanomaterials have the ability to return to their original shape after deformation when exposed to an external stimulus such as heat or light. They are commonly used in biomedical implants, actuators, robotics, and self-deploying structures in aerospace engineering (Xia et al. 2021).

2.1.4 Multifunctional Nanomaterials

These nanomaterials integrate multiple functionalities, such as conductivity, magnetism, biocompatibility, and responsiveness, into a single material. They are widely used in biosensors, wearable electronics, energy storage devices, and flexible displays (Kyriakides et al. 2021).

2.2 Classification Based on Composition

2.2.1 Carbon-Based Nanomaterials

Carbon-based nanomaterials exhibit exceptional mechanical, electrical, and thermal properties due to their unique structures. Graphene is a two-dimensional nanomaterial composed of a single layer of carbon atoms arranged in a honeycomb lattice. It is renowned for its exceptional mechanical strength, outstanding electrical conductivity, high thermal conductivity, and remarkable flexibility. These unique properties make graphene highly suitable for a wide range of advanced applications, including biosensors, where its conductivity enhances signal detection; flexible electronics, due to its bendable yet robust nature; and energy storage devices such as supercapacitors and batteries, where it contributes to faster charge-discharge cycles and improved capacity (Wei and Yang 2019). Carbon nanotubes (CNTs), on the other hand, are cylindrical nanostructures made of rolled graphene sheets. They exhibit extraordinary mechanical strength and electrical conductivity, which make them highly valuable in the development of lightweight composite materials for the aerospace and automotive industries. CNTs are also utilized in nanoelectronics and targeted drug delivery systems, where their high surface area and ability to penetrate biological membranes enable efficient therapeutic delivery (Rashko et al. 2022). Fullerenes are another class of carbon-based nanomaterials, consisting of spherical or ellipsoidal carbon molecules, such as the well-known C_{60} buckyball. Their unique electronic and structural properties render them useful in a variety of applications, including solar cells, where they function as efficient electron acceptors; drug carriers, due to their capacity to encapsulate therapeutic agents; and antioxidant therapies, where they scavenge reactive oxygen species and mitigate oxidative stress (Dinadayalane and Leszczynski 2012). Collectively, these carbon-based nanomaterials have revolutionized numerous fields, offering promising solutions in nanomedicine, electronics, and renewable energy as seen in (Figure 2).

2.2.2 Metal-Based Nanomaterials

Metal-based nanomaterials include nanoparticles composed of metals or metal oxides, offering excellent antimicrobial, catalytic, and optical properties. Gold nanoparticles (AuNPs) are widely studied and utilized due to their unique optical, electronic, and biocompatible properties. They are especially valuable in biomedical applications such as targeted drug delivery, where their surface can be easily functionalized with therapeutic agents, allowing for precise delivery to diseased tissues. In cancer treatment, AuNPs play a significant role in photothermal therapy, where they absorb light and convert it into heat to selectively destroy cancer cells. Additionally, their excellent conductivity and surface plasmon resonance properties make them highly effective in biosensors for disease diagnosis and monitoring (Amina and Guo 2020). Silver nanoparticles (AgNPs) are renowned for their potent antimicrobial activity, which stems from their ability to disrupt microbial cell membranes and interfere with metabolic processes. As a result, they are extensively used in wound dressings to prevent infections, in water purification systems to eliminate pathogens, and in antimicrobial coatings for medical devices and consumer products (Zahoor et al. 2021). Titanium dioxide (TiO_2) nanoparticles possess strong photocatalytic activity and high stability, making them ideal for environmental and cosmetic applications. In the presence of UV light, TiO_2 nanoparticles can degrade organic pollutants, which is leveraged in self-cleaning surfaces and air-purifying coatings. Moreover, their ability to absorb and scatter ultraviolet radiation makes them effective as UV-blocking agents in sunscreens, offering protection against harmful solar radiation without penetrating the skin (Anaya-Esparza et al. 2020). These metal-based nanoparticles continue to advance the fields of medicine, environmental science, and consumer safety through their versatile functionalities.

2.2.3 Polymer-Based Nanomaterials

Polymer-based nanomaterials are an important class of nanostructured materials composed of synthetic or natural polymers, engineered for applications requiring biocompatibility, flexibility,

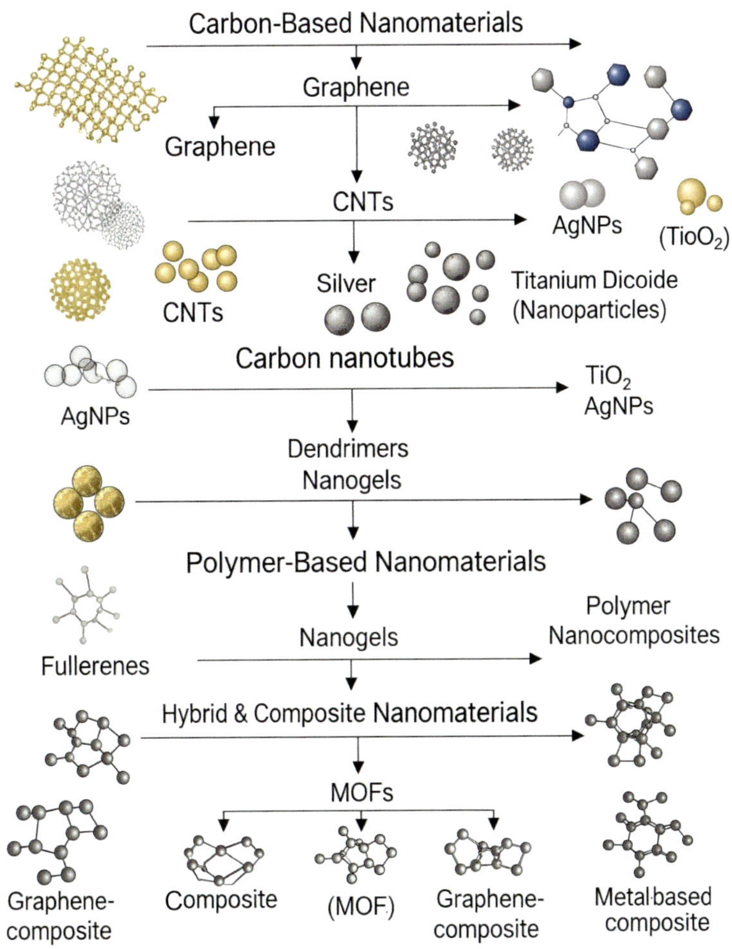

Figure 2: Classification Based on Composition.

and tunable physical or chemical properties. Among the most prominent types are dendrimers, which are highly branched, tree-like polymeric nanostructures. Their well-defined architecture and multiple surface functional groups make them ideal candidates for drug delivery, gene therapy, and catalytic applications, as they can encapsulate or bind various therapeutic agents and release them in a controlled manner (Palmerston Mendes et al. 2017). Another significant type is nanogels—hydrophilic polymer networks capable of absorbing large amounts of water while maintaining their structural integrity. Their high water content and soft consistency mimic natural tissues, making them particularly suitable for drug delivery, where they can respond to environmental stimuli, and in tissue engineering for promoting cell growth and repair (Dehchani et al. 2024). Polymer nanocomposites represent another innovative category, where polymers are reinforced with various nanoparticles such as clay, carbon nanotubes, or metal oxides. This combination enhances mechanical strength, thermal stability, and barrier properties, enabling their widespread use in fields like food packaging, automotive parts, and biomedical devices such as prosthetics and implants (Hassan et al. 2021). Together, these polymer-based nanomaterials offer multifunctional solutions across industries, contributing significantly to advances in healthcare, engineering, and material science.

2.2.4 Hybrid and Composite Nanomaterials

Hybrid and composite nanomaterials combine two or more types of nanomaterials to achieve superior mechanical, electrical, and thermal properties. These materials enhance the functionality of individual components and are widely used in energy storage, electronics, coatings, and biomedicine. Metal-Organic Frameworks (MOFs) are a class of crystalline, porous materials formed by the coordination of metal ions or clusters with organic ligands. Their highly tunable structures, large surface areas, and adjustable pore sizes make them exceptionally versatile for a range of applications. MOFs are widely used in gas storage, particularly for hydrogen and carbon dioxide, due to their ability to adsorb gases at high capacities. Additionally, they serve as efficient catalysts in chemical reactions because of their active metal centers and accessible pores. In the biomedical field, MOFs have emerged as promising drug delivery systems, as their porous structures can encapsulate therapeutic molecules and release them in a controlled manner, improving treatment efficacy and reducing side effects (Lawson et al. 2021). Another innovative material class is graphene-based composites, which integrate graphene with polymers, metals, or ceramics to significantly enhance their physical and chemical properties. Graphene contributes exceptional electrical conductivity, mechanical strength, and flexibility to these composites, making them ideal for use in high-performance applications such as battery electrodes, supercapacitors, and flexible electronic devices. By improving the structural integrity and conductive pathways of traditional materials, graphene-based composites enable the development of next-generation energy storage systems and wearable electronics with superior durability and efficiency.

The classification of smart nanomaterials highlights their diverse functionalities and compositions, making them highly versatile for applications in medicine, electronics, environmental sustainability, and industrial engineering. Stimuli-responsive, self-healing, and shape-memory nanomaterials are transforming the way materials interact with their surroundings, while carbon-based, metal-based, polymer-based, and hybrid nanomaterials continue to advance technological innovations (Khare et al. 2024). As research progresses, these materials are expected to play a crucial role in nanomedicine, smart coatings, sustainable energy solutions, and next-generation electronic devices. However, further studies on toxicity, large-scale production, and regulatory frameworks are essential to ensure their safe and widespread adoption in various industries (Díez-Pascual and Rahdar 2022).

3. Properties of Smart Nanomaterials

Smart nanomaterials possess unique and highly tunable properties that make them suitable for a wide range of applications in biomedicine, environmental science, energy storage, and electronics. These materials exhibit stimuli-responsive behavior, high surface area-to-volume ratio, and exceptional mechanical, electrical, thermal, and optical properties. Understanding their characteristics is essential for designing functional, adaptive, and sustainable nanomaterials as seen in Figure 3 (Su and Song 2021).

3.1 Structural Properties

3.1.1 High Surface Area-to-Volume Ratio

One of the most fundamental and advantageous properties of smart nanomaterials is their exceptionally high surface area-to-volume ratio. This characteristic greatly enhances their reactivity, adsorption capacity, and catalytic efficiency, making them highly effective in a wide range of applications. In catalysis, for example, many chemical reactions occur at the surface of a material. The increased surface area provided by nanomaterials allows more active sites for these reactions, thereby significantly improving reaction rates and overall catalytic performance. In drug delivery, this high surface area enables nanocarriers to adsorb or bind a larger number of drug molecules, facilitating efficient transport and targeted release within the body, which enhances therapeutic outcomes while minimizing side effects. Similarly, in the field of sensing, the heightened surface area leads to increased interaction between the sensor and analytes, resulting in greater sensitivity and faster detection of biological or chemical signals. This makes smart nanomaterials highly

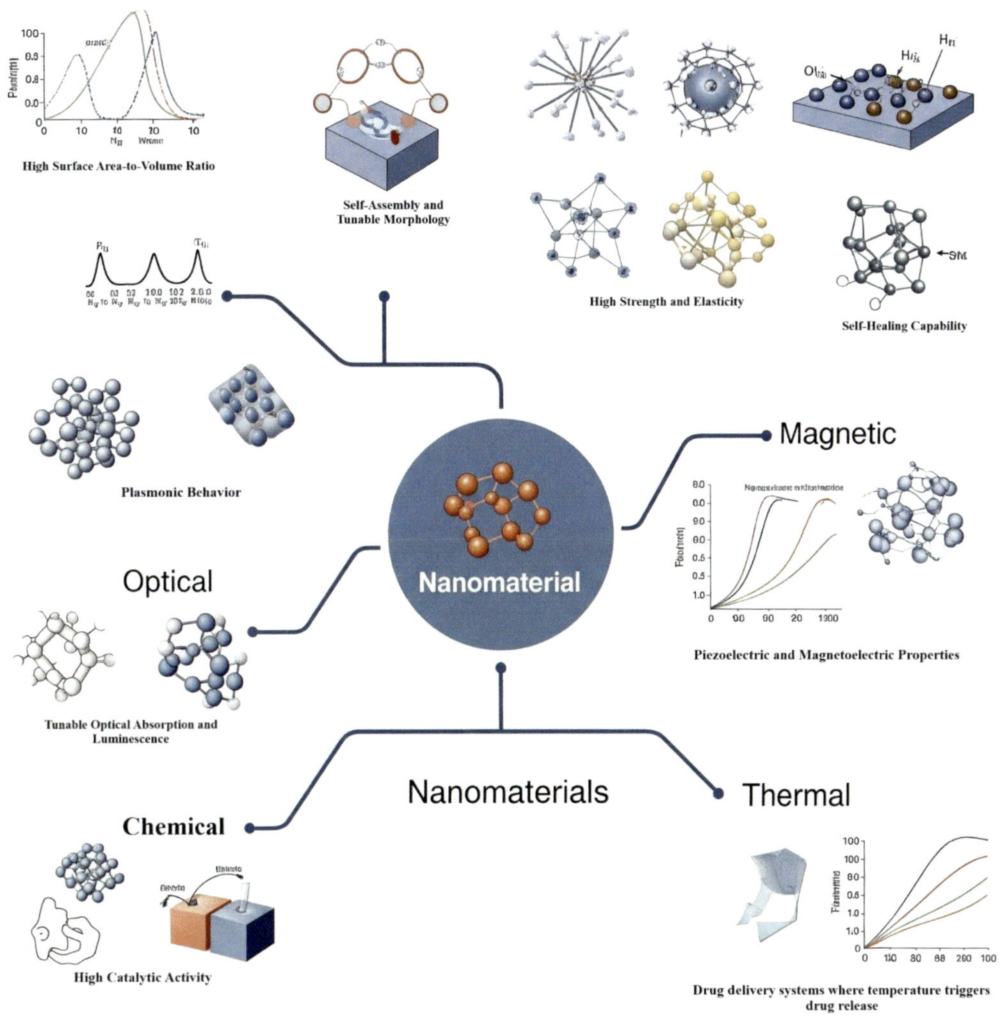

Figure 3: Properties of Nanomaterials.

suitable for developing next-generation biosensors and diagnostic tools with improved precision and responsiveness (Sable et al. 2024).

3.1.2 Self-Assembly and Tunable Morphology

Smart nanomaterials possess the remarkable ability to self-assemble into various nanostructures, such as nanotubes, nanospheres, nanorods, and nanowires, depending on their chemical composition, molecular interactions, and synthesis methods. This self-assembly behavior allows precise control over their morphology, which in turn influences their physical and functional properties. The tunable morphology of these materials plays a critical role in enhancing their performance across diverse applications. For instance, in coatings and composite materials, specific nanostructures like nanorods or nanowires contribute to improved mechanical strength, flexibility, and durability. In electronic and energy storage devices, well-ordered structures facilitate efficient charge transport and conductivity, making them ideal for use in batteries, supercapacitors, and photovoltaic systems. In biomedical applications, morphology is crucial for controlling drug release kinetics; nanospheres or nanocapsules can be designed to release therapeutic agents in a sustained or stimuli-responsive manner, thereby increasing treatment precision and minimizing side effects. The ability of smart nanomaterials to adopt specific shapes tailored to application needs makes them highly versatile and effective in advancing modern technology and medicine (Li et al. 2021).

3.1.3 Shape-Memory Effect

Certain smart nanomaterials exhibit a remarkable property known as the shape-memory effect, which allows them to return to their original shape after deformation when exposed to specific external stimuli such as heat, light, or an electric field. This reversible transformation is made possible by their unique molecular or crystalline structure, which can "memorize" and revert to a predefined shape. This property has opened new avenues in multiple high-tech fields. In biomedical applications, shape-memory nanomaterials are used in implants that can adapt to physiological conditions within the body—such as temperature or pH—enhancing patient comfort and functionality, particularly in stents, sutures, and orthopedic devices. In the field of soft robotics and actuators, these materials enable adaptive and flexible movements, mimicking natural muscle functions and allowing for the development of responsive, lightweight robotic systems. In aerospace engineering, shape-memory materials are utilized in components that require dynamic shape control, such as deployable structures or aerodynamic surfaces that adjust during flight to optimize performance and fuel efficiency. The ability of these nanomaterials to undergo controlled shape changes makes them highly valuable in applications demanding precision, adaptability, and smart responsiveness (Dayyoub et al. 2022).

3.2 Mechanical Properties

3.2.1 High Strength and Elasticity

Smart nanomaterials, particularly carbon-based ones such as graphene and carbon nanotubes (CNTs), are renowned for their exceptional tensile strength, elasticity, and lightweight nature. These properties stem from their unique atomic structures, which provide remarkable mechanical performance even at the nanoscale. When incorporated into structural composites, these materials significantly enhance durability and resistance to mechanical stress, making them ideal for use in aerospace, automotive, and construction industries. In the realm of wearable electronics, the high flexibility and strength of graphene and CNTs contribute to the development of stretchable, bendable devices that maintain performance under repeated movement and deformation. Additionally, their impact-resistant nature makes them suitable for use in protective coatings and films, where they help absorb shock and prevent damage from physical impacts. The integration of these advanced carbon-based nanomaterials into various systems offers a powerful combination of strength, flexibility, and resilience, pushing the boundaries of innovation in multiple sectors (Islam et al. 2022).

3.2.2 Self-Healing Capability

Self-healing smart nanomaterials are designed with embedded microcapsules or specialized polymeric networks that can release healing agents automatically when mechanical damage occurs. This innovative feature allows the material to repair itself without external intervention, significantly extending its functional lifespan. In the automotive industry, self-healing coatings are employed to autonomously repair surface scratches, preserving the aesthetic and protective qualities of vehicle exteriors without the need for costly maintenance. In the biomedical field, these materials contribute to the long-term durability of implants and prosthetics by healing microcracks or wear-induced damage within the body, thereby reducing the risk of implant failure and the need for surgical replacements. Additionally, in the electronics sector, self-healing materials are used in circuits and flexible devices to restore electrical conductivity after damage, ensuring uninterrupted performance and enhancing device reliability. The ability of these materials to autonomously detect and repair damage makes them highly valuable in creating more resilient, low-maintenance systems across various advanced technological applications (Shojaei and Khasraghi 2021).

3.3 Thermal Properties

3.3.1 High Thermal Conductivity

Nanomaterials such as graphene, carbon nanotubes (CNTs), and metal nanowires are known for their exceptional thermal conductivity, which makes them highly effective in managing heat in various technological applications. Their ability to rapidly transfer heat stems from their unique atomic structures and high surface area, enabling efficient phonon transport. In microelectronics, these nanomaterials are used to dissipate heat generated by densely packed components, preventing overheating and improving device performance and longevity. In energy storage systems, such as batteries and supercapacitors, effective thermal management using these materials helps maintain stable operation, enhances safety, and extends the lifespan of the devices. Additionally, in industrial applications, nanomaterials serve as advanced coolants and heat transfer agents, improving the efficiency of thermal processes and reducing energy consumption. The integration of thermally conductive nanomaterials into modern technologies represents a critical advancement in maintaining performance, safety, and durability under high thermal loads (Chen and Chen 2020).

3.3.2 Thermoresponsive Behavior

Certain polymers and metal oxides possess the unique ability to undergo reversible phase transitions—such as changes in state or crystallinity—when exposed to specific temperatures. This thermal responsiveness makes them highly valuable in a range of advanced applications. In drug delivery systems, these materials can be engineered to release therapeutic agents in response to body heat or externally applied temperature changes, allowing for controlled and targeted drug release. In the textile industry, smart fabrics incorporating such materials can dynamically regulate body temperature by altering their structure in response to environmental conditions, enhancing wearer comfort and energy efficiency. Additionally, these phase-change materials (PCMs) are widely used in thermal energy storage systems, where they absorb, store, and release heat during phase transitions. This enables efficient temperature control in buildings, electronics, and renewable energy systems. The versatility and tunability of these thermally responsive materials make them ideal for applications that require adaptive behavior and energy efficiency (Abdah et al. 2020).

3.4 Electrical and Electronic Properties

3.4.1 High Electrical Conductivity

Many smart nanomaterials, including graphene and silver nanowires, are distinguished by their exceptional electrical conductivity, which is critical for the advancement of next-generation electronic and biomedical technologies. Their superior electron mobility and conductivity, even at the nanoscale, make them ideal for use in flexible and stretchable electronics, such as wearable sensors and foldable displays, where traditional rigid conductors fall short. In biomedical applications, these materials are incorporated into biosensors and implantable medical devices, allowing for accurate real-time monitoring of physiological signals with high sensitivity and biocompatibility. Furthermore, their excellent conductive and reflective properties make them effective in electromagnetic interference (EMI) shielding, protecting sensitive electronic components from external electromagnetic radiation and ensuring device reliability. The combination of flexibility, biocompatibility, and electrical performance enables these smart nanomaterials to play a pivotal role in shaping the future of smart, responsive, and high-performance technologies (Tan et al. 2020).

3.4.2 Piezoelectric and Magnetoelectric Properties

Piezoelectric smart nanomaterials have the unique ability to generate an electric charge when exposed to mechanical stress or deformation. This property makes them highly valuable in energy-harvesting technologies, particularly in the development of wearable devices that convert body movements into usable electrical energy, promoting self-powered electronics. In biomedical and industrial settings, these materials are employed in highly sensitive sensors that can detect pressure, vibration, or physiological signals, enabling real-time monitoring and diagnostics. Additionally, their tactile responsiveness is leveraged in robotic systems, where they serve as key components in artificial skin and touch-sensitive interfaces, enhancing the ability of robots to interact with their environment.

Magnetoelectric nanomaterials, on the other hand, exhibit a coupled response to magnetic and electric fields, allowing them to convert magnetic signals into electrical ones and vice versa. This dual functionality enables their application in targeted drug delivery systems, where magnetic fields guide drug-loaded nanoparticles to specific sites, and in magnetic hyperthermia therapy for cancer treatment, where localized heating destroys tumor cells. They also play a critical role in data storage and memory devices by enabling efficient information encoding and retrieval. Furthermore, these materials are used in smart actuators and soft robotics, where they enable controlled movement and adaptive responses to external magnetic stimuli. The multifunctionality of piezoelectric and magnetoelectric nanomaterials makes them integral to the advancement of intelligent, responsive, and energy-efficient systems (Ebrahimi et al. 2020).

3.5 Optical Properties

3.5.1 Tunable Optical Absorption and Luminescence

Smart nanomaterials possess the remarkable ability to absorb, reflect, or emit light at specific wavelengths, a property that is highly dependent on their size, shape, and surface structure. This tunable optical behavior is fundamental to numerous advanced applications. In the field of medicine, these materials are utilized in light-sensitive drug delivery systems such as photodynamic therapy, where they generate reactive oxygen species upon light exposure to selectively destroy cancer cells without harming surrounding healthy tissue. In renewable energy, smart nanomaterials enhance the efficiency of solar cells by optimizing light absorption across a broad spectrum, thereby improving energy conversion rates. Additionally, their ability to emit light under specific conditions is leveraged in fluorescent biosensors used for medical diagnostics. These biosensors enable sensitive and specific detection of biological markers, aiding in early disease diagnosis and monitoring. The optical versatility of smart nanomaterials continues to drive innovation in healthcare, energy, and environmental technologies (Kang et al. 2020).

3.5.2 Plasmonic Behavior

Certain metallic nanoparticles, particularly those composed of gold and silver, exhibit a phenomenon known as surface plasmon resonance (SPR). This occurs when conduction electrons on the nanoparticle surface oscillate in resonance with incident light, resulting in a strong enhancement of the local electromagnetic field. This unique optical property significantly boosts the interaction of these nanoparticles with light, making them highly valuable in various biomedical and technological applications. In biosensing, SPR enables the development of ultra-sensitive sensors capable of detecting minute concentrations of biomarkers, which is essential for early disease diagnosis and monitoring. In cancer treatment, gold and silver nanoparticles are employed in photothermal therapy, where SPR-induced light absorption leads to localized heating that selectively destroys cancer cells with minimal damage to surrounding tissues. Furthermore, SPR enhances contrast in advanced imaging techniques such as optical coherence tomography and photoacoustic imaging, allowing for more precise and detailed visualization of biological structures. The SPR effect thus plays a pivotal role in expanding the capabilities of nanomedicine and diagnostic technologies (Philip and Kumar 2022).

3.6 Chemical Properties

3.6.1 High Catalytic Activity

Due to their high surface area and reactivity, smart nanomaterials function as efficient catalysts in water purification and pollutant degradation, fuel cells and hydrogen production, and industrial chemical reactions (Namdeti et al. 2025).

3.6.2 Selective Permeability and Adsorption

Nanoporous materials, such as metal-organic frameworks (MOFs) and zeolites, are characterized by their high surface area and precisely tunable pore sizes, which enable selective adsorption of molecules. This property makes them highly effective for various advanced applications. In environmental technology, they are widely used for gas storage and separation, particularly in capturing carbon dioxide (CO_2) from industrial emissions, contributing to climate change mitigation. In biomedicine, their porous structure allows for controlled drug delivery and release, ensuring targeted and sustained therapeutic effects. Additionally, nanoporous materials play a crucial role in water purification by selectively adsorbing and filtering out toxic substances, heavy metals, and organic pollutants. Their multifunctionality and adaptability make them indispensable in sustainable and precision-driven technologies (Singh et al. 2020).

3.7 Biological and Biocompatibility Properties

3.7.1 Biodegradability and Non-Toxicity

Smart nanomaterials designed for biomedical applications must be biodegradable and non-toxic to prevent adverse effects. Examples include chitosan-based nanoparticles for wound healing, liposome-encapsulated drugs for targeted therapy, and biodegradable polymeric scaffolds for tissue engineering (Campora and Ghersi 2022).

3.7.2 Antimicrobial Activity

Certain metal nanoparticles, notably silver and copper nanoparticles, are renowned for their potent antimicrobial properties, which make them highly effective in combating a wide range of bacteria, viruses, and fungi. These nanoparticles are extensively used in medical coatings to prevent infections on surgical instruments, implants, and hospital surfaces, thereby reducing healthcare-associated infections. In the food industry, they are incorporated into packaging materials to inhibit microbial growth, significantly enhancing the shelf life and safety of perishable products. Additionally, silver and copper nanoparticles play a vital role in water purification systems, where their antimicrobial action helps eliminate harmful pathogens, ensuring safe and clean drinking water. The broad-spectrum antimicrobial efficacy and versatility of these metal nanoparticles make them invaluable in healthcare, food safety, and environmental applications (Crisan et al. 2021).

3.8 Environmental Stability and Sustainability

3.8.1 Photocatalytic Activity

Photocatalytic nanomaterials, including titanium dioxide (TiO_2) and zinc oxide (ZnO), harness light energy to initiate chemical reactions that break down harmful pollutants and organic contaminants. When exposed to ultraviolet or visible light, these materials generate reactive oxygen species capable of degrading pollutants, making them highly effective for environmental remediation. They are widely used in air and water purification systems to remove toxins, bacteria, and volatile organic compounds, ensuring cleaner and safer environments. Additionally, their photocatalytic properties are applied in self-cleaning surfaces, such as antimicrobial coatings for hospital walls, which reduce the spread of infections by continuously decomposing organic matter. Furthermore, these nanomaterials contribute to solar energy harvesting technologies, enhancing green energy solutions by converting light into usable power with high efficiency (Puri and Gupta 2023).

3.8.2 Eco-Friendly and Renewable Resource Utilization

Developing bio-based smart nanomaterials promotes sustainability by minimizing environmental impact through the use of renewable and biodegradable resources. Cellulose nanofibers, derived from plant biomass, are widely used in creating biodegradable packaging materials that reduce plastic waste and pollution. Protein-based nanoparticles, sourced from natural proteins, offer biocompatible and efficient platforms for targeted drug delivery, enhancing therapeutic outcomes while minimizing side effects. Additionally, algae-derived nanomaterials provide eco-friendly options for energy storage devices, such as supercapacitors and batteries, due to their abundant availability and renewable nature. These bio-based nanomaterials combine functionality with environmental responsibility, supporting the advancement of green nanotechnology. Smart nanomaterials exhibit a diverse range of structural, mechanical, thermal, electrical, optical, chemical, biological, and environmental properties that make them highly valuable in modern scientific and industrial applications. Their stimuli-responsive nature, high surface area, and multifunctionality provide enhanced performance, adaptability, and sustainability across multiple domains, including medicine, electronics, energy, and environmental science (Patani et al. 2023). As research progresses, smart nanomaterials will continue to drive innovation, offering next-generation solutions for complex challenges in healthcare, energy, and sustainable development (Shah et al. 2022).

4. Synthesis and Fabrication Techniques

Smart nanomaterials exhibit stimuli-responsive behavior, enhanced mechanical properties, and unique electrical, optical, and biological functionalities that make them highly valuable in applications such as biomedicine, energy storage, environmental science, and electronics. To harness these properties effectively, precise synthesis and fabrication techniques are essential. Nanomaterials can be synthesized using top-down and bottom-up approaches, each offering distinct advantages depending on the desired size, shape, composition, and functionality of the nanostructures (Puri and Gupta 2023) as seen in Table 1.

Table 1: Comparison of Top-Down vs. Bottom-Up Approaches.

Approach	Advantages	Disadvantages	References
Top-Down	Precise control over dimensions, suitable for large production.	High energy consumption, limitations in feature size, possible defects.	Puri and Gupta (2023)
Bottom-Up	High purity, atomic-level control, cost-effective.	Time-consuming, difficult to scale up.	Shah et al. (2022)
Hybrid	Combines advantages of both approaches.	Complex processing steps, requires advanced techniques.	Namdeti et al. (2025)

4.1 Classification of Synthesis Techniques

Synthesis methods for smart nanomaterials are broadly categorized into three main approaches. Top-down approaches involve breaking down bulk materials into nanoscale structures using techniques like milling, lithography, and etching, offering precise size control but sometimes causing defects. Bottom-up approaches build nanomaterials by assembling atoms or molecules through chemical vapor deposition, self-assembly, or sol-gel processes, enabling better control over composition and structure at the atomic level. Hybrid approaches combine both top-down and bottom-up techniques, leveraging the advantages of each to achieve superior control over nanomaterial size, shape, and functionality, optimizing their properties for specific applications.

4.2 Top-Down Approaches

In top-down synthesis, bulk materials are physically or chemically reduced to nanoscale dimensions using mechanical, thermal, or chemical methods (Puri and Gupta 2023).

4.2.1 Mechanical Milling (Ball Milling)

High-energy ball milling is a top-down synthesis method that uses mechanical grinding to break bulk materials into nanoparticles. It is commonly employed for producing metallic and ceramic nanomaterials due to its scalability and cost-effectiveness. This technique enables the production of nanomaterials with controlled particle size, making it suitable for large-scale applications. However, it has certain limitations, such as the potential introduction of impurities from the milling media and equipment. Additionally, it offers limited control over the shape and uniformity of the nanoparticles, which can affect their performance in specific applications (Ramezani and Neitzert 2012).

4.2.2 Lithography (Photolithography, Electron-Beam Lithography, Nanoimprint Lithography)

Photolithography, electron-beam lithography (EBL), and nanoimprint lithography (NIL) are top-down nanofabrication techniques used to create precise nanostructures. Photolithography employs UV light and masks to etch patterns on substrates, while EBL uses focused electron beams for higher-resolution nanoscale patterning. NIL involves pressing a mold onto a material to imprint nanostructures. These methods are widely applied in semiconductor device fabrication and biosensor development due to their precision. However, they are often costly and complex, with resolution limits imposed by the wavelength of light in photolithography or beam precision in EBL, impacting scalability and throughput (Ramezani and Neitzert 2012).

4.2.3 Etching Techniques (Wet and Dry Etching)

Wet etching and dry etching are common top-down techniques used to selectively remove materials at the nanoscale. Wet etching involves chemical solutions that dissolve specific materials, offering simplicity and cost-effectiveness. Dry etching, on the other hand, uses plasma or reactive gases to precisely etch materials, enabling finer control and anisotropic patterns. These methods are extensively applied in fabricating microelectronics and microelectromechanical systems (MEMS) devices, where precise material removal is critical. However, both etching techniques may cause surface damage or roughness, potentially affecting device performance and reliability, thus requiring careful optimization of process parameters (Shah et al. 2022).

4.3 Bottom-Up Approaches

In bottom-up synthesis, nanomaterials are built atom by atom or molecule by molecule using chemical and physical self-assembly techniques.

4.3.1 Chemical Vapor Deposition (CVD) and Atomic Layer Deposition (ALD)

Chemical Vapor Deposition (CVD) and Atomic Layer Deposition (ALD) are widely used bottom-up synthesis techniques for fabricating nanomaterials. CVD involves gaseous precursors that chemically react or decompose on a substrate surface, forming nanomaterials such as carbon nanotubes (CNTs), graphene, and thin-film coatings. ALD deposits materials one atomic layer at a time through alternating chemical reactions, ensuring precise thickness control. Both methods offer high purity, excellent uniformity, and scalability suitable for industrial production. However, they require high temperatures and carefully controlled environments, which can increase complexity and operational costs (Shah et al. 2022).

4.3.2 Sol-Gel Method

The sol-gel process involves the hydrolysis and condensation of metal alkoxides to form a gel-like network, which is subsequently dried and calcined to produce nanomaterials. This versatile method is widely used to synthesize ceramic nanoparticles, metal oxides, and silica-based nanomaterials. The sol-gel technique offers several advantages, including a simple and low-temperature process that allows precise control over the composition and homogeneity of the final product. However, it can be time-consuming and may lead to particle agglomeration, which can affect the uniformity and performance of the synthesized nanomaterials (Shah et al. 2022).

4.3.3 Hydrothermal and Solvothermal Synthesis

Hydrothermal synthesis involves using high-pressure and high-temperature aqueous or organic solvents to grow nanostructures in a controlled environment. This method is commonly applied to synthesize metal oxides, carbon nanomaterials, and quantum dots with excellent purity and crystallinity. Its ability to produce highly crystalline nanomaterials makes it suitable for large-scale production in various industrial applications. However, hydrothermal synthesis requires specialized autoclave equipment to maintain the necessary pressure and temperature conditions, which can increase operational complexity and cost (Ebrahimi et al. 2020).

4.3.4 Green Synthesis (Biogenic Synthesis)

Green synthesis of nanomaterials utilizes plant extracts, bacteria, fungi, or biomolecules to produce nanomaterials through environmentally friendly processes. This method is widely applied to create biodegradable and biocompatible nanomaterials for drug delivery systems, as well as antimicrobial silver nanoparticles. The primary advantages of green synthesis include its non-toxic nature, sustainability, and cost-effectiveness, making it an attractive alternative to conventional chemical methods. However, challenges remain in optimizing these biological processes for large-scale production, as variations in biological sources can affect consistency and yield (Ebrahimi et al. 2020).

4.4 Hybrid and Advanced Synthesis Techniques

4.4.1 Electrospinning

Electrospinning uses electrostatic forces to draw polymer solutions into fine nanofibers, producing materials with high surface area and porosity. This technique is widely applied in creating tissue engineering scaffolds, wound dressings, and filtration membranes due to its ability to mimic the extracellular matrix and provide effective barriers. Electrospinning is cost-effective and scalable, making it suitable for industrial production. However, its main limitation is that it is restricted to producing fiber-like structures, which may not be suitable for applications requiring other nanomaterial shapes or morphologies (Namdeti et al. 2025).

4.4.2 Self-Assembly and Directed Assembly

Self-assembly is a bottom-up approach where nanoparticles spontaneously organize into highly ordered structures under specific environmental conditions such as temperature, pH, or solvent polarity. This method is extensively used in applications like DNA nanotechnology, targeted drug carriers, and nanophotonics, where precise control over the arrangement and functionality of nanomaterials is crucial. The key advantage of self-assembly is its ability to achieve highly precise and complex structures at the nanoscale. However, it requires strictly controlled conditions to ensure consistent and reproducible results, which can limit its practicality in large-scale manufacturing (Namdeti et al. 2025).

4.4.3 3D Printing of Nanomaterials

Nano-ink printing is an additive manufacturing technique that uses inks containing functional nanoparticles to fabricate nanostructured materials with high precision. This method enables the production of custom-designed biomedical implants, flexible electronics, and sensitive sensors, allowing complex geometries and tailored properties. One of its major advantages is the high design flexibility, making it suitable for personalized and intricate applications. However, the technique faces limitations due to the restricted range of compatible materials available for nano-inks, which can constrain the diversity of achievable nanomaterial properties and applications (Namdeti et al. 2025).

5. Emerging Trends and Future Prospects

The synthesis and fabrication of smart nanomaterials involve diverse physical, chemical, and biological methods tailored to achieve precise control over size, shape, composition, and functionalities. Top-down techniques ensure scalability and mechanical strength, while bottom-up methods provide high purity and atomic-level precision. Hybrid and emerging approaches such as 3D printing and electrospinning are paving the way for next-generation applications in healthcare, electronics, energy, and sustainable materials. As nanotechnology advances, automation, AI-driven design, and eco-friendly synthesis will drive the future development of smart nanomaterials, ensuring their widespread adoption across scientific and industrial sectors (Chen and Chen 2020).

6. Applications of Smart Nanomaterials

Smart nanomaterials have revolutionized various industries due to their stimuli-responsive properties, superior mechanical strength, enhanced electrical conductivity, and tunable optical characteristics as seen in (Table 2). Their unique functionalities make them indispensable in fields such as biomedicine, energy storage, environmental science, electronics, and food technology. Below is an in-depth discussion of their applications across different domains (Husen and Siddiqi 2023) as seen in (Figure 4).

6.1 Biomedical and Healthcare Applications

Targeted Drug Delivery: Smart nanomaterials such as liposomes, dendrimers, polymeric nanoparticles, and metallic nanoparticles are widely used in targeted drug delivery systems. These nanoparticles encapsulate drugs and release them at specific sites in response to external stimuli like pH, temperature, or magnetic fields. Example: pH-sensitive nanoparticles release cancer drugs in acidic tumor environments, minimizing side effects, and reducing toxicity while enhancing therapeutic efficacy (Hassan et al. 2020).

Regenerative Medicine and Tissue Engineering: Nanomaterials such as nanofibers, carbon nanotubes (CNTs), and hydroxyapatite nanoparticles play a significant role in tissue regeneration. These materials mimic the extracellular matrix, promoting cell adhesion, proliferation, and differentiation. Example: Graphene-based scaffolds support bone and neural tissue regeneration, enhancing the healing process in orthopedic and neural injuries (Hassan et al. 2020).

6.2 Energy and Environmental Applications

Energy Storage and Conversion: Nanomaterials enhance the performance of batteries, supercapacitors, and fuel cells by improving conductivity, charge storage capacity, and efficiency. Example: Graphene-based electrodes improve lithium-ion battery performance, increasing storage

Table 2: Smart Nanomaterials and Their Application.

Smart Nanomaterial	Properties/Features	Applications	References
Carbon Nanotubes (CNTs)	High strength, conductivity, flexibility	Drug delivery systems, biosensors	Jha et al. (2020)
Quantum Dots	Tunable fluorescence, high photostability	Bioimaging, cancer diagnostics	Mohammadi et al. (2022)
Gold Nanoparticles (AuNPs)	Biocompatible, modifiable surface, optical properties	Targeted drug delivery, photothermal therapy	Badir et al. (2025)
Silver Nanoparticles (AgNPs)	Antimicrobial, conductive	Wound dressings, coatings, water treatment	Siritongsuk et al. (2022)
Magnetic Nanoparticles (e.g. Fe3O4)	Magnetic responsiveness, modifiable surface	MRI contrast agents, targeted drug delivery	Rahman (2023)
Zinc Oxide Nanoparticles (ZnO NPs)	UV-blocking, antibacterial, semiconducting	Sunscreens, antimicrobial coatings	Badir et al. (2025)
Graphene Oxide (GO)	High surface area, good conductivity	Tissue engineering, biosensors	Jha et al. (2020)
Silica Nanoparticles	Biocompatibility, porous structure	Controlled drug release, imaging agents	Rahman (2023)
Stimuli-responsive Polymers	Respond to pH, temperature, light, etc.	Smart drug delivery systems	Jha et al. (2020)
Nanoclay	High aspect ratio, swelling behavior	Barrier materials, drug delivery	Mohammadi et al. (2022)
Shape Memory Nanomaterials	Return to original shape on heating	Smart textiles, biomedical stents	Badir et al. (2025)
Self-healing Nanomaterials	Autonomous repair mechanisms	Coatings, structural composites	Mohammadi et al. (2022)
Metal-Organic Frameworks (MOFs)	High porosity, tunable chemical functionality	Gas storage, drug encapsulation	Rahman (2023)
Nanozymes	Enzyme-mimicking nanomaterials	Biosensing, cancer therapy	Jha et al. (2020)
TiO_2 Nanoparticles	Photocatalytic, UV absorption	Water purification, self-cleaning surfaces	Badir et al. (2025)

capacity and lifespan. This enables the development of longer-lasting and faster-charging batteries for electric vehicles and portable electronics (Hassan et al. 2020).

Solar Energy Harvesting: Smart nanomaterials boost solar energy efficiency by enhancing light absorption and electron transport in solar cells. Example: Perovskite quantum dots improve photovoltaic cell efficiency, increasing solar energy conversion rates and leading to more sustainable energy solutions (Hassan et al. 2020).

Water Purification and Filtration: Nanomaterials like carbon nanotubes, titanium dioxide, and silver nanoparticles are used for water filtration and purification. Example: Nanoporous membranes remove heavy metals and pathogens from drinking water, providing access to clean water in remote and underdeveloped regions (Hassan et al. 2020).

Air Pollution Control: Catalytic nanomaterials like titanium dioxide (TiO_2) nanoparticles help in air purification by breaking down pollutants. TiO_2 coatings on buildings reduce airborne pollutants through photocatalytic reactions, improving air quality in urban environments (Shah et al. 2022).

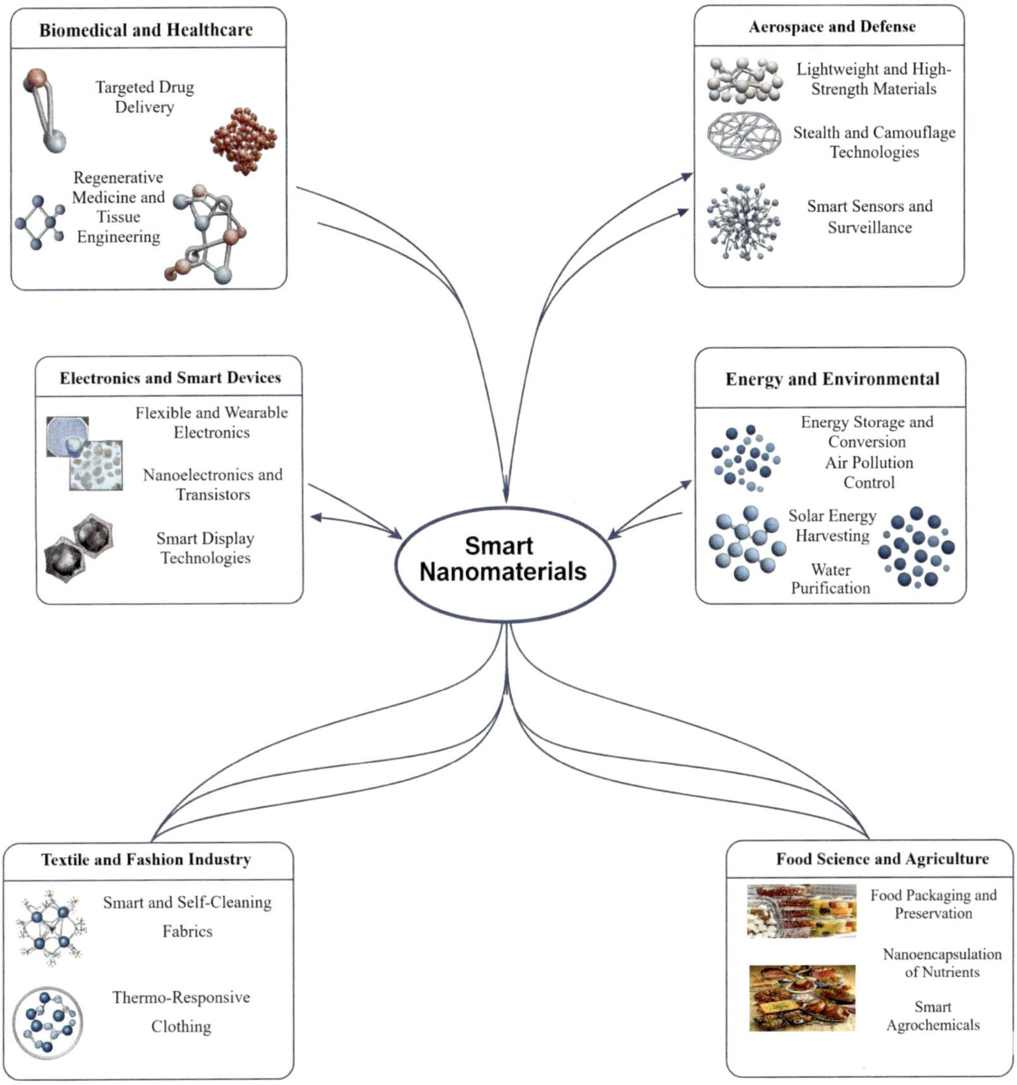

Figure 4: Applications of Smart Nanomaterials.

6.3 Electronics and Smart Devices

Flexible and Wearable Electronics: Nanomaterials enable the development of flexible, lightweight, and stretchable electronic devices. Example: Graphene-based sensors in smartwatches monitor health metrics like heart rate and glucose levels, enhancing user comfort and improving real-time health monitoring (Shah et al. 2022).

Nanoelectronics and Transistors: Semiconductor nanomaterials like quantum dots and carbon nanotubes are used in next-generation transistors and processors. Silicon nanowires improve the efficiency of transistors in modern microprocessors, reducing power consumption while increasing processing speeds in computers and smartphones (Shah et al. 2022).

Smart Display Technologies: Quantum dot nanomaterials enhance brightness, contrast, and energy efficiency in displays. Example: Quantum-dot LED (QLED) screens in high-resolution televisions provide superior image quality with lower power consumption (Triana et al. 2022).

6.4 Aerospace and Defense Applications

Lightweight and High-Strength Materials: Nanocomposites provide superior strength-to-weight ratios, enhancing aircraft and spacecraft performance. Carbon nanotube-reinforced polymers are used in aerospace components. Reduces fuel consumption and enhances durability (Triana et al. 2022).

Stealth and Camouflage Technologies: Nanomaterials with adaptive optical properties enable next-generation camouflage coatings. Metamaterial-based cloaking devices bend light around objects. Enhances stealth capabilities in military applications (Triana et al. 2022).

Smart Sensors and Surveillance: Nanomaterial-based sensors improve threat detection and surveillance systems. Nanowire-based infrared detectors enhance night vision capabilities. Improves security and military intelligence operations (Kumar and Kumari 2021).

6.5 Food Science and Agriculture

Food Packaging and Preservation: Smart nanomaterials enhance food shelf life, safety, and quality by preventing microbial contamination. Silver nanoparticle-infused packaging films prevent bacterial growth. Reduces food waste and enhances consumer safety (Kumar and Kumari 2021).

Nanoencapsulation of Nutrients: Nanomaterials encapsulate bioactive compounds, protecting them from degradation and improving absorption. Nano-encapsulated vitamins in functional foods enhance nutrient bioavailability. Improves nutritional intake and food fortification strategies (Kumar and Kumari 2021).

Smart Agrochemicals: Nanoparticle-based fertilizers and pesticides increase efficiency and reduce environmental impact. Slow-release nanopesticides minimize soil contamination. Enhances crop yield while reducing chemical runoff (Kumar and Kumari 2021).

6.6 Textile and Fashion Industry

Smart and Self-Cleaning Fabrics: Nanocoating makes textiles water-repellent, stain-resistant, and antimicrobial. Nano-TiO_2-coated fabrics break down stains under sunlight. Reduces the need for frequent washing and extends fabric lifespan (Singh et al. 2024b).

Thermo-Responsive Clothing: Smart nanomaterials adapt to temperature changes, providing comfort and energy savings. Phase-change nanomaterials in winter jackets retain body heat. Enhances thermal regulation and wearability (Prasad et al. 2023).

7. Future Prospects and Challenges

While smart nanomaterials offer numerous advantages, scalability, cost-effectiveness, and safety concerns remain challenges. Future research will focus on:

- Sustainable and green nanotechnology for reducing environmental impact.
- AI and nanomaterial simulations for precise material design.
- Biodegradable nanomaterials to prevent long-term ecological damage.

Smart nanomaterials are transforming industries, from biomedicine to electronics, aerospace, agriculture, and food science. Their unique stimuli-responsive properties and superior mechanical, electrical, and optical features enable groundbreaking advancements. However, continued research is needed to address scalability, environmental impact, and regulatory challenges to fully harness their potential for a sustainable future (Prasad et al. 2023).

8. Challenges and Limitations

Smart nanomaterials have gained significant attention for their potential applications in diverse fields such as medicine, energy, environment, and electronics. However, despite their promising properties and functionalities, several challenges and limitations hinder their widespread adoption and commercialization. These challenges range from safety concerns and regulatory issues to high production costs and scalability problems. This section provides an in-depth discussion of the major obstacles facing smart nanomaterials and the possible ways to overcome them (Thangudu 2020).

Safety and Toxicological Concerns: One of the most pressing challenges associated with smart nanomaterials is their potential toxicity to humans, animals, and the environment. Due to their nanoscale size and high surface reactivity, these materials can easily enter biological systems and interact with cells, tissues, and organs, leading to unforeseen health effects (Prasad et al. 2023).

Biological Toxicity: Nanoparticles can penetrate biological barriers such as the blood-brain barrier and accumulate in vital organs. Studies suggest that certain nanomaterials, such as carbon nanotubes and some metal oxides, may cause oxidative stress, DNA damage, and cytotoxicity. The long-term effects of exposure to nanomaterials remain unclear, necessitating further research into their biodegradability and bioaccumulation (Zhou et al. 2018).

Environmental Toxicity: The release of nanomaterials into the environment through industrial waste, product degradation, or accidental spills raises concerns about their persistence and impact on ecosystems. Heavy metal nanoparticles like silver and zinc oxide can disrupt aquatic and terrestrial organisms. Lack of proper disposal and recycling methods for nanomaterials can contribute to long-term pollution (Zhou et al. 2018).

Regulatory and Ethical Issues: Due to limited data on nanotoxicity, regulatory frameworks for smart nanomaterials are still evolving. Different countries have varying safety guidelines, creating challenges for global commercialization. The ethical implications of using nanomaterials in human enhancement, such as in medicine and military applications, also require attention (Zhou et al. 2018).

High Production Costs and Scalability Issues: Despite their superior properties, the production of smart nanomaterials is expensive and technologically demanding. Several factors contribute to prohibitive costs, limiting their industrial-scale adoption (Zhou et al. 2018).

Expensive Raw Materials and Fabrication Techniques: Many smart nanomaterials require rare or expensive precursor materials, such as gold, platinum, and graphene derivatives. High-tech synthesis methods like chemical vapor deposition (CVD), laser ablation, and atomic layer deposition demand significant investment in specialized equipment and infrastructure. The low yield of many fabrication methods makes large-scale production economically unfeasible (Jha et al. 2021).

Scalability and Mass Production Challenges: Nanomaterials often exhibit batch-to-batch variations, making it difficult to maintain consistency in large-scale manufacturing. Issues like particle aggregation, dispersion stability, and contamination arise when scaling up from laboratory to industrial production. Some smart nanomaterials lose their unique properties when synthesized in bulk quantities, reducing their effectiveness (Yu et al. 2024).

Energy-Intensive Processing: Many nanomaterial synthesis techniques consume considerable amounts of energy, making them environmentally unsustainable. Alternative green synthesis methods, such as bio-inspired nanomaterials and eco-friendly solvents, are being explored but require further optimization (Mazari et al. 2021).

Stability, Degradation, and Shelf Life: Smart nanomaterials must maintain their functionality over time. However, they often face degradation, instability, or undesired interactions with their surroundings (Mazari et al. 2021).

Chemical and Thermal Stability: Many nanomaterials are sensitive to humidity, temperature, and UV radiation, leading to degradation. Oxidation and aggregation of nanoparticles can reduce their effectiveness in applications such as drug delivery and coatings (Yu et al. 2024).

Long-Term Performance in Real-World Conditions: Some smart nanomaterials lose their stimuli-responsive properties over time, affecting their reliability. External factors like mechanical stress, exposure to chemicals, and biological fluids can alter their performance (Mazari et al. 2021).

Storage and Handling: Nanomaterials require specialized storage conditions to prevent degradation and agglomeration. Ensuring long shelf life without compromising effectiveness remains a major hurdle in commercial applications (Yu et al. 2024).

Integration into Existing Systems and Compatibility Issues: For nanomaterials to be widely used, they must be seamlessly integrated into existing technologies and manufacturing systems. However, this integration poses significant challenges.

Compatibility with Conventional Materials: Many nanomaterials require modifications to blend with traditional materials used in electronics, coatings, and biomedical devices. Interfacial adhesion and compatibility issues may arise when incorporating nanomaterials into polymers, metals, or ceramics (Mazari et al. 2021).

Complex Fabrication and Processing Requirements: Industries require cost-effective and scalable processes to incorporate nanomaterials into products. Advanced manufacturing techniques such as 3D printing and nanolithography offer solutions but require further development (Malik et al. 2023).

Consumer Acceptance and Market Penetration: Many industries and consumers remain hesitant to adopt nanotechnology due to safety concerns and unfamiliarity. Clear labelling, education, and transparent risk assessments are needed to build trust and market confidence (Malik et al. 2023).

9. Conclusion and Future Remarks

Smart nanomaterials represent a revolutionary advancement in science and technology, offering a vast array of applications that enhance efficiency, sustainability, and innovation across various industries. These materials, characterized by their nanoscale dimensions and unique physicochemical properties, have significantly transformed fields such as medicine, electronics, energy storage, and environmental remediation. Their ability to respond dynamically to external stimuli—such as temperature, pH, light, and magnetic fields—sets them apart from conventional materials and enables their integration into cutting-edge technologies.

References

Abdah, M. A. A. M., Azman, N. H. N., Kulandaivalu, S., and Sulaiman, Y. (2020). Review of the use of transition-metal-oxide and conducting polymer-based fibres for high-performance supercapacitors. *Materials & Design*, 186: 108199.

Alayande, A. B., Goh, K., Son, M., Kim, C. M., Chae, K. J., Kang, Y. et al. (2020). Recent progress in one-and two-dimensional nanomaterial-based electro-responsive membranes: versatile and smart applications from fouling mitigation to tuning mass transport. *Membranes*, 11(1): 5.

Amina, S. J., and Guo, B. (2020). A review on the synthesis and functionalization of gold nanoparticles as a drug delivery vehicle. *International journal of nanomedicine*, 9823–9857.

Anaya-Esparza, L. M., Villagrán-de la Mora, Z., Ruvalcaba-Gómez, J. M., Romero-Toledo, R., Sandoval-Contreras, T., Aguilera-Aguirre, S. et al. (2020). Use of titanium dioxide (TiO2) nanoparticles as reinforcement agent of polysaccharide-based materials. *Processes*, 8(11): 1395.

Badir, A., Refki, S., and Sekkat, Z. 2025. Utilizing Gold Nanoparticles in Plasmonic Photothermal Therapy for Cancer Treatment. *Heliyon*.

Behera, A., and Behera, A. 2022. Nanomaterials. *Advanced Materials: An Introduction to Modern Materials Science*, 77–125.

Campora, S., and Ghersi, G. 2022. Recent developments and applications of smart nanoparticles in biomedicine. *Nanotechnology Reviews*, 11(1): 2595–2631.

Chaikittisilp, W., Yamauchi, Y., and Ariga, K. (2022). Material evolution with nanotechnology, nanoarchitectonics, and materials informatics: what will be the next paradigm shift in nanoporous materials? *Advanced Materials*, 34(7): 2107212.

Chen, X. K., and Chen, K. Q. (2020). Thermal transport of carbon nanomaterials. *Journal of Physics: Condensed Matter*, 32(15): 153002.

Crisan, M. C., Teodora, M., and Lucian, M. (2021). Copper nanoparticles: Synthesis and characterization, physiology, toxicity, and antimicrobial applications. *Applied Sciences*, 12(1): 141.

Dayyoub, T., Maksimkin, A. V., Filippova, O. V., Tcherdyntsev, V. V., and Telyshev, D. V. (2022). Shape memory polymers as smart materials: a review. *Polymers*, 14(17): 3511.

Dehchani, A. J., Jafari, A., and Shahi, F. (2024). Nanogels in Biomedical Engineering: Revolutionizing Drug Delivery, Tissue Engineering, and Bioimaging. *Polymers for Advanced Technologies*, 35(10): e6595.

Díez-Pascual, A. M., and Rahdar, A. (2022). Graphene-based polymer composites for flexible electronic applications. *Micromachines*, 13(7): 1123.

Dinadayalane, T. C., and Leszczynski, J. (2012). Fundamental structural, electronic, and chemical properties of carbon nanostructures: graphene, fullerenes, carbon nanotubes, and their derivatives. *Handbook of computational chemistry*, 793–867.

Ebrahimi, F., Hosseini, S. H. S., and Singhal, A. (2020). A comprehensive review on the modeling of smart piezoelectric nanostructures. *Structural Engineering and Mechanics*, 74(5): 611–633.

Hassan, S. S., Kamel, A. H., Hashem, H. M., and Bary, E. A. (2020). Drug delivery systems between metal, liposome, and polymer-based nanomedicine: a review. *Eur Chem Bull*, 9(3): 91–102.

Hassan, T., Salam, A., Khan, A., Khan, S. U., Khanzada, H., Wasim, M. et al. (2021) Functional nanocomposites and their potential applications: A review. *Journal of Polymer Research*, 28(2): 36.

Husen, A., and Siddiqi, K. S. (Eds.). (2023). *Advances in smart nanomaterials and their applications*. Elsevier.

Islam, M. H., Afroj, S., Uddin, M. A., Andreeva, D. V., Novoselov, K. S., and Karim, N. (2022). Graphene and CNT-based smart fiber-reinforced composites: a review. *Advanced Functional Materials*, 32(40): 2205723.

Jha, R., Singh, A., Sharma, P. K., and Fuloria, N. K. (2020). Smart carbon nanotubes for drug delivery system: A comprehensive study. *Journal of Drug Delivery Science and Technology*, 58: 101811.

Jha, R., Singh, A., Sharma, P. K., Porwal, O., and Fuloria, N. K. (2021). Graphene-based nanomaterial system: A boon in the era of smart nanocarriers. *Journal of Pharmaceutical Investigation*, 51: 245–280.

Kang, J., Zheng, C., Khan, K., Tareen, A. K., Aslam, M., and Wang, B. (2020). Two dimensional nanomaterials-enabled smart light regulation technologies: Recent advances and developments. *Optik*, 220: 165191.

Khan, Y., Sadia, H., Ali Shah, S. Z., Khan, M. N., Shah, A. A., Ullah, N. et al. (2022). Classification, synthetic, and characterization approaches to nanoparticles, and their applications in various fields of nanotechnology: A review. *Catalysts*, 12(11): 1386.

Khare, N., Patani, A., Khare, P., and Singh, S. (2024). Recent Updates on the Use of Smart Nanostructures for Food Packaging Applications. In *Metal and Metal-Oxide Based Nanomaterials: Synthesis, Agricultural, Biomedical and Environmental Interventions* (pp. 133–155). Singapore: Springer Nature Singapore.

Konstantopoulos, G., Koumoulos, E. P., and Charitidis, C. A. (2022). Digital innovation enabled nanomaterial manufacturing; machine learning strategies and green perspectives. *Nanomaterials*, 12(15): 2646.

Kumar, S., and Kumari, P. (2021). Flexible Nano Smart sensors. *Nanosensors for Smart Manufacturing*, 199.

Kyriakides, T. R., Raj, A., Tseng, T. H., Xiao, H., Nguyen, R., Mohammed, F. S. et al. (2021). Biocompatibility of nanomaterials and their immunological properties. *Biomedical Materials*, 16(4): 042005.

Lawson, H. D., Walton, S. P., and Chan, C. (2021). Metal–organic frameworks for drug delivery: a design perspective. *ACS Applied Materials & Interfaces*, 13(6): 7004–7020.

Li, Z., Fan, Q., and Yin, Y. (2021). Colloidal self-assembly approaches to smart nanostructured materials. *Chemical Reviews*, 122(5): 4976–5067.

Malik, S., Muhammad, K., and Waheed, Y. (2023). Nanotechnology: A revolution in modern industry. *Molecules*, 28(2): 661.

Mancardi, G., Mikolajczyk, A., Annapoorani, V. K., Bahl, A., Blekos, K., Burk, J. et al. (2023). A computational view on nanomaterial intrinsic and extrinsic features for nanosafety and sustainability. *Materials Today*, 67: 344–370.

Mazari, S. A., Ali, E., Abro, R., Khan, F. S. A., Ahmed, I., Ahmed, M. et al. (2021). Nanomaterials: Applications, waste-handling, environmental toxicities, and future challenges–A review. *Journal of Environmental Chemical Engineering*, 9(2): 105028.

Mohammadi, R., Naderi-Manesh, H., Farzin, L., Vaezi, Z., Ayarri, N., Samandari, L. et al. (2022). Fluorescence sensing and imaging with carbon-based quantum dots for early diagnosis of cancer: A review. *Journal of Pharmaceutical and Biomedical Analysis*, 212: 114628.

Namdeti, R., Al-Kathiri, D. S. M. S., Rao, G. B., Khan, A. H., and Alshehri, H. A. M. (2025). A Critical Review of Smart Materials for Efficient Water Purification: Towards Sustainable Clean Water Solutions. *Journal of Membrane Science and Research*, 11(1).

Palmerston Mendes, L., Pan, J., and Torchilin, V. P. (2017). Dendrimers as nanocarriers for nucleic acid and drug delivery in cancer therapy. *Molecules*, 22(9): 1401.

Papadakis, C. M., Niebuur, B. J., and Schulte, A. (2023). Thermoresponsive Polymers under Pressure with a Focus on Poly (N-isopropylacrylamide)(PNIPAM). *Langmuir*, 40(1): 1–20.

Patani, A., Patel, A., Prajapati, D., Khare, N., and Singh, S. (2023). Various agriculture crop plant-based bioactive compounds and their use in nanomaterial synthesis and applications. pp. 223–241. *In: Nanomaterials from Agricultural and Horticultural Products*. Singapore: Springer Nature Singapore.

Philip, A., and Kumar, A. R. (2022). The performance enhancement of surface plasmon resonance optical sensors using nanomaterials: A review. *Coordination Chemistry Reviews*, 458: 214424.

Prasad, S. R., Kumbhar, V. B., and Prasad, N. R. (2023). Applications of Nanotechnology in Textile: A Review. *ES Food & Agroforestry*, 15: 1019.

Puri, N., and Gupta, A. (2023). Water remediation using titanium and zinc oxide nanomaterials through disinfection and photo catalysis process: A review. *Environmental Research*, 227: 115786.

Rahman, M. (2023). Magnetic resonance imaging and iron-oxide nanoparticles in the era of personalized medicine. *Nanotheranostics*, 7(4): 424.

Ramezani, M., and Neitzert, T. (2012). Mechanical milling of aluminum powder using planetary ball milling process. *Journal of Achievements in Materials and Manufacturing Engineering*, 55(2): 790–798.

Rashko, M. N., Hamad, S. M., Barzinjy, A. A., and Hamad, A. H. (2022). Mechanical properties of carbon nanotubes (CNTs): a review. *Eurasian Journal of Science and Engineering*, 8(2): 54–68.

Sable, H., Kumar, V., Singh, V., Rustagi, S., Chahal, S., and Chaudhary, V. (2024). Strategically engineering advanced nanomaterials for heavy-metal remediation from wastewater. *Coordination Chemistry Reviews*, 518: 216079.

Shah, M. A., Pirzada, B. M., Price, G., Shibiru, A. L., and Qurashi, A. (2022). Applications of nanotechnology in smart textile industry: A critical review. *Journal of Advanced Research*, 38: 55–75.

Shojaei, A., and Khasraghi, S. S. (2021). Self-healing and self-sensing smart polymer composites. pp. 307–357. *In: Composite Materials*. Elsevier.

Singh, B. K., Lee, S., and Na, K. (2020). An overview on metal-related catalysts: metal oxides, nanoporous metals and supported metal nanoparticles on metal organic frameworks and zeolites. *Rare Metals*, 39: 751–766.

Singh. N. B., Kumar, B., Usman, U. L., and Susan, M. A. B. H. (2024a). Nano revolution: exploring the frontier nanomaterials in science, technology and society. *Nano-Structure & Nano-Objects*, 39: 101299.

Singh, N., Madhu, A., and Yadav, M. (2024b). Nano-finishing in Apparel Textiles. pp. 415–440. *In: Nanotechnology in Textile Finishing: Advancements and Applications*. Singapore: Springer Nature Singapore.

Singh, S., and Melnik, R. (2022). Coupled multiphysics modelling of sensors for chemical, biomedical, and environmental applications with focus on smart materials and low-dimensional nanostructures. *Chemosensors*, 10(5): 157.

Siritongsuk, P., Thammawithan, S., Srichaiyapol, O., Nasompag, S., Pongha, S., Daduang, S. et al. (2022). Synthesis and application of AgNPs-Chitosan composite as a self-disinfecting coating in water-based polyurethane. *Coatings*, 12(12): 1832.

Soares, P. I., Romao, J., Matos, R., Silva, J. C., and Borges, J. P. (2021). Design and engineering of magneto-responsive devices for cancer theranostics: Nano to macro perspective. *Progress in Materials Science*, 116: 100742.

Su, M., and Song, Y. (2021). Printable smart materials and devices: strategies and applications. *Chemical Reviews*, 122(5): 5144–5164.

Tan, D., Jiang, C., Li, Q., Bi, S., and Song, J. (2020). Silver nanowire networks with preparations and applications: a review. *Journal of Materials Science: Materials in Electronics*, 31: 15669–15696.

Thangudu, S. (2020). Next generation nanomaterials: smart nanomaterials, significance, and biomedical applications. *Applications of Nanomaterials in Human Health*, 287–312.

Triana, M. A., Hsiang, E. L., Zhang, C., Dong, Y., and Wu, S. T. (2022). Luminescent nanomaterials for energy-efficient display and healthcare. *ACS Energy Letters*, 7(3): 1001–1020.

Wei, Y., and Yang, R. (2019). Nanomechanics of graphene. *National Science Review*, 6(2): 324–348.

Xia, Y., He, Y., Zhang, F., Liu, Y., and Leng, J. (2021). A review of shape memory polymers and composites: mechanisms, materials, and applications. *Advanced Materials*, 33(6): 2000713.

Yan, N., Zheng, Z., Liu, Y., Jiang, X., Wu, J., Feng, M. et al. (2022). Photo-responsive shape memory polymer composites enabled by doping with biomass-derived carbon nanomaterials. *Nano Research*, 15: 1383–1392.

Yu, M., Mathew, A., Liu, D., Chen, Y., Wu, J., Zhang, Y. et al. (2024). Microfluidics for Formulation and Scale-Up Production of Nanoparticles for Biopharma Industry. pp. 395–420. *In: Microfluidics in Pharmaceutical Sciences: Formulation, Drug Delivery, Screening, and Diagnostics*. Cham: Springer Nature Switzerland.

Zahoor, M., Nazir, N., Iftikhar, M., Naz, S., Zekker, I., Burlakovs, J. et al. (2021). A review on silver nanoparticles: Classification, various methods of synthesis, and their potential roles in biomedical applications and water treatment. *Water*, 13(16): 2216.

Zhai, L., Narkar, A., and Ahn, K. (2020). Self-healing polymers with nanomaterials and nanostructures. *Nano Today*, 30: 100826.

Zhang, Q., Zhang, Y., Wan, Y., Carvalho, W., Hu, L., and Serpe, M. J. (2021). Stimuli-responsive polymers for sensing and reacting to environmental conditions. *Progress in Polymer Science*, 116: 101386.

Zhou, Y., Peng, Z., Seven, E. S., and Leblanc, R. M. (2018). Crossing the blood-brain barrier with nanoparticles. *Journal of Controlled Release*, 270: 290–303.

CHAPTER-2

Recent Advances in Smart Nanomaterials for Environmental Remediation

Apurva Sharma,[1] *Pankaj Kumar,*[1,2,*] *Snigdha Singh,*[1]
Virendra Kumar Yadav,[3] *Shivraj Gangadhar Wanale*[4]
and *Vedpathak Mangesh Mohan*[5]

1. Introduction

The increasing burden of environmental pollution—ranging from contaminated water bodies and degraded soils to polluted air—poses significant risks to ecological systems and human health (Tariq et al. 2025). Traditional remediation techniques, while effective to a certain extent, often fall short in terms of efficiency, selectivity, cost-effectiveness, and environmental sustainability (Ningthoujam et al. 2022). The rapid advancement in nanotechnology has opened new avenues for addressing these limitations through the development and application of smart nanomaterials, which possess unique physicochemical properties and responsive behaviours.

Environmental remediation encompasses a broad spectrum of strategies aimed at removing, neutralizing, or containing pollutants to restore natural ecosystems. However, several persistent challenges hinder the success of conventional remediation approaches (Mondal and Sharma 2016). These include non-selectivity of adsorbents, which reduces efficiency in complex environmental matrices, slow reaction kinetics and limited degradation capacity of traditional catalysts, secondary pollution arising from chemical reagents or incomplete treatment processes, inaccessibility of pollutants in micro- or nanoscale domains within soil and sediments and high energy and operational costs, particularly in large-scale or remote applications (Anani et al. 2022). With the growing urgency to address pollution in an environmentally benign and cost-effective manner, there is a pressing need for next-generation materials that can offer multifunctionality, adaptability, and enhanced performance under dynamic environmental conditions (Cerra and Fratoddi 2024).

[1] Department of Earth and Environmental Sciences, Parul Institute of Applied Sciences, Parul University, Vadodara, Gujarat, 391760, India. ORCID: 0000-0003-3493-8016
[2] Faculty of Earth Science, University Malaysia Kelantan Jeli Campus, 17600 Jeli, Kelantan Malaysia.
[3] Marwadi University Research Center, Department of Microbiology, Faculty of Sciences, Marwadi University, Rajkot, Gujarat 360003, India.
[4] Department of Chemistry, Netaji Subhashchandra Bose Art's, Commerce and Science College, Nanded, 431601, Maharashtra, India.
[5] School of Technology, Department of Civil Engineering, Sanjay Ghodawat University, Kolhapur, 416 118 Maharashtra, India.
* Corresponding author: pankaj.kumar25135@paruluniversity.ac.in

DOI: 10.1201/9781003558804-2

1.1 Emergence of Smart Nanomaterials

Smart nanomaterials, often defined by their ability to respond to external stimuli such as pH, temperature, light, magnetic fields, or redox conditions, have emerged as promising tools in environmental remediation (Yilmaz-Sercinoglu et al. 2023). These materials exhibit *size-dependent properties*, *high surface area*, and *tunable reactivity*, enabling targeted and efficient interaction with a wide range of contaminants (Mitra and Saha 2025).

The evolution of these materials from passive to *stimuli-responsive systems* has marked a change in basic assumptions in remediation technologies. For example, photocatalytic nanomaterials can harness solar energy for pollutant degradation, while magnetic nanoparticles allow easy recovery and reuse through magnetic separation (Iqbal et al. 2022). Additionally, hybrid and functionalised nanostructures offer tailored selectivity towards specific contaminants, thereby enhancing remediation efficacy while minimising collateral environmental impact (Chidichimo et al. 2013).

This chapter aims to explore the recent advances in the design, synthesis, and application of smart nanomaterials for environmental remediation. The core objectives are to present a comprehensive classification of smart nanomaterials used in air, water, and soil remediation; highlight state-of-the-art synthesis techniques and functionalisation methods that enhance material performance; explain the mechanistic pathways through which these materials interact with pollutants; discuss real-world applications, including case studies and pilot projects, demonstrating the practical relevance of smart nanomaterials; address ecotoxicological risks, life-cycle considerations, and regulatory perspectives; and identify emerging trends and future directions to guide further research and industrial adoption (Yadav et al. 2021). Through this discussion, the chapter intends to provide a foundational yet forward-looking perspective on how smart nanomaterials can contribute to sustainable environmental management and pollution abatement in the 21st century. A summary of the role of nanomaterials in environmental remediation is shown in Figure 1.

2. Classification of Smart Nanomaterials

Smart nanomaterials represent a versatile class of engineered materials at the nanoscale (1–100 nm), characterised by their ability to respond to environmental stimuli, exhibit selective reactivity, and enhance pollutant remediation efficiency. Their classification is generally based on their functional properties, mode of activation, and environmental compatibility. This section provides an overview of four major classes of smart nanomaterials with significant relevance to environmental remediation. Smart nanomaterials for environmental remediation have been summarised in Table 1.

2.1 Stimuli-Responsive Nanomaterials

Stimuli-responsive nanomaterials, also known as "intelligent" or "adaptive" nanomaterials, are designed to change their physical or chemical properties in response to specific environmental triggers (Elsaid et al. 2023). These triggers include pH (e.g., pH-sensitive polymers that release active agents in acidic or basic conditions), temperature (e.g., thermo-responsive gels that expand or collapse), light (e.g., photochromic materials that alter their structure under UV/visible light), redox potential (e.g., nanomaterials that activate under reducing or oxidising conditions), and magnetic and electric fields (Aghababai Beni and Jabbari 2022).

Such materials offer on-demand activation, ensuring controlled release or targeted action at contaminated sites. For instance, pH-responsive nanocarriers can deliver oxidants precisely in acidic environments like mining-impacted soils or landfill leachate zones. This targeted behaviour minimises material wastage and secondary environmental impact. Examples include poly(N-isopropylacrylamide)-coated nanoparticles for temperature-triggered contaminant release; pH-responsive mesoporous silica nanoparticles loaded with heavy metal chelators; and redox-sensitive polymeric micelles for organic pollutant degradation (Roy et al. 2021).

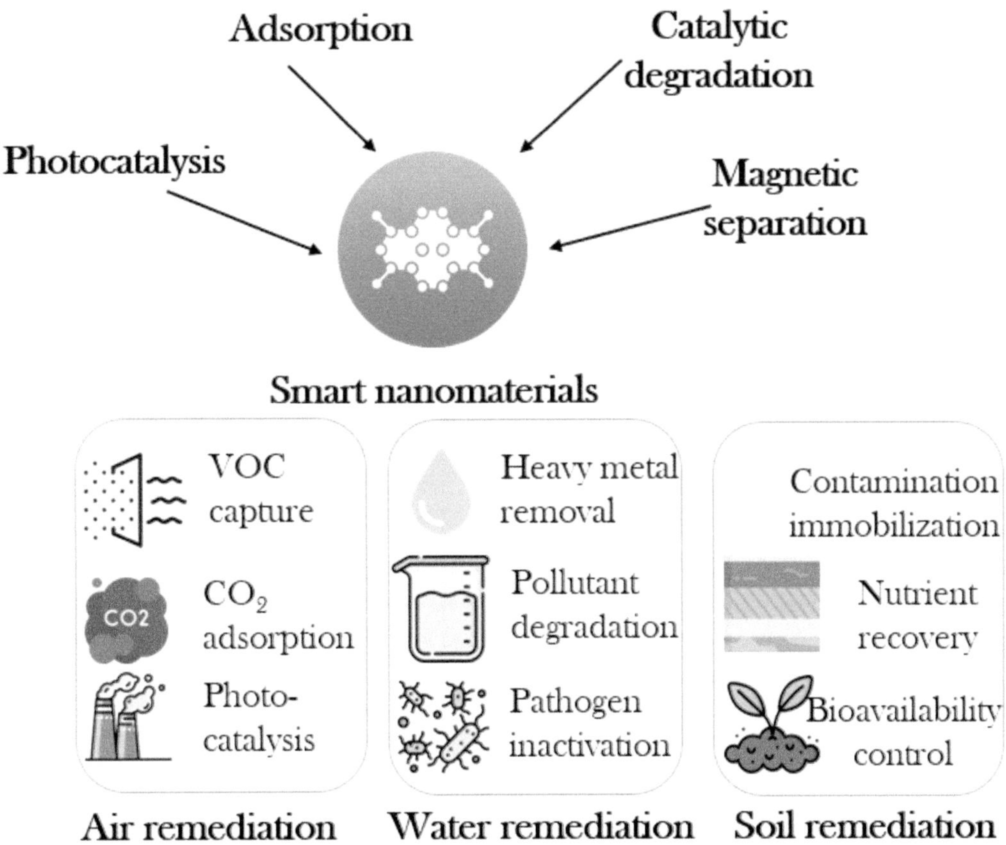

Figure 1: Overview of role of smart nanomaterials in environmental remediation.

2.2 Magnetic Nanomaterials

Magnetic nanomaterials, primarily composed of iron oxides such as magnetite (Fe_3O_4) or maghemite (γ-Fe_2O_3), have garnered attention for their dual functionality: high reactivity towards contaminants and easy retrieval via external magnetic fields (Umar et al. 2013). These properties enable efficient removal of pollutants such as heavy metals, dyes, and pathogens; separation and recovery after treatment, reducing the cost and environmental impact of remediation efforts; and reusability, improving sustainability and lowering material consumption (Raja et al. 2023).

Surface modification with functional groups (e.g., amines, thiols, carboxylates) can enhance the selectivity and binding capacity of magnetic nanoparticles (Sable et al. 2024). Additionally, magnetic nanocomposites can be coupled with photocatalytic or adsorptive components to achieve multifunctionality.

2.3 Photocatalytic Nanomaterials

Photocatalytic nanomaterials leverage the energy from light (UV or visible) to initiate redox reactions that degrade organic pollutants and destroy pathogens. Upon light absorption, these materials generate electron–hole pairs, which produce reactive oxygen species (ROS) such as hydroxyl radicals and superoxide anions. These ROS are highly effective in breaking down complex and persistent contaminants (Amin et al. 2021). The most widely used photocatalyst is titanium dioxide (TiO_2), though others like zinc oxide (ZnO), graphitic carbon nitride (g-C_3N_4), and silver-

Table 1: Types of smart nanomaterials.

Type of Smart Nanomaterial	Key Properties	Mechanism of Action	Target Pollutants	Environmental Applications	Advantages	Limitations	Reference
Stimuli-Responsive Nanomaterials	pH, temperature, redox, or light-sensitive behaviour; reversible response to stimuli	Release or capture of contaminants in response to environmental changes	Organic dyes, heavy metals, pharmaceutical residues	Targeted release in water treatment; responsive sensors	High selectivity, controllable behaviour	Environmental stability concerns; potential toxicity	(Farzin et al. 2024)
Magnetic Nanomaterials (e.g. Fe_3O_4, $MnFe_2O_4$)	Superparamagnetism; high surface area; ease of separation using external magnetic fields	Adsorption and separation under magnetic field	Heavy metals (Pb, As, Cr), oils, dyes, microplastics	Wastewater treatment; oil spill clean-up; soil remediation	Easy recovery, reusable, high adsorption capacity	Agglomeration risk; surface oxidation issues	(S. Liu et al. 2020)
Photocatalytic Nanomaterials (e.g. TiO_2, ZnO, g-C_3N_4)	Photoreactivity under UV/visible light; electron-hole pair generation; oxidative potential	Photocatalytic degradation; advanced oxidation processes (AOPs)	Organic pollutants (dyes, pesticides), VOCs, pathogens	Water disinfection, air purification, wastewater treatment	High degradation efficiency; renewable energy use	Light penetration limits; potential metal leaching	(Tabasum et al. 2023)
Biodegradable and Green Nanomaterials	Environmentally benign; derived from natural sources; biocompatible	Biodegradation, adsorption, or enzyme-mimicking activity	Organic waste, pharmaceuticals, emerging contaminants	Green synthesis routes; eco-friendly remediation	Low toxicity; sustainable production	Limited long-term stability: lower reactivity compared to metal-based nanomaterials	(Fuku et al. 2023)
Hybrid and Composite Nanomaterials	Tailored multifunctional properties; synergistic effects of combined materials	Dual/tri-mode action (adsorption + photocatalysis + magnetic separation)	Mixed pollutants, heavy metals, organics	Multifunctional remediation (water, soil, air)	Enhanced performance; multiple functionalities	Complex synthesis; cost-intensive	(Su 2017)
Metal-Organic Frameworks (MOFs)	High porosity, tunable surface chemistry, large surface area	Adsorption, catalysis, and selective trapping of contaminants	Gases (CO_2, CH_4), VOCs, heavy metals	Air purification, gas separation, water treatment	High adsorption capacity, selective removal	Moisture sensitivity, scalability issues	(Dong et al. 2023)
Carbon-Based Nanomaterials (CNTs, Graphene, Graphene Oxide)	High surface area, electrical conductivity, hydrophobicity	Adsorption, electron transfer, catalytic degradation	Organic dyes, VOCs, heavy metals	Water and air filtration; sensors for pollutant detection	High adsorption, robust structure	Difficult recovery, aggregation tendencies	(Nayak and Pathan 2023)

based nanomaterials are also gaining attention (Corsi et al. 2023). Recent advances include doping with metals or non-metals to extend light absorption into the visible range; nanocomposites with carbon-based materials (e.g., graphene, CNTs) for enhanced charge separation; and immobilisation on substrates for better recovery and reuse (Ata et al. 2019).

2.4 Biodegradable and Green Nanomaterials

With growing environmental concerns over nanomaterial toxicity and persistence, there is a shift towards developing biodegradable and eco-friendly nanomaterials (Das et al. 2015). These materials are synthesised from renewable resources, are often non-toxic, and degrade naturally after use, minimising ecological disruption. Key sources and types of green nanomaterials include biopolymers (e.g., chitosan, cellulose, starch, alginate), plant extracts, and biodegradable carriers, which can be used for controlled release of remediation agents. Such nanomaterials are especially valuable for applications in sensitive ecosystems like wetlands, agricultural soils, and groundwater zones. Additionally, green synthesis approaches reduce the need for hazardous solvents and energy-intensive procedures (Ayeleru et al. 2025).

3. Synthesis and Functionalization Techniques

The performance of smart nanomaterials in environmental remediation is strongly influenced by their synthesis methods and functionalisation strategies. These factors determine their size, morphology, surface properties, stability, and reactivity. To enhance specificity, responsiveness, and environmental safety, researchers employ a variety of synthesis routes and surface engineering approaches (Chang et al. 2018). This section discusses three prominent strategies: green synthesis, surface modification and doping, and the development of hybrid/composite nanostructures.

3.1 Green Synthesis Approaches

Green synthesis emphasises the production of nanomaterials using environmentally benign solvents, renewable biological agents, and energy-efficient processes. Unlike traditional chemical synthesis, which often involves toxic reducing agents and harsh conditions, green synthesis aligns with the principles of green chemistry, minimising ecological risks and health hazards (Nazarzadeh Zare et al. 2021). Different methods for green synthesis of nanomaterials have been shown in Table 2. Greenly synthesised nanomaterials are ideal for remediation in ecologically sensitive zones (e.g., wetlands, agricultural fields), where toxicity and bioaccumulation concerns are significant (Guerra et al. 2018).

3.2 Surface Modification and Doping

Surface functionalisation refers to the alteration of nanomaterial surfaces by attaching specific chemical groups, polymers, or molecules to enhance their selectivity, stability, and interaction with target pollutants (Palchoudhury et al. 2020). Doping, on the other hand, involves the incorporation of foreign atoms into the crystal lattice of nanomaterials to tune their electrical, magnetic, or photocatalytic properties (Al-Khayri et al. 2025). Surface modification increases binding affinity for specific pollutants (e.g., heavy metals, dyes, antibiotics), improves dispersibility in aqueous environments, and enhances biocompatibility or reduces aggregation. Amino ($-NH_2$), carboxyl (–COOH), thiol (–SH), and hydroxyl (–OH) groups, along with polymers like polyethylene glycol (PEG), polyacrylic acid (PAA), and polyethyleneimine (PEI), are the common functional groups and agents used for this process (Neouze and Schubert 2008). Modified or doped nanomaterials demonstrate greater efficiency in degrading organic pollutants, selective adsorption of toxic ions, and enhanced photocatalytic activity under natural light conditions.

Table 2: Green synthesis of nanomaterials for environmental remediation.

Synthesis Method	Reducing/ Stabilizing Agents	Source	Nanomaterials Produced	Advantages	Limitations	Examples/ Applications	References
Plant Extract-Mediated Synthesis	Polyphenols, flavonoids, alkaloids, terpenoids	Leaves, flowers, roots, fruits (e.g. neem, tea, aloe)	AgNPs, AuNPs, ZnO, TiO_2	Eco-friendly, renewable, abundant bio-resources; no harsh chemicals	Variability in size and shape; scalability challenges	AgNPs for antibacterial coatings; AuNPs for sensing	(Dejen et al. 2023)
Microbial Synthesis	Enzymes, proteins, metabolites	Bacteria (e.g. *Bacillus*, *Pseudomonas*), fungi (*Aspergillus*, *Fusarium*)	AgNPs, AuNPs, ZnO, SeNPs	Biocompatible, mild reaction conditions, low energy input	Slow reaction kinetics; risk of contamination	SeNPs as antioxidants; ZnO for wastewater treatment	
Algae-Mediated Synthesis	Polysaccharides, proteins, pigments (chlorophyll, carotenoids)	Microalgae (e.g. *Chlorella*, *Spirulina*), macroalgae (e.g. *Sargassum*)	AgNPs, AuNPs, CuO	Fast, large-scale potential, eco-friendly	Purification challenges; limited mechanistic understanding	AgNPs for antimicrobial activity; CuO for dye degradation	
Biopolymer-Assisted Synthesis	Chitosan, cellulose, starch, alginate	Agricultural/industrial waste biomass	Metal nanoparticles (Ag, Au), oxides (ZnO, Fe_3O_4)	Low toxicity, stabilizing agents inherently biocompatible	Lower reactivity compared to chemical methods	Chitosan-ZnO for heavy metal removal; alginate-Fe_3O_4 for magnetic separation	
Green Chemical Reduction (using bio-reducing agents)	Ascorbic acid, glucose, citric acid	Fruits (e.g. lemon, orange), sugars	AgNPs, AuNPs, PdNPs	Mild conditions, low energy input, rapid synthesis	Limited control over nanoparticle uniformity	AgNPs for water disinfection; PdNPs for catalysis	(Jadhav et al. 2024)
Green Sol-Gel Method	Plant extracts, biopolymers, sugars	Renewable bio-resources	SiO_2, TiO_2, Fe_2O_3 nanoparticles	Lower energy demand, improved surface functionalization	Requires post-synthesis calcination for stability	TiO_2 for photocatalysis; Fe_2O_3 for arsenic removal	
Green Co-Precipitation	Plant extracts, biomolecules	Biomass, plant materials	Fe_3O_4, ZnO, CuO	Scalable, simple process, high yield	Particle agglomeration; limited particle size control	Fe_3O_4 for magnetic separation; CuO for wastewater treatment	(Dejen et al. 2023)

3.3 Hybrid and Composite Nanostructures

Hybrid and composite nanomaterials are formed by combining two or more nanostructures to exploit synergistic properties, thereby overcoming limitations of individual components. These materials offer multifunctionality, improved mechanical stability, and enhanced environmental performance. Core–shell structures, wherein a nanoparticle core is enclosed by another material layer (e.g., $Fe_3O_4@SiO_2$); heterojunctions, which are interfaces of two semiconductors to improve charge transfer (e.g., TiO_2/g-C_3N_4); and matrix-based composites, which include nanoparticles embedded in a biopolymer or carbonaceous matrix for structural support and improved dispersion, are the different types of composites that can be used (Sperling and Parak 2010).

4. Mechanisms of Action in Environmental Remediation

Smart nanomaterials exhibit multiple mechanisms to remove or neutralise pollutants from the environment. Their effectiveness stems from their high surface area, tunable surface properties, and multifunctional behaviour, allowing them to adsorb, degrade, transform, or recover contaminants in complex matrices. This section explores four major mechanisms—adsorption; catalytic degradation and redox reactions; photocatalysis and advanced oxidation processes (AOPs); and magnetic separation and recovery—that underpin the remediation capabilities of smart nanomaterials (Figure 2).

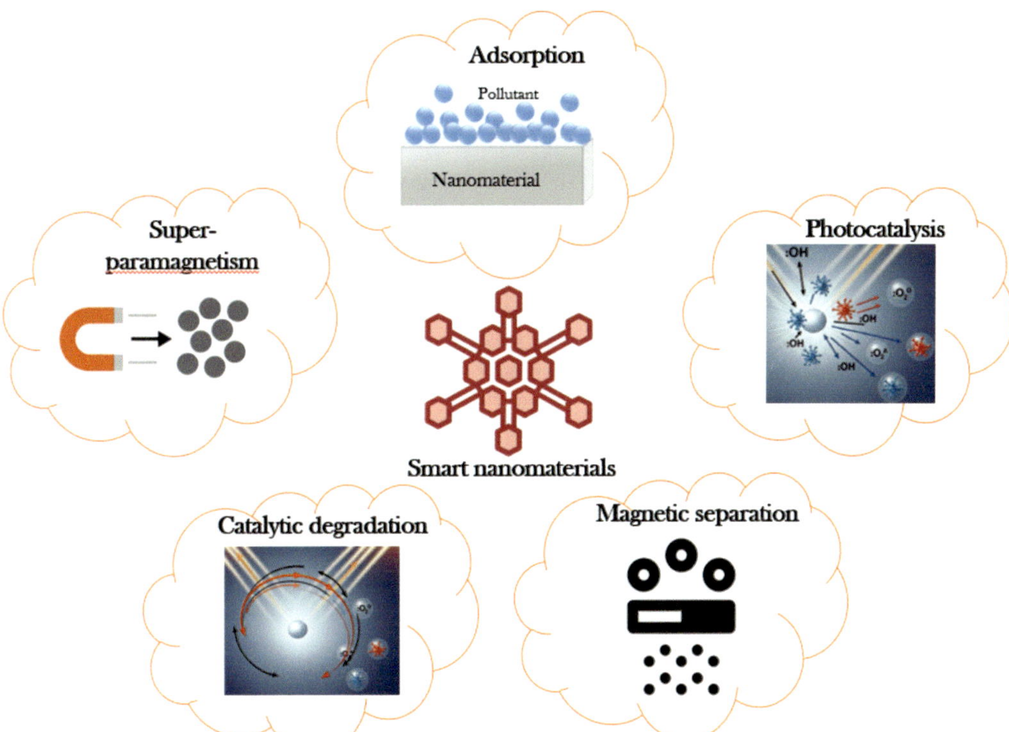

Figure 2: Mechanisms of action of smart nanomaterials.

4.1 Adsorption

Adsorption is one of the most widely used mechanisms for pollutant removal, wherein contaminants adhere to the surface of nanomaterials due to physical or chemical interactions. Smart nanomaterials, due to their high surface-to-volume ratios and tailored surface chemistry, can exhibit superior adsorption capabilities (Ahmad et al. 2022).

4.2 Catalytic Degradation and Redox Reactions

Some nanomaterials can catalyse redox reactions that break down harmful pollutants into less toxic or inert substances. These reactions are especially useful in treating POPs, pesticides, and industrial effluents. They involve electron transfer between the pollutant and nanomaterial surface and can occur under ambient conditions or be activated by co-catalysts, pH, or oxidants (Mondal and Palit 2022).

4.3 Photocatalysis and Advanced Oxidation Processes (AOPs)

Photocatalysis involves the activation of a nanomaterial by light (usually UV or visible), generating reactive species (e.g., hydroxyl radicals •OH, superoxide O_2^-) that oxidise organic pollutants. This is a core technique in advanced oxidation processes (AOPs), aimed at complete mineralisation of pollutants. Upon light irradiation, a semiconductor photocatalyst (e.g., TiO_2, ZnO) becomes excited, promoting electrons from the valence band to the conduction band. The resulting electron–hole pairs initiate redox reactions that degrade contaminants (Raskar et al. 2024).

4.4 Magnetic Separation and Recovery

Magnetic nanomaterials, especially those incorporating iron oxides (Fe_3O_4, γ-Fe_2O_3), provide a practical advantage through their easy separation using external magnetic fields. This not only simplifies the treatment process but also enhances the reusability and cost-efficiency of nanomaterials. Their key features include rapid and selective recovery of nanomaterials post-treatment; usefulness in batch or continuous flow systems; and the ability to be regenerated and reused for multiple cycles (Javed et al. 2020).

5. Applications in Water and Wastewater Treatment

Water pollution remains one of the most critical environmental challenges worldwide. Contaminants such as heavy metals, organic pollutants, pathogens, and oil-based compounds severely impact human health and aquatic ecosystems. Smart nanomaterials offer innovative, efficient, and often multifunctional solutions to treat these diverse pollutants. Their high surface area, selective reactivity, tunable functionalities, and regenerative capabilities make them excellent candidates for advanced water and wastewater treatment technologies (Wang et al. 2015). This section outlines the major applications of smart nanomaterials in the removal and remediation of contaminants in aqueous systems (Table 3).

5.1 Heavy Metal Removal

Heavy metals such as lead (Pb), cadmium (Cd), chromium (Cr), arsenic (As), and mercury (Hg) are persistent, toxic, and non-biodegradable. Conventional treatment methods often fall short in completely removing these elements or are cost-prohibitive. Smart nanomaterials, particularly those with functionalised surfaces or redox activity, have demonstrated excellent performance in adsorbing, reducing, or precipitating heavy metals from contaminated water (Ghanghas et al. 2025).

Table 3: Smart nanomaterial applications in water treatment.

Nanomaterial Type	Target Pollutant(s)	Mechanism of Action	Performance/ Removal Efficiency	Advantages	Limitations	Reference
Magnetic Nanoparticles (Fe_3O_4, $MnFe_2O_4$)	Heavy metals (Pb^{2+}, Cd^{2+}, As^{5+}), dyes, oils	Adsorption + magnetic separation	$Pb^{2+} > 95\%$, dyes $> 90\%$	Easy recovery, reusable, fast kinetics	Surface fouling; aggregation risk	(Rosaline et al. 2024)
Photocatalytic Nanomaterials (TiO_2, ZnO, g-C_3N_4)	Organic pollutants (dyes, phenols, pesticides)	Photocatalytic degradation under UV/visible light	Methylene blue $> 98\%$ within 2 hours	Renewable energy (solar-driven), high degradation	Light penetration limitations; nanoparticle leaching	
Carbon-Based Nanomaterials (CNTs, Graphene Oxide)	Organic dyes (methylene blue, rhodamine B), heavy metals	π-π interactions, hydrophobic adsorption, electron transfer	Dyes $> 90\%$; Pb^{2+} ~85%	High surface area; multifunctional; rapid uptake	Recovery challenges; potential ecotoxicity	(Nayak and Pathan 2023)
Biodegradable/Green Nanomaterials (Chitosan NPs, Starch-Capped AgNPs)	Heavy metals, pathogens, emerging contaminants	Adsorption, antimicrobial activity	Pathogens $> 99\%$; $Cr^{6+} > 85\%$	Eco-friendly; biodegradable; low toxicity	Stability concerns; limited lifespan in water	(Fuku et al. 2023)
Stimuli-Responsive Nanomaterials (pH, temperature-sensitive)	Heavy metals, dyes, pharmaceutical residues	pH-triggered adsorption/desorption; selective capture	$Pb^{2+} > 90\%$ at pH 5; dyes $> 80\%$	Selectivity; controllable binding/ release	Environmental stability; high synthesis cost	(Plazas-Tuttle et al. 2015)
Hybrid/Composite Nanomaterials (Fe_3O_4/TiO_2, ZnO/Graphene)	Mixed pollutants (heavy metals + organics)	Combined adsorption, photocatalysis, magnetic separation	Mixed pollutant removal $> 90\%$	Multifunctionality; enhanced removal performance	Complex synthesis; cost-intensive	(Su 2017)
Metal-Organic Frameworks (MOFs)	Heavy metals, antibiotics, dyes, VOCs	Adsorption via high surface area and porosity	Antibiotics $> 85\%$; dyes $> 90\%$	High selectivity; tunable structures	Sensitive to moisture; scalability issues	(Dong et al. 2023)

5.2 Pathogen Inactivation

The presence of bacteria, viruses, and protozoa in water sources poses serious health threats, especially in regions lacking adequate sanitation. Smart nanomaterials have emerged as powerful antimicrobial agents owing to their ability to disrupt cell membranes, generate reactive oxygen species (ROS), and interfere with microbial metabolism (Sheoran et al. 2024).

5.3 Oil Spill Cleanup and Desalination

Oil spills and saline water intrusion are increasingly common due to industrial activities, shipping, and climate change. Smart nanomaterials can be tailored for hydrophobic, oleophilic, or ion-selective properties, enabling their use in oil–water separation and desalination technologies (Liu et al. 2014).

Oil Spill Cleanup:

1. Superhydrophobic magnetic nanocomposites (e.g., Fe_3O_4 coated with silica or carbon) can selectively adsorb oil while allowing easy magnetic retrieval.
2. Graphene-based sponges and aerogels exhibit excellent oil absorption capacities and reusability.
3. Biodegradable nanomaterials minimize secondary pollution risks after cleanup.

Desalination:

1. Nanostructured membranes with embedded nanoparticles (e.g., carbon nanotubes, metal oxides) enhance salt rejection and water flux.
2. Capacitive deionization using graphene-based electrodes allows selective ion removal with low energy input.
3. Photothermal nanomaterials convert solar energy into localized heat for solar distillation and membrane distillation processes.

6. Applications in Soil Remediation

Soil contamination poses a significant threat to agricultural productivity, food safety, groundwater quality, and ecosystem health. Pollutants such as heavy metals, hydrocarbons, pesticides, industrial chemicals, and persistent organic pollutants (POPs) accumulate in soil due to industrial emissions, improper waste disposal, excessive use of agrochemicals, and mining activities. Traditional soil remediation techniques often suffer from limitations such as excessive cost, long treatment times, and ecological disruption. Smart nanomaterials have emerged as a promising alternative, offering high reactivity, selectivity, mobility, and environmental compatibility for *in-situ* and *ex-situ* remediation strategies. This section discusses their applications in immobilising contaminants, enhancing soil health, and scaling up to field applications (S. Liu et al. 2021).

6.1 Immobilization of Contaminants

One of the most effective strategies for soil remediation is the immobilisation of pollutants, which reduces their mobility, leachability, and bioavailability, thereby preventing them from entering food chains or groundwater systems. Smart nanomaterials, owing to their large surface area, ion-exchange capacity, and tailored functional groups, can strongly bind to soil contaminants (Chauhan et al. 2025).

6.2 Bioavailability Control and Nutrient Recovery

Smart nanomaterials can regulate the bioavailability of nutrients and pollutants in soil, thereby facilitating both remediation and restoration of soil health. These materials can enhance nutrient recovery, stimulate microbial activity, and promote plant growth, making them ideal for integrated soil rehabilitation (Gautam and Chattopadhyaya 2016).

6.3 Field-Scale Applications and Pilot Studies

Despite the promising results at laboratory scale, the real-world application of nanotechnology in soil remediation requires validation through field trials and pilot studies. Several recent initiatives have demonstrated the feasibility, cost-effectiveness, and environmental safety of smart nanomaterials at larger scales (Mensah et al. 2021).

Case Studies and Field Examples:

- A field pilot using nZVI and biochar nanocomposites in mining-impacted soil showed a 70–90% reduction in lead and cadmium mobility over six months.
- Magnetite nanoparticles applied in contaminated agricultural land successfully immobilized arsenic, reducing its uptake by rice crops.
- A combination of nanoclay and phytoremediation was employed in petroleum-contaminated fields, enhancing hydrocarbon degradation and biomass production.
- Field-scale injection of nano-hydroxyapatite near e-waste recycling zones led to reduced lead bioavailability and improved microbial diversity.

Challenges and Considerations:

- Nanomaterial stability and aging in varied soil matrices.
- Toxicological assessments for long-term soil and groundwater safety.
- Cost-benefit analysis for large-scale implementation.
- Regulatory compliance and public acceptance.

7. Air Pollution Control using Smart Nanomaterials

Air pollution remains one of the most critical environmental challenges globally, with severe implications for human health, climate change, and ecological systems. Key pollutants include volatile organic compounds (VOCs), greenhouse gases (GHGs) like CO_2 and CH_4, nitrogen oxides (NOx), sulphur dioxide (SO_2), and fine particulate matter (PM2.5 and PM10). Traditional air pollution control technologies often fall short in efficiency, selectivity, and adaptability to dynamic environmental conditions. Smart nanomaterials, owing to their tunable surface properties, high surface-to-volume ratios, and functional responsiveness to environmental stimuli, have emerged as advanced tools for efficient and targeted air purification (Tariq et al. 2025). This section explores how these materials contribute to the control and remediation of air pollutants through innovative mechanisms (Table 4).

7.1 Removal of Volatile Organic Compounds (VOCs)

VOCs such as benzene, formaldehyde, toluene, and xylene are emitted from industrial activities, solvents, vehicular exhausts, and household products. These compounds are not only toxic but also contribute to ground-level ozone and smog formation. Smart nanomaterials can effectively adsorb or catalytically degrade VOCs under ambient conditions.

Nanotechnology offers a range of materials and mechanisms for the effective removal of VOCs. Photocatalytic materials such as titanium dioxide (TiO_2), zinc oxide (ZnO), and doped semiconductor nanoparticles are activated by UV or visible light, initiating oxidation reactions that break down VOCs into less harmful substances like carbon dioxide and water (Shayegan et al. 2018). Metal–organic frameworks with high porosity and selectivity are well-suited for trapping low-concentration VOCs, while functionalised carbon-based nanomaterials, including graphene oxide

Table 4: Smart Nanomaterials for Air Pollution Mitigation.

Nanomaterial Type	Target Air Pollutants	Mechanism of Action	Performance/Removal Efficiency	Advantages	Limitations	References
Photocatalytic Nanomaterials (TiO_2, ZnO, g-C_3N_4)	VOCs (formaldehyde, toluene), NOx, SOx, CO, PM	Photocatalytic oxidation under light (UV/visible)	VOCs: >90%; NOx: ~80%	Solar-driven; broad-spectrum activity	UV light dependence (TiO_2); nanoparticle leaching	(Zhao et al. 2020)
Metal-Organic Frameworks (MOFs)	VOCs, CO_2, SOx, NOx, benzene	Physical adsorption; selective binding via porous structures	CO_2: ~95%; VOCs: ~85%	High surface area; tunable selectivity	Sensitivity to humidity; regeneration challenges	(Mammadova et al. 2022)
Carbon-Based Nanomaterials (Graphene, CNTs, Carbon Dots)	VOCs, CO_2, $PM_{2.5}$, SOx	Adsorption, electron transfer, filtration enhancement	CO_2: 80–90%; $PM_{2.5}$: ~95%	High surface area; reusable; functionalization options	Aggregation risk; potential toxicity	(Zhao et al. 2020)
Magnetic Nanomaterials (Fe_3O_4, $CoFe_2O_4$)	$PM_{2.5}$, VOCs, CO_2	Adsorption, magnetic recovery, electron transfer	$PM_{2.5}$: >90%; VOCs: ~80%	Easy recovery; reusability	Surface oxidation; cost of magnetic separation	(Latef et al. 2025)
Stimuli-Responsive Nanomaterials	VOCs, NOx, SOx, CO_2	pH/temperature/light-triggered adsorption & release	VOCs: ~85%; NOx: ~80%	Selectivity; on-demand capture & release	Limited field studies; scalability issues	
Bio-Inspired Nanomaterials (Enzyme-functionalized NPs)	VOCs, NOx, CO_2	Catalytic breakdown via biomimetic pathways	VOCs: ~90%; NOx: ~75%	Selectivity; eco-friendly	Stability of enzymes; cost	(Sharma et al. 2022)
Hybrid Nanomaterials (TiO_2/Graphene, MOF/Carbon)	Mixed pollutants (VOCs + CO_2 + PM)	Combined adsorption, photocatalysis, filtration	VOCs: >90%; CO_2: ~85%; $PM_{2.5}$: >95%	Multifunctional; synergistic effects	Complex synthesis; expensive	

and carbon nanotubes (CNTs), exhibit a strong affinity towards aromatic VOCs due to their oxygen-containing functional groups (Xie et al. 2022). Additionally, stimuli-responsive nanomaterials can self-regenerate after VOC adsorption through light- or heat-induced desorption, enabling prolonged and efficient use (Nawaz et al. 2024).

These advanced nanomaterials are being applied in various air purification technologies, including indoor air purifiers, smart HVAC systems, and VOC filters in industrial exhausts. They are also integrated into coatings for self-cleaning surfaces in hospitals, public buildings, and other sensitive environments, contributing to improved air quality and reduced health risks from VOC exposure.

7.2 Capture of Greenhouse Gases and Particulate Matter

Mitigating greenhouse gas emissions (CO_2, CH_4, N_2O) and controlling particulate matter (PM) is essential for addressing climate change and improving air quality. Smart nanomaterials offer innovative solutions for the selective capture, conversion, and separation of these pollutants.

Nanomaterials are playing a pivotal role in greenhouse gas capture technologies, offering innovative solutions for reducing atmospheric CO_2 levels. Amine-functionalised nanomaterials, such as silica or polymeric nanocomposites, chemically bind CO_2 with high selectivity, while photocatalytic nanomaterials can harness solar energy to convert CO_2 into valuable chemicals like methanol and formic acid. Metal–organic frameworks and zeolites, with their precisely engineered pore sizes, further enhance CO_2 capture by selectively adsorbing it over other gases (Ashley et al. 2012). Additionally, magnetic nanocomposites offer the advantage of easy recovery and regeneration, making the capture process more efficient and economically viable (Sun et al. 2024).

In the field of particulate matter (PM) filtration, nanomaterials are transforming air purification strategies. Nanofibre membranes and electrospun mats are highly effective in trapping fine particles such as PM2.5 and ultrafine particulates, while still maintaining air permeability for practical use in filtration systems (Xiao et al. 2018). Advanced smart coatings on filters can further degrade PM-bound organic pollutants or bacteria under light irradiation, providing a multifunctional approach to air quality management (H. Liu et al. 2020). Electrostatic nanomaterials, with their charge-responsive surfaces, enhance the capture of airborne particles in dynamic environments, offering robust performance even under variable conditions (Ji et al. 2023). These innovations highlight the potential of nanotechnology in addressing pressing environmental challenges such as air pollution and climate change.

7.3 Smart Filters and Coatings

Smart filters and coatings represent a dynamic category of nanotechnology-enabled air purification systems that adapt to environmental stimuli such as light, humidity, temperature, and pollutant concentration.

Nanotechnology has enabled the development of advanced materials with diverse functions tailored for environmental and health applications. Photo-responsive coatings, incorporating materials like titanium dioxide (TiO_2) or graphitic carbon nitride (g-C_3N_4), can degrade airborne pathogens and pollutants upon exposure to light, offering a passive and energy-efficient approach to air purification (Njema and Kibet 2024). Thermo-responsive polymers, on the other hand, can release or capture pollutants depending on the surrounding temperature, enabling dynamic control over contaminant capture based on environmental conditions (Musarurwa and Tavengwa 2022). Nanocomposite coatings infused with silver or copper nanoparticles demonstrate potent antimicrobial and antifungal properties, making them ideal for use in medical settings, hospitals, and cleanroom environments where sterility is critical (Ealias and Saravanakumar 2017). Additionally, electrospun nanofibre filters, embedded with functional nanoparticles, provide multi-layer filtration capabilities and can incorporate self-cleaning features, ensuring sustained performance and reduced maintenance

(Le et al. 2024). These advancements highlight the versatility of nanomaterials in creating multifunctional systems for pollution control and public health protection. Nanomaterials have found diverse applications in enhancing air quality and environmental safety. They are increasingly used in air conditioning systems, automotive cabins, face masks, and industrial air scrubbers for effective filtration of pollutants. Self-decontaminating wall paints and glass coatings, particularly in hospitals, schools, and public buildings, help maintain sterile and hygienic environments. Additionally, wearable air purifiers with integrated nanofibre membranes offer personal exposure reduction, enabling protection from airborne pollutants in everyday settings.

8. Toxicity, Environmental Impact, and Risk Assessment

While smart nanomaterials hold immense promise for environmental remediation, their increased use raises important concerns regarding their toxicity, ecotoxicological impacts, and long-term environmental behaviour. Nanomaterials can interact with biological systems in unpredictable ways due to their nanoscale size, high reactivity, and surface area, potentially leading to unintended toxicological effects. Thus, a thorough understanding of their risk profile and environmental footprint is essential for sustainable deployment. This section explores the key challenges and current approaches in assessing and mitigating the toxicity and environmental risks associated with smart nanomaterials.

8.1 Ecotoxicological Concerns

Smart nanomaterials can enter ecosystems through various pathways—during manufacturing, application, disposal, or accidental release. Once in the environment, they may interact with soil, water, and living organisms, leading to ecotoxic effects that vary depending on their composition, shape, surface chemistry, and dosage.

Key environmental and ecological concerns have emerged regarding the use of engineered nanomaterials. Aquatic toxicity is a significant issue, as nanomaterials such as silver nanoparticles (AgNPs), zinc oxide (ZnO), and titanium dioxide (TiO_2) can exhibit toxic effects on aquatic organisms like algae, daphnia, and fish, potentially disrupting food chains and altering aquatic ecosystem dynamics (Sibiya et al. 2022). Bioaccumulation and biomagnification are also critical concerns, as certain nanomaterials may accumulate in the tissues of organisms and transfer across trophic levels, posing risks to both wildlife and humans (Isibor et al. 2024). In soils, exposure to nanomaterials can lead to microbial disruption, altering microbial diversity, enzymatic activity, nutrient cycling, and soil fertility, which are vital for ecosystem stability (Simonin and Richaume 2015). Furthermore, the generation of reactive oxygen species (ROS) by many nanoparticles can damage cellular components such as DNA, proteins, and lipids in plants and animals, leading to oxidative stress and cytotoxicity (Horie and Tabei 2021). For instance, studies have shown that carbon nanotubes can cause respiratory toxicity in mammals and inhibit root elongation in plants, while metal oxide nanoparticles, depending on concentration and exposure routes, can negatively affect cell viability, photosynthesis, and reproduction in aquatic organisms. These findings underscore the importance of understanding and mitigating the potential ecological risks associated with nanomaterial use.

8.2 Life Cycle Assessment (LCA)

Life Cycle Assessment is a systematic method for evaluating the environmental impacts of a product or technology throughout its entire lifespan, from raw material extraction to disposal. LCA is crucial in understanding whether the environmental benefits of smart nanomaterials outweigh their production and end-of-life impacts.

LCA of smart nanomaterials encompasses several critical components that determine their overall environmental impact (Nizam et al. 2021). The first stage, raw material extraction and synthesis, involves assessing energy consumption and emissions generated during nanoparticle production, particularly through methods like hydrothermal or chemical synthesis (Hischier and Walser 2012). The usage phase evaluates the balance between benefits, such as pollutant removal efficiency in environmental remediation, and the associated energy or resource inputs required to achieve these outcomes (Gallagher et al. 2017). Disposal and recycling present additional challenges, as nanomaterials may persist in landfills or wastewater, raising concerns about long-term environmental risks and highlighting the need for effective recovery or degradation strategies. Finally, transport and packaging contribute to the overall carbon footprint, with emissions generated through logistics, storage, and distribution of nanomaterials requiring careful consideration to minimise their environmental burden (Salahshoori et al. 2024).

8.3 Regulatory Guidelines and Safe Use Practices

As the application of nanotechnology expands, regulatory bodies worldwide are working to develop guidelines and frameworks for the safe manufacture, use, and disposal of nanomaterials. OECD guidelines, the European Union's REACH regulation, the US EPA and FDA, and India's DST and MoEFCC are some regulatory organisations that have initiated frameworks for nanotechnology safety and risk management in industrial applications and provide guidance on environmental and health assessments for engineered nanomaterials in commercial use.

9. Recent Innovations and Case Studies

The development of smart nanomaterials has advanced rapidly in recent years, moving beyond laboratory-scale proof-of-concept studies to tangible applications in real-world environmental remediation challenges. This transition is marked by innovations in synthesis, functionalisation, and integration techniques, coupled with pilot-scale and field deployments demonstrating effectiveness and scalability. This section highlights key innovations and presents illustrative case studies showcasing the practical impact of smart nanomaterials in diverse environmental settings, including applications in developing countries and large-scale industrial and municipal contexts.

9.1 Lab-to-Field Translations

One of the most critical phases in the development of smart nanomaterials is the successful translation of laboratory findings into field-applicable technologies. This involves scaling up production while maintaining material quality, ensuring operational stability under variable environmental conditions, and addressing cost-effectiveness.

Recent advancements in nanomaterials have led to the development of innovative and scalable green synthesis methods, enabling the eco-friendly and cost-effective production of nanomaterials using plant extracts, microbial synthesis, and biopolymer templates. Such techniques are now viable for bulk manufacturing, offering a sustainable alternative to conventional methods (Ying et al. 2022). Furthermore, the emergence of modular nanocomposite systems, which combine magnetic, photocatalytic, and adsorptive functionalities, has enhanced the multifunctionality of these materials. This modular approach not only facilitates diverse remediation applications but also allows for easy recovery and reuse of nanomaterials, improving their practicality in real-world scenarios (Javaid et al. 2022). Additionally, the development of smart, stimuli-responsive nanomaterials that dynamically adjust their surface charge, catalytic activity, or other properties in response to environmental changes has significantly improved the efficiency and adaptability of nanomaterials in heterogeneous field conditions. These advancements collectively contribute to the growing potential of nanotechnology in addressing environmental challenges through sustainable and efficient solutions.

Case Examples:

- **Photocatalytic nanomaterials for dye degradation:** TiO_2-based nanostructures, optimized in the lab, have been successfully integrated into solar reactors deployed in textile wastewater treatment plants, achieving high degradation rates with solar energy.
- **Magnetic nano-adsorbents for heavy metal removal:** Lab-scale magnetic iron oxide nanoparticles coated with biopolymers have been scaled for *in-situ* treatment of arsenic-contaminated groundwater, allowing rapid adsorption and magnetic separation onsite.

9.2 Real-World Applications in Developing Countries

Developing countries face unique environmental remediation challenges due to limited infrastructure, high pollutant loads, and resource constraints. Smart nanomaterials offer tailored solutions that are often low-cost, easy to deploy, and compatible with existing technologies.

Key Applications:

- **Arsenic and fluoride removal from groundwater:** Nanomaterials such as iron oxide-coated sand and nZVI have been used successfully in community-scale water purification units in countries like India and Bangladesh (Patel et al. 2022, Zhu et al. 2023).
- **Decentralized wastewater treatment:** Composite nanomaterial membranes and photocatalysts incorporated into low-maintenance reactors help treat domestic and industrial wastewater in areas lacking centralized sewage systems (Iyyappan et al. 2024).
- **Agricultural soil remediation:** Biochar-supported nanoparticles enhance nutrient availability and immobilize pesticides and heavy metals, promoting sustainable farming practices in contaminated regions (Arabzadeh Nosratabad et al. 2024).

Case Studies:

- In rural India, magnetic nanocomposite filters have been deployed to reduce lead and cadmium in drinking water, providing affordable and user-friendly solutions to remote communities (Patel et al. 2022).
- A pilot project in Kenya utilized solar-driven photocatalytic reactors with smart nanomaterials to treat hospital wastewater, significantly reducing organic load and pathogen content without reliance on electricity (Constantin et al. 2024).

9.3 Industrial and Municipal Use Cases

Beyond small-scale applications, industrial and municipal sectors are increasingly adopting smart nanomaterials to meet stricter environmental regulations and improve sustainability.

Nanomaterial-based solutions are emerging as powerful tools for addressing environmental challenges across various industrial sectors. In wastewater treatment, nanomaterial-based catalysts and adsorbents have demonstrated exceptional efficiency in degrading complex pollutants such as dyes, phenols, and heavy metals. This advancement enables industries, including the textile and chemical sectors, to progress towards zero liquid discharge goals, reducing environmental impact (Bassey and Bassey 2024). In air pollution control, smart nano-coatings applied to filters capture VOCs and particulate matter from industrial stacks, offering enhanced efficiency and self-cleaning capabilities that extend the lifespan of air filtration systems. Furthermore, nanotechnology plays a critical role in soil remediation at contaminated industrial sites, where nano-remediation strategies—such as the use of nZVI and metal oxide nanoparticles—are employed to detoxify sites contaminated with chlorinated solvents and heavy metals, restoring the health of ecosystems and reducing human exposure to hazardous substances (Mech et al. 2022). Together, these innovative approaches highlight the transformative potential of nanotechnology in environmental management and industrial sustainability (Gottardo et al. 2021).

Nanomaterials have revolutionised several aspects of water and waste management, offering advanced solutions for environmental challenges. In drinking water treatment plants, nanomaterial-enhanced filtration membranes are increasingly used to improve the removal of emerging contaminants, including pharmaceuticals and endocrine disruptors, ensuring safer drinking water for communities. For landfill leachate treatment, magnetic nano-adsorbents have proven highly effective in recovering heavy metals and organic pollutants, thereby reducing the environmental impact of landfill sites and facilitating resource recovery (Gupta et al. 2020). Additionally, in sludge management, the application of nanomaterial catalysts accelerates biodegradation processes while simultaneously minimising odour emissions, enhancing the overall efficiency and sustainability of sludge treatment operations (Aich et al. 2014). Together, these advancements underscore the transformative potential of nanotechnology in addressing complex environmental issues in water and waste sectors.

Case Examples:

- The use of graphene oxide-based membranes in a municipal wastewater treatment plant in Europe demonstrated superior removal of micropollutants compared to conventional membranes (Januário et al. 2021).
- An industrial pilot implemented magnetic nanocomposites for heavy metal capture from electroplating wastewater, achieving cost-effective metal recovery and water reuse (Waqas and Ahmad 2024).

10. Challenges and Future Perspectives

Despite the remarkable progress and promising applications of smart nanomaterials in environmental remediation, several challenges remain that must be addressed to unlock their full potential and ensure sustainable, responsible deployment. This section discusses key hurdles related to scaling up and commercialisation, the imperative of aligning nanomaterial technologies with circular economy principles, and emerging trends that will shape the future design of smart nanomaterials.

10.1 Scale-Up and Commercialization

One of the foremost challenges in advancing smart nanomaterials from laboratory research to widespread practical use is the scale-up of synthesis and fabrication processes while maintaining material quality, reproducibility, and cost-effectiveness.

Key Challenges:

- Manufacturing Consistency: Nanomaterial properties—such as size, morphology, surface functionality—critically influence their performance. Achieving uniformity at large scale is technically complex and can impact efficacy.
- Cost of Production: Many synthesis techniques, especially those involving high-purity precursors or sophisticated functionalization, remain expensive. Developing affordable green synthesis and continuous manufacturing methods is essential.
- Material Stability: Nanomaterials must maintain functionality over prolonged periods and under diverse environmental conditions, resisting aggregation, fouling, or degradation.
- Safety and Handling: Commercialization requires protocols for safe production, storage, transportation, and disposal to prevent occupational and environmental hazards.

- Regulatory Approval: Navigating regulatory frameworks and obtaining approvals based on comprehensive risk assessments can delay market entry.

Opportunities:

- Development of continuous flow reactors and automation in nanomaterial synthesis to enhance throughput and quality control.
- Scale-up of bio-based synthesis approaches to reduce costs and environmental footprint.
- Public-private partnerships and government incentives to support pilot-scale demonstrations and commercialization efforts.

10.2 Integration with Circular Economy and Sustainable Practices

Sustainability is a critical lens through which future nanomaterial development must be viewed, emphasizing resource efficiency, recyclability, and minimal environmental footprint.

Integration Challenges:

- Nanomaterial Recovery and Reuse: Post-remediation recovery of nanomaterials, especially after use in water or soil, is often difficult. Lost nanoparticles may pose environmental risks or represent wasted resources.
- End-of-Life Management: Strategies for degradation, recycling, or safe disposal of nanomaterials are still underdeveloped.
- Life Cycle Impacts: Complete life cycle assessments (LCA) need to be incorporated early in design to avoid shifting burdens from remediation to production or disposal phases.

Sustainable Practices:

- Designing biodegradable and eco-friendly nanomaterials that break down into harmless products after use.
- Embedding nanomaterials within recyclable composite matrices to facilitate recovery.
- Employing waste valorisation approaches, such as using agricultural residues or industrial by-products as raw materials for green synthesis.

10.3 Future Trends in Smart Nanomaterial Design

The next generation of smart nanomaterials is expected to be shaped by advances in materials science, nanotechnology, and digital technologies, fostering multifunctionality, adaptability, and eco-compatibility. Some of these future trends are discussed in this section.

Multifunctional nanomaterials are smart nanomaterials formed by the integration of combined properties—e.g., magnetic, photocatalytic, and adsorptive—into single materials for enhanced remediation efficiency (Reddy Esanakula et al. 2025). Bioinspired and biomimetic designs draw inspiration from natural systems, enabling the development of nanomaterials that exhibit selective pollutant targeting, self-healing capabilities, or adaptive responses to environmental stimuli (Srinivas et al. 2025). At the nano-bio interface, the constructive interaction between nanomaterials and microorganisms is harnessed to enhance remediation strategies, such as nanomaterial-augmented bioremediation processes. Smart sensors and responsive systems integrate sensing capabilities within nanomaterials, allowing for real-time monitoring of pollutants and controlled activation of remediation functions (Husen and Salahuddin Siddiqi 2023). AI and ML further support the field by enabling predictive modelling of nanomaterial behaviour, optimising synthesis methods, and tailoring designs for specific environmental applications (Onamade et al. 2025). Underpinning these innovations is the commitment to circular and sustainable design principles, ensuring that materials are developed with reusability, recyclability, and biodegradability in mind, aligning with the broader goals of sustainability and environmental stewardship (Sohail et al. 2019).

11. Conclusion

Smart nanomaterials represent a cutting-edge frontier in environmental remediation, offering innovative solutions to some of the most pressing pollution challenges faced worldwide. Their unique size-dependent properties, tunable surface chemistry, and responsiveness to external stimuli enable highly efficient and selective removal of contaminants from water, soil, and air. This versatility positions them as powerful tools in addressing heavy metals, organic pollutants, pathogens, and even complex mixtures of contaminants. The reviewed classifications—stimuli-responsive, magnetic, photocatalytic, and biodegradable nanomaterials—demonstrate the breadth of design strategies currently employed to tailor materials for specific environmental applications. Advances in green synthesis and functionalisation have made the production of these materials more sustainable and scalable, yet significant challenges remain. Commercial-scale manufacturing must overcome issues of reproducibility, cost, and material stability without compromising environmental safety. Moreover, the potential ecotoxicological impacts of nanomaterials necessitate rigorous life cycle assessments and the development of regulatory frameworks to ensure safe deployment. Integration within circular economy models is essential to maximise resource efficiency, promote recycling, and minimise waste generation associated with nanomaterial use. Looking ahead, the future of smart nanomaterials lies in the development of multifunctional, adaptive systems that combine remediation with real-time sensing and self-regulation. Bioinspired designs and nano-bio interfaces offer exciting pathways for synergistic remediation approaches, while digital technologies such as artificial intelligence promise to revolutionise material design and application optimisation. Continued interdisciplinary collaboration among material scientists, environmental engineers, toxicologists, and policymakers is crucial to overcoming existing barriers. In conclusion, smart nanomaterials hold transformative potential to enhance environmental remediation technologies significantly. By addressing current limitations and embracing sustainable innovation, they can contribute substantially to cleaner ecosystems and healthier communities globally.

References

Aghababai Beni, A., and Jabbari, H. (2022). Nanomaterials for environmental applications. *Results in Engineering*, 15: 100467. https://doi.org/10.1016/J.RINENG.2022.100467.

Ahmad, F., Salem-Bekhit, M. M., Khan, F., Alshehri, S., Khan, A., Ghoneim, M. M. et al. (2022). Unique properties of surface-functionalized nanoparticles for bio-application: functionalization mechanisms and importance in application. *Nanomaterials,* 12(8): 1333. https://doi.org/10.3390/NANO12081333.

Aich, N., Plazas-Tuttle, J., Lead, J. R., and Saleh, N. B. (2014). A critical review of nanohybrids: synthesis, applications and environmental implications. *Environmental Chemistry*, 11(6): 609–623. https://doi.org/10.1071/EN14127.

Amin, S., Solangi, A. R., Hassan, D., Hussain, N., Ahmed, J., and Baksh, H. (2021). Recent trends in development of nanomaterials based green analytical methods for environmental remediation. *Current Analytical Chemistry*, 17(4): 438–448. https://doi.org/10.2174/1573411016666200319100707.

Anani, O. A., Inobeme, A., Osarenotor, O., Olisaka, F. N., Aidonojie, P. A., Olatunji, E. O. et al. (2022). Application of microorganisms as biofactories to produce biogenic nanoparticles for environmental cleanup: currents advances and challenges. *Current Nanoscience*, 19(6): 770–782. https://doi.org/10.2174/1573413719666221219164613/CITE/REFWORKS.

Arabzadeh Nosratabad, N., Yan, Q., Cai, Z., and Wan, C. (2024). Exploring nanomaterial-modified biochar for environmental remediation applications. *Heliyon*, 10(18): e37123. https://doi.org/10.1016/J.HELIYON.2024.E37123.

Ashley, M., Magiera, C., Ramidi, P., Blackburn, G., Scott, T. G., Gupta, R. et al. (2012). Nanomaterials and processes for carbon capture and conversion into useful by-products for a sustainable energy future. *Greenhouse Gases: Science and Technology*, 2(6): 419–444. https://doi.org/10.1002/GHG.1317.

Ata, S., Tabassum, A., Bibi, I., Majid, F., Sultan, M., Ghafoor, S. et al. (2019). Lead remediation using smart materials. A review. *Zeitschrift Fur Physikalische Chemie*, 233(10): 137–1409. https://doi.org/10.1515/ZPCH-2018-1205/ASSET/GRAPHIC/J_ZPCH-2018-1205_FIG_004.JPG.

Ayeleru, O. O., Sadare, O. O., Idris, A. O., Pandey, S., and Olubambi, P. A. (2025). Introduction to smart nanomaterials for environmental remediation. *Smart Nanomaterials for Environmental Applications*, 3–28. https://doi.org/10.1016/B978-0-443-21794-4.00008-9.

Bassey, K. E., and Bassey, K. E. (2024). From waste to wonder: Developing engineered nanomaterials for multifaceted applications. *GSC Advanced Research and Reviews*, 2024(03): 109–123. https://doi.org/10.30574/gscarr.2024.20.3.0326.

Cerra, S., and Fratoddi, I. (2024). Nanoscale Solutions: The Transformative Applications of Functionalized Nanomaterials in Environmental Remediation. *Applied Nano,* 5(1): 14–19. https://doi.org/10.3390/APPLNANO5010002.

Chang, J., Zhang, L., and Wang, P. (2018). Intelligent environmental nanomaterials. *Environmental Science: Nano*, 5(4): 811–836. https://doi.org/10.1039/C7EN00760D.

Chauhan, P., Pandit, J., Harkhani, K., Tripathi, A., Kumar, N., Patel, M. V. et al. (2025). Nanomaterials a promising approach for environmental restoration. *Nanotechnology for Environmental Engineering*, 10(1): 1–18. https://doi.org/10.1007/S41204-025-00411-1.

Chidichimo, G., Cupelli, D., De Filpo, G., Formoso, P., and Nicoletta, F. P. (2013). Nanoparticles as a Smart Technology for Remediation. *Sustainable Development in Chemical Engineering: Innovative Technologies*, 297–348. https://doi.org/10.1002/9781118629703.CH12.

Constantin, M. A., Constantin, L. A., Ionescu, I. A., Nicolescu, C. M., Bumbac, M., and Tiron, O. (2024). Performance of a Solar-Driven Photocatalytic Membrane Reactor for Municipal Wastewater Treatment. *Processes*, 12(3): 617. https://doi.org/10.3390/PR12030617.

Corsi, I., Venditti, I., Trotta, F., and Punta, C. (2023). Environmental safety of nanotechnologies: The eco-design of manufactured nanomaterials for environmental remediation. *Science of The Total Environment*, 864: 161181. https://doi.org/10.1016/J.SCITOTENV.2022.161181.

Das, S., Sen, B., and Debnath, N. (2015). Recent trends in nanomaterials applications in environmental monitoring and remediation. *Environmental Science and Pollution Research*, 22(23): 18333–18344. https://doi.org/10.1007/S11356-015-5491-6/METRICS.

Dejen, K. D., Sabir, F. K., Ananda Murthy, H. C., Ayanie, G. T., Shume, M. S., and Bekele, E. T. (2023). Green Synthesis of Nanomaterials for Environmental Remediation. *Green Nanoremediation: Sustainable Management of Environmental Pollution*, 27–65. https://doi.org/10.1007/978-3-031-30558-0_2.

Dong, A., Chen, D., Li, Q., and Qian, J. (2023). Metal-Organic Frameworks for Greenhouse Gas Applications. *Small*, 19(10): 2201550. https://doi.org/10.1002/SMLL.202201550.

Ealias, A. M., and Saravanakumar, M. P. (2017). A review on the classification, characterisation, synthesis of nanoparticles and their application. *IOP Conference Series: Materials Science and Engineering*, 263(3): 032019. https://doi.org/10.1088/1757-899X/263/3/032019.

Elsaid, K., Olabi, A. G., Abdel-Wahab, A., Elkamel, A., Alami, A. H., Inayat, A. et al. (2023). Membrane processes for environmental remediation of nanomaterials: Potentials and challenges. *Science of The Total Environment*, 879: 162569. https://doi.org/10.1016/J.SCITOTENV.2023.162569.

Farzin, M. A., Naghib, S. M., and Rabiee, N. (2024). Bio-inspired and biomimetic composites based on biodegradable polymers for sensing applications with emphasis on early diagnosis of cancer. *Chemical Engineering Journal*, 493: 152445. https://doi.org/10.1016/J.CEJ.2024.152445.

Fuku, X., Dyosiba, X., and Iftikhar, F. J. (2023). Green prepared nanomaterials from various biodegradable wastes and their application in energy. *Nano-Structures & Nano-Objects*, 35: 100997. https://doi.org/10.1016/J.NANOSO.2023.100997.

Gallagher, M. J., Allen, C., Buchman, J. T., Qiu, T. A., Clement, P. L., Krause, M. O. P. et al. (2017). Research highlights: applications of life-cycle assessment as a tool for characterizing environmental impacts of engineered nanomaterials. *Environmental Science: Nano*, 4(2): 276–281. https://doi.org/10.1039/C7EN90005H.

Gautam, R. K., and Chattopadhyaya, M. C. (2016). Advanced Nanomaterials for Wastewater Remediation. *Advanced Nanomaterials for Wastewater Remediation*, 1–414. https://doi.org/10.1201/9781315368108/Advanced-Nanomaterials-Wastewater-Remediation-Ravindra-Kumar-Gautam-Mahesh-Chandra-Chattopadhyaya.

Ghanghas, P., Anita, Yadav, A., and Poonia, K. (2025). Advantages of nanomaterials for environmental remediation. *Nanomaterials in Environmental Remediation*, 93–108. https://doi.org/10.1201/9781003484189-6.

Gottardo, S., Mech, A., Drbohlavová, J., Małyska, A., Bøwadt, S., Riego Sintes, J. et al. (2021). Towards safe and sustainable innovation in nanotechnology: State-of-play for smart nanomaterials. *NanoImpact*, 21: 100297. https://doi.org/10.1016/J.IMPACT.2021.100297.

Guerra, F. D., Attia, M. F., Whitehead, D. C., and Alexis, F. (2018). Nanotechnology for Environmental Remediation: Materials and Applications. *Molecules*, 23(7): 1760. https://doi.org/10.3390/MOLECULES23071760.

Gupta, A., Kumar, V., Ahmed, S., and Gautam, S. (2020). Impact of Nanotechnology in the Development of Smart Cities. *Lecture Notes in Civil Engineering*, 58: 845–857. https://doi.org/10.1007/978-981-15-2545-2_68.

Hischier, R., and Walser, T. (2012). Life cycle assessment of engineered nanomaterials: State of the art and strategies to overcome existing gaps. *Science of The Total Environment*, 425: 271–282. https://doi.org/10.1016/J.SCITOTENV.2012.03.001.

Horie, M., and Tabei, Y. (2021). Role of oxidative stress in nanoparticles toxicity. *Free Radical Research*, 55(4): 331–342. https://doi.org/10.1080/10715762.2020.1859108.

Husen, A., and Salahuddin Siddiqi, K. (2023). *Advances in Smart Nanomaterials and their Applications*. Elsevier. https://books.google.co.in/books?hl=en&lr=&id=2aCSEAAAQBAJ&oi=fnd&pg=PP1&dq=Future+Trends+in+

Smart+Nanomaterial+Design+for+environmental+remediation&ots=hzUJ6fvi8x&sig=EftDzIXkXI2Q3Mq MapdHyYK8zL8&redir_esc=y#v=onepage&q&f=false.

Iqbal, H. M. N., Bilal, M., and Rodriguez-Couto, S. (2022). Smart nanohybrid constructs: Concept and designing for environmental remediation. *Chemosphere*, 301: 134616. https://doi.org/10.1016/J.CHEMOSPHERE.2022.134616.

Isibor, P. O., Oyegbade, S. A., Oni, J. G., Ahmed, W. W., Abiodun, E. O., and Oyewole, O. A. (2024). Nanoparticles in Food Chains: Bioaccumulation and Trophic Transfer. *Environmental Nanotoxicology*, 203–233. https://doi.org/10.1007/978-3-031-54154-4_11.

Iyyappan, J., Gaddala, B., Gnanasekaran, R., Gopinath, M., Yuvaraj, D., and Kumar, V. (2024). Critical review on wastewater treatment using photo catalytic advanced oxidation process: Role of photocatalytic materials, reactor design and kinetics. *Case Studies in Chemical and Environmental Engineering*, 9: 100599. https://doi.org/10.1016/J.CSCEE.2023.100599.

Jadhav, S., Yadav, H., Sankhla, M. S., Sonone, S. S., Parihar, K., and Singh, P. (2024). Green-Synthesis of Nanomaterials for Environmental Remediation. *Nano-Phytoremediation and Environmental Pollution*, 84–96. https://doi.org/10.1201/9781003186298-5/Green-Synthesis-Nanomaterials-Environmental-Remediation-Swapnali-Jadhav-Himanshu-Yadav-Mahipal-Singh-Sankhla-Swaroop-Sonone-Kapil-Parihar-Prashant-Singh.

Januário, E. F. D., Vidovix, T. B., Beluci, N. de C. L., Paixão, R. M., Silva, L. H. B. R. da., Homem, N. C. et al. (2021). Advanced graphene oxide-based membranes as a potential alternative for dyes removal: A review. *Science of The Total Environment*, 789: 147957. https://doi.org/10.1016/J.SCITOTENV.2021.147957.

Javaid, A., Imran, M., Latif, S., Hussain, N., and Bilal, M. (2022). Functionalized magnetic nanostructured composites and hybrids for photocatalytic elimination of pharmaceuticals and personal care products. *Science of The Total Environment*, 849: 157683. https://doi.org/10.1016/J.SCITOTENV.2022.157683.

Javed, R., Zia, M., Naz, S., Aisida, S. O., Ain, N. ul., and Ao, Q. (2020). Role of capping agents in the application of nanoparticles in biomedicine and environmental remediation: recent trends and future prospects. *Journal of Nanobiotechnology*, 18(1): 1–15. https://doi.org/10.1186/S12951-020-00704-4.

Ji, X., Huang, J., Teng, L., Li, S., Li, X., Cai, W. et al. (2023). Advances in particulate matter filtration: Materials, performance, and application. *Green Energy & Environment*, 8(3): 673–697. https://doi.org/10.1016/J.GEE.2022.03.012.

Latef, A. A. H. A., Zayed, E. M., and Omar, A. A. (Eds.). (2025). *Bioremediation and Nanotechnology for Climate Change Mitigation*. Springer. Retrieved 2 June 2025, from https://books.google.co.in/books?hl=en&lr=&id=rSRaEQAAQBAJ&oi=fnd&pg=PA193&dq=Photocatalytic+Nanomaterials+for+air+pollution+mitigation&ots=AAGqfo6U69&sig=RtvqsY6_MwAUqnX3EWf05cBctTI&redir_esc=y#v=onepage&q=Photocatalytic%20Nanomaterials%20for%20air%20pollution%20mitigation&f=false.

Le, B., Omran, N., Elnabawy, E., Hassanin, A. H., Mahmoud, K., Shehata, N. et al. (2024). Exploring advances in nanofiber-based face masks: a comprehensive review of mechanical, electrostatic, and antimicrobial functionality filtration for the removal of airborne particulate matter and pathogens. *Emergent Materials*, 7(3): 765–800. https://doi.org/10.1007/S42247-023-00622-9/FIGURES/22.

Liu, H., Cao, C., Huang, J., Chen, Z., Chen, G., and Lai, Y. (2020). Progress on particulate matter filtration technology: basic concepts, advanced materials, and performances. *Nanoscale*, 12(2): 437–453. https://doi.org/10.1039/C9NR08851B.

Liu, S., Low, Z. X., Xie, Z., and Wang, H. (2021). TEMPO-Oxidized Cellulose Nanofibers: A Renewable Nanomaterial for Environmental and Energy Applications. *Advanced Materials Technologies*, 6(7): 2001180. https://doi.org/10.1002/ADMT.202001180.

Liu, S., Yu, B., Wang, S., Shen, Y., and Cong, H. (2020). Preparation, surface functionalization and application of Fe3O4 magnetic nanoparticles. *Advances in Colloid and Interface Science*, 281: 102165. https://doi.org/10.1016/J.CIS.2020.102165.

Liu, Y., Jiao, Y., Zhang, Z., Qu, F., Umar, A., and Wu, X. (2014). Hierarchical SnO2 nanostructures made of intermingled ultrathin nanosheets for environmental remediation, smart gas sensor, and supercapacitor applications. *ACS Applied Materials and Interfaces*, 6(3): 2174–2184. https://doi.org/10.1021/AM405301V/SUPPL_FILE/AM405301V_SI_001.PDF.

M. Al-Khayri, J., Anju T. R., and Jain Shri Mohan. (2025). *Nanomaterial Green Synthesis*. 2. https://doi.org/10.1007/978-3-031-84643-4.

Mammadova, S., Nasibova, A., Khalilov, R., Mehraliyeva, S., Valiyeva, M., Gojayev, A. et al. (2022). Nanomaterials application in air pollution remediation. *Eurasian Chemical Communications*, 4(2): 160–166.

Mech, A., Gottardo, S., Amenta, V., Amodio, A., Belz, S., Bøwadt, S. et al. (2022). Safe- and sustainable-by-design: The case of Smart Nanomaterials. A perspective based on a European workshop. *Regulatory Toxicology and Pharmacology*, 128: 105093. https://doi.org/10.1016/J.YRTPH.2021.105093.

Mensah, M. B., Lewis, D. J., Boadi, N. O., and Awudza, J. A. M. (2021). Heavy metal pollution and the role of inorganic nanomaterials in environmental remediation. *Royal Society Open Science*, 8(10). https://doi.org/10.1098/RSOS.201485.

Mitra, A. K., and Saha, R. (2025). Functionalized Nanomaterials : Properties to Perspective Applications in Environmental Remediation. *Nanomaterials in Environmental Remediation*, 127–152. https://doi.org/10.1201/9781003484189-8.

Mondal, K., and Sharma, A. (2016). Recent advances in the synthesis and application of photocatalytic metal–metal oxide core–shell nanoparticles for environmental remediation and their recycling process. *RSC Advances*, 6(87): 83589–83612. https://doi.org/10.1039/C6RA18102C.

Mondal, S., and Palit, D. (2022). Prospects and implementation of nanotechnology in environmental remediation and clean up. *Natural Resources Conservation and Advances for Sustainability*, 271–287. https://doi.org/10.1016/B978-0-12-822976-7.00020-X.

Musarurwa, H., and Tavengwa, N. T. (2022). Thermo-responsive polymers and advances in their applications in separation science. *Microchemical Journal*, 179: 107554. https://doi.org/10.1016/J.MICROC.2022.107554.

Nawaz, F., Ali, M., Ahmad, S., Yong, Y., Rahman, S., Naseem, M. et al. (2024). Carbon based nanocomposites, surface functionalization as a promising material for VOCs (volatile organic compounds) treatment. *Chemosphere*, 364: 143014. https://doi.org/10.1016/J.CHEMOSPHERE.2024.143014.

Nayak, T., and Pathan, A. (2023). Environmental remediation and application of carbon-based nanomaterials in the treatment of heavy metal-contaminated water: A review. *Materials Today: Proceedings*, 92: 1659–1670. https://doi.org/10.1016/J.MATPR.2023.06.227.

Nazarzadeh Zare, E., Mudhoo, A., Ali Khan, M., Otero, M., Bundhoo, Z. M. A., Patel, M. et al. (2021). Smart Adsorbents for Aquatic Environmental Remediation. *Small*, 17(34): 2007840. https://doi.org/10.1002/SMLL.202007840.

Neouze, M. A., and Schubert, U. (2008). Surface modification and functionalization of metal and metal oxide nanoparticles by organic ligands. *Monatshefte Fur Chemie*, 139(3): 183–195. https://doi.org/10.1007/S00706-007-0775-2/METRICS.

Ningthoujam, R., Singh, Y. D., Babu, P. J., Tirkey, A., Pradhan, S., and Sarma, M. (2022). Nanocatalyst in remediating environmental pollutants. *Chemical Physics Impact*, 4: 100064. https://doi.org/10.1016/J.CHPHI.2022.100064.

Nizam, N. U. M., Hanafiah, M. M., and Woon, K. S. (2021). A Content Review of Life Cycle Assessment of Nanomaterials: Current Practices, Challenges, and Future Prospects. *Nanomaterials*, 11(12): 3324. https://doi.org/10.3390/NANO11123324.

Njema, G. G., and Kibet, J. K. (2024). A review of novel materials for nano-photocatalytic and optoelectronic applications: Recent perspectives, water splitting and environmental remediation. *Progress in Engineering Science*, 1(4): 100018. https://doi.org/10.1016/J.PES.2024.100018.

Onamade, A. O., Asaju, O. A., and Adewumi, B. J. (2025). Challenges and future prospects of smart nanomaterials for environmental remediation. *Smart Nanomaterials for Environmental Applications*, 757–780. https://doi.org/10.1016/B978-0-443-21794-4.00013-2.

Palchoudhury, S., Aich, N., and Zhou, Z. (2020). Advances in Smart Nanomaterials: Environmental Perspective. *Journal of Nanomaterials*, 2020. https://doi.org/10.1155/2020/6715765.

Patel, R., Shah, D., Shah, S., and Shah, M. (2022). Green nanomaterials for removal of arsenic and fluoride contamination from wastewater. *Materials Today: Proceedings*, 62(P13): 7318–7323. https://doi.org/10.1016/J.MATPR.2022.05.100.

Plazas-Tuttle, J., Rowles, L. S., Chen, H., Bisesi, J. H., Sabo-Attwood, T., and Saleh, N. B. (2015). Dynamism of Stimuli-Responsive Nanohybrids: Environmental Implications. *Nanomaterials*, 5(2): 1102–1123. https://doi.org/10.3390/NANO5021102.

Raja, M. A., Ahmad, M. A., Daniyal, M., and Husen, A. (2023). Management of wastewater and other environmental issues using smart nanomaterials. *Advances in Smart Nanomaterials and Their Applications*, 489–503. https://doi.org/10.1016/B978-0-323-99546-7.00003-3.

Raskar, N. D., Dake, D. V., Mane, V. A., Sonpir, R. B., Khawal, H. A., Mote, V. D. et al. (2024). Nanoengineered reduced graphene oxide-Fe doped α-MnO2: A multifunctional smart material for energy storage and environmental remediation. *Journal of Energy Storage*, 86: 111206. https://doi.org/10.1016/J.EST.2024.111206.

Reddy Esanakula, J., Reddy Gondi, K., Chaithanya, K., and Reddy, K. K. (2025). Applications of Magneto-responsive Smart Materials in Environmental Nanotechnology: Opportunities and Challenges. *Journal of Environmental Review Article Nanotechnology J. Environ. Nanotechnol*, 14(1): 491–504. https://doi.org/10.13074/jent.2025.03.2511170.

Rosaline, D. R., Keerthana, C., Vasthi, S. S., Rubini, S. H., Ratna Monica, J. H., Manikandan, A. et al. (2024). Magnetic nanoparticles and nanocomposites for the applications of photocatalytic degradation of organic dyes. *Magnetic Nanoparticles and Polymer Nanocomposites*, 459–497. https://doi.org/10.1016/B978-0-323-85748-2.00018-9.

Roy, A., Sharma, A., Yadav, S., Jule, L. T., and Krishnaraj, R. (2021). Nanomaterials for Remediation of Environmental Pollutants. *Bioinorganic Chemistry and Applications*, 2021(1): 1764647. https://doi.org/10.1155/2021/1764647.

Sable, H., Kumar, V., Singh, V., Rustagi, S., Chahal, S., and Chaudhary, V. (2024). Strategically engineering advanced nanomaterials for heavy-metal remediation from wastewater. *Coordination Chemistry Reviews*, 518: 216079. https://doi.org/10.1016/J.CCR.2024.216079.

Salahshoori, I., Yazdanbakhsh, A., Namayandeh Jorabchi, M., Kazemabadi, F. Z., Khonakdar, H. A., and Mohammadi, A. H. (2024). Recent advances and applications of stimuli-responsive nanomaterials for water treatment: A comprehensive review. *Advances in Colloid and Interface Science*, 333: 103304. https://doi.org/10.1016/J.CIS.2024.103304.

Sharma, K., Tyagi, S., Vikal, S., Devi, A., Gautam, Y. K., and Singh, B. P. (2022). Green Nanomaterials for Remediation of Environmental Air Pollution. *Handbook of Green and Sustainable Nanotechnology*, 1–26. https://doi.org/10.1007/978-3-030-69023-6_66-1.

Shayegan, Z., Lee, C. S., and Haghighat, F. (2018). TiO2 photocatalyst for removal of volatile organic compounds in gas phase – A review. *Chemical Engineering Journal*, 334: 2408–2439. https://doi.org/10.1016/J.CEJ.2017.09.153.

Sheoran, A. R., Lakra, N., Luhach, A., Saharan, B. S., Debnath, N., and Sharma, P. (2024). Nano-engineered Solutions for Sustainable Environmental Cleanup. *BioNanoScience*, 14(4): 4083–4107. https://doi.org/10.1007/S12668-024-01370-8.

Sibiya, A., Jeyavani, J., Santhanam, P., Preetham, E., Freitas, R., and Vaseeharan, B. (2022). Comparative evaluation on the toxic effect of silver (Ag) and zinc oxide (ZnO) nanoparticles on different trophic levels in aquatic ecosystems: A review. *Journal of Applied Toxicology*, 42(12): 1890–1900. https://doi.org/10.1002/JAT.4310.

Simonin, M., and Richaume, A. (2015). Impact of engineered nanoparticles on the activity, abundance, and diversity of soil microbial communities: a review. *Environmental Science and Pollution Research*, 22(18): 13710–13723. https://doi.org/10.1007/S11356-015-4171-X/METRICS.

Sohail, M. I., Waris, A. A., Ayub, M. A., Usman, M., Zia ur Rehman, M., Sabir, M. et al (2019). Environmental application of nanomaterials: A promise to sustainable future. *Comprehensive Analytical Chemistry*, 87: 1–54. https://doi.org/10.1016/BS.COAC.2019.10.002.

Sperling, R. A., and Parak, W. J. (2010). Surface modification, functionalization and bioconjugation of colloidal inorganic nanoparticles. *Philosophical Transactions of the Royal Society A: Mathematical, Physical and Engineering Sciences*, 368(1915): 1333–1383. https://doi.org/10.1098/RSTA.2009.0273.

Srinivas, A., Chandran, K. R., Nehra, S., and Kumar, D. (2025). Future Aspects in the Field of Bio-nanomaterials Toward Environmental Assessment. *Bio-Nanomaterials in Environmental Remediation: Industrial Applications*, 295–317. https://doi.org/10.1002/9783527848546.CH12.

Su, C. (2017). Environmental implications and applications of engineered nanoscale magnetite and its hybrid nanocomposites: A review of recent literature. *Journal of Hazardous Materials*, 322: 48–84. https://doi.org/10.1016/J.JHAZMAT.2016.06.060.

Sun, Y., Jiang, Y., Wei, H., Zhang, Z., Irshad, S., Liu, X. X. et al. (2024). Nano-enabled strategies for greenhouse gases emission mitigation: a comprehensive review. *Nano Today*, 57: 102378. https://doi.org/10.1016/J.NANTOD.2024.102378.

Tabasum, S., Rani, S., Thakur, A., Sindhu, M., Singh Maan, K., Vijay Patel, U. et al. (2023). Photo-catalytic Detoxification of Chlorpyrifos Pesticide From the Aquatic Environment Using g-C3N4 Doped With GO Nano-composite. *AIP Conference Proceedings*, *2800*(1). https://doi.org/10.1063/5.0163717/2910077.

Tariq, N., Afzal, W., Rehman, A. ur, Nadeem, S., Durrani, F., Ullah, A. et al. (2025). Nanotechnology in Environmental Remediation: A Sustainable Frontier. *Journal of Medical & Health Sciences Review*, 2(2). https://doi.org/10.62019/RJSVWH72.

Umar, A., Akhtar, M. S., Dar, G. N., and Baskoutas, S. (2013). Low-temperature synthesis of α-Fe2O3 hexagonal nanoparticles for environmental remediation and smart sensor applications. *Talanta*, 116: 1060–1066. https://doi.org/10.1016/J.TALANTA.2013.08.026.

Wang, Z., Guo, J., Ma, J., and Shao, L. (2015). Highly regenerable alkali-resistant magnetic nanoparticles inspired by mussels for rapid selective dye removal offer high-efficiency environmental remediation. *Journal of Materials Chemistry A*, 3(39): 19960–19968. https://doi.org/10.1039/C5TA04840K.

Waqas, M., and Ahmad, H. (2024). Trapping of heavy metal ions from electroplating wastewater with phosphorylated double-shelled hollow spheres. *Chemosphere*, 350: 140968. https://doi.org/10.1016/J.CHEMOSPHERE.2023.140968.

Xiao, J., Liang, J., Zhang, C., Tao, Y., Ling, G. W., and Yang, Q. H. (2018). Advanced Materials for Capturing Particulate Matter: Progress and Perspectives. *Small Methods*, 2(7): 1800012. https://doi.org/10.1002/SMTD.201800012.

Xie, Y., Lyu, S., Zhang, Y., and Cai, C. (2022). Adsorption and Degradation of Volatile Organic Compounds by Metal–Organic Frameworks (MOFs): A Review. *Materials*, 15(21): 7727. https://doi.org/10.3390/MA15217727.

Yadav, N., Garg, V. K., Chhillar, A. K., and Rana, J. S. (2021). Detection and remediation of pollutants to maintain ecosustainability employing nanotechnology: A review. *Chemosphere*, 280: 130792. https://doi.org/10.1016/J.CHEMOSPHERE.2021.130792.

Yilmaz-Sercinoglu, Z., Durmaz-Sam, S., Ulucan-Karnak, F., and Kuru, C. İ. (2023). Smart Nanomaterials for Bioremediation and Biodegradation: Laboratory and On-Site Applications and Future Trends. *Sustainable Nanomaterials for Biosystems Engineering*, 341–367. https://doi.org/10.1201/9781003333517-16.

Ying, S., Guan, Z., Ofoegbu, P. C., Clubb, P., Rico, C., He, F. et al. (2022). Green synthesis of nanoparticles: Current developments and limitations. *Environmental Technology & Innovation*, 26: 102336. https://doi.org/10.1016/J.ETI.2022.102336.

Zhao, Q., Li, X., Zhou, Q., Wang, D., and Xu, H. (2020). Nanomaterials Developed for Removing Air Pollutants. *Advanced Nanomaterials for Pollutant Sensing and Environmental Catalysis*, 203–247. https://doi.org/10.1016/B978-0-12-814796-2.00006-X.

Zhu, W., Kang, J., Zhang, D., Tian, M., and Meng, X. (2023). Extraction of andradite from flotation tailings as a potential material for As(III) removal from simulated wastewater. *Journal of Water Process Engineering*, 54. https://doi.org/10.1016/j.jwpe.2023.103988.

CHAPTER-3

Green Synthesis of Nanomaterials

Unlocking Sustainable Solutions for Environmental and Industrial Challenges

Hemant S. Tarkas,[1,†] *Jaywant P. Sonawane,*[2,†] *Vijay S. Baviskar,*[1]
Tejas B. Chaudhari,[3] *Leena P. Shirsath*[3] and *Sandip P. Patil*[3,*]

1. Introduction

Nanotechnology is a rapidly evolving field that focuses on manipulating materials at the nanoscale to develop innovative solutions across a range of industries. The importance of nanomaterials, particularly in applications like environmental remediation, biomedicine, and energy storage, has driven the exploration of various synthesis methods. Conventional synthesis methods often pose limitations such as toxicity and environmental harm, prompting the rise of green synthesis techniques. Green synthesis, an eco-friendly alternative, leverages natural resources and processes to minimise environmental impact while achieving high-efficiency nanomaterial production. This approach, grounded in the principles of green chemistry, offers numerous advantages over traditional methods, such as reduced waste and the use of renewable, natural reducing and stabilising agents. Biological, chemical, and physical mechanisms of green synthesis have enabled the creation of various nanomaterials like metallic nanoparticles, metal oxides, carbon-based materials, and hybrid composites. These green-synthesised nanomaterials have found diverse applications in environmental remediation, biomedicine, energy production, agriculture, and industry. Despite the promising potential, challenges related to toxicity, scalability, and the need for further research into the mechanisms of synthesis remain. Some future perspectives focus on advancing the field and addressing these concerns through innovation and interdisciplinary approaches.

[1] Department of Applied Sciences and Humanities, R. C. Patel Institute of Technology, Shirpur-425405, Maharashtra, India. https://orcid.org/0000-0002-5313-2504, https://orcid.org/0000-0003-1580-4303.
[2] Department of Chemistry, R. C. Patel Arts, Commerce and Science College, Shirpur-425405, Maharashtra, India. https://orcid.org/0000-0002-8745-5379.
[3] Department of Microbiology and Biotechnology, R. C. Patel Arts, Commerce and Science College, Shirpur-425405, Maharashtra, India. https://orcid.org/0000-0002-9859-020X.
† Authors contributed equally as the first author
* Corresponding author: patilsandip3@gmail.com

DOI: 10.1201/9781003558804-3

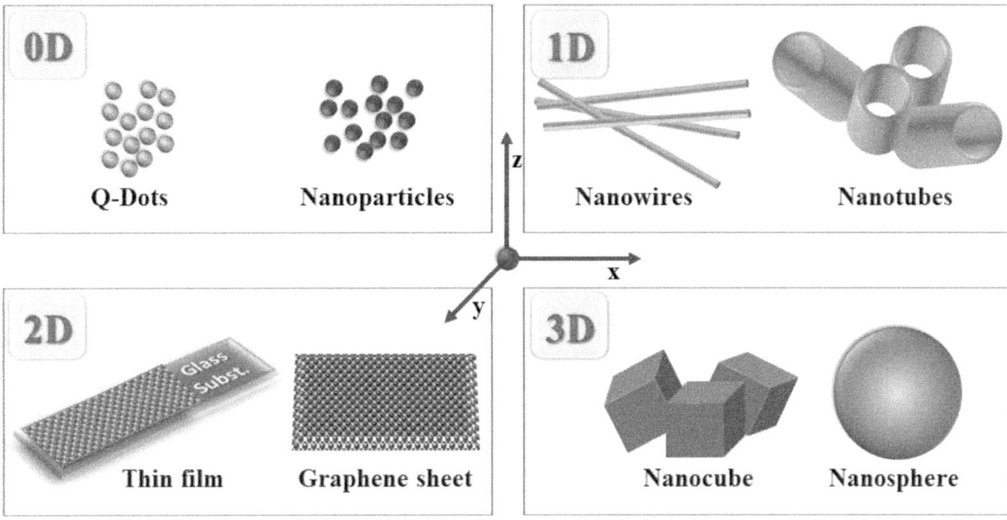

Figure 1: Structural classifications of nanomaterials as 0D, 1D, 2D and 3D.

1.1 Overview of Nanotechnology

Advanced nanotechnology has the potential to transform many fields by supporting the creation of new materials with novel properties (Bhushan 2017). The word 'nano' originates from the Greek word for tiny and first came into use in the early 1900s. Nobel laureate Richard Zsigmondy deployed a microscope for analysing gold colloids in 1925, using the term nanometre (nm) to define particle size. In the 21st century, it has emerged as a key tool to address challenges across diverse sectors. Nanotechnology refers to practices at the nanoscale, where matter is manipulated at atomic and molecular levels, typically between 1 to 100 nm (Bhushan 2017, Anajwala et al. 2010). Nanomaterials are materials with at least one dimension between 1–100 nm that possess superior properties like high surface area, chemical reactivity, and quantum effects, which make them ideal for various applications. Recent discoveries at the nanoscale promise to tackle present and future challenges (Bhushan 2015). Traditional nanoparticle synthesis methods, including chemical and physical methods, often involve toxic reagents and complex, energy-intensive equipment. This raises concerns about their environmental impact, leading to the rise of green nanotechnology, which is primarily focused on eco-friendly techniques for nanomaterial production (Pellegrini et al. 2016).

According to orientation and structural components in the range of 1 to 100 nm, nanomaterials are classified into four broad categories: 0D, 1D, 2D, and 3D, as shown in Figure 1.

Zero-Dimensional (0D) nanomaterials like quantum dots (Q-Dots) and nanoparticles have three dimensions in the nanoscale (typically less than 100 nm), and their properties are strongly influenced by their size. One-Dimensional (1D) materials like nanowires, nanotubes, and nanorods feature one dimension in the nanoscale, and their unique properties arise from the restricted direction of growth. Two-Dimensional (2D) materials like graphene, nanoplates, and thin films have two nanoscale dimensions that offer unique electronic and mechanical properties, which are highly valuable in electronics and sensor applications. Moreover, Three-Dimensional (3D) materials are structured at the nanoscale in all three dimensions—for example, nanocubes and nanospheres—and may include bulk nanostructures. These bulk nanostructures include nanostructured surfaces or nanoporous materials that exhibit high surface area and distinctive catalytic properties (Li et al. 2025).

1.2 Importance of Nanomaterials

Nanomaterials are significant because of their unique properties, which differ drastically from those of bulk materials. Properties such as small size, high surface area, and enhanced reactivity make them ideal for a wide range of applications, such as in electronics, medicine, environmental remediation, and energy storage. In electronics, nanomaterials enable the development of smaller, faster, and more efficient devices. Moreover, in medicine, they are used for targeted drug delivery and diagnostic imaging (Li et al. 2025). Their ability to interact with biological systems at the molecular level opens new possibilities in disease treatment and prevention (Zhao et al. 2024). Furthermore, nanomaterials play a crucial role in improving energy efficiency, for example, through more effective solar cells (Phogat et al. 2024) and batteries (Vindhyasarumi et al. 2025), and in environmental protection by enabling more efficient water and air purification (Meenakshi et al. 2025). The versatile nature of nanomaterials positions them at the forefront of technological advancements across multiple industries (Zhao et al. 2024).

1.3 Conventional Synthesis Methods: Limitations and Environmental Impact

Conventional chemical nanoparticle synthesis methods, including laser ablation and sol-gel, along with physical methods like ball milling and chemical vapour deposition (CVD), are widely used due to their ability to precisely control nanoparticle size and morphology. However, these methods come with several limitations and significant environmental impacts. Typically, they involve the use of toxic chemicals, elevated temperature processing, and energy-intensive conditions. CVD requires hazardous precursor gases and solvents, which can lead to harmful emissions and chemical waste (Arora et al. 2025). The energy consumption involved in methods like laser ablation and high-temperature synthesis also contributes to a high carbon footprint, further exacerbating environmental concerns (Al-Harbi et al. 2024). Additionally, these methods often produce unwanted by-products that require careful handling and disposal, leading to potential contamination and waste management challenges. As a result, the environmental sustainability of conventional chemical and physical synthesis methods has been questioned, driving the search for greener, more eco-friendly alternatives (Deka et al. 2025).

1.4 Green Synthesis: An Eco-friendly Alternative

Green synthesis of nanomaterials offers an eco-friendly alternative to traditional methods by utilising natural resources and processes that minimise environmental impact (Deka et al. 2025). This approach relies on renewable, non-toxic materials like plant extracts, biopolymers, and microorganisms that reduce, stabilise, and fabricate nanoparticles, thereby avoiding the use of harmful chemicals and reducing waste (Roy et al. 2025). Green synthesis can be classified into physical, chemical, and biological methods. Chemical green synthesis involves the use of non-toxic, natural compounds or biomolecules as reducing agents, while physical methods often utilise low-energy processes like ultrasonic or microwave-assisted techniques (Mir et al. 2025). Biological methods include plant-mediated synthesis, where plant extracts act as reducing agents, and microbial synthesis, which uses microbes to produce nanoparticles (Patil et al. 2019, Roy et al. 2025). These methods not only offer environmentally safe production routes but also tend to be cost-effective and scalable, making green synthesis a promising solution for sustainable nanomaterial production. Owing to the metal ion reduction capability of microorganisms, metal nanostructures with a variety of dimensions and morphologies can be easily synthesised (Mehra et al. 2025). On the other hand, in plant extract-based synthesis, parts of plants like leaves, seeds, fruits, and flowers are employed for the synthesis of nanostructured materials (Fatima et al. 2024). These parts inherently contain phytochemicals, including polyphenols, flavonoids, terpenoids, and alkaloids, which help reduce metal ions, resulting in metal nanostructures. Hence, this method is sometimes referred to as the phytochemical synthesis method of nanostructure production. Both microbial synthesis and plant extract (or phytochemical synthesis) methods are scalable and significantly cost-effective, making them among the most promising methods of nanoparticle synthesis to tackle future challenges and applications (Mehra et al. 2025).

2. Principles of Green Synthesis of Nanomaterials

2.1 Concept of Green Synthesis

Production of nanomaterials using green alternative is an eco-friendly and sustainable method that utilises biological sources like plants, fungi, bacteria, algae, and biomolecules to produce nanoparticles (Alsaiari et al. 2023). Unlike conventional physical and chemical methods that often involve hazardous by-products, toxic solvents, high energy consumption, green synthesis highlights biocompatibility, cost-effectiveness, and environmental safety. Green synthesis includes the use of natural and biological agents that act as both reducing and stabilising agents, facilitating the conversion of metal ions into nanoparticles under mild conditions (Vaseghi et al. 2018). Plant extracts contain high amounts of phytochemicals like flavonoids, alkaloids, tannins, and polyphenols, and hence they are commonly used because of their ability to efficiently reduce metal ions. Similarly, microorganisms such as bacteria and fungi produce enzymes and metabolites that mediate nanoparticle formation. This method is widely applied in synthesising metal nanoparticles (silver, gold, and zinc oxide), metal oxide nanoparticles, and carbon-based nanomaterials. The synthesised nanoparticles exhibit enhanced biocompatibility and are extensively used in biomedicine, environmental remediation, agriculture, and catalysis. Silver nanoparticles produced using plant extracts show remarkable antimicrobial properties, making them useful in wound healing and antibacterial coatings (Omran and Baek 2022). However, challenges remain related to scalability, reproducibility, and precise control over nanoparticle size and shape. Advancements in nanobiotechnology continue to refine this method, making it a promising avenue for sustainable nanomaterial development.

2.2 Green Chemistry Principles

One of the key principles in green chemistry of nanomaterials is waste prevention, which ensures the generation of minimal by-products and hazardous substances during synthesis. Atom economy further supports this by maximising the conversion of reactants into the final nanoparticle product, reducing material wastage. Instead of using toxic reagents, green synthesis emphasises less hazardous chemical processes, relying on natural reducing and stabilising agents such as plant extracts, bacteria, and fungi. The principle of designing safer chemicals ensures that nanoparticles are effective yet non-toxic to humans and the environment (Soeteman-Hernández et al. 2020). Similarly, safer solvents and auxiliaries like water or ethanol replace hazardous organic solvents. Energy conservation is also crucial, with energy efficiency promoting ambient temperature and pressure reactions to reduce the carbon footprint of nanoparticle synthesis. To ensure sustainability, renewable feedstocks like plant-based compounds and microbial enzymes are used, aligning with bio-based nanomaterial production (Osterberg et al. 2023). The principle of reducing derivatives minimises unnecessary modifications and processing steps, making synthesis more straightforward and resource-efficient. Catalysis plays a key role by employing biological catalysts like enzymes to enhance reaction efficiency and selectivity. Additionally, green nanoparticles should be biodegradable to ensure they break down into non-toxic components after use (Bellanthudawa et al. 2023). Real-time studies for pollution prevention help monitor and control synthesis processes, reducing environmental risks. Safer chemistry for accident prevention minimises the likelihood of toxic spills, explosions, or hazardous reactions, making nanoparticle production safer and more sustainable (Asmatulu et al. 2024).

2.3 Advantages of Green Synthesis Over Conventional Methods

Green synthesis is an eco-friendly and sustainable approach that offers significant advantages over conventional physical and chemical methods for nanoparticle production. One of the primary benefits is its environmental safety, as it reduces the use of toxic chemicals and hazardous solvents, thereby reducing harmful by-products and ensuring biocompatibility. Unlike traditional methods that require high temperatures, pressure, and energy-intensive processes, green synthesis operates under mild and ambient conditions, leading to lower energy consumption and cost-effectiveness. Another advantage is the use of biological entities like plants, bacteria, fungi, and algae that act as natural reducing and stabilising agents (Hasan et al. 2018). These biological resources contain metabolites like flavonoids, proteins, and enzymes that facilitate nanoparticle formation in a non-toxic and sustainable manner. This not only makes the process safer but also allows for the production of nanoparticles with enhanced stability and functionality. Furthermore, green synthesis provides better biocompatibility and biodegradability, making the nanoparticles suitable for applications in medicine, drug delivery, and agriculture without posing risks to human health and the environment (Saratale et al. 2018). Additionally, green synthesis ensures cost-effectiveness and scalability by utilising easily available and renewable resources. Conventional methods often require expensive precursors, hazardous reagents, and sophisticated instrumentation, whereas green synthesis leverages biological materials that are abundant and inexpensive (Jain et al. 2024). Despite challenges in achieving precise control over size, shape, and reproducibility, ongoing research in nanobiotechnology is addressing these issues.

2.4 Role of Natural Reducing and Stabilizing Agents

Bio-based reducing and stabilising agents derived from plants, fungi, bacteria, and algae offer an eco-friendly alternative to conventional chemical reagents.

Natural compounds such as polyphenols, flavonoids, alkaloids, terpenoids, proteins, and sugars present in biological extracts act as reducing agents. They donate electrons to metal ions, transforming them into nanoparticles. For example, plant-derived polyphenols and flavonoids efficiently reduce silver and gold ions, leading to the formation of stable nanoparticles with controlled size and morphology (Can 2020). Similarly, microbial enzymes and metabolites such as NADH-dependent reductases and organic acids play a role in the bioreduction process (Qu et al. 2018).

After reduction, nanoparticles tend to aggregate due to high surface energy. Natural biomolecules such as proteins, polysaccharides, and surfactants act as capping agents, preventing aggregation and enhancing stability. Functional groups such as hydroxyl, carboxyl, and amine in these molecules interact with nanoparticle surfaces, providing steric or electrostatic stabilisation. For instance, proteins from fungi and bacterial exopolysaccharides form protective layers around nanoparticles, improving dispersion and biocompatibility (Mouro et al. 2024).

The dual role of natural reducing and stabilising agents enhances the eco-friendliness, biocompatibility, and functional properties of nanoparticles, making them suitable for biomedical, environmental, and catalytic applications (Devi et al. 2024). This green approach aligns with sustainable nanotechnology by reducing toxicity and environmental hazards.

2.5 Key Environmental and Economic Benefits

The production of nanoparticles through sustainable methods offers numerous environmental and economic benefits. Environmentally, green synthesis of nanoparticles eliminates the need for toxic chemicals and energy-intensive processes, thereby reducing pollution and minimizing hazardous waste generation (Osman et al. 2024). The utilization of biological resources like plant extracts, bacteria and fungi promote eco-friendly practices by reducing chemical waste and supporting the circular economy. Additionally, nanoparticles synthesized using green methods are often biodegradable and biocompatible, ensuring minimal long-term environmental impact (Afonso et al. 2024). Since these processes occur under ambient conditions, energy consumption and carbon emissions are significantly lower compared to conventional methods. Furthermore, nanoparticles play a crucial role in wastewater treatment by effectively removing heavy metals and organic pollutants, thereby contributing to environmental remediation and pollution control (Latif et al. 2020).

From an economic perspective, green synthesis of nanoparticles is highly cost-effective as it relies on readily available biological materials instead of expensive chemicals and high-energy equipment. The low-cost production process makes large-scale manufacturing more feasible and ensures affordability for industries like medicine, agriculture, and environmental remediation (Lewoyehu et al. 2021). The growing market for sustainable nanotechnology also drives economic growth and encourages investment in research and development. Additionally, the expanding field of nanotechnology creates job opportunities in research, manufacturing, and product development that foster employment generation. Industries that adopt sustainable nanoparticle production benefit from enhanced regulatory compliance, corporate social responsibility, and long-term sustainability (Khatoon et al. 2025).

3. Mechanisms of Green Nanomaterials Synthesis

3.1 Biological Methods

3.1.1 Plant-Mediated Synthesis

Plant-mediated synthesis of nanomaterials is a green and eco-friendly approach that utilises plant extracts to produce nanoparticles. This method is an alternative to conventional physical and chemical synthesis techniques that involve toxic chemicals, high energy consumption, and hazardous by-products. Plants serve as natural biofactories that provide a rich source of phytochemicals such as flavonoids, alkaloids, tannins, phenols, and terpenoids that act as reducing, capping, and stabilising agents during nanoparticle formation (Kulkarni et al. 2023). The process of plant-mediated synthesis involves mixing a metal salt solution with plant extracts, which triggers the reduction of metal ions into nanoparticles under ambient conditions. Various metals and metal oxide nanoparticles like silver (Ag), gold (Au), zinc oxide (ZnO), and copper oxide (CuO) nanoparticles have been successfully synthesised using plant-based methods (Letchumanan et al. 2021). The size, shape, and stability of the nanoparticles depend on factors such as plant species, extract concentration, reaction temperature, and pH. Plant-mediated nanoparticles exhibit excellent biocompatibility, which makes them suitable for various applications like medicine, agriculture, and environmental remediation. They have demonstrated strong antimicrobial, anticancer, and antioxidant properties, which make them valuable in drug delivery, wound healing, and biosensing. In agriculture, these nanoparticles enhance plant growth and protect against pathogens. Additionally, they are employed in various environmental applications like wastewater treatment and pollutant degradation (Singh et al. 2019). However, challenges like scalability, reproducibility, and precise control over nanoparticle characteristics require further research and optimisation to enhance industrial applications (Harish et al. 2023).

3.1.2 Microbial Synthesis

Microorganisms act as biological factories that reduce metal ions into stable nanomaterials through enzymatic and metabolic processes. This method is a sustainable alternative to conventional physical and chemical synthesis of nanoparticles. Microorganisms mediate nanoparticle synthesis by secreting enzymes, proteins, and metabolites that facilitate metal ion reduction and nanoparticle stabilisation. Bacterial species like *Pseudomonas aeruginosa*, *Bacillus subtilis* and *Escherichia coli* are known to produce silver, gold, and zinc oxide nanoparticles (Fazaa 2022). Similarly, fungi like *Aspergillus niger* and *Fusarium oxysporum* have demonstrated efficient nanoparticle biosynthesis due to their high metal tolerance and extracellular enzyme production (Adeleke et al. 2024, Patil et al. 2021). Algae and yeast also contribute to nanoparticle synthesis by absorbing and transforming metal ions into stable nanostructures. Microbial synthesis of nanomaterials offers various advantages like cost-effectiveness, eco-friendliness, and enhanced biocompatibility. These biosynthesised nanoparticles exhibit excellent antimicrobial, catalytic, and biomedical properties, which make them valuable in drug delivery, biosensing, wastewater treatment, and agricultural applications.

Additionally, microbial synthesis enables precise control over nanoparticle size, shape, and stability, which is crucial for various industrial applications (Hulkoti and Taranath 2014, Patil et al. 2023).

Despite its benefits, challenges like process scalability, optimisation of microbial culture conditions, and nanoparticle recovery need further research. Advances in nanobiotechnology and genetic engineering are expected to enhance microbial synthesis efficiency, which makes it a key technology for sustainable nanomaterial production.

3.1.3 Enzyme-Mediated Synthesis

Enzyme-mediated synthesis offers precise control over nanoparticle size, shape, and functionality while operating under mild reaction conditions. Unlike traditional chemical and physical methods, enzyme-mediated synthesis minimises the need for toxic reagents and high-energy inputs, hence making it an environmentally friendly and biocompatible alternative. Enzymes such as oxidoreductases (laccase, peroxidase, and nitrate reductase) and hydrolases play a crucial role in the synthesis of nanoparticles. These enzymes catalyse the reduction of metal ions into their nanoparticle form while simultaneously stabilising them by preventing aggregation. Nitrate reductase from fungi and bacteria is widely used to synthesise silver and gold nanoparticles due to its ability to facilitate electron transfer. Laccase is a copper-containing oxidase that is employed in the synthesis of metal oxide nanoparticles, which contributes to applications in biomedicine and environmental remediation (Rudakiya and Gupte 2019, Patil et al. 2024).

This enzyme-mediated approach is highly advantageous for biomedical applications as it produces biocompatible nanoparticles with enhanced stability and functionality (Sharifi et al. 2020, Patil and Chaudhari 2017). Additionally, it is widely applied in biosensing, drug delivery, and antimicrobial coatings. The process aligns with green chemistry principles that reduce environmental toxicity and ensure sustainability. However, challenges like enzyme stability, scalability, and cost-effective production need further research and optimisation. With advancements in nanobiotechnology, enzyme-mediated synthesis is emerging as a promising technique for developing functional nanomaterials in a sustainable and controlled manner (Shirsath et al. 2018).

3.2 Chemical and Physical Methods

The chemical processes involved in assembling nanomaterials particle by particle are known as bottom-up approaches in nanotechnology. These methods allow precise control over size, shape, and composition. One widely used technique is chemical vapour deposition (CVD) (Machac et al. 2020). In CVD, gaseous precursors undergo reactions to form nanostructures like carbon nanotubes. The sol-gel process involves hydrolysis and polymerisation of precursors to create nanomaterials like metal oxides (Dong et al. 2020). In precipitation methods, controlled chemical reactions in solution lead to nanoparticle formation. Electrodeposition enables the growth of nanostructures by reducing metal ions onto a substrate. Self-assembly is another crucial approach where molecules arrange into ordered nanostructures due to intermolecular forces. Additionally, atomic or molecular condensation allows controlled deposition of atoms onto surfaces, which forms precise nanostructures. These bottom-up methods are widely used in nanotechnology for applications in medicine, electronics, and catalysis, offering innovative solutions for material design at the nanoscale.

3.3 Green Synthesis Conditions

Adsorbate type, catalyst efficiency, pH, temperature, extract concentration, particle size, and synthesis technique are some of the variables that affect the synthesis of nanoparticles. Studies indicate that nanoparticles are highly dynamic, with their properties evolving over time and in response to environmental conditions, which in turn influence their behaviour and potential applications.

3.3.1 pH

In the environmentally friendly synthesis of nanoparticles, the pH level is a crucial component. Since the pH levels in the final material are found to influence the magnitude and texture of the generated materials, changing the fluid's pH can effectively control the nanoparticles' nature and scope (Armendariz et al. 2004).

3.3.2 Temperature

Temperature stands as a key factor influencing nanoparticle synthesis across all three methods. Green synthesis of nanoparticles occurs at ambient temperatures or below 100°C. The properties of the final nanoparticles are significantly influenced by the temperature of the reaction medium (Rai et al. 2006).

3.3.3 Pressure

Pressure plays a crucial part in the production of nanoparticles since it directly influences the particle size as well as nature. The formation process is impacted by the pressure exerted on the reaction medium; research indicates that biological agents reduce metal ions more quickly in atmospheric pressure conditions (Tran et al. 2013).

3.3.4 Time

Several factors influence the properties and abilities of green-synthesised nanoparticles, including the length of incubation, the synthesis method, exposure to light, and preservation conditions. Over time, these nanoparticles may undergo changes such as aggregation, shrinkage, growth, or degradation, which can impact their stability and future applications (Kuchibhatla et al. 2012).

3.3.5 Particle Shape and Size

One important characteristic that defines the properties of nanoparticles is their size. Research has indicated that the melting point of nanoparticles drops as they approach the nanometre scale. Furthermore, differently configured nanoparticles have comparable energy levels, which make shape change easier. The kind of energy used to analyse nanoparticles can also affect how they appear. The morphology and fluid behaviour of synthesised nanoparticles have a significant impact on their chemical composition (Akbari et al. 2011).

4. Types of Nanomaterials Synthesized by Green Methods

4.1 Metallic Nanoparticles

Green methods offer environmentally friendly alternatives to traditional methods and are utilised increasingly to synthesise various types of nanomaterials (Mehra et al. 2025). These include carbon-based nanomaterials like graphene and carbon nanotubes, as well as metallic nanoparticles like copper (Cu), silver (Ag), and gold (Au). To minimise toxicity and their impact on the environment, these green synthesis methods make use of microbes, plant extracts, and environmentally friendly solvents.

4.1.1 Gold Nanoparticles (Au-Np)

Gold has fascinated humans for its shine, non-reactivity, and stability, making it a highly sought-after noble metal. In green synthesis of gold nanoparticles, ionic gold species like Gold Chloride ($AuCl_3$), Gold Nitrate ($Au(NO_3)_3$), and Chloroauric Acid ($HAuCl_4$) are typically used (Mehra et al. 2025). While gold is non-reactive, these compounds are formed using highly corrosive and costly reagents. In order to minimise material losses, efficient, cost-effective methods are essential for nanoparticle synthesis. Green synthesis offers a reliable and economical alternative, reducing the use of toxic chemicals while preserving gold's unique properties.

a. Phytochemical Production of Au-Nps

The green phytochemical synthesis of Au-Nps via plant extract has gained significant attention due to its advancements (Fatima et al. 2024). The process begins with selecting the plant part containing phytochemicals, which is cleaned and crushed using sterilised instruments in a solvent medium, typically water. These phytochemicals act as reducing or capping agents during synthesis. Parameters like pH, extract concentration, and temperature influence the dimension-dependent properties of Au-Nps. Characterisation techniques such as UV-visible spectroscopy, FTIR, XRD, and TEM can evaluate these properties, revealing significant potential in biomedical diagnostics, imaging, drug delivery, and therapies (Fatima et al. 2024, Fouda et al. 2022).

In phytochemical synthesis of Au-Nps, chloroauric acid or gold chloride ($HAuCl_4$) is commonly used to prepare Au^+ ionic solutions. This is done by dissolving gold dust or pieces into aqua regia, a mixture of nitric and hydrochloric acid. The $HAuCl_4$ powder is then mixed with plant extracts under optimised conditions to reduce gold ions and form pure Au-Nps. Various plants, including *Azadirachta indica*, *Camellia sinensis*, and *Aloe vera*, are rich in bioactive compounds like terpenoids, flavonoids, and polyphenols that help in the reduction of gold ions and the production of nanoparticles (Wang et al. 2015). Research has shown that gold nanoparticles produced from *Cinnamomum verum* and *Cinnamomum cassia* are rod-shaped, while most nanoparticles are spherical, ranging from 5 to 60 nm in size (Wang et al. 2015). For instance, *Eucalyptus* leaf extract produces spherical nanoparticles in the 5–30 nm range, which are ideal for antimicrobial sensor development. These green synthesis methods offer customisable nanoparticle sizes and shapes, making them suitable for drug delivery, sensing, and catalysis. Additionally, Au-Nps demonstrate antimicrobial (Thangamani et al. 2019), antioxidant (Chen et al. 2021), and cytotoxic potential against cancer cells, with studies showing that they induce mitochondrial dysfunction and caspase activation for anticancer activity. Despite these promising results, further optimisation and functionalisation strategies are necessary to enhance their biomedical applications (Perveen et al. 2021). Researchers have also reported the synthesis of Au-Nps using various plant extracts, including *Jasminum auriculatum*, *Simarouba glauca*, and *Glycyrrhzia glabra*, demonstrating antimicrobial, antioxidant, and anticancer activities (Karnwal et al. 2024).

b. Microbial Production of Au-Nps

Microbial green synthesis of gold nanoparticles (Au-Np) utilises microorganisms like bacteria, fungi, and algae as reducing agents. These microbes release energy or electrons during metabolic activity, which interact with Au^+ ions, resulting in the formation of neutral Au nanoparticles. Bacteria play a crucial role in biosynthesising various nanomaterials through intracellular and extracellular biosynthesis. In intracellular synthesis, the process occurs within microbial cells, typically in the periplasmic space or cytoplasm, and involves bacterial enzymes reducing the metal ions (Figure 2a). In contrast, extracellular biosynthesis occurs outside the microbial cells, in the growth medium or cell wall, with bacteria secreting enzymes or metabolites for reduction (Figure. 2b) (Chowdhury et al. 2022, Martinaga et al. 2024). Although intracellular synthesis is more complex and slower, it often results in nanoparticles with uniform dimensions. Extracellular synthesis is faster but less controlled, although optimisation can achieve desired outcomes. Common bacteria used for

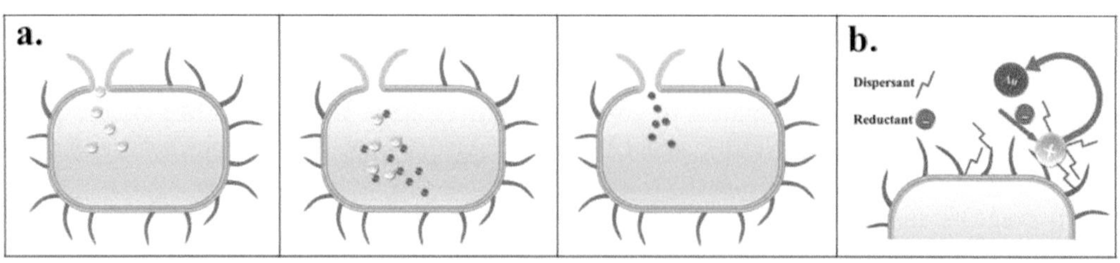

Figure 2: **a.** Intracellular and **b.** Extracellular microbial synthesis of Au-Nps.

intracellular synthesis include *Escherichia coli* and *Pseudomonas aeruginosa*, while *Bacillus cereus* and *Klebsiella pneumoniae* are used in extracellular methods (Kalishwaralal et al. 2010).

Trichoderma saturnisporum was used to produce violet spherical Au-Np (8–30 nm) with antioxidant, cytotoxic, antibacterial, and antibiofilm properties (Abu-Elghait et al. 2025). *Fusarium pseudonygamai* TB–13c also yielded multifunctional Au-Np (Soliman et al. 2024). Fungal synthesis offers high efficiency, but controlling particle properties remains a challenge (Xu et al. 2024). Endophytic fungus *Phoma* sp. produced spherical Au-Np (10–100 nm) (Soltani et al. 2022).

Microorganisms play a vital role in the green synthesis of nanoparticles through intracellular and extracellular mechanisms. In intracellular synthesis, microbes absorb metal ions that are reduced inside the cytoplasm or periplasmic space by enzymes and biomolecules (Chowdhury et al. 2022). This method is observed in bacteria like *Bacillus* sp. and fungi like *Fusarium oxysporum* that produce silver and gold nanoparticles. However, extracting these nanoparticles requires cell lysis, which makes this process complicated. In extracellular synthesis, microbes secrete enzymes or biomolecules that reduce metal ions outside the cell, forming nanoparticles in the surrounding medium. This method, especially seen in *Pseudomonas* sp., *Aspergillus* sp., and actinomycetes, further simplifies the recovery of nanoparticles. The extracellular approach is often preferred due to its ease of purification and scalability. Microbial nanoparticle synthesis is eco-friendly, cost-effective, and has applications in medicine, catalysis, and environmental remediation (Ahmad et al. 2003). This makes it a promising alternative to conventional chemical and physical nanoparticle synthesis methods.

4.1.2 Silver Nanoparticles (Ag-NPs)

Silver nanoparticles have garnered significant attention for their unique properties and potential in various fields such as electronics, environmental remediation, and medicine. Traditional synthesis methods often rely on high-energy processes and hazardous chemicals, which can harm both the environment and human health (Yadha et al. 2024). In contrast, green synthesis provides an eco-friendly alternative, using plant extracts, microbes, and other biocompatible agents to reduce silver ions into nanoparticles. Several techniques like X-ray diffraction (XRD), transmission electron microscopy (TEM), and UV-visible spectroscopy are commonly used to analyse Ag-NPs, ensuring uniform size and stability. These nanoparticles show promising biomedical applications, particularly due to their antibacterial properties (Hairil et al. 2024). Green synthesis not only minimises environmental impact but also facilitates large-scale production of Ag-NPs with controllable characteristics for various industrial uses.

a. Phytochemical Production of Ag-Nps

In green synthesis, several phytochemicals, particularly phenolic compounds, reduce silver ions (Ag^{+}) to produce silver nanoparticles (AgNPs). Phenolic compounds, found in fruits, leaves, stems, and seeds, include flavonoids like rutin, kaempferol, and quercetin (Figure 3). These compounds have antioxidant properties and can reduce Ag^{+} ions to Ag-Np. Due to their ability to stabilise nanoparticles and chelate metal ions, they help control the synthesis process by adjusting key parameters for nanoparticle formation.

Recent studies have explored the green synthesis of silver nanoparticles (Ag-Np) using various plant extracts and compounds. Rutin-3–O-glycoside, isolated from *Carica papaya*, was used in a green metal-reducing method to produce therapeutic Ag-Nps (Yadha et al. 2024). Flavonoids like rutin, quercetin, and kaempferol have been successfully used to synthesise stable, 50–90 nm Ag-Nps with antibacterial and anti-inflammatory properties. Additionally, kaempferol-capped Ag-Nps have been shown to inhibit methicillin-resistant *Staphylococcus aureus* and have potential for liver cancer therapy (Alyami et al. 2022). Saponins and terpenes from plants such as *Pterocarpus santalinus* and *Ocimum tenuiflorum* also act as reducing agents for Ag-Np production, resulting in spherical particles that are effective against pathogens (Segura et al. 2020). Green synthesis using sugars like fructose and lactose has also been explored, producing nanoparticles with antibacterial properties

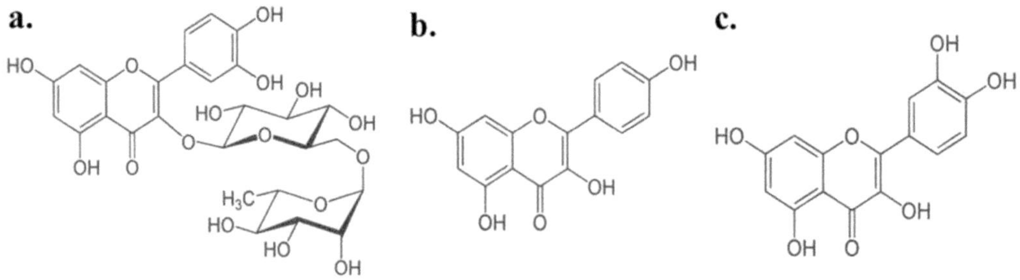

Figure 3: Molecular structure of flavonoids **a.** rutini **b.** kaempferol and c. quercetin.

(Arshadi et al. 2018). These environmentally friendly synthesis methods are cost-effective, energy-efficient, and beneficial for large-scale applications in industries such as catalysis and medicine.

b. Microbial Production of Ag-Nps

Recent studies on the extracellular biosynthesis of silver nanoparticles (AgNPs) have shown promising results, using microorganisms like *Klebsiella pneumoniae*, *Micrococcus luteus*, and *Enterobacter aerogenes* (Patel et al. 2024). These biogenic AgNPs exhibit diverse morphologies and sizes, with applications in environmental remediation, such as dye removal, and in medicine for their antibacterial properties. Factors like phytochemicals, pH, and reaction conditions influence AgNP synthesis, optimising production (Vanlalveni et al. 2024).

In 2023, Soliman et al. (2023) demonstrated the green synthesis of Ag-Np using *Trichoderma saturnisporum* fungi, obtaining the highest Ag-Np growth at 1 millimole of $AgNO_3$, 28°C, and 72 hours of incubation. After dehydration at 150°C for 48 hours, the brown-coloured Ag-Np was characterised, revealing spherical particles ranging from 10 to 70 nm. These nanoparticles showed potential in antibacterial, antibiofilm, cytotoxicity, and antioxidant activities. Fungus-mediated biosynthesis is recognised for its cost-effectiveness and environmental benefits. Several fungi, including *Fusarium* (Rai et al. 2021) and *Penicillium* (Barkhade 2018), produce Ag-Np extracellularly, with temperature and pH significantly affecting particle size and dispersity. This green approach could offer sustainable alternatives in health and agriculture for antibacterial and anticancer applications.

4.1.3 Copper Nanoparticles (Cu-Nps)

a. Phytochemical Production of Cu-Nps

The synthesis of copper nanoparticles follows a similar mechanism to that of gold (Au) and silver (Ag) nanoparticles through phytochemical reduction. Copper salts like copper (II) sulphate ($CuSO_4$), copper (II) chloride ($CuCl_2$), and copper (II) nitrate ($Cu(NO_3)_2$) serve as sources of Cu^{2+} ions, which are reduced by plant-derived phytochemicals to form Cu-NPs. The green synthesis of Cu-NPs has garnered significant attention due to its enhanced biological compatibility and environmental sustainability. These Cu-NPs have demonstrated a wide range of applications, including antiviral, antibacterial, antifungal, and anticancer properties (Rakshit et al. 2023). Studies have shown that Cu-NPs can efficiently detect hexavalent chromium, break down synthetic dyes, and inhibit cancer cell proliferation. Plant extracts from species like *Wrightia tinctoria*, *Blumea balsamifera*, and *Acanthospermum hispidum* have been used to synthesise Cu-NPs with sizes ranging from 15 to 55 nm, exhibiting notable biological activities (Rajagopal et al. 2021). Flavonoids, phenols, and tannins in these plants act as reducing as well as stabilising agents. This green synthesis method is a promising alternative for biomedicine, environmental remediation, and nanotechnology, with further research needed to unlock its full potential.

b. Microbial Production of Cu-Nps

Biosynthesis of copper nanoparticles (Cu-NPs) using bacteria provides an eco-friendly alternative to traditional methods. Two studies examined this process with different bacterial strains. *Shewanella loihica* PV-4 produced Cu-NPs (10–16 nm) exhibiting strong antibacterial properties against *E. coli* through reactive oxygen species generation (Lv et al. 2018). Meanwhile, *Pseudomonas silesiensis* strain A3 generated Cu-NPs (87.1 nm), with optimal biosynthesis achieved using 2% blackstrap sugar cane molasses as a carbon source, yielding Cu-NPs of 66.12 nm. These studies emphasise bacterial biosynthesis for antimicrobial Cu-NPs.

Recent research has explored the use of various microbes in copper nanoparticle (Cu-NP) production, offering cost-effective and eco-friendly alternatives to traditional methods. Bacteria like *Shewanella oneidensis* (Kimber et al. 2018) and fungi such as *Aspergillus niger* (Hussain et al. 2023) and *Trichoderma virens* (Dorjee et al. 2024) have been successfully used for synthesis. These Cu-NPs, ranging from 8–500 nm, demonstrate significant antibacterial, antidiabetic, and anticancer activities. Characterisation techniques like FTIR, TEM, SEM, and UV-Vis spectroscopy aid in their analysis (Dorjee et al. 2024).

4.2 Metal Oxide Nanoparticles

Green synthesis of metallic nanoparticles uses eco-friendly techniques to convert metal ions into nanoparticles through natural resources like microbes, plant extracts, and biopolymers. This approach offers a sustainable alternative to conventional synthesis that often involves hazardous chemicals, high energy use, and harmful by-products. Green synthesis is cost-effective, scalable, and environmentally friendly. It often produces nanoparticles with desirable qualities, such as high stability and biocompatibility, suitable for various applications. The process involves metal salt preparation, introducing a green reducing agent, nanoparticle formation, and purification.

Green synthesis of metal oxide nanoparticles, like TiO_2, ZnO, Fe_2O_3, and SnO_2, uses natural resources like plant extracts and microorganisms. This eco-friendly approach produces non-toxic, biodegradable, and biocompatible materials, ideal for biomedical, sensor, and catalytic applications.

4.2.1 Titanium Dioxide (TiO_2)

Titanium dioxide nanoparticles are gaining widespread attention for their exceptional photocatalytic properties, stability, and large surface area, making them highly versatile across various industries (Hamrayev et al. 2024). They are particularly effective in environmental remediation, degrading pollutants in air and water via photocatalysis. TiO_2-Np also enhance dye-sensitised solar cells and support photocatalytic water splitting for hydrogen production. Additionally, TiO_2 is used in sunscreens and cosmetics for UV protection and as an antibacterial agent in medical equipment and food packaging. In sensors and diagnostics, TiO_2 serves biomolecular sensing and gas detection roles. Recent research focuses on improving TiO_2 photocatalytic efficiency and utilising eco-friendly green synthesis methods, enhancing their application potential (Inamdar et al. 2024). Recent investigations have explored the green synthesis of TiO_2 nanoparticles from plant extracts (Eddy et al. 2024). Plant secondary metabolites, particularly polyphenols, enhance the stability and effectiveness of TiO_2-Np (Eddy et al. 2024). These green-synthesised nanoparticles possess unique properties, making them suitable for applications like antibacterial treatments, organic pollutant removal, and photocatalysis. The technique, utilising bioactive substances such as proteins and phytochemicals, is scalable and holds promise for environmental applications like water purification and self-cleaning surfaces.

Over the past decade, various microbes have been explored for the synthesis of TiO_2 nanoparticles. Fungi like *Aspergillus flavus* have produced spherical TiO_2 nanoparticles (62–74 nm) with antibacterial properties against *E. coli*. Bacteria such as *Lactobacillus* sp. and *Saccharomyces cerevisiae* have also synthesised TiO_2 nanoparticles ranging from 8–35 nm (Jha et al. 2009). *Aeromonas hydrophila* produced TiO_2 nanoparticles (40.50 nm) with strong antibacterial activity against *S. aureus* and *S. pyogenes* (Jayaseelan et al. 2013), demonstrating their potential for environmental and antimicrobial applications.

4.2.2 Zinc Oxide (ZnO)

ZnO nanoparticles synthesised via green methods offer enhanced biocompatibility, improved antimicrobial activity, and higher surface area, making them ideal for applications in environmental remediation, biomedical fields, sensors, and energy production. Plant extracts, bacteria, and fungi serve as eco-friendly reducing agents, replacing traditional synthesis methods and resulting in nanoparticles with tailored properties.

Recent studies on the green synthesis of zinc oxide nanoparticles using plant extracts have highlighted their potential in various applications like medicine and environmental remediation. Plants like *Tinospora cordifolia* and *Acacia modesta* have been used to synthesise ZnO NPs with notable antioxidant, antibacterial, and photocatalytic properties (Jayaseelan et al. 2024). Phytochemicals, such as flavonoids and phenolic compounds, act as reducing agents, making the process eco-friendly (Al-darwesh et al. 2024). These plant-based ZnO NPs also show enhanced biocompatibility and reduced toxicity, which makes them suitable for biomedical applications. Other studies have reported the successful synthesis of ZnO NPs from plants like *Morinda citrifolia*, *Catharanthus roseus*, and *Fagonia cretica*, which exhibit antibacterial, antidiabetic, and anticancer activities (Greeshma et al. 2023). Characterisation techniques like XRD and SEM confirm the size and composition of these nanoparticles. In addition to plant-based methods, microbial biosynthesis using bacteria like *Pseudomonas stutzeri* (Murali et al. 2023) and probiotics such as *Lactobacillus fermentum* (Shanmugam et al. 2023) has been explored. These bacteria utilise biosurfactants as reducing agents to synthesise ZnO NPs with promising antimicrobial properties (Shanmugam et al. 2023). The green synthesis of ZnO NPs offers a sustainable, cost-effective, and environmentally friendly alternative to traditional chemical methods, with applications in medicine, agriculture, and environmental protection.

4.2.3 Iron Oxide (Fe_2O_3)

Green synthesis of iron oxide nanoparticles (Fe_2O_3-NPs) has gained significant attention as an eco-friendly alternative to traditional chemical approaches (Aldossari et al. 2023). This process involves reducing iron salts such as iron chloride or iron sulphate using plant extracts, which serve as both reducing and stabilising agents. Plants like coriander, rosemary, and mint have been shown to effectively reduce these salts to Fe_2O_3 nanoparticles, which are typically spherical or cuboidal in shape with sizes ranging from 25 to 70 nm. Recent studies have confirmed the successful synthesis of Fe_2O_3 NPs using plant extracts, with characterisation techniques such as X-ray diffraction (XRD), scanning electron microscopy (SEM), and Fourier transform infrared spectroscopy (FTIR) revealing their formation and properties (Sridevi et al. 2023). These green-synthesised Fe_2O_3 NPs exhibit a variety of promising applications, including antiglycation activity, cytotoxic effects against glioblastoma cells, and photocatalytic degradation of industrial dyes. Notably, Fe_2O_3 NPs derived from plant extracts show enhanced antioxidant potential when compared to chemically synthesised counterparts (Kumar et al. 2023). Further modification with polymers like polyvinylpyrrolidone can enhance their photocatalytic and antioxidant properties. Plants such as *Ocimum sanctum*, *Bombax malabaricum*, and *Madhuca indica* (Kumari et al. 2023) have also been used in the synthesis of Fe_2O_3 NPs, which have demonstrated antibacterial, anticancer, and anti-inflammatory effects. These nanoparticles offer significant potential for both biomedical and environmental applications, particularly in pollutant degradation. Additionally, recent research has explored the use of microorganisms like marine *Streptomyces* sp. (Attea et al. 2024) and soil bacteria such as *Bacillus* and *Klebsiella* species for the biosynthesis of iron oxide nanoparticles (Daramola et al. 2024). These biogenic NPs have shown antibacterial and antifungal properties, highlighting the versatility of green synthesis in producing environmentally friendly, cost-effective nanomaterials. Further studies involving fungi like *Beauveria bassiana* (Neama et al. 2024) and iron–manganese oxide nanoparticles from *Aspergillus niger* (Kamal et al. 2023) continue to show promise for their use in environmental remediation and antimicrobial treatments, further supporting the sustainability and potential of green synthesis in nanomaterial production.

4.3 Carbon-Based Nanomaterials

Carbon-based nanomaterials like carbon nanotubes, carbon quantum dots, and graphene possess exceptional physical, chemical, and mechanical properties that make them valuable for environmental pollution challenges (Dutta et al. 2022). These materials are widely used in energy storage, tissue engineering, drug delivery, and biosensors (Qasim et al. 2023). Green synthesis methods using renewable biomass offer sustainable alternatives to traditional fabrication, eliminating toxic catalysts and enhancing environmental benefits. This eco-friendly approach, rooted in green chemistry, promotes the use of waste materials, making carbon nanoparticles more sustainable in biomedical and environmental applications (Korah et al. 2021). Green-synthesised carbon-based nanomaterials, like graphene quantum dots, are also being explored for water treatment and environmental remediation, contributing to sustainable technological advancements. These developments highlight the promising role of carbon nanomaterials in both addressing environmental concerns and advancing biomedical technologies.

4.3.1 Carbon Nanotubes (CNTs)

Carbon nanotubes (CNTs) are cylindrical nanomaterials composed of carbon atoms arranged in a hexagonal lattice and are known for their remarkable mechanical, electrical, and thermal properties. With their high strength-to-weight ratio, electrical conductivity, and large surface area, CNTs have found applications in diverse fields like electronics, energy storage, environmental remediation, and biomedicine. CNTs are classified into single-walled (SWCNTs) and multi-walled (MWCNTs), each with distinct characteristics suitable for different uses. Traditionally, CNTs are synthesised using chemical methods, but recent research has focused on green synthesis approaches using renewable biomass, plant extracts, microorganisms, or agricultural waste, offering eco-friendly, scalable alternatives that reduce environmental impact. These green methods not only minimise the use of toxic catalysts but also promote the use of readily available biomass as precursors, making the process more sustainable and cost-effective (Qasim et al. 2023). Walnut shells have been utilised for CNT synthesis through pyrolysis, providing a solution to waste disposal while producing high-purity multi-walled CNTs, ideal for applications like gas adsorption and waste purification (Roquia et al. 2021). Microbial biosynthesis of nanoparticles, including CNTs, by bacteria like *Shewanella* and *Geobacter*, further highlights the potential of using microorganisms to produce eco-friendly nanoparticles for applications in energy conversion, medicine, and environmental remediation (Koul et al. 2021). Similarly, fungi, known for their high biomass production and metal tolerance, offer another sustainable alternative for nanoparticle synthesis, with the added benefit of imparting antimicrobial properties (Chauhan et al. 2023). These advances point towards the future potential of green-synthesised CNTs in addressing both environmental and technological challenges.

4.3.2 Graphene

Graphene is a single layer of carbon atoms arranged in a 2D honeycomb lattice (Figure 4), celebrated for its exceptional properties like high electrical conductivity, mechanical strength, and thermal conductivity. Since its discovery, graphene has garnered immense interest for potential applications in diverse fields, including electronics, energy storage, biomedicine, and nanotechnology. Its unique structure facilitates rapid electron movement, making it ideal for use in transistors, sensors, and flexible electronics. Additionally, graphene's large surface area and efficient heat conduction make it a promising material for batteries, supercapacitors, and water filtration systems. However, challenges remain in large-scale production and integration into commercial products. While graphene's properties position it as a game-changer for numerous industries, its practical use still

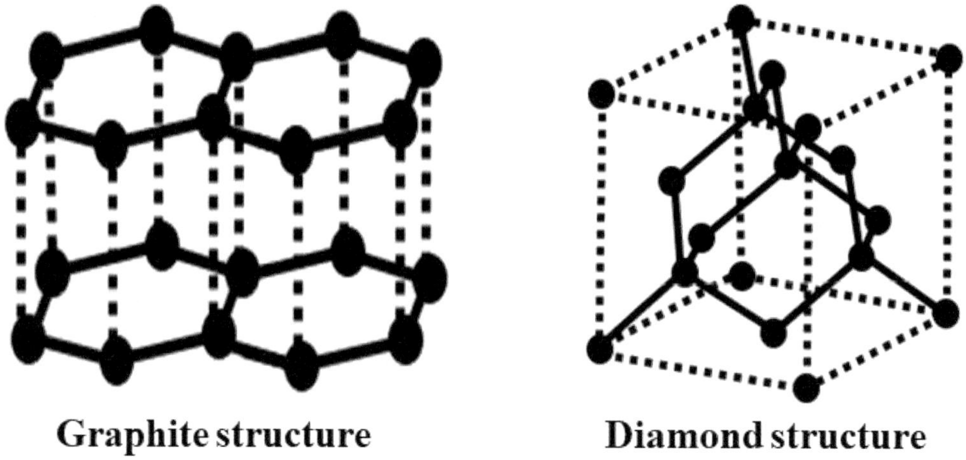

Figure 4: Carbon allotropes as graphene and diamond structure.

requires further development. In contrast, diamond, with its three-dimensional tetrahedral structure, is known for its unmatched hardness, making it ideal for cutting tools and electronics, but the flexibility and conductivity of graphene present a unique set of opportunities in advanced technology.

Green synthesis of graphene refers to environmentally friendly methods using renewable resources and non-toxic reagents that offer a sustainable alternative to traditional techniques that involve hazardous chemicals and high-energy processes. One common approach utilises plant-based materials, such as plant extracts, as reducing agents to convert carbon sources like biomass or organic waste into graphene or graphene oxide. Natural compounds in plant extracts, like flavonoids and phenolic acids, can reduce graphite oxide to graphene oxide or reduced graphene oxide. Additionally, microorganisms like bacteria and fungi can be used to produce graphene in controlled environments, providing low-cost and eco-friendly solutions for large-scale production.

Recent studies have focused on plant extracts as reducing agents for graphene oxide (GO), with successful applications using extracts from eucalyptus bark, ginger, garlic, and onion (Rattan et al. 2020, Manchala et al. 2019). These green methods offer advantages such as simplicity, cost-effectiveness, and reduced environmental impact compared to traditional methods using harsh chemicals like hydrazine. Rai et al. (2020) also explored using substances like L-cysteine and ascorbic acid for the reduction of GO, supporting more sustainable graphene production.

Characterisation techniques such as XRD, FTIR, Raman spectroscopy, and TEM have confirmed the successful reduction of GO using plant-based reducing agents. The resulting biosynthesised graphene has shown potential applications, including antibacterial properties and high-performance supercapacitors (Manchala et al. 2019, Khanam et al. 2019). A study by Kadyan et al. (2024) demonstrated the green synthesis of graphene quantum dots (GQDs) using *Mangifera indica* leaves, showing strong antioxidant and antimicrobial activity. This approach offers an eco-friendly and efficient solution for combating antibiotic resistance and microbial infections, positioning green synthesis methods as a sustainable alternative for graphene production.

4.4 Hybrid and Composite Nanomaterials

Hybrid and composite nanomaterials are sophisticated materials created by combining various nanoparticles or materials to enhance their properties and enable unique functionalities not found in individual components. These materials exhibit exceptional mechanical, electrical, thermal, and chemical properties, making them valuable in industries such as electronics, energy storage, environmental remediation, and biomedicine. Hybrid nanomaterials typically involve inorganic nanoparticles, such as metals, metal oxides, or ceramics, combined with organic molecules, polymers, or carbon-based materials like graphene and carbon nanotubes. By combining materials with complementary properties, hybrid nanomaterials can enhance electrical conductivity, mechanical strength, flexibility, and bioactivity.

Green synthesis methods for hybrid nanomaterials utilise renewable, non-toxic resources and mild conditions, reducing environmental impact (Devi et al. 2020). Natural molecules like flavonoids, polyphenols, and proteins from plant extracts, alongside enzymes and metabolites from microorganisms, help reduce metal ions and stabilise nanoparticles (Sabouri et al. 2023). These methods offer advantages over traditional techniques, including reduced reliance on harmful chemicals and lower energy consumption. They also tend to be simpler, cost-effective, and scalable for large-scale production.

Phytochemical synthesis refers to using plant-derived compounds as reducing agents, stabilisers, or capping agents in the creation of nanomaterials (Hasan et al. 2023). The combination of silver nanoparticles with carbon-based materials, like graphene oxide, can merge the antimicrobial properties of silver with the electrical conductivity of graphene (Aththanayaka et al. 2023). Similarly, composite nanomaterials involve embedding nanoparticles within matrix materials like polymers, which are stabilised by phytochemicals from plant extracts (Dehghan et al. 2022). These materials have applications in drug delivery, sensors, and packaging. Studies highlight the green synthesis of nanocomposites with diverse applications in photocatalysis, antimicrobial activity, and environmental remediation, proving the potential of such eco-friendly methods for advancing nanotechnology across various sectors.

5. Applications of Green-Synthesized Nanomaterials

5.1 Environmental Applications

Green-synthesised nanoparticles' biocompatibility and controlled release capabilities adapt them for diverse applications (Rashwan et al. 2022a, b). These nanoparticles have enormous potential for use in tumour therapy (Devanesan et al. 2021), systems for delivering drugs (Fang et al. 2019), and medical imaging in the biomedical field. Additionally, they are utilised in agriculture to play a crucial role in protecting crops, supplying nutrients, and supporting plant development. These applications help minimise negative environmental impacts by fostering sustainability and enhancing farming practices. Moreover, green nanoparticles have proven effective in environmental remediation like air pollution management, antimicrobial treatments, and the purification of soil and water (Jain et al. 2021). Their ability to eliminate pollutants and enhance restoration treatments makes them a part of a cleaner and greener environment.

5.1.1 Water Treatment and Purification

The size-dependent characteristics, high reactivity, functionalisation, and significant surface area of nanomaterials make them appropriate for wastewater treatment (Thangavelu et al. 2022). Nanotechnology-based wastewater treatment technologies aim to improve water disinfection, decontamination, safe wastewater reuse, and seawater desalination (Kim et al. 2019). This covers the elimination of contaminants by biosorption and nano-adsorption, the identification of pollution by nanoscale sensors, and a variety of filtration processes like electro-dialysis, nanofiltration, reverse osmosis, and ultrafiltration (Kumar et al. 2014). When compared to bulk particles, nanoparticles exhibit remarkable optical, electrical, and magnetic properties due to their significantly larger surface area and high surface energy.

Numerous potential uses in wastewater treatment are made possible by the special and improved qualities of materials at the nanoscale (Jain et al. 2021). When components change from being bulky to being nanostructured, their characteristics and functions may also change (Jiang et al. 2018). Materials' properties depend on two things: the length scale of the material and how the atoms or molecules are arranged.

Under the influence of visible, infrared, or ultraviolet light, the rate of chemical reaction can be altered (Ahmed and Ogulata 2022). A substance that absorbs light and helps the reacting substances undergo chemical transformation is called a photocatalyst. Nanostructured semiconductor materials are the best photocatalysts due to the abundance of electrons and holes generated by light that are present on the surface. Photocatalysis can be used to remove both carbon-based and mineral contaminants like heavy metal ions and microorganisms (Kajitvichyanukul et al. 2022).

Nanofiltration is another suggested method for sewage purification. This specific kind of membrane technique may have additional uses for membranes (Religa et al. 2011). Membrane separation is a force-driven process. Clean water can pass through but harmful substances like organic compounds, nutrients, suspended particles, microorganisms, inorganic metal ions and oxygen-depleting contaminants are effectively filtered out (Theodorakopoulos et al. 2023).

Nanosorbents have been extensively utilised in the treatment of wastewater to remove organic dyes and metals from aqueous solutions (Gautam and Tiwari 2020). The unique characteristics of nanoparticles are their small size and large surface area relative to their volume. Because they are readily adaptable and biocompatible, they are especially well suited for specific kinds of contaminants (Ahmed et al. 2022). At the nanoscale, metal oxides exhibit enhanced adsorption compared to their bulk counterparts due to the formation of metal–ligand precipitates.

5.1.2 Air Pollution Control

One of the biggest issues the world is currently dealing with is contaminated air, which is worsening every year and has serious and catastrophic effects on the planet and its ecosystem. Both natural and synthetic sources contribute to air pollution, but it seems that human-caused sources influence air pollution more than natural ones. The World Health Organization states that sulphur dioxide, nitrogen dioxide, carbon monoxide, particulate matter, and surface ozone are the primary airborne pollutants. An emerging technology that could provide a temporary solution and help withstand air pollution is green nanotechnology. Owing to their high reactivity and large specific surface areas, green nanomaterials act as efficient adsorbents, catalysts, and sensing agents. Nanomaterials can more quickly catalyse reactions or adsorb large amounts of air pollutants due to their extensive surface area and energy. As a result, energy consumption in the degradation process is low, and it may also assist in reducing pollutant release. This section clarifies future trends in this field as well as the use of green nanomaterials for air pollution detection and remediation.

5.2 Biomedical Applications

5.2.1 Antimicrobial and Antiviral Agents

Nanoparticles have emerged as powerful antimicrobial and antiviral agents in biomedical applications due to their unique physicochemical properties like high surface area, enhanced reactivity, and the ability to disrupt microbial structures. Metal-based nanoparticles such as silver (AgNPs), gold (AuNPs), copper oxide (CuO), and zinc oxide (ZnO) exhibit broad-spectrum antimicrobial activity by generating reactive oxygen species (ROS) that disrupt cell membranes, interfere with enzyme functions, and inhibit DNA replication (Sharma et al. 2024). Silver nanoparticles are widely used in wound dressings, coatings for medical devices, and antimicrobial textiles due to their potent bactericidal properties against drug-resistant pathogens.

In antiviral applications, nanoparticles act by preventing virus attachment, entry, or replication within host cells (Milovanovic et al. 2017). Gold and silver nanoparticles functionalised with biomolecules can block viral surface proteins, inhibiting interactions with host cell receptors. Silver nanoparticles have shown effectiveness against influenza, HIV, and coronaviruses by disrupting viral envelopes and altering viral genetic material. Additionally, nanocarriers are employed for targeted drug delivery that enhances the efficacy of antiviral therapies while reducing side effects. Nanoparticles offer significant advantages like prolonged antimicrobial activity, reduced antibiotic resistance development, and improved drug bioavailability (Raza et al. 2019). However, challenges related to cytotoxicity, stability, and controlled release need further investigation for safe clinical applications. Ongoing research in nanomedicine aims to optimise nanoparticle formulations for enhanced biocompatibility and targeted therapeutic effects, making them promising candidates for combating infectious diseases in modern healthcare.

5.2.2 Drug Delivery Systems

Nanoparticles have revolutionised drug delivery systems in biomedical applications by offering targeted, efficient, and controlled release of therapeutic agents. Their small size, large surface area, and tunable physicochemical properties enable enhanced bioavailability, prolonged circulation time, and reduced systemic toxicity. Distinct types of nanoparticles, including liposomes, polymeric nanoparticles, metallic nanoparticles, dendrimers, and lipid-based carriers, are designed to improve drug solubility, stability, and site-specific delivery (Bhatia and Bhatia 2016). Functionalisation with ligands such as antibodies, peptides, or aptamers allows nanoparticles to selectively bind to diseased cells, minimising off-target effects and enhancing therapeutic efficacy.

In cancer therapy, nanoparticles facilitate the targeted delivery of chemotherapeutic agents, reducing adverse effects on healthy tissues while improving drug accumulation at tumour sites through enhanced permeability and retention (EPR) effects. Additionally, they are used for gene delivery in genetic disorders, enabling the transport of nucleic acids such as siRNA and mRNA into cells with high efficiency. Nanoparticles also play a vital role in vaccine delivery, as seen in mRNA COVID-19 vaccines in which lipid nanoparticles protect and transport the genetic material into host cells for an immune response (Hou et al. 2021). Furthermore, nanoparticles are utilised in antimicrobial therapy, overcoming drug resistance by enhancing the penetration of antibiotics into bacterial biofilms (Singh et al. 2022). Despite their advantages, challenges such as toxicity, immunogenicity, and large-scale production need further investigation. Advances in nanomedicine continue to optimise nanoparticle formulations to ensure safer and more effective drug delivery systems. With continuous research, nanoparticle-based drug delivery holds immense potential for personalised medicine and improved therapeutic outcomes in various diseases.

5.2.3 Bioimaging and Biosensors

Nanoparticles have revolutionised biomedical applications, particularly in bioimaging and biosensing, due to their unique optical, electrical, and magnetic properties. In bioimaging, nanoparticles enhance contrast and provide high-resolution imaging for disease diagnosis and monitoring. Quantum dots, gold nanoparticles, and iron oxide nanoparticles are widely used in magnetic resonance imaging (MRI), fluorescence, and computed tomography (CT). Quantum dots have tunable fluorescence that enables real-time tracking of biomolecules and cancer cells. On the other hand, gold nanoparticles enhance contrast in CT scans. Superparamagnetic iron oxide nanoparticles improve MRI sensitivity and are used for targeted imaging of tumours and neural disorders. These nanoparticles can also be functionalised with biomolecules to enhance specificity, enabling early disease detection. In biosensing, nanoparticles serve as highly sensitive platforms for detecting biomolecules, pathogens, and toxins. Gold nanoparticles and carbon-based nanomaterials like graphene and carbon nanotubes are commonly integrated into biosensors to enhance signal transduction. Plasmonic biosensors, based on gold nanoparticles, detect biomolecular interactions through surface plasmon resonance (SPR), allowing real-time monitoring of disease markers (Singh et al. 2023). Similarly, enzyme-functionalised nanoparticles improve the sensitivity of electrochemical biosensors for glucose and cancer biomarker detection (Wang et al. 2020). Nanoparticle-based biosensors provide rapid, cost-effective, and highly sensitive diagnostics, essential for early disease detection and personalised medicine. Despite challenges such as toxicity and biocompatibility, ongoing advancements in nanotechnology continue to drive the development of safer and more efficient nanoparticle-based biomedical tools.

5.3 Energy Applications

The globe is increasingly concerned about an impending energy crisis due to the limited availability of conventional energy resources and the rising daily energy consumption (Padhiari et al. 2022). Efforts are underway worldwide to find alternatives and develop efficient electronic devices and sustainable energy sources, with the manufacturing of green nanomaterials emerging as an appealing solution (Padhiari et al. 2022). Green synthesis processes utilise natural resources without causing harmful effects, making them more environmentally friendly than traditional methods (Sarkar et al. 2019). Furthermore, nanomaterials synthesised this way are better suited for current energy applications. Energy harvesting applications are typically categorised into two main areas: energy conversion and energy storage. In energy conversion devices, such as fuel cells (Hassaan et al. 2023), hybrid solar cells, photoelectrochemical cells (El Golli et al. 2021), and dye-sensitised solar cells (Ali et al. 2023), green-synthesised nanomaterials play an active role. Similar innovative applications of green-synthesised nanomaterials have also been observed in energy storage systems.

5.3.1 Photocatalysis for Solar Energy Conversion

Usually, catalysis is known as a substance that enhances chemical reactions, while photocatalysis accelerates photoreactions. In solar energy conversion, a photocatalytic reactor or photocatalyst is used to convert solar energy into useful forms, such as hydrogen or synthetic fuels, by using semiconductor materials to capture sunlight and drive chemical reactions. When exposed to light, the photocatalyst absorbs photons, creating electron-hole pairs that interact with reactants like carbon dioxide or water to produce energy-rich molecules, such as hydrogen through water splitting or carbon-based fuels via CO_2 reduction. This process mimics natural photosynthesis and provides a sustainable energy source. Additionally, photocatalysts can decompose dyes and facilitate nitrogen reduction to yield ammonia in some applications.

The effectiveness of a photocatalytic system depends on factors like light absorption capacity, charge carrier dynamics, and surface reactivity, making material selection crucial. For photocatalysis in solar energy conversion, materials must have a bandgap between 1.5 and 3 eV, high photostability, and the ability to separate and transport charge carriers. They should also have a high surface area for better reactant interaction, catalyse reactions like water splitting or CO_2 reduction, and be environmentally friendly, cost-effective, and scalable. Green synthesis methods offer an eco-friendly approach to producing photocatalytic materials using renewable or non-toxic processes, minimising environmental impact. Green-synthesised materials like TiO_2, metal oxides, nanoparticles, and carbon-based composites, made from plant extracts or natural precursors, often show enhanced efficiency, better charge carrier separation, and greater stability under light exposure. This sustainable approach supports the development of solar energy technologies such as hydrogen production and CO_2 reduction.

Green synthesis of nanomaterials for photocatalysis in solar energy conversion has gained significant attention due to its eco-friendly, cost-effective nature, presenting a sustainable alternative to traditional chemical methods (Padhiari et al. 2022, Sarkar et al. 2019). Various approaches, including plant-mediated, biological, and environmentally friendly chemical methods, have been explored to synthesise photocatalysts like ZnO, TiO_2, metal sulphides, and g-C_3N_4-based materials (Ali et al. 2023). These green-synthesised nanomaterials have demonstrated promising applications in dye-sensitised solar cells and the photocatalytic degradation of organic pollutants. ZnO nanostructures synthesised using garlic bulb extract showed similar photocatalytic activity to chemically synthesised ZnO in degrading methylene blue under concentrated sunlight (El Golli et al. 2021). Combining green synthesis with solar concentration technology offers a cost-effective solution for large-scale photocatalytic wastewater remediation, reducing both costs and environmental impact. Recent reviews emphasise the benefits of green synthesis methods, such as cost-effectiveness and environmental friendliness, particularly in wastewater treatment. While

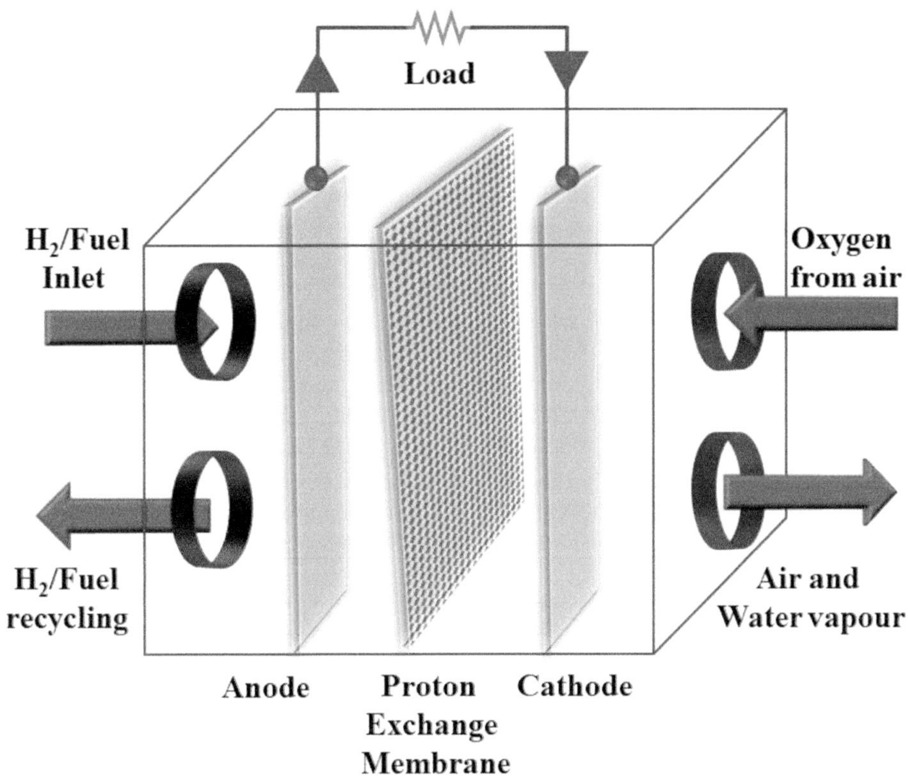

Figure 5: Fuel cell device and its working mechanism.

these methods hold great promise, further research is needed to optimise their potential for future applications, especially in energy storage and environmental remediation (Gupta et al. 2023).

5.3.2 Fuel Cells and Energy Storage

A fuel cell is a device that converts chemical energy from fuel directly into electrical energy through an electrochemical reaction. It works similarly to a battery but has the key difference that, as long as you supply the fuel and oxygen, it will continue generating electricity. Figure 5 schematically illustrates the structure of fuel cell devices and their working mechanism.

A fuel cell works by first supplying hydrogen gas to the anode (negative side). It is divided into two parts, protons (H^+) and electrons (e^-). The protons move through an electrolyte, a substance that allows ions like H^+ to pass but blocks electrons. The electrons travel through an external circuit, generating electricity, and then reach the cathode (positive side), where they combine with oxygen and the protons to form water (H_2O) and heat. The key by-products of this process are water and heat, making fuel cells a clean and efficient energy source, particularly when hydrogen is used as the fuel (Kausar et al. 2023).

Conductivity, stability, ionic conductivity, low gas permeability, and mechanical strength are crucial factors when choosing materials for fuel cell design in order to ensure long-term performance. Furthermore, to guarantee long-term viability and successful energy conversion, cost-effectiveness, recyclability, and environmental sustainability are essential, particularly for large-scale production (Bayat et al. 2023). Green-synthesised material that meets all the criteria required for fuel cell application is sustainable for utilisation in the future (El-Esawy et al. 2024).

Green nanocomposites, which offer improved effectiveness and ecological benefits, are substitutes for the electrodes and electrolytes in microbial fuel cells (MFCs). These materials improve features including power density and efficiency, replacing the role of platinum-based electrodes. To fully explore their potential in sustainable fuel cell technologies, more study is necessary, as there are still challenges with their widespread application (Kausar et al. 2023). Recent research has explored the potential of green nanoparticles in various fields, including energy production and biological applications. Biogenic platinum nanoparticles (Pt NPs) synthesised using green methods have shown promise in cytotoxicity tests, antibacterial activities, and methanol oxidation reactions for clean energy generation (Bayat et al. 2023). Green nanoparticles offer advantages in biomedical applications, crop improvement, and energy storage devices like batteries and supercapacitors (El-Esawy et al. 2024). Nanomaterials have also been utilised in fuel cell components to enhance efficiency. For example, bacterial nanocellulose (BNC) has been used as a sustainable support for platinum nanoparticles, with Pt/BNC catalysts demonstrating high electrocatalytic performance in methanol oxidation (Ponjavic et al. 2022). Sustainable techniques for nanomaterial synthesis have also been investigated for preventing corrosion and improving microbial fuel cells (MFCs). Green synthesis methods have been employed to produce platinum nanoparticles for oxygen reduction reactions (Rosas-Medellín et al. 2021). Biomimetically synthesised gold and silver nanoparticles from weed extracts have significantly enhanced MFC performance, increasing coulombic efficiency and power density by over 200% (Abbasi et al. 2021). These advancements in green nanomaterials offer promising solutions for energy production and environmental sustainability.

5.3.3 Hydrogen Generation and storage

Hydrogen generation and storage are crucial to transitioning to a sustainable, low-carbon energy future. As a clean fuel, hydrogen can be used across various applications, including transportation, electricity generation, and industrial processes, providing a viable alternative to fossil fuels. The main methods for hydrogen production are water electrolysis, which uses electricity—ideally from renewable sources like wind or solar—to split water into hydrogen and oxygen. This process is environmentally friendly and produces zero emissions when powered by green electricity. Another common method is steam methane reforming, which extracts hydrogen from natural gas but results in significant carbon emissions, making it less sustainable unless paired with carbon capture technologies. Additionally, methods like biomass gasification and photoelectrochemical water splitting are being explored for hydrogen generation from renewable sources. A major challenge in hydrogen energy, however, lies in its storage. Hydrogen has a low energy density by volume, which makes efficient storage difficult. Common storage methods include high-pressure gas tanks, cryogenic liquid storage, and solid-state storage using metal hydrides or carbon-based materials. Each method has its own challenges, such as the cost and weight of high-pressure tanks, the energy-intensive nature of cryogenic storage, and the early-stage development of solid-state storage. Recent advancements in nanomaterials and green-synthesised nanoparticles show promise in improving storage efficiency and reducing costs. Innovations in these areas are essential for the widespread adoption of hydrogen energy, positioning it as a key element in clean, sustainable energy systems.

5.4 Agricultural Applications

Despite being essential to meeting the growing demand for nutrition worldwide, current farming practices have detrimental effects on the environment. The growth of farming practices has led to widespread soil contamination by various harmful substances, such as toxic metals and chemical fertilisers (Jakhar et al. 2022). These pollutants pose serious risks to both environments and social well-being, while also lowering soil reproduction via the food chain (Chu et al. 2023). Investigating innovative and environmentally friendly restoration methods that aid in restoring soil quality and reducing the effects of soil pollution is essential considering these challenges. A major advancement in solving the intricate issues facing the world's food production system is the application of nanotechnology in agriculture (He et al. 2019). The widespread application of nanomaterials in soil wellness, nutrition oversight, precision farming, and insect control presents outstanding chances to adopt efficient and environmentally friendly agricultural methods (Adeyemi et al. 2022).

Recently, there has been an increase in the acknowledgement of nanotechnology as an innovative and promising method to address the challenges associated with soil pollution in farming settings. The unique properties of nanomaterials, such as their large surface area, their nature, and capacity to alter physical and chemical characteristics, make them particularly invaluable for soil remediation. Nanotechnology has the potential to significantly change traditional soil remediation techniques by offering more accurate, efficient, and environmentally friendly means of reducing the effects of agricultural contamination. This comprehensive section's goal is to examine and gather the most current information on the use of nanotechnology in cleaning up contaminated agricultural soils. Nanotechnology involves the deliberate management of materials at the nanoscale and offers unprecedented opportunities for the creation and implementation of tailored solutions for soil remediation.

5.4.1 Nano-Fertilisers

Nanofertilisers are environmentally friendly materials that can give plants vital nourishment, promote plant development, and increase crop production. To promote plant growth, biofertilisers—also referred to as bacterial fertilisers or microbial cultures—use live or dormant microorganisms. These biofertilisers are divided into groups like nitrogen fixation, phosphate support, phosphate growth, beneficial nutrients, and Rhizobacteria that promote plant growth (El-Ghamry et al. 2018). According to their nutrient contents, functions, and consistency, nanofertilisers can be broadly classified into three groups: (i) nanoparticulates, which include carbon nanotubes, SiO_2 and TiO_2 and which aid in plant growth; (ii) micronutrients, which include Cu, Zn, Fe, Mo, Ni and Mn; and (iii) macronutrients, which include P, Ca, N and K. Nanofertilisers can promote plant growth through various mechanisms (El-Ghamry et al. 2018).

Accordingly, brinjal and tomatoes can benefit from the use of onion-derived silver nanoparticles as a nanofertiliser (de França Bettencourt et al. 2020). To promote the growth of grain plants, they also created a biofertiliser that contains microbes with chemical auxin complexes. The results showed that both single and combined oxides enhanced grain plant root development, new weight, and seed germination. Nanofertilisers offer several benefits in contrast to conventional fertilisers, such as enhanced efficiency from the targeted supply of vital nutrients to plants and a reduction in environmental consequences owing to the lower quantities of fertiliser needed. This technology has the capacity to optimise crop production while also minimising the ecological impacts associated with fertilisers. The benefits of nanofertilisers encompass their elevated nutrient density, gradual nutrient release, and enhanced absorption by plants (Table 1).

To sum up, green nanomaterials provide a good substitute for nanofertilisers, encouraging plant development, seed germination, and total weight gain. In addition to starting several biochemical processes within the plant, such as enhancing its immune response and encouraging the production of auxin, nanofertilisers help plants better manage a variety of stressors, including drought, diseases, and pests. Additionally, nanofertilisers improve the soil's fertility and structure by making nutrients more soluble, which encourages the growth of healthy plants. The processes used to make these eco-friendly fertilisers are simple and quick. Enhancement of plant growth can be ascribed to several distinct mechanisms.

5.4.2 Pest Control and Plant Protection

Pests pose significant biotic challenges in agriculture and food production. To tackle these issues, humans are continually exploring new control methods. Conventional methods, like integrated pest control, are not working, and chemical pesticides have detrimental effects on both humans and animals. They also reduce soil fertility, increase pest resistance, eliminate natural predators, pollute the environment, reduce biodiversity, and endanger human health. Implementing nanotechnology in pest control can serve as a promising alternative to the challenges associated with conventional

Table 1: Pros and cons of nanofertilizers compared to traditional fertilizers.

Characteristics	Nano Fertilizers	Traditional Fertilizers
Nutrient absorption effectiveness	Increases the yield of fertilizer and improves nutrient uptake by plants while preserving fertilizer.	Its mass materials are difficult for plants to absorb, which reduces their effectiveness.
Controlled relief mechanisms	Encasing combined with an epoxy resin, oil and sulphur coating enables precise control of nutrient release.	An excessive discharge can lead to toxicity, disrupting the green equilibrium.
Nutrient dissolution and distribution	Improves the distribution and solubility of insoluble mineral components in the soil, increasing plant accessibility.	Due to larger particle size and reduced solubility, plants have less access to these nutrients.
Duration of effective release	Augments and extends the rate at which plants acquire nutrients.	When delivered, essential nutrients for plants often become lost as insoluble salts.
Minimal fertilizer application required	Reduces the amount of nutrients lost as a result of movement, water runoff and evaporation.	Evaporation, water runoff and travel all result in significant fertilizer loss.

chemical pesticides. Greater efficacy against microbes and low toxicity of nanocides to non-target species are two benefits of employing nanotechnology to control insect pests, underscoring the previously reported low level of environmental harm. It is also more ecologically sustainable because it provides targeted, selective, and extended release of formulated nanomaterials. As a result, using nanotechnology to manage insect pests is seen as both environmentally friendly and a successful method of control in green agriculture.

5.4.3 Soil Nutrient Management

Soil is an essential resource for human existence, crucial to producing food, fibre, and fuel. Healthy soils possess high fertility, diverse ecosystems, and robust physical properties, which are vital for producing nutritious food. When soil suffers degradation, its productivity diminishes, necessitating restoration efforts. Eco-friendly nanoparticles derived from plants and microbes contribute to soil recovery at the nanoscale level. The interaction between nanoparticles, plants, and microbes is important for sustaining soil fertility by regulating nutrients, agrochemicals, and contaminants, with its dynamics shaped by numerous factors involved in soil rehabilitation.

Nanoparticles (NPs) enhance the delivery of nutrients in plants due to their broad surface, high binding ability, and carefully regulated dispersion mechanisms. Food security and sustainable agriculture are supported by the effective and targeted nutrient supply that is ensured by encapsulation in nano-carriers, which shields nutrients from environmental stress (Nair et al. 2011). All things considered, nanofertilisers improve crop yield by boosting photosynthesis, encouraging plant growth hormones, and optimising nutrient utilisation (Rui et al. 2016). Additionally, nanofertilisers enhance photosynthesis by improving light absorption and boosting chlorophyll content. Titanium oxide nanoparticles (TiO_2-NPs) significantly increase water and nutrient uptake, scavenge reactive oxygen species (ROS), and enhance photosynthesis, as seen in a 73% biomass increase in spinach over 30 days (Noji et al. 2011). NPs also influence enzyme activity and the electron transport system, improving plant physiology. They can be absorbed through immediate foliar spraying, dusted on seeds, grains, and plant roots. While promising, further research is needed on the role of NPs in photosynthesis across different plant species (Anusuya and Sathiyabama 2015).

As an alternative, nanotechnology continues to be at the forefront of improving food production and availability by putting forward best practices for nutrient and soil fertility management. It has the potential to completely transform agriculture because NP characteristics can be changed based on predetermined goals, they have demonstrated an effective carrier system for delivering nutrients to the intended locations gradually and steadily, they are less toxic than other chemicals, and they are simple to process and engineer.

5.5 Industrial Applications

Green synthesis of nanoparticles utilises natural resources like plant extracts, microorganisms, or biopolymers, reducing reliance on harmful chemicals. This sustainable approach produces nanoparticles with unique properties such as high surface area and biocompatibility. These eco-friendly nanoparticles find applications across various industries, including medicine, environmental remediation, and energy storage.

5.5.1 Catalysis and Chemical Reactions

Green-synthesised nanoparticles are gaining attention in catalysis and industrial chemical reactions due to their eco-friendly production and excellent catalytic properties (Gupta et al. 2023). Produced using natural resources like plant extracts or microorganisms, they offer a sustainable alternative to traditional methods involving toxic chemicals. These nanoparticles provide enhanced catalytic activity, selectivity, and stability, making them ideal for processes like biomass conversion, wastewater treatment, and fine chemical production. Their small size and large surface area enable efficient reactions under milder conditions, reducing energy consumption and carbon footprint. Moreover, they can be recycled, promoting sustainability in industrial applications (Nguyen et al. 2022).

Green-synthesised nanoparticles are emerging as highly effective catalysts for various chemical processes, offering a sustainable alternative to traditional methods that often rely on harmful chemicals and energy-intensive processes. These nanoparticles are produced using natural materials like plant extracts, algae, or microorganisms, which significantly reduces the environmental impact (Nasrollahzadeh et al. 2021). Their small size, large surface area, and high reactivity make them particularly valuable in enhancing reactions in industrial applications such as wastewater treatment, energy production, and fine chemical synthesis. Green-synthesised nanoparticles, particularly those made from noble metals, are commonly used in reactions like hydrogenation, oxidation, and polymerisation, improving catalytic efficiency, selectivity, and recyclability. This leads to cost savings and enhances environmental sustainability. Furthermore, the green synthesis process itself offers benefits like reduced pollution, lower costs, and greater safety for both humans and the environment (Gómez-López et al. 2020). By utilising renewable resources like agricultural waste, green synthesis minimises the carbon footprint and avoids toxic chemicals. Research has also demonstrated that these nanoparticles improve properties like thermal and electrical conductivity, and they show potential in applications ranging from energy storage and biological labelling to cancer therapy. Despite challenges such as toxicity concerns, the potential of green-synthesised nanoparticles to revolutionise various industries remains immense, offering cleaner, more efficient, and sustainable solutions for a variety of sectors, particularly in environmental catalysis, where they address issues like pollution remediation and waste treatment.

5.5.2 Nanocomposites in Coatings and Textiles

Green-synthesised nanoparticles are becoming increasingly popular in the textile industry due to their ability to enhance fabric properties while being environmentally friendly. Nanoparticles like silver, titanium dioxide, and zinc oxide, synthesised using plant extracts or microorganisms, provide textiles with antimicrobial, self-cleaning, and UV-protective features (Gautam et al. 2022). Silver nanoparticles help prevent bacterial growth and reduce odour in clothing, especially in sportswear and medical textiles. Titanium dioxide nanoparticles make fabrics hydrophobic, stain-resistant, and UV-protective, improving durability and comfort. This eco-friendly production method aligns with the demand for sustainable textiles, offering a non-toxic, environmentally conscious alternative to traditional treatments.

Green-synthesised nanoparticles are increasingly used in antimicrobial textiles to enhance fabric performance while promoting environmental sustainability (Swielam et al. 2024). Nanoparticles like silver, copper, and zinc oxide, derived from plant extracts or microorganisms, possess strong antimicrobial properties, preventing bacterial, fungal, and pathogen growth on fabrics (Li et al. 2014). Silver nanoparticles, for example, are commonly used in sportswear and medical textiles due to their ability to inhibit odour-causing bacteria. These nanoparticles not only improve hygiene but also extend fabric lifespan by reducing microbial degradation (Ahmed et al. 2022). The green synthesis process ensures that these nanoparticles are free from toxic chemicals, making textiles safer and more eco-friendly. Additionally, antimicrobial nanocomposites like chitosan and zinc oxide nanoparticles offer sustainable solutions for recycling textile waste (Patil et al. 2020). Innovations like hybrid nanostructures, such as silver on zinc oxide nanowires, further enhance antimicrobial performance and hold promise for applications in wound dressings, water treatment, and air filtration (Amibo et al. 2022). This growing use of green nanotechnology reflects the demand for sustainable, high-performance materials in textiles.

5.5.3 Food Packaging and Preservation

Green-synthesised nanoparticles, such as silver, zinc oxide, and titanium dioxide, are gaining popularity in the food and packaging industries for their ability to enhance food safety, extend shelf life, and offer sustainable alternatives to traditional materials. Produced through eco-friendly methods like plant extracts or microorganisms, these nanoparticles possess antimicrobial properties that help prevent spoilage and contamination (Kumar et al. 2020). They also improve the mechanical and barrier properties of packaging, providing better resistance to moisture, oxygen, and light, thus preserving food freshness. Additionally, these nanoparticles serve as natural preservatives and nutrient delivery systems, enhancing food stability and reducing the need for harmful chemicals.

These nanoparticles are increasingly used in food preservation, providing a sustainable alternative to conventional methods that rely on synthetic chemicals. Silver, copper, and zinc oxide nanoparticles, synthesised through green methods, exhibit strong antimicrobial properties that inhibit the growth of bacteria and fungi. Silver nanoparticles, in particular, are used in food packaging to extend the shelf life of perishable goods like meats, dairy, and produce. Nanoparticles derived from agricultural waste, such as cellulose and pectin, also offer antimicrobial benefits and biodegradability (Pradhan et al. 2023).

In addition, these nanoparticles enhance food quality and nutritional content by improving the absorption of nutrients like vitamins and minerals. This provides a safer and more sustainable alternative to synthetic additives. Green-synthesised nanoparticles also play a key role in nutrient fortification, making essential nutrients like iron, zinc, and calcium more accessible for absorption.

In food packaging, nanotechnology is being used to create antimicrobial nanocomposites that improve food safety and shelf life. Green-synthesised nanoparticles, such as silver, are incorporated into packaging materials to prevent spoilage and reduce the need for chemical preservatives (Flores-Contreras et al. 2024). However, challenges remain, such as concerns over toxicity and the need for further research on long-term effects. Despite this, green nanotechnology continues to offer promising solutions for safer, more sustainable food packaging.

6. Challenges and Future Perspectives in Green Synthesis

Green synthesis, a method that uses environmentally friendly approaches to produce nanoparticles and other materials, has gained significant attention in recent years. This approach leverages natural processes, such as plant extracts and microorganisms, to replace harmful chemical methods traditionally used in industries like materials science, pharmaceuticals, and electronics. While green synthesis offers promising benefits such as reduced toxicity, sustainability, and biocompatibility, it also faces numerous challenges that hinder its widespread industrial adoption. These challenges include issues related to scalability, consistency, cost-effectiveness, and environmental impact. As researchers continue to explore ways to optimise these methods, the future of green synthesis holds enormous potential for more sustainable and eco-friendly industrial practices. Addressing these challenges and refining the techniques could lead to a broader application of green synthesis methods across various industries.

6.1 Limitations of Current Green Synthesis Approaches

Green synthesis, particularly through phytochemical and microbial methods, has garnered significant attention for producing nanomaterials and chemical compounds in an environmentally friendly manner. These methods align with sustainable chemistry principles by minimising the use of hazardous chemicals, reducing energy consumption, and lowering environmental impact (Ying et al. 2022). Phytochemical synthesis, which utilises plant extracts like flavonoids, alkaloids, and phenolics to reduce metal ions and form nanoparticles, is cost-effective, biocompatible, and eco-friendly. However, it has limitations such as variability in extract composition, which can lead to inconsistent nanoparticle quality and size, complicating scalability (Shaikh et al. 2021). The slow reaction time and difficulty in controlling particle size and shape further hinder large-scale production. Additionally, extracting and purifying plant materials is time-consuming and expensive, and large-scale harvesting can have negative environmental impacts. Microbial synthesis, using bacteria, fungi, or algae, offers an eco-friendly, scalable alternative that produces highly specific nanoparticles. However, microbial synthesis faces challenges such as long production times, unpredictable biosynthesis pathways, and potential contamination (Carmona et al. 2023). Moreover, the need for specialised infrastructure and concerns about genetically modified organisms (GMOs) add complexity and cost, while regulatory and ethical issues may limit its use in fields like food and medicine. Despite their potential, both phytochemical and microbial methods require further optimisation to become commercially viable at scale.

6.2 Future Research Directions and Emerging Trends

The field of nanomaterials is rapidly expanding, with ongoing research revealing new properties and applications. Emerging trends in sustainability, energy, healthcare, electronics, and environmental remediation are expected to shape the future, offering transformative solutions to global challenges.

The future of nanomaterials research holds immense promise across multiple sectors, addressing global challenges and revolutionising industries like energy, healthcare, and environmental sustainability. In the area of Sustainable Nanomaterials and Green Nanotechnology, researchers are focusing on low-impact synthesis techniques using renewable resources like plant extracts, microorganisms, and sustainable solvents, alongside waste-to-resource approaches that convert agricultural or industrial waste into nanomaterial precursors. Biodegradable nanomaterials from natural polymers are being developed to reduce ecological footprints, while the recycling and reuse of nanomaterials is an emerging field aimed at supporting a circular economy. In energy harvesting and storage, nanomaterials are improving solar cell efficiency with materials like quantum dots, graphene, and perovskites, and advancing battery technologies by enhancing energy density and charging times. Supercapacitors and hybrid systems are also under development, alongside nanogenerators that harvest energy from mechanical movements (Ramadoss et al. 2024). Nanomaterials in Healthcare are transforming drug delivery, diagnostics, regenerative medicine, and cancer therapy, with targeted drug delivery systems and stimuli-responsive nanocarriers for personalised medicine. Nanomaterials in Electronics are enabling breakthroughs in quantum computing, flexible electronics, and energy-efficient devices. Finally, Environmental Remediation technologies are employing nanomaterials for water purification, air pollution control, and soil clean-up, offering innovative solutions for sustainability (Asghar et al. 2024). The growing focus on sustainability, biocompatibility, and high-performance systems requires continued interdisciplinary collaboration, ensuring nanomaterials will play a significant role in shaping the technological landscape of the future, providing scalable, safe, and cost-effective solutions to pressing global issues.

References

Abbasi, S. A., and Patnaik, P. (2021). Biomimetically generated nanoparticles in boosting the performance of microbial fuel cells. *Nature Environment and Pollution Technology*, 20(5): 1881–1886.

Abu-Elghait, M., Soliman, M. K., Azab, M. S., and Salem, S. S. (2025). Response surface methodology: Optimization of myco-synthesized gold and silver nanoparticles by *Trichoderma saturnisporum*. *Biomass Conversion and Biorefinery*, 15: 4211–4224.

Adeleke, B. S., Olowe, O. M., Ayilara, M. S., Fasusi, O. A., Omotayo, O. P., Fadiji, A. E. et al. (2024). Biosynthesis of nanoparticles using microorganisms: A focus on endophytic fungi. *Heliyon*, 10(21): 39636.

Adeyemi, J. O., Oriola, A. O., Onwudiwe, D. C., and Oyedeji, A. O. (2022). Plant extracts mediated metal-based nanoparticles: synthesis and biological applications. *Biomolecules*, 12(5): 627.

Afonso, I. S., Cardoso, B., Nobrega, G., Minas, G., Ribeiro, J. E., and Lima, R. A. (2024). Green synthesis of nanoparticles from olive oil waste for environmental and health applications: A review. *Journal of Environmental Chemical Engineering*: 114022.

Ahmad, A., Senapati, S., Khan, M. I., Kumar, R., Ramani, R., Srinivas, V. et al. (2003). Intracellular synthesis of gold nanoparticles by a novel alkalotolerant actinomycete, *Rhodococcus* species. *Nanotechnology*, 14(7): 824.

Ahmed, S. F., Mofijur, M., Ahmed, B., Mehnaz, T., Mehejabin, F., Maliat, D. et al. (2022). Nanomaterials as a sustainable choice for treating wastewater. *Environmental Research*, 214: 113807.

Ahmed, T., and Ogulata, R. T. (2022). A review on silver nanoparticles-green synthesis, antimicrobial action, and application in textiles. *Journal of Natural Fibers*, 19(14): 8463–8484.

Akbari, B., Tavandashti, M. P., and Zandrahimi, M. (2011). Particle size characterization of nanoparticles–a practical approach. *Iranian Journal of Materials Science and Engineering*, 8(2): 48–56.

Al-darwesh, M. Y., Ibrahim, S. S., and Mohammed, M. A. (2024). A review on plant extract mediated green synthesis of zinc oxide nanoparticles and their biomedical applications. *Results in Chemistry*: 101368.

Aldossari, S. M., Rehman, L. U., Ahmad, I., Aslam, M., Fozia, F., Mohany, M. et al. (2023). Phytosynthesized Iron Oxide Nanoparticles Using Aqueous Extract of *Saccharum arundinaceum* (Hardy Sugar Cane), their Characterizations, Antiglycation, and Cytotoxic Activities. *ACS omega*, 8(44): 41214-41222.

Al-Harbi, N., and Abd-Elrahman, N. K. (2024). Physical methods for preparation of nanomaterials, their characterization, and applications: a review. *Journal of Umm Al-Qura University for Applied Sciences*: 1–22.

Ali, J., Bibi, S., Jatoi, W. B., Tuzen, M., Jakhrani, M. A., Feng, X. et al. (2023). Green synthesized zinc oxide nanostructures and their applications in dye-sensitized solar cells and photocatalysis: A review. *Materials Today Communications*: 106840.

Alsaiari, N. S., Alzahrani, F. M., Amari, A., Osman, H., Harharah, H. N., Elboughdiri, N. et al. (2023). Plant and microbial approaches as green methods for the synthesis of nanomaterials: synthesis, applications, and future perspectives. *Molecules*, 28(1): 463.

Alyami, N. M., Alyami, H. M., and Almeer, R. (2022). Using green biosynthesized kaempferol-coated sliver nanoparticles to inhibit cancer cells growth: An in vitro study using hepatocellular carcinoma (HepG2). *Cancer Nanotechnology*, 13(1): 26.

Amibo, T. A., Beyan, S. M., Mustefa, M., Sundramurthy, V. P., and Bayu, A. B. (2022). Development of nanocomposite based antimicrobial cotton fabrics impregnated by nano SiO_2 loaded AgNPs derived from eragrostis teff straw. *Materials Research Innovations*, 26(7): 405–414.

Anajwala, C. C., Jani, G. K., and Swamy, S. V. (2010). Current trends of nanotechnology for cancer therapy. *International Journal of Pharmaceutical Sciences and Nanotechnology*, 3(3): 1043–1056.

Anusuya, S., and Sathiyabama, M. (2015). Foliar application of β-D-glucan nanoparticles to control rhizome rot disease of turmeric. *International Journal of Biological Macromolecules*, 72: 1205–1212.

Armendariz, V., Herrera, I., Peralta-Videa, J. R., Jose-Yacaman, M., Troiani, H., Santiago, P. et al. (2004). Size controlled gold nanoparticle formation by Avena sativa biomass: use of plants in nanobiotechnology. *Journal of Nanoparticle Research*, 6: 377–382.

Arora, K., Ledwani, L., and Komal. (2025). A Comprehensive Review on the Synthesis and Therapeutic Potential of Cobalt Ferrite ($CoFe_2O_4$) Nanoparticles. *Chemistry Select*, 10(1): e202404136.

Arshadi, E., Sedaghat, S., and Moradi, O. (2018). Green synthesis and characterization of silver nanoparticles using fructose. *Asian Journal of Green Chemistry*, 2(1): 41–50.

Asghar, N., Hussain, A., Nguyen, D. A., Ali, S., Hussain, I., Junejo, A. et al. (2024). Advancement in nanomaterials for environmental pollutants remediation: a systematic review on bibliometrics analysis, material types, synthesis pathways, and related mechanisms. *Journal of Nanobiotechnology*, 22(1): 26.

Asmatulu, R., Khan, W. S., and Asmatulu, E. (Eds.). (2024). Nanotechnology safety. Elsevier.

Aththanayaka, S., Thiripuranathar, G., and Ekanayake, S. (2023). Microwave-assisted phytogenic $Ag/Ag_2O/ZnO$ nanocomposite as a replacement of Ag/Ag_2O and ZnO nanoparticles: A comparative antioxidant study. *Surfaces and Interfaces*, 43: 103547.

Attea, S. A., Ghareeb, M. A., Kelany, A. K., Elhakim, H. K., Allemailem, K. S., Bukhari, S. I. et al. (2024). Biosynthesis of Iron Oxide Nanoparticles by Marine *Streptomyces* sp. SMGL39 with Antibiofilm Activity: *In vitro* and *in silico* study. *Molecules*, 29(19): 4784.

Barkhade, T. (2018). Extracellular Biosynthesis of Silver Nanoparticles Using Fungus *Penicillium* Species. *International Journal of Research-Granthaalayah*, 6(1): 277–283.

Bayat, R., Akin, M., Yilmaz, B., Bekmezci, M., Bayrakci, M., and Sen, F. (2023). Biogenic platinum-based nanoparticles: Synthesis, characterization and their applications for cell cytotoxic, antibacterial effect, and direct alcohol fuel cells. *Chemical Engineering Journal Advances*, 14: 100471.

Bayda, S., Adeel, M., Tuccinardi, T., Cordani, M., and Rizzolio, F. (2019). The history of nanoscience and nanotechnology: from chemical–physical applications to nanomedicine. *Molecules*, 25(1): 112.

Bellanthudawa, B. K. A., Nawalage, N. M. S. K., Handapangoda, H. M. A. K., Suvendran, S., Wijayasenarathne, K. A. S. H., Rathnasuriya, M. L. D. et al. (2023). A perspective on biodegradable and non-biodegradable nanoparticles in industrial sectors: applications, challenges, and future prospects. *Nanotechnology for Environmental Engineering*, 8(4): 975–1013.

Bhatia, S., and Bhatia, S. (2016). Nanoparticle types, classification, characterization, fabrication methods and drug delivery applications. *Natural Polymer Drug Delivery Systems: Nanoparticles, Plants, and Algae*, 33–93.

Bhushan, B. (2015). Governance, policy, and legislation of nanotechnology: a perspective. *Microsystem Technologies*, 21: 1137–1155.

Bhushan, B. (Ed.) (2017). Introduction to Nano Technology. Springer Handbook of Petroleum Technology: 1–19.

Can, M. (2020). Green gold nanoparticles from plant-derived materials: An overview of the reaction synthesis types, conditions, and applications. *Reviews in Chemical Engineering*, 36(7): 859–877.

Carmona, M., Poblete-Castro, I., Rai, M., and Turner, R. J. (2023). Opportunities and obstacles in microbial synthesis of metal nanoparticles. *Microbial Biotechnology*, 16(5): 871–876.

Chauhan, A., Anand, J., Parkash, V., and Rai, N. (2023). Biogenic synthesis: A sustainable approach for nanoparticles synthesis mediated by fungi. *Inorganic and Nano-Metal Chemistry*, 53(5): 460–473.

Chowdhury, N. K., Choudhury, R., Gogoi, B., Chang, C. M., and Pandey, R. P. (2022). Microbial synthesis of gold nanoparticles and their application. *Current Drug Targets*, 23(7): 752–760.

Chu, S., Xian, L., Lai, C., Yang, W., Wang, J., Long, M. et al. (2023). Migration and risks of potentially toxic elements from sewage sludge applied to acid forest soil. *Journal of Forestry Research*, 34(6): 2011–2026.

Daramola, O. B., George, R. C., Torimiro, N., and Olajide, A. A. (2024). Insights on the synthesis of iron-oxide nanoparticles and the detection of iron-reducing genes from soil microbes. *Colloids and Surfaces C: Environmental Aspects*, 2: 100025.

de França Bettencourt, G. M., Degenhardt, J., Torres, L. A. Z., de Andrade Tanobe, V. O., and Soccol, C. R. (2020). Green biosynthesis of single and bimetallic nanoparticles of iron and manganese using bacterial auxin complex to act as plant bio-fertilizer. *Biocatalysis and Agricultural Biotechnology*, 30: 101822.

Dehghan, Z., Ranjbar, M., Govahi, M., and Khakdan, F. (2022). Green synthesis of Ag/Fe_3O_4 nanocomposite utilizing *Eryngium planum L.* leaf extract and its potential applications in medicine. *Journal of Drug Delivery Science and Technology*, 67: 102941.

Deka, K., Nongbet, R. D., Das, K., Saikia, P., Kaur, S., Talukder, A. et al. (2025). Understanding the mechanism underlying the green synthesis of Metallic nanoparticles using plant extract (s) with special reference to Silver, Gold, Copper and Zinc oxide nanoparticles. *Hybrid Advances*: 100399.

Devanesan, S., and AlSalhi, M. S. (2021). Green synthesis of silver nanoparticles using the flower extract of Abelmoschus esculentus for cytotoxicity and antimicrobial studies. *International Journal of Nanomedicine*: 3343–3356.

Devi KR, S., Mathew, S., Rajan, R., Georgekutty, J., Pinheiro, D., and U, A. (2020). Synthesis and characterization of $CeO_2/Bi_2O_3/gC_3N_4$ ternary Z-scheme nanocomposite. *International Journal of Applied Ceramic Technology*, 17(5): 2346–2356.

Devi, L., Kushwaha, P., Ansari, T. M., Kumar, A., and Rao, A. (2024). Recent trends in biologically synthesized metal nanoparticles and their biomedical applications: a review. *Biological Trace Element Research*, 202(7): 3383–3399.

Dong, Y., Du, X. Q., Liang, P., and Man, X. L. (2020). One-pot solvothermal method to fabricate 1D-VS4 nanowires as anode materials for lithium-ion batteries. *Inorganic Chemistry Communications*, 115: 107883.

Dorjee, L., Kamil, D., Kumar, R., Khokhar, M. K., Aggarwal, R., and Gogoi, R. (2024). Synthesis of promising copper nanoparticles utilizing biocontrol agents, *Trichoderma virens* and *Chaetomium globosum*. *The Indian Journal of Agricultural Sciences*, 94(1): 061–067.

Duran, N., Marcato, P. D., Durán, M., Yadav, A., Gade, A., and Rai, M. (2011). Mechanistic aspects in the biogenic synthesis of extracellular metal nanoparticles by peptides, bacteria, fungi, and plants. *Applied Microbiology and Biotechnology*, 90: 1609–1624.

Eddy, D. R., Rahmawati, D., Permana, M. D., Takei, T., Noviyanti, A. R., and Rahayu, I. (2024). A review of recent developments in green synthesis of TiO_2 nanoparticles using plant extract: Synthesis, characterization and photocatalytic activity. *Inorganic Chemistry Communications*: 112531.

El Golli, A., Fendrich, M., Bazzanella, N., Dridi, C., Miotello, A., and Orlandi, M. (2021). Wastewater remediation with ZnO photocatalysts: green synthesis and solar concentration as an economically and environmentally viable route to application. *Journal of Environmental Management*, 286: 112226.

El-Ghamry, A., Mosa, A. A., Alshaal, T., and El-Ramady, H. (2018). Nanofertilizers vs. biofertilizers: new insights. *Environment, Biodiversity and Soil Security*, 2(2018): 51–72.

Fang, C., Ma, Z., Chen, L., Li, H., Jiang, C., and Zhang, W. (2019). Biosynthesis of gold nanoparticles, characterization and their loading with zonisamide as a novel drug delivery system for the treatment of acute spinal cord injury. *Journal of Photochemistry and Photobiology B: Biology*, 190: 72–75.

Fatima, Z., Saleem, R., Khan, R. R. M., Liaqat, M., Pervaiz, M., Saeed, Z. et al. (2024). Green Synthesis, Properties, and Biomedical Potential of Gold Nanoparticles: A Comprehensive Review. *Biocatalysis and Agricultural Biotechnology*, 103271.

Fazaa, S. A. (2022). Eco-friendly synthesis of zinc oxide nanoparticles using *Bacillus subtilis*, characterization and antibacterial potential against *Staphylococcus aureus* associated with cardiac catheterization. *Iranian Journal of Catalysis*, 12(2).

Flores-Contreras, E. A., González-González, R. B., Pablo Pizaña-Aranda, J. J., Parra-Arroyo, L., Rodríguez-Aguayo, A. A., Iñiguez-Moreno, M. et al. (2024). Agricultural waste as a sustainable source for nanoparticle synthesis and their antimicrobial properties for food preservation. *Frontiers in Nanotechnology*, 6: 1346069.

Fouda, A., Eid, A. M., Guibal, E., Hamza, M. F., Hassan, S. E. D., Alkhalifah, D. H. M. et al. (2022). Green synthesis of gold nanoparticles by aqueous extract of Zingiber officinale: characterization and insight into antimicrobial, antioxidant, and in vitro cytotoxic activities. *Applied Sciences*, 12(24): 12879.

Gautam, R. K., and Tiwari, I. (2020). Humic acid functionalized magnetic nanomaterials for remediation of dye wastewater under ultrasonication: Application in real water samples, recycling and reuse of nanosorbents. *Chemosphere*, 245: 125553.

Gautam, Y. K., Sharma, K., Tyagi, S., Kumar, A., and Singh, B. P. (2022). Applications of green nanomaterials in coatings. *Green Nanomaterials for Industrial Applications*: 107–152.

Gómez-López, P., Puente-Santiago, A., Castro-Beltrán, A., do Nascimento, L. A. S., Balu, A. M., Luque, R. et al. (2020). Nanomaterials and catalysis for green chemistry. *Current Opinion in Green and Sustainable Chemistry*, 24: 48–55.

González-Melendi, P., Fernández-Pacheco, R., Coronado, M. J., Corredor, E., Testillano, P. S., Risueño, M. C. et al. (2008). Nanoparticles as smart treatment-delivery systems in plants: assessment of different techniques of microscopy for their visualization in plant tissues. *Annals of Botany*, 101(1): 187–195.

Greeshma, K. P., and Thamizselvi, R. (2023). Phytogenic synthesis of ZnO nanoparticles from *Catharanthus Roseus* and *Morinda Citrifolia* leaf extract and its promising multifunctional biological applications. *Journal of Drug Delivery Science and Technology*, 87: 104785.

Gupta, D., Boora, A., Thakur, A., and Gupta, T. K. (2023). Green and sustainable synthesis of nanomaterials: recent advancements and limitations. *Environmental Research*, 231: 116316.

Hairil Anuar, A. H., Abd Ghafar, S. A., Hanafiah, R. M., Lim, V., and Mohd Pazli, N. F. A. (2024). Critical Evaluation of Green Synthesized Silver Nanoparticles-Kaempferol for Antibacterial Activity Against Methicillin-Resistant *Staphylococcus aureus*. *International Journal of Nanomedicine*: 1339–1350.

Hamrayev, H., Korpayev, S., and Shameli, K. (2024). Advances in synthesis techniques and environmental applications of TiO_2 nanoparticles for wastewater treatment: a review. *Journal of Research in Nanoscience and Nanotechnology*, 12(1): 1–24.

Harish, V., Ansari, M. M., Tewari, D., Yadav, A. B., Sharma, N., Bawarig, S. et al. (2023). Cutting-edge advances in tailoring size, shape, and functionality of nanoparticles and nanostructures: A review. *Journal of the Taiwan Institute of Chemical Engineers*, 149: 105010.

Hasan, G. G., Mohammed, H. A., Althamthami, M., Khelef, A., Laouini, S. E., and Meneceur, S. (2023). Synergistic effect of novel biosynthesis SnO_2@ Fe_3O_4 nanocomposite: A comprehensive study of its photocatalytic of Dyes & antibiotics, antibacterial, and antimutagenic activities. *Journal of Photochemistry and Photobiology A: Chemistry*, 443: 114874.

Hasan, M., Ullah, I., Zulfiqar, H., Naeem, K., Iqbal, A., Gul, H. et al. (2018). Biological entities as chemical reactors for synthesis of nanomaterials: Progress, challenges and future perspective. *Materials Today Chemistry*, 8: 13–28.

Hassaan, M. A., El-Nemr, M. A., Elkatory, M. R., Ragab, S., Niculescu, V. C., and El Nemr, A. (2023). Principles of photocatalysts and their different applications: a review. *Topics in Current Chemistry*, 381(6): 31.

Hou, X., Zaks, T., Langer, R., and Dong, Y. (2021). Lipid nanoparticles for mRNA delivery. *Nature Reviews Materials*, 6(12): 1078–1094.

Hulkoti, N. I., and Taranath, T. C. (2014). Biosynthesis of nanoparticles using microbes—a review. *Colloids and surfaces B: Biointerfaces*, 121: 474–483.

Hussain, S. B., Aslam, M., Aslam, A., Noor, S., and Zubair, M. (2023). Production of copper nanoparticles by Aspergillus niger and evaluation of their biomedical applications. *Asian Journal of Biochemistry, Genetics and Molecular Biology*, 15(2): 46–59.

Inamdar, A. K., Inamdar, S. A., Birajdar, C. T., Bhale, J., Rajmane, S. V., Shinde, B. H. et al. (2024). Applications of Titanium Dioxide (TiO_2) Nanoparticles in Photocatalysis. *Science of Advanced Materials*, 16(7): 757–771.

Jain, K., Takuli, A., Gupta, T. K., and Gupta, D. (2024). Rethinking nanoparticle synthesis: a sustainable approach vs. traditional methods. *Chemistry–An Asian Journal*, 19(21): 202400701.

Jain, R., Mendiratta, S., Kumar, L., and Srivastava, A. (2021). Green synthesis of iron nanoparticles using *Artocarpus heterophyllus* peel extract and their application as a heterogeneous Fenton-like catalyst for the degradation of Fuchsin Basic dye. *Current Research in Green and Sustainable Chemistry*, 4: 100086.

Jakhar, A. M., Aziz, I., Kaleri, A. R., Hasnain, M., Haider, G., Ma, J. et al. (2022). Nano-fertilizers: A sustainable technology for improving crop nutrition and food security. *NanoImpact*, 27: 100411.

Jayaseelan, C., Rahuman, A. A., Roopan, S. M., Kirthi, A. V., Venkatesan, J., Kim, S. K. et al. (2013). Biological approach to synthesize TiO_2 nanoparticles using Aeromonas hydrophila and its antibacterial activity. *Spectrochimica Acta Part A: Molecular and Biomolecular Spectroscopy*, 107: 82–89.

Jha, A. K., Prasad, K., and Kulkarni, A. R. (2009). Synthesis of TiO_2 nanoparticles using microorganisms. *Colloids and Surfaces B: Biointerfaces*, 71(2): 226–229.

Jiang, Y., Peng, Z., Zhang, S., Li, F., Liu, Z., Zhang, J. et al. (2018). Facile in-situ Solvothermal Method to synthesize double shell $ZnIn_2S_4$ nanosheets/TiO_2 hollow nanosphere with enhanced photocatalytic activities. *Ceramics International*, 44(6): 6115–6126.

Kadyan, P., Thillai Arasu, P., and Kataria, S. K. (2024). Graphene Quantum dots: Green Synthesis, characterization, and antioxidant and antimicrobial potential. *International Journal of Biomaterials*, 2024(1): 2626006.

Kajitvichyanukul, P., Nguyen, V. H., Boonupara, T., Thi, L. A. P., Watcharenwong, A., Sumitsawan, S. et al. (2022). Challenges and effectiveness of nanotechnology-based photocatalysis for pesticides-contaminated water: a review. *Environmental Research*, 212: 113336.

Karnwal, A., Kumar Sachan, R. S., Devgon, I., Devgon, J., Pant, G., Panchpuri, M. et al. (2024). Gold nanoparticles in nanobiotechnology: from synthesis to biosensing applications. *ACS omega*, 9(28): 29966–29982.

Khanam, P. N., and Hasan, A. (2019). Biosynthesis and characterization of graphene by using non-toxic reducing agent from Allium Cepa extract: Anti-bacterial properties. *International Journal of Biological Macromolecules*, 126: 151–158.

Khatoon, U. T., and Velidandi, A. (2025). An Overview on the Role of Government Initiatives in Nanotechnology Innovation for Sustainable Economic Development and Research Progress. *Sustainability*, 17(3): 1250.

Kim, M., Jee, S. C., Shinde, S. K., Mistry, B. M., Saratale, R. G., Saratale, G. D. et al. (2019). Green-synthesis of anisotropic peptone-silver nanoparticles and its potential application as anti-bacterial agent. *Polymers*, 11(2): 271.

Kimber, R. L., Lewis, E. A., Parmeggiani, F., Smith, K., Bagshaw, H., Starborg, T. et al. (2018). Biosynthesis and characterization of copper nanoparticles using *Shewanella oneidensis*: application for click chemistry. *Small*, 14(10): 1703145.

Korah, B. K., Plathanam, N. J., Chacko, A. R., Punnoose, M. S., Abraham, T., and Mathew, B. (2021). Green Synthesized Carbon-Based Nanomaterials: Synthesis and Properties. In Renewable Materials and Green Technology Products. Apple *Academic Press*: 1–53.

Koul, B., Poonia, A. K., Yadav, D., and Jin, J. O. (2021). Microbe-mediated biosynthesis of nanoparticles: Applications and future prospects. *Biomolecules*, 11(6): 886.

Kuchibhatla, S. V., Karakoti, A. S., Baer, D. R., Samudrala, S., Engelhard, M. H., Amonette, J. E. et al. (2012). Influence of aging and environment on nanoparticle chemistry: implication to confinement effects in nanoceria. *The Journal of Physical Chemistry C*, 116(26): 14108–14114.

Kulkarni, D., Sherkar, R., Shirsathe, C., Sonwane, R., Varpe, N., Shelke, S. et al. (2023). Biofabrication of nanoparticles: sources, synthesis, and biomedical applications. *Frontiers in Bioengineering and Biotechnology*, 11: 1159193.

Kumar, P., Kumar, S., and Thakur, N. (2023). *Azadirachta indica* and polyvinylpyrrolidone encapsulated Fe_2O_3 nanoparticles to enhance the photocatalytic and antioxidant activity. *Inorganic Chemistry Communications*, 155: 111084.

Kumar, S., Ahlawat, W., Bhanjana, G., Heydarifard, S., Nazhad, M. M., and Dilbaghi, N. (2014). Nanotechnology-based water treatment strategies. *Journal of nanoscience and nanotechnology*, 14(2): 1838–1858.

Kumar, S., Mudai, A., Roy, B., Basumatary, I. B., Mukherjee, A., and Dutta, J. (2020). Biodegradable hybrid nanocomposite of chitosan/gelatin and green synthesized zinc oxide nanoparticles for food packaging. *Foods*, 9(9): 1143.

Kumari, T., Phogat, D., and Shukla, V. (2023). Exploring the multipotentiality of plant extracts for the green synthesis of iron nanoparticles: A study of adsorption capacity and dye degradation efficiency. *Environmental Research*, 229: 116025.

Latif, A., Sheng, D., Sun, K., Si, Y., Azeem, M., Abbas, A. et al. (2020). Remediation of heavy metals polluted environment using Fe-based nanoparticles: Mechanisms, influencing factors, and environmental implications. *Environmental Pollution*, 264: 114728.

Letchumanan, D., Sok, S. P., Ibrahim, S., Nagoor, N. H., and Arshad, N. M. (2021). Plant-based biosynthesis of copper/copper oxide nanoparticles: an update on their applications in biomedicine, mechanisms, and toxicity. *Biomolecules*, 11(4): 564.

Lewoyehu, M. (2021). Comprehensive review on synthesis and application of activated carbon from agricultural residues for the remediation of venomous pollutants in wastewater. *Journal of Analytical and Applied Pyrolysis*, 159: 105279.

Li, W., and Zhao, D. (2013). An overview of the synthesis of ordered mesoporous materials. *Chemical Communications*, 49(10): 943–946.

Li, X., Aftab, S., Mukhtar, M., Kabir, F., Khan, M. F., Hegazy, H. H. et al. (2025). Exploring nanoscale perovskite materials for next-generation photodetectors: a comprehensive review and future directions. *Nano-Micro Letters*, 17(1): 28.

Li, Z., Tang, H., Yuan, W., Song, W., Niu, Y., Yan, L. et al. (2014). Ag nanoparticle–ZnO nanowire hybrid nanostructures as enhanced and robust antimicrobial textiles via a green chemical approach. *Nanotechnology*, 25(14): 145702.

Lv, Q., Zhang, B., Xing, X., Zhao, Y., Cai, R., Wang, W. et al. (2018). Biosynthesis of copper nanoparticles using *Shewanella loihica* PV-4 with antibacterial activity: novel approach and mechanisms investigation. *Journal of Hazardous Materials*, 347: 141–149.

Machac, P., Cichon, S., Lapcak, L., and Fekete, L. (2020). Graphene prepared by chemical vapour deposition process. *Graphene Technology*, 5: 9–17.

Manchala, S., Tandava, V. S. R. K., Jampaiah, D., Bhargava, S. K., and Shanker, V. (2019). Novel and highly efficient strategy for the green synthesis of soluble graphene by aqueous polyphenol extracts of *eucalyptus* bark and its applications in high-performance supercapacitors. *ACS Sustainable Chemistry & Engineering*, 7(13): 11612–11620.

Meenakshi, M., Bhaskar, R., Kumar, S. A., and Kumar, R. S. (2025). A Concise Review on Magnetic Nanoparticles: Their Properties, Types, Synthetic Methods, and Current Trending Applications. *Current Nanoscience*, 21(1): 2–17.

Mehra, V., Kumar, S., Tamang, A. M., and Chandraker, S. K. (2025). Green Synthesis of Gold Nanoparticles (AuNPs) by Using Plant Extract and Their Biological Application: A Review. *BioNanoScience*, 15(1): 1–20.

Milovanovic, M., Arsenijevic, A., Milovanovic, J., Kanjevac, T., and Arsenijevic, N. (2017). Nanoparticles in antiviral therapy. *Antimicrobial Nanoarchitectonics*: 383–410.

Mir, M. A., and Banik, B. K. (2025). Synthesis of Schiff base ligands under environment friendly conditions: A systematic review. *Inorganic Chemistry Communications*: 113987.

Moreira, N. F., Narciso-da-Rocha, C., Polo-López, M. I., Pastrana-Martínez, L. M., Faria, J. L., Manaia, C. M. et al. (2018). Solar treatment (H_2O_2, TiO_2-P25 and GO-TiO_2 photocatalysis, photo-Fenton) of organic micropollutants, human pathogen indicators, antibiotic resistant bacteria and related genes in urban wastewater. *Water Research*, 135: 195–206.

Mouro, C., Gomes, A. P., and Gouveia, I. C. (2024). Microbial exopolysaccharides: structure, diversity, applications, and future frontiers in sustainable functional materials. *Polysaccharides*, 5(3): 241–287.

Muñoz-García, J., Vázquez, L., Cuerno, R., Sánchez-García, J. A., Castro, M., Gago, R. et al. (2009). Toward functional nanomaterials. *ZM Wang. Springer, Dordrecht*: 323.

Murali, M., Gowtham, H. G., Shilpa, N., Singh, S. B., Aiyaz, M., Sayyed, R. Z. et al. (2023). Zinc oxide nanoparticles prepared through microbial mediated synthesis for therapeutic applications: A possible alternative for plants. *Frontiers in Microbiology*, 14: 1227951.

Nair, R., Poulose, A. C., Nagaoka, Y., Yoshida, Y., Maekawa, T., and Kumar, D. S. (2011). Uptake of FITC labelled silica nanoparticles and quantum dots by rice seedlings: effects on seed germination and their potential as biolabels for plants. *Journal of Fluorescence*, 21: 2057–2068.

Nasrollahzadeh, M., Sajjadi, M., Iravani, S., and Varma, R. S. (2021). Green-synthesized nanocatalysts and nanomaterials for water treatment: Current challenges and future perspectives. *Journal of Hazardous Materials*, 401: 123401.

Neama, K. T., and Alzubaidy, M. H. J. (2024). Biosynthesis of Iron Oxide Nanoparticles by Fungi *Beauveria bassiania* and Study of their Structural and Optical Properties. *Wasit Journal for Pure sciences*, 3(3): 353–364.

Nguyen, N. T. T., Nguyen, L. M., Nguyen, T. T. T., Liew, R. K., Nguyen, D. T. C., and Van Tran, T. (2022). Recent advances on botanical biosynthesis of nanoparticles for catalytic, water treatment and agricultural applications: A review. *Science of the Total Environment*, 827: 154160.

Noji, T., Kamidaki, C., Kawakami, K., Shen, J. R., Kajino, T., Fukushima, Y. et al. (2011). Photosynthetic oxygen evolution in mesoporous silica material: adsorption of photosystem II reaction centre complex into 23nm nanopores in SBA. *Langmuir*, 27(2): 705–713.

Omran, B. A., and Baek, K. H. (2022). Valorization of agro-industrial biowaste to green nanomaterials for wastewater treatment: Approaching green chemistry and circular economy principles. *Journal of Environmental Management*, 311: 114806.

Osman, A. I., Zhang, Y., Farghali, M., Rashwan, A. K., Eltaweil, A. S., Abd El-Monaem, E. M. et al. (2024). Synthesis of green nanoparticles for energy, biomedical, environmental, agricultural, and food applications: A review. *Environmental Chemistry Letters*, 22(2): 841–887.

Osterberg, M., Henn, K. A., Farooq, M., and Valle-Delgado, J. J. (2023). Biobased nanomaterials- the role of interfacial interactions for advanced materials. *Chemical Reviews*, 123(5): 2200–2241.

Padhiari, S., Mishra, M., and Hota, G. (2022). Green synthesis of nanomaterials for photocatalytic application. Design, Fabrication, and Characterization of Multifunctional *Nanomaterials*: 373–398.

Patel, B., Yadav, V. K., Desai, R., Patel, S., Amari, A., Choudhary, N. et al. (2024). Bacteriogenic synthesis of morphologically diverse silver nanoparticles and their assessment for methyl orange dye removal and antimicrobial activity. *PeerJ*, 12: e17328.

Patil, A. H., Jadhav, S. A., Gurav, K. D., Waghmare, S. R., Patil, G. D., Jadhav, V. D. et al. (2020). Single step green process for the preparation of antimicrobial nanotextiles by wet chemical and sonochemical methods. *The Journal of the Textile Institute*, 111(9): 1380–1388.

Patil, S. P., Chaudhari, T. B., Dalal, K. S., Phirke, N. V., and Chaudhari, B. L. (2024). Laccase-Mediated Valorization of Lignocellulosic Biomass: An Eco-friendly Approach Towards the Sustainable Environment. In *Industrial Microbiology and Biotechnology: A New Horizon of the Microbial World*. Singapore: *Springer Nature* Singapore: 635–661.

Patil, S. P., Dalal, K. S., Shirsath, L. P., and Chaudhari, B. L. (2023). Microbial Hyaluronidase: Its Production, Purification and Applications. In: *Industrial Microbiology and Biotechnology: Emerging concepts in Microbial Technology* (pp. 473–497). Singapore: Springer Nature Singapore.

Patil, S. P., Shirsath, L. P., and Chaudhari, B. L. (2021). A halotolerant hyaluronidase from newly isolated *Brevibacterium halotolerans* DC1: Purification and characterization. *International Journal of Biological Macromolecules*, 166: 839–850.

Patil, S., and Chaudhari, B. (2017). A simple, rapid and sensitive plate assay for detection of microbial hyaluronidase activity. *Journal of Basic Microbiology*, 57(4): 358–361.

Patil, S., Bhadane, B., Shirsath, L., Patil, R., and Chaudhari, B. (2019). Steroidal fraction of *Carissa carandas* L. inhibits microbial hyaluronidase activity by mixed inhibition mechanism. *Preparative Biochemistry and Biotechnology*, 49(3): 298–306.

Pellegrini, G., Bonanni, V., Campo, G., PINEIDER, F., Sangregorio, C., de Julián Fernández, C. et al. (2016). Magnetoplasmonics. *In Encyclopedia of Nanotechnology*: 1879–1903.

Phogat, P., Jha, R., and Singh, S. (2024). Harnessing ZnO morphologies in energy application and sustainable development. *PhysicaScripta*, 99(10): 102004.

Ponjavic, M., Stevanovic, S., Nikodinovic-Runic, J., Jeremic, S., Cosovic, V. R., and Maksimovic, V. (2022). Bacterial nanocellulose as green support of platinum nanoparticles for effective methanol oxidation. *International Journal of Biological Macromolecules*, 223: 1474–1484.

Pradhan, D., Jaiswal, A. K., and Jaiswal, S. (2023). Nanocellulose based green nanocomposites: characteristics and application in primary food packaging. *Food Reviews International*, 39(9): 7148–7179.

Qasim, M., Clarkson, A. N., and Hinkley, S. F. (2023). Green synthesis of carbon nanoparticles (CNPs) from biomass for biomedical applications. *International Journal of Molecular Sciences*, 24(2): 1023.

Qu, G., Guo, J., Yang, D., and Sun, Z. (2018). Biocatalysis of carboxylic acid reductases: phylogenesis, catalytic mechanism and potential applications. *Green Chemistry*, 20(4): 777–792.

Rai, A., Singh, A., Ahmad, A., and Sastry, M. (2006). Role of halide ions and temperature on the morphology of biologically synthesized gold nanotriangles. *Langmuir*, 22(2): 736–741.

Rai, M., Bonde, S., Golinska, P., Trzcińska-Wencel, J., Gade, A., Abd-Elsalam, K. A. et al. (2021). Fusarium as a novel fungus for the synthesis of nanoparticles: mechanism and applications. *Journal of Fungi*, 7(2): 139.

Rai, S., Bhujel, R., Biswas, J., and Swain, B. P. (2020, November). A green approach for synthesis of graphene. *AIP Conference Proceedings* (Vol. 2273, No. 1). AIP Publishing.

Rajagopal, G., Nivetha, A., Sundar, M., Panneerselvam, T., Murugesan, S., Parasuraman, P. et al. (2021). Mixed phytochemicals mediated synthesis of copper nanoparticles for anticancer and larvicidal applications. *Heliyon*, 7(6).

Rajakumar, G., Rahuman, A. A., Roopan, S. M., Khanna, V. G., Elango, G., Kamaraj, C. et al. (2012). Fungus-mediated biosynthesis and characterization of TiO2 nanoparticles and their activity against pathogenic bacteria. *Spectrochimica Acta Part A: Molecular and Biomolecular Spectroscopy*, 91: 23–29.

Rakshit, S., Jana, P. C., and Kamilya, T. (2023). Green synthesis of copper nanoparticles by using plant extracts and their biomedical applications–an extensive review. *Current Nanomaterials*, 8(2): 110–125.

Ramadoss, J., and Sonachalam, A. (2024). Energy harvesting and storage applications for hybrid nanomaterials. pp. 119–137. *In:* Technological Applications of Nano-Hybrid Composites. IGI Global.

Rashwan, A. K., Karim, N., Xu, Y., Hanafy, N. A., Li, B., Mehanni, A. H. E. et al. (2023). An updated and comprehensive review on the potential health effects of curcumin-encapsulated micro/nanoparticles. *Critical Reviews in Food Science and Nutrition*, 63(29): 9731–9751.

Rashwan, A. K., Younes, H. A., Karim, N., Taha, E. M., Mozafari, M. R., Tawfeuk, H. Z. et al. (2022). Health properties of bioactive food compounds-loaded micro and nano-encapsulation systems: a review. *SVU-International Journal of Agricultural Sciences*, 4(2): 178–203.

Rattan, S., Kumar, S., and Goswamy, J. K. (2020). Graphene oxide reduction using green chemistry. *Materials Today: Proceedings*, 26: 3327–3331.

Raza, A., Sime, F. B., Cabot, P. J., Maqbool, F., Roberts, J. A., and Falconer, J. R. (2019). Solid nanoparticles for oral antimicrobial drug delivery: A review. *Drug Discovery Today*, 24(3): 858–866.

Religa, P., Kowalik, A., and Gierycz, P. (2011). Application of nanofiltration for chromium concentration in the tannery wastewater. *Journal of Hazardous Materials*, 186(1): 288–292.

Roquia, A., Khalfan Hamed Alhashmi, A., and Hamed Abdullah Alhasmi, B. (2021). Synthesis and characterisation of carbon nanotubes from waste of Juglans regia (walnut) shells. *Fullerenes, Nanotubes and Carbon Nanostructures*, 29(11): 860–867.

Rosas-Medellín, D., Pérez-Salcedo, K. Y., Morales-Acosta, D., Rodríguez-Varela, F. J., and Escobar, B. (2021). Green synthesis of Pt nanoparticles and their application in the oxygen reduction reaction. *Journal of Materials Research*, 36: 4131–4140.

Roy, N., Nivedya, T., Paira, P., and Chakrabarty, R. (2025). Selenium-based nanomaterials: green and conventional synthesis methods, applications, and advances in dye degradation. *RSC advances*, 15(4): 3008–3025.

Rudakiya, D. M., and Gupte, A. (2019). Laccases in nano-biotechnology: recent trends and advanced applications. *Recent Advances in Biotechnology*: 39–62.

Rui, M., Ma, C., Hao, Y., Guo, J., Rui, Y., Tang, X. et al. (2016). Iron oxide nanoparticles as a potential iron fertilizer for peanut (Arachis hypogaea). *Frontiers in plant science*, 7: 815.

Sabouri, Z., Oskuee, R. K., Sabouri, S., Moghaddas, S. S. T. H., Samarghandian, S., Abdulabbas, H. S. et al. (2023). Phytoextract-mediated synthesis of Ag-doped ZnO–MgO–CaO nanocomposite using *Ocimum Basilicum* L. seeds extract as a highly efficient photocatalyst and evaluation of their biological effects. *Ceramics International*, 49(12): 20989–20997.

Saratale, R. G., Saratale, G. D., Shin, H. S., Jacob, J. M., Pugazhendhi, A., Bhaisare, M. et al. (2018). New insights on the green synthesis of metallic nanoparticles using plant and waste biomaterials: current knowledge, their agricultural and environmental applications. *Environmental Science and Pollution Research*, 25: 10164–10183.

Sarkar, S., Sarkar, S., and Bhattacharjee, C. (2019). Green Synthesis of Novel Photocatalysts. *Nanophotocatalysis and Environmental Applications: Materials and Technology*, 241–261.

Segura, R., Vásquez, G., Colson, E., Gerbaux, P., Frischmon, C., Nesic, A. et al. (2020). Phytostimulant properties of highly stable silver nanoparticles obtained with saponin extract from *Chenopodium quinoa*. *Journal of the Science of Food and Agriculture*, 100(13): 4987–4994.

Shaikh, W. A., Chakraborty, S., Owens, G., and Islam, R. U. (2021). A review of the phytochemical mediated synthesis of AgNP (silver nanoparticle): The wonder particle of the past decade. *Applied Nanoscience*, 11(11): 2625–2660.

Shanmugam, R., Munusamy, T., Jayakodi, S., Al-Ghanim, K. A., Nicoletti, M., Sachivkina, N. et al. (2023). Probiotic-bacteria (*Lactobacillus fermentum*)-wrapped zinc oxide nanoparticles: Biosynthesis, characterization, and antibacterial activity. *Fermentation*, 9(5): 413.

Sharifi, M., Sohrabi, M. J., Hosseinali, S. H., Hasan, A., Kani, P. H., Talaei, A. J. et al. (2020). Enzyme immobilization onto the nanomaterials: Application in enzyme stability and prodrug-activated cancer therapy. *International Journal of Biological Macromolecules*, 143: 665–676.

Sharma, S. D., Gupta, V., Sain, D., Rahi, R. K., Yadav, M., and Neelam, D. K. (2024). Metal And Metal Oxide Nanoparticles: Unveiling the Mechanisms of Antimicrobial Activity. Hunan Daxue Xuebao. *Journal of Hunan University Natural Sciences*, 74–94.

Shirsath, L. P., Patil, S. P., and Patil, U. K. (2018). Incidence of leaf spot disease on cotton caused by *Curvularia verruculosa* and role of its hydrolytic enzymes in pathogenesis. *Physiology and Molecular Biology of Plants*, 24(4): 711–714.

Singh, A. K., Anwar, M., Pradhan, R., Ashar, M. S., Rai, N., and Dey, S. (2023). Surface plasmon resonance based-optical biosensor: Emerging diagnostic tool for early detection of diseases. *Journal of Biophotonics*, 16(7): 202200380.

Singh, A., Amod, A., Pandey, P., Bose, P., Pingali, M. S., Shivalkar, S. et al. (2022). Bacterial biofilm infections, their resistance to antibiotics therapy and current treatment strategies. *Biomedical Materials*, 17(2): 022003.

Singh, S., Kumar, V., Romero, R., Sharma, K., and Singh, J. (2019). Applications of nanoparticles in wastewater treatment. *Nanobiotechnology in bioformulations*: 395–418.

Soeteman-Hernández, L. G., Blab, G. A., Carattino, A., Dekker, F., Dekkers, S., van der Linden, M. et al. (2020). Challenges of implementing nano-specific safety and safe-by-design principles in academia. *NanoImpact*, 19: 100243.

Soliman, M. K., Abu-Elghait, M., Salem, S. S., and Azab, M. S. (2024). Multifunctional properties of silver and gold nanoparticles synthesis by *Fusarium pseudonygamai*. *Biomass Conversion and Biorefinery*, 14(22): 28253–28270.

Soliman, M. K., Salem, S. S., Abu-Elghait, M., and Azab, M. S. (2023). Biosynthesis of silver and gold nanoparticles and their efficacy towards antibacterial, antibiofilm, cytotoxicity, and antioxidant activities. *Applied Biochemistry and Biotechnology*, 195(2): 1158–1183.

Soltani Nejad, M., Samandari Najafabadi, N., Aghighi, S., Pakina, E., and Zargar, M. (2022). Evaluation of *Phoma sp.* biomass as an endophytic fungus for synthesis of extracellular gold nanoparticles with antibacterial and antifungal properties. *Molecules*, 27(4): 1181.

Sridevi, H., Bhat, M. R., Kumar, P. S., Kumar, N. M., and Selvaraj, R. (2023). Structural characterization of cuboidal α-Fe_2O_3 nanoparticles synthesized by a facile approach. *Applied Nanoscience*, 13(8): 5605–5613.

Swielam, E. M., Hussien, Z. M., and Hasanin, M. S. (2024). Design, characterizations, and antimicrobial activity of sustainable home furnishing-based waste fabric treated using biobased nanocomposite. *Bioresources and Bioprocessing*, 11(1): 75.

Szabó, Z., Volk, J., Fülöp, E., Deák, A., and Bársony, I. (2013). Regular ZnO nanopillar arrays by nanosphere photolithography. *Photonics and Nanostructures-Fundamentals and Applications*, 11(1): 1–7.

Thangavelu, L., and Veeraragavan, G. R. (2022). A Survey on Nanotechnology-Based Bioremediation of Wastewater. *Bioinorganic Chemistry and Applications*, 2022(1): 5063177.

Theodorakopoulos, G. V., Arfanis, M. K., Sánchez Pérez, J. A., Agüera, A., Cadena Aponte, F. X., Markellou, E. et al. (2023). Novel pilot-scale photocatalytic nanofiltration reactor for agricultural wastewater treatment. *Membranes*, 13(2): 202.

Thirugnanasambandan, T., Karuppaiah, C., Sankar, B. V., and Gopinath, S. C. (2024). Insights and potentials of two-dimensional black phosphorous-based solar cells. *Physica Scripta*, 99(5): 052002.

Tran, Q. H., and Le, A. T. (2013). Silver nanoparticles: synthesis, properties, toxicology, applications and perspectives. *Advances in Natural Sciences: Nanoscience and Nanotechnology*, 4(3): 033001.

Vanlalveni, C., Ralte, V., Zohmingliana, H., Das, S., Anal, J. M. H., Lallianrawna, S. et al. (2024). A review of microbes mediated biosynthesis of silver nanoparticles and their enhanced antimicrobial activities. *Heliyon*: 10(11).

Vaseghi, Z., Nematollahzadeh, A., and Tavakoli, O. (2018). Green methods for the synthesis of metal nanoparticles using biogenic reducing agents: a review. *Reviews in Chemical Engineering*, 34(4): 529–559.

Vindhyasarumi, A., Saritha, A., Sethulekshmi, A. S., and Joseph, K. (2025). Advancements in carbon microsphere-based nanocomposites: synthesis, properties and multifaceted applications: a comprehensive review. *Carbon Letters*: 1–30.

Wang, C. F., Sun, X. Y., Su, M., Wang, Y. P., and Lv, Y. K. (2020). Electrochemical biosensors based on antibody, nucleic acid and enzyme functionalized graphene for the detection of disease-related biomolecules. *Analyst*, 145(5): 1550–1562.

Wang, T., Jiao, Y., Chai, Q., and Yu, X. (2015). Gold nanoparticles: synthesis and biological applications. *Nano Life*, 5(03): 1542007.

Xu, F., Li, Y., Zhao, X., Liu, G., Pang, B., Liao, N. et al. (2024). Diversity of fungus-mediated synthesis of gold nanoparticles: properties, mechanisms, challenges, and solving methods. *Critical Reviews in Biotechnology*, 44(5): 924–940.

Yacamán, M. J., Ascencio, J. A., Liu, H. B., and Gardea-Torresdey, J. (2001). Structure shape and stability of nanometric sized particles. *Journal of Vacuum Science & Technology B: Microelectronics and Nanometer Structures Processing, Measurement, and Phenomena*, 19(4): 1091–1103.

Yadha, H., Kolure, R., Thakur, S., Mandava, K., and Boddu, S. (2024). QBD approach for green synthesis of Rutin silver nanoparticles-screening for antioxidant, anticancer and anticlastogenic potential. *Heliyon*: 10(20).

Ying, S., Guan, Z., Ofoegbu, P. C., Clubb, P., Rico, C., He, F. et al. (2022). Green synthesis of nanoparticles: Current developments and limitations. *Environmental Technology & Innovation*, 26: 102336.

Zhao, L. X., Chen, L. L., Cheng, D., Wu, T. Y., Fan, Y. G., and Wang, Z. Y. (2024). Potential application prospects of biomolecule-modified two-dimensional chiral nanomaterials in biomedicine. *ACS Biomaterials Science & Engineering*, 10(4): 2022–2040.

CHAPTER-4

Advanced Nanomaterial-Based Hydrogels for Environmental Pollution Remediation

Ianca Oliveira Borges,[1,2,*] *Aline de Fátima Silva Lago,*[1] *Rafael Carvalho do Lago,*[1] *Pareshkumar G. Moradeeya*[3] and *Gustavo Henrique Denzin Tonoli*[1]

1. Introduction

The explosive increase in population and swift worldwide expansion have resulted in a high rate of environmental pollution, a topic of various policies and scientific investigations (Duta et al. 2021). Due to the severe health problems for humans and animals associated with environmental contamination, this topic has become a global concern. Even in small concentrations, toxic pollutants in outdoor and indoor environments, such as gases, bioaerosols, oxides, pharmaceuticals, pesticides, microorganisms (MOs), metals, among others (Almomani et al. 2020), are significant threats that can cause severe consequences for human health and the environment (González-Martín et al. 2021). According to data from the World Health Organization (WHO), around 91% of the global population lives in regions where pollution levels surpass acceptable limits (Ju et al. 2021). The pollution issue negatively affects socioeconomic development, making it crucial to remediate concentrations of all contaminants. It is imperative to use new technologies that offer sustainability, reliability, and high quality in the economic, social, and environmental spheres, focusing on complying with existing norms and regulations (Almomani et al. 2020).

Considering the issue of pollution, new intelligent materials on different scales are constantly being developed. The goal is to improve treatment efficiency and solve pollution-related problems through coordinated ecological development among scientists, politicians, industry, and the public sector (Mu et al. 2020). Additionally, it is crucial that the development of new materials be carried out in a multidisciplinary manner, as the advanced optimization of nanomaterials requires profound knowledge, focusing on low-cost solutions (Rando et al. 2024). This comprehensive approach is crucial for developing cost-effective, eco-friendly, and pioneering technologies for contaminant removal, ensuring that these solutions are more inclusive and efficient.

[1] Department of Forest Sciences, Federal University of Lavras, P.O. Box 3037, Lavras, 37200-900, Minas Gerais, Brazil. https://orcid.org/0000-0002-4254-7851, https://orcid.org/0000-0002-8558-5217, https://orcid.org/0000-0002-6502-8974
[2] Department of Geosciences and Natural Resource Management, University of Copenhagen, Denmark.
[3] School of Chemical and Metallurgical Engineering, University of the Witwatersrand, Johannesburg, South Africa. https://orcid.org/0000-0003-2780-2532
* Corresponding author: iancaborges@hotmail.com, https://orcid.org/0000-0003-4370-4705

DOI: 10.1201/9781003558804-4

Nanomaterials (NMs) are materials characterized by having an internal structure or surface that falls within the nanoscale range, typically between 1 and 100 nanometers. Thanks to their increased surface area relative to volume and the influence of quantum effects, NMs demonstrate unique physical and chemical attributes that are markedly different from those of larger-scale materials (Rando et al. 2024). NMs can be classified into several structural dimensions: zero-dimensional (0D), with three axes (x, y, z); one-dimensional (1D), limited to two dimensions; two-dimensional (2D), limited to one direction; and three-dimensional (3D), which encompass structures larger than 100 nm. Despite the different dimensions, the importance of NMs lies in their versatility and ability to provide innovative solutions in various fields (Agshar et al. 2024).

Nanomaterials (NMs) are excellent adsorbents, such as hydrogels, and serve as effective catalysts for environmental remediation. Their unique characteristics include high chemical reactivity, a large adsorption surface area, lower operating temperature, and active atomicity, which make them highly efficient in addressing environmental contaminants. The compact size of nanoparticles enhances the ability of surface atoms to adsorb and react, stabilizing the charge. Because of their small size, NMs have highly reactive surfaces, capable of effectively adsorbing contaminants and degrading pollutants. One of the main advantages of utilizing NMs for remediation is their cost-efficiency, which greatly aids in achieving sustainable economic development. The advantages of nano-adsorbents go beyond cost-effectiveness, including higher efficiency, reliability, and ease of use. These properties make nanomaterials an innovative and promising choice for addressing environmental pollution problems more effectively and sustainably (Sable et al. 2024).

Hydrogels consist of cross-linked, 3D hydrophilic networks that can preserve substantial levels of water. They display customizable biodegradability, biocompatibility, acute environmental detection capabilities, and impressive mechanical properties (Zhang and Huang 2021). Their capability to conserve a vast volume of water is due to their three-dimensional structure, giving them a gel-like look and behavior (Priya et al. 2024). Hydrogels boast a range of impressive traits, such as being biodegradable, biocompatible, hydrophilic, superabsorbent, viscoelastic, and soft. Additionally, they respond to different factors, including heat, electric and magnetic fields, biological molecules, and electrolytic strength. This versatility significantly expands their potential uses (Siepmann et al. 2011). The traits of hydrogels can be adapted by modifying the type and organization of their monomers, and also by adjusting the methods of their preparation.

Hydrogels fall into two primary groups: natural and synthetic polymers. Natural polymers include substances such as agarose (Bilal et al. 2019), alginate (Hao et al. 2017), cellulose (Isobe et al. 2018), and chitosan (Ouyang et al. 2018). These substances are plentiful in nature and possess advantages such as biochemical properties and biocompatibility with other materials (Pena et al. 2018). Conversely, synthetic polymers are produced through the polymerization of artificial monomers, including polyvinyl alcohol (Zhang et al. 2019), polyethylene glycol (Cruz-Acuna et al. 2018), and peptides (Mondal et al. 2020), a synthetic biopolymer. Hydrogels are being developed using synthetic or natural polymers by way of physical or covalent cross-linking methods. These materials' properties can be precisely adjusted to meet a wide range of application needs (Kasinski et al. 2020).

As a result, hydrogels can be synthesized through multiple techniques, including chemical bonding of monomers, physical bonding via alterations in pH or temperature, and combining organic or artificial polymers (Priya et al. 2024). These approaches enable scientists to develop hydrogels with tailored properties for different applications, particularly in pollutant remediation. Nevertheless, conventional hydrogels face difficulties in simultaneously achieving electrical conductance, robust adherence, and solid mechanical properties. Consequently, crafting hydrogels that can effectively bond to a variety of surfaces while preserving their unique properties continues to pose a significant challenge.

2. Properties of Hydrogel

The attributes of a hydrogel differ based on the materials incorporated and their intended uses. For pollutant remediation, an optimal hydrogel should exhibit high adsorption and swelling capabilities, responsiveness to temperature and pH variations, and biodegradability.

Adsorption capability of hydrogels is closely tied to their swelling behavior. As hydrogels swell, they markedly expand their porosity and surface area, which enhances their capability to detain and retain contaminants like organic compounds and heavy metals (Zhang and Huang 2021). During the swelling process, functional groups carboxyl (–COOH) and hydroxyl (–OH) become more accessible for chemical and physical interactions with contaminants (Asghar et al. 2024). This not only enhances adsorption efficiency but also allows more uniform distribution of pollutants within the hydrogel structure. Additionally, swelling directly affects the diffusion of pollutants, allowing them to move more easily through the expanded polymer network.

Hydrogels are interlinked polymers featuring 3D frameworks that can expand multiple times and absorb water when submerged in an aqueous medium (Mahinroosta et al. 2018). Various factors influence the swelling kinetics and equilibrium in hydrogels, including the crosslinking ratio; highly cross-linked polymer networks exhibit reduced swelling, whereas less cross-linked networks swell more. The polymer's chemical composition significantly influences the swelling behavior of hydrogels. Polymers with hydrophilic groups tend to swell more extensively compared to those with hydrophobic groups. Moreover, pH and temperature significantly modify the swelling behavior of hydrogels (Bashir et al. 2020). Thermo-responsive hydrogels experience volumetric changes when exposed to temperature fluctuations, expanding below their critical solution temperature and contracting above it. Conversely, pH-responsive hydrogels react to changes in pH by either swelling or shrinking (Hu et al. 2015). These hydrogels contain basic and ionizable acidic groups within their polymer structure, which enables them to either uptake or release protons based on the surrounding pH. At higher pH values, acidic groups within the polymer lose their protons, while at lower pH values, primary groups gain protons (Darban et al. 2022).

Stimuli-reactive hydrogels have the unique capability to react to a range of external factors, including pH, magnetic fields, light, temperature, and ionic strength (Ozay et al. 2011). Depending on their application, these materials can shift from solution to gel. There are three main categories of stimuli-responsive hydrogels (Figure 1). Chemical factors include solvent composition, ionic strength, electrochemical fields, and pH levels. Physical stimuli encompass magnetic fields, light, ultrasound, temperature, mechanical force, and electric fields. Lastly, biological factors involve antigens, glucose, ligands, enzymes, and other biological molecules (Rittikulsittichai et al. 2016). Multi-responsive hydrogels can react to two or more of these stimuli simultaneously, greatly enhancing their versatility across diverse applications (Darban et al. 2022).

The biodegradability of hydrogels refers to their ability to decompose into harmless and non-toxic final products by the activity of microorganisms, including fungi and bacteria (Darban et al. 2022). This process is influenced by the functional groups in the hydrogel's framework and the synthesis techniques used (Darban et al. 2022). Degradation entails the dissolution and hydrolysis of biological components within the hydrogel, leading to the formation of safer end products. Various parameters, including the chemical structure—where hydrophilic and natural groups facilitate degradation—affect the breakdown of polymers (both synthetic and natural) in the hydrogel. Additionally, the synthesis method and materials impact the hydrogel's biodegradability, while environmental conditions (pH and temperature) also play a role in polymer degradation through solubilization (Zhang and Huang 2021). Ensuring biodegradability is crucial for the ecological sustainability of hydrogels, minimizing their environmental impact after disposal.

The production of hydrogels using natural biodegradable materials or those with biological compatibility is an effective approach to mitigating the effects of pollution. Before large-scale application, it is crucial to conduct comprehensive risk assessments. Large-scale green synthesis can significantly reduce costs, representing a sustainable approach (Sable et al. 2024). Additionally,

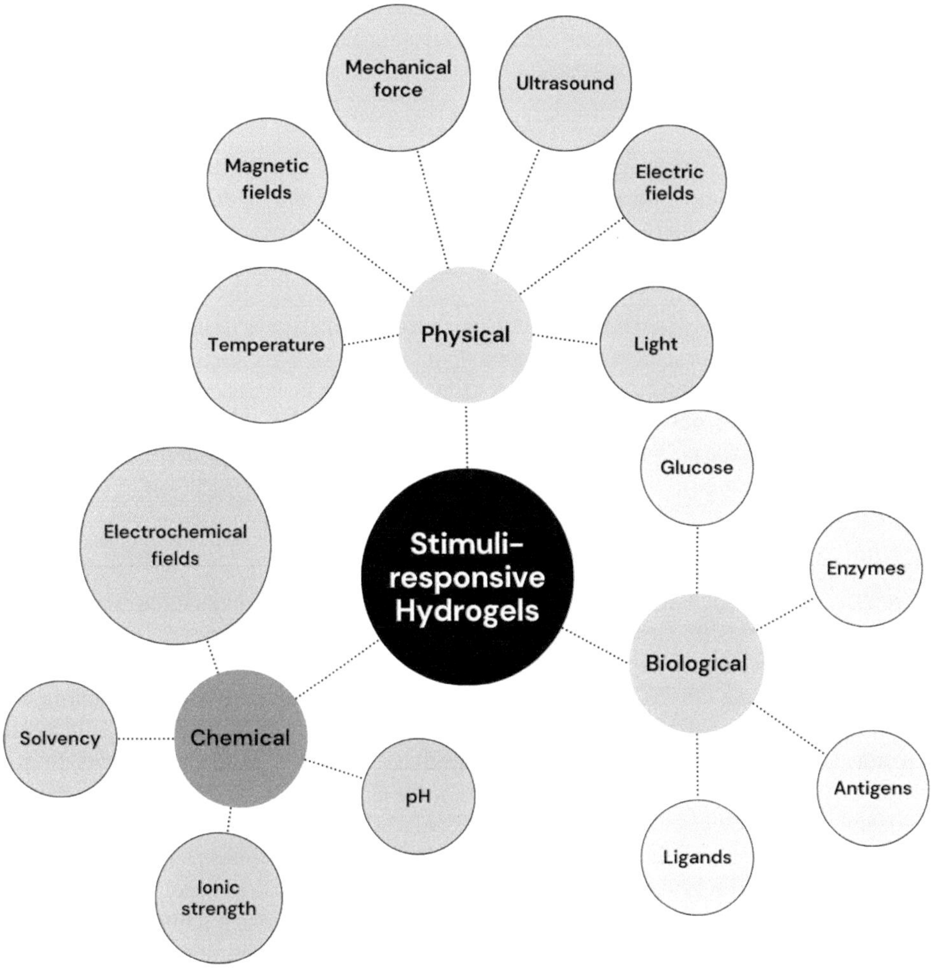

Figure 1: Response of hydrogel to different stimuli.

the use of biocompatible and environmentally acceptable materials, such as nanomaterials, can reduce toxicity and increase stability. Investing in efficient and economically viable techniques for the collection and reuse of nanomaterials is essential to ensure their sustained use in a sustainable manner (Karnwal and Malik 2024).

3. Synthesis of Hydrogel

Hydrogels hold immense potential due to their extensive surface area, strong adsorption ability, excellent solubility in organic solvents, and ease of fabrication. They are also cost-effective, biodegradable, recyclable, and offer enhanced functionality. Additionally, the notable hydrophilicity of cellulose-based hydrogels further enhances their suitability for environmental remediation (Dai et al. 2020). However, the performance of these materials heavily relies on their physical and chemical attributes (Malik et al. 2017). Therefore, the foremost and pivotal phase in establishing a successful adsorption system involves synthesizing a hydrogel with materials specifically tailored for the intended application.

The synthesis of hydrogels involves essential components such as monomers, initiators, and crosslinkers. Various monomers, including acrylamide, polyvinyl alcohol, polyvinylpyrrolidone, acrylic acid, and 2-dimethylaminoethyl methacrylate (DMAEM), possess different attributes related to pollutant remediation capacity, physical strength, and mechanical resistance (Darban et al. 2022). Crosslinking agents are essential in the synthesis of hydrogels, forming a three-dimensional polymer matrix and stabilizing the interaction between functional monomers and target molecules, which is crucial for remediation efforts. As a result, crosslinkers affect the hydrophilic or hydrophobic properties of the polymers, their selectivity, mechanical stability, and morphology (Yan and Row 2006).

The synthesis of hydrogels can be achieved via two main approaches: chemical and physical crosslinking. Each method imparts unique characteristics to the hydrogels due to the varying nature of the polymer chain crosslinks. Ionizing energy, such as gamma rays, X-rays, high-energy ultraviolet rays, accelerated electrons, and ion beams, is instrumental in these polymerization and crosslinking reactions, as it possesses sufficient energy to disrupt chemical bonds (Darban et al. 2022). In polymerization processes, ionizing radiation is utilized to begin the process, efficiently transforming monomers into elongated polymer links. This method is lauded for its speed and effectiveness. Additionally, ionizing radiation is critical in crosslinking, where it precisely forms chemical linkages between polymer links to create a robust 3D system. This crosslinking significantly enhances the polymer material's mechanical characteristics, stability, and other desired attributes (Priya et al. 2024).

3.1 Physical Cross-Linking

Hydrogels formed via physical crosslinking exhibit non-covalent interactions between polymer chains. Instead, they rely on physical forces like hydrogen bonds, hydrophobic forces, coordination bonds, and Van der Waals forces. This type of crosslinking can be undone under specific circumstances, allowing the hydrogel to undergo structural alterations without disrupting covalent bonds, unlike chemical crosslinking. Consequently, hydrogels formed through physical crosslinking demonstrate heightened responsiveness to various external factors such as pH, ionic strength, or temperature. They possess unique characteristics, such as the ability to "self-heal," where the gel can reassemble itself after being disrupted (Priya et al. 2024).

Hydrogels created via these physical interactions are known as physical gels. They are distinguished by their high-water content, sensitivity to environmental conditions, and the ability to reversibly react to temperature modifications. These hydrogels have a limited lifespan, typically enduring only a handful of days up to one month under biological conditions, making them ideal for short-term drug delivery applications. They are considered safe for use because the gelation process avoids the need for toxic covalent crosslinking agents. The physical methods used to prepare hydrogels offer advantages like thermal changeability and elevated water reactivity (Lin et al. 2020). Hydrogels can be synthesized from chitosan in combination with small anionic compounds, such as sulfates, phosphates, and citrates of metals like platinum, palladium, and molybdenum. The efficiency of synthesizing these hydrogels is affected by factors such as the anion's charge and scale, and the degree of deacetylation in the chitosan (Priya et al. 2024).

3.1.1 Freeze-Thaw Synthesis

One physical approach to synthesizing hydrogels is through the crystallization process using the freeze-thaw method. This technique involves repeated cycles of freezing and thawing, which induce the formation of crystalline regions within the hydrogel matrix (Zhang et al. 2013). Crystallization is achieved through the freezing of small molecular weight dissolved substances or major solvents, which increases the density of polymers. This process decreases the distance between polymer chains, enabling them to align and form a structured 3D network (Wang and Wang 2016). The repeated freeze-thaw process in hydrogels results in the formation of pores. This occurs as the melting crystals during the thawing phase leave behind voids, creating a porous network within the hydrogel (Guan et al. 2014). The mechanical characteristics of freeze-thaw hydrogels can be precisely customized by varying the density of polymers, the temperature at which freezing occurs, the lengths of the freezing and thawing periods, and the total number of freeze-thaw cycles performed (Wang and Wang 2016). Hydrogels synthesized through this method exhibit greater elasticity compared to those produced by chemical methods, attracting global interest (Qi et al. 2015). For example, poly(vinyl alcohol) and carboxymethylcellulose hydrogels are created using the freeze-thaw method and are employed in environmental remediation to retain metals (nickel, copper, zinc and silver) (Wang and Wang 2016).

3.1.2 Self-Assembling Synthesis

Self-assembling occurs when individual monomer units autonomously align and arrange themselves into fibers through non-covalent bonds (Yang et al. 2007). When these fibers remain adequately dissolved in a liquid, they entangle and effectively immobilize the solvent, forming a 3D structure. Non-covalent interactions provide stability to the hydrogel structure, making them softer compared to those formed by covalent crosslinking (Cui et al. 2010). These interactions confer notable benefits to self-assembled hydrogels, including resilience to environmental changes and the ability to repair themselves (Rajbhandary and Nilsson 2016).

3.1.3 Instantaneous Gelation Synthesis

Instant gelation is another method for the rapid blending of hydrogels (Feng et al. 2010). Zhou et al. (2014) developed magnetic hydrogel spheres using chitosan, specifically designed for environmental remediation by adsorbing lead ions (Pb^{2+}). These spheres showed an impressive adsorption capacity of 171.0 mg/g and could be revitalized, retaining 90% of their adsorption efficiency even after four cycles.

3.1.4 Ionotropic Gelation Synthesis

Ionotropic gelation facilitates the creation of hydrogels by forming nanoparticles and microparticles via electrostatic attractions between two ionic entities, with one being a polymer. This method allows for precise control under optimal conditions (Pedroso-Santana and Fleitas-Salazar 2020). In their study, Guerrero et al. (2020) synthesized hydrogel spheres using sodium alginate and hydroxypropyl cellulose in varied proportions of 50:50, 75:25, and 100:0 to assess their effectiveness in removing lead ions (Pb^{2+}). Among the tested ratios, the 75:25 composition demonstrated superior adsorption capacity over the other ratios, achieving 47.72 mg/g after 3 hours of contact. This composition also exhibited an impressive adsorption rate of 95.45%.

3.1.5 Inverse Emulsion Synthesis

By utilizing the inverse emulsion technique, a "water-in-oil" emulsion is achieved, where a water-soluble monomer is distributed within an uninterrupted oil medium, such as paraffin oil. This process involves the implementation of a suitable stabilizing agent to maintain the emulsion's stability. The process then undergoes a phase change in a coagulation solution, which results in the monomer being released and a porous membrane being formed. This method is advantageous for creating fine powders. Adjusting the reaction parameters allows for the control of particle size to achieve the desired results (Darban et al. 2022). Klinpituksa and Kosaiyakanon (2017) investigated the properties of a highly absorbent hydrogel, which was created using acrylic acid and carboxymethylcellulose through the inverse emulsion polymerization technique. They employed N,N'-methylene bisacrylamide as the crosslinker and potassium persulfate as the initiator. The resulting hydrogel demonstrated an impressive swelling capacity, reaching 544.95 g/g in deionized water and 44.0 g/g in a 0.9% NaCl solution.

3.2 Chemical Cross-Linking

In chemical crosslinking, the formation of networks is significantly easier to manage, especially when compared to those produced through physical crosslinking. This ease of control stems from the fact that their preparation and application do not rely on the pH level. Crosslinking typically occurs via chemical reactions, such as polymerization or reactions initiated by crosslinking agents. Covalent bonds are pivotal in constructing polymer chains within chemically crosslinked hydrogels. These bonds provide substantial strength and stability, forming a three-dimensional network of interlinked polymer chains (Priya et al. 2024). This structure endows chemically crosslinked hydrogels with enhanced robustness and resistance to environmental fluctuations. Therefore, these hydrogels are typically characterized by exceptional mechanical resilience and prolonged durability. The subsequent sections will detail various methodologies for synthesizing hydrogels through chemical modifications.

3.2.1 Free Radical Polymerization

Hydrogels are synthesized through free radical polymerization, utilizing N,N'-methylene bisacrylamide as a crosslinking catalyst. This procedure encompasses three main phases: initiation, polymer chain extension, and termination. In the initial phase, free radicals are produced by an initiator, such as ammonium persulfate, potassium persulfate, or by external stimuli like temperature, light, redox reactions, ultraviolet, or gamma radiation (Dil and Sadeghi 2018). Next, free radicals combine with a monomer, forming a monomer radical. This newly formed monomer radical then engages with additional monomers present in the solution, leading to the creation of polymer chains. During this propagation phase, the crosslinking agent is incorporated, forming a 3D structure. The final stage, termination, occurs through either disproportionation or combination. In combination, two growing chains merge to form a longer polymer chain. In the process of disproportionation, hydrogen shifts from the end of one elongating network to another. This transfer produces a polymer with one terminal group that is unsaturated and another that is saturated. To speed up the process, an accelerator like N,N,N',N'-tetramethylenediamine can be incorporated into the reaction process (Khan and Lo 2016). Shah et al. (2018) examined a superabsorbent hydrogel using acrylic acid and acrylamide in a free radical polymerization process. Their study focused on pollution remediation by removing multiple metals (nickel, cadmium, cobalt, and copper) from liquid solutions.

3.2.2 High Energy Irradiation

The formation of hydrogels through high-energy radiation requires the use of electron beams, ultraviolet light, and other types of radiation at room or lower temperatures, without needing catalysts, initiators, or crosslinking agents (Delbecq et al. 2013). This method outperforms advanced chemical initiation processes for hydrogel formation by eliminating by-product generation (Yang et al. 2015). In this technique, crosslinking density is dictated by the irradiation duration and dose. Prominent natural and synthetic polymers employed in this synthesis include alginate, gelatin, polyvinyl alcohol, chitosan, hyaluronic acid, carboxymethyl cellulose, and polyethylene glycol (Gaharwar et al. 2013). It is crucial to highlight that this crosslinking approach resembles free radical polymerization in hydrocarbons, encompassing three stages: initiation, polymer chain propagation, and termination. Upon exposure to irradiation, the reaction solution generates a hydroxyl free radical, which subsequently gives rise to a free radical monomer. The synthesis advances as the network reaches the pivotal gelation stage (Darban et al. 2022). The study by Maziad et al. (2015) assessed hydrogels made from polyacrylic acid/polyvinyl for water purification using gamma irradiation. The results demonstrated extraction capacities of 150 mg/g for nickel ions (Ni^{2+}), 155 mg/g for cobalt ions (Co^{2+}), and 193 mg/g for copper ions (Cu^{2+}). Additionally, the findings showed that the hydrogel expanded by 273%.

3.2.3 Reaction of Functional Groups

The formation of bonds between crosslinkers and the functional groups in polymer molecules is essential for the reaction. Hydrophilic groups, such as hydroxyl (–OH) and amines ($–NH_2$), interact with crosslinking facilitators like glutaraldehyde, which contains aldehyde functional groups. This interaction results in aldol products through covalent bonding. The creation of hydrogels from polymers containing hydroxyl groups requires specific conditions, such as a suppressor (methanol), low pH levels, and elevated temperatures. Nevertheless, protein-based hydrogels do not require these specific conditions (Pakdel and Peighambardoust 2018). Conversely, polymers with ester functional groups undergo chemical crosslinking through condensation processes in the presence of a crosslinking agent (Tran et al. 2017).

3.2.4 Grafting Reactions

This method involves grafting hydrophilic moieties such as NH_2 (amino), $CONH_2$ (acylamino), COOH (carboxyl), and SO_3H (sulfonic) onto the hydrogel's surface (Godiya et al. 2019). Adding these functional groups enhances the efficiency of remediation (adsorption or desorption) and the selectivity for specific heavy metals. Consequently, this leads to increased hydrophilicity, total active adsorption areas, and surface charge (Aly and El-Bisi 2018). Qi et al. (2019) engineered an innovative hydrogel derived from salecan-based polysaccharides. This was achieved through graft copolymerization, incorporating sodium vinyl sulfonate and acrylamide. The primary goal was effectively decontaminating lead ions (Pb^{2+}) from wastewater.

3.3 Irradiation-Based Cross Linking

Crosslinking through irradiation presents an attractive method for synthesizing hydrogels, particularly in situations where swift gelation and affordability are critical. By utilizing light-responsive UV irradiation and functional groups, scientists can quickly fabricate hydrogels, enabling numerous applications (Kausar 2020). However, certain prerequisites and potential limitations must be considered, such as ensuring the chosen light-sensitive groups are suitable for the intended application and the hydrogel's sensitivity to environmental factors like light and temperature (Priya et al. 2024).

Radiation-induced crosslinking, leveraging light-responsive functional groups, represents a highly efficient approach for hydrogel synthesis. This approach provides benefits compared to conventional chemical techniques, making it an attractive alternative (Nguyen et al. 2019). Notable benefits include rapid production, as hydrogels can be formed quickly using UV light-sensitive functional groups, resulting in a swifter process, especially when contrasted with chemical crosslinking, which requires prolonged reaction times. Additionally, this technique can reduce production costs by minimizing the reliance on costly crosslinking agents or catalysts (Singh and Singh 2018).

4. Functional Mechanisms of Hydrogel for Pollution Remediation

For a detailed and visual understanding of the functional mechanisms of hydrogel for pollution remediation, the reader is encouraged to closely examine Figure 2. This figure illustrates the efficient application of hydrogel, highlighting key processes and the interaction between components that make this technological solution so effective. Understanding these mechanisms is essential to appreciate the innovation and positive environmental impact that hydrogel remediation can provide.

4.1 Adsorption

During the production of adsorptive hydrogels, a polymer network is formed, enabling the incorporation of various functional groups, such as hydroxyl, amide, carboxyl, sulfonic acid, and amino. These functional groups readily form hydrogen bonds, either with water or pollutants (Badsha et al. 2021). The functional moieties within the hydrogel act as active regions, facilitating the storage of pollutants within the polymer network (Badsha et al. 2021).

During the adsorption phase, pollutants are retained in the solid phase (hydrogel) through physical interactions involving weak forces such as hydrophobic interactions, π-π interactions, electrostatic interactions, van der Waals forces, and hydrogen bonds (Seida and Tokuyama 2022). Adsorption can also occur through chemical interactions (chemisorption), which involve strong

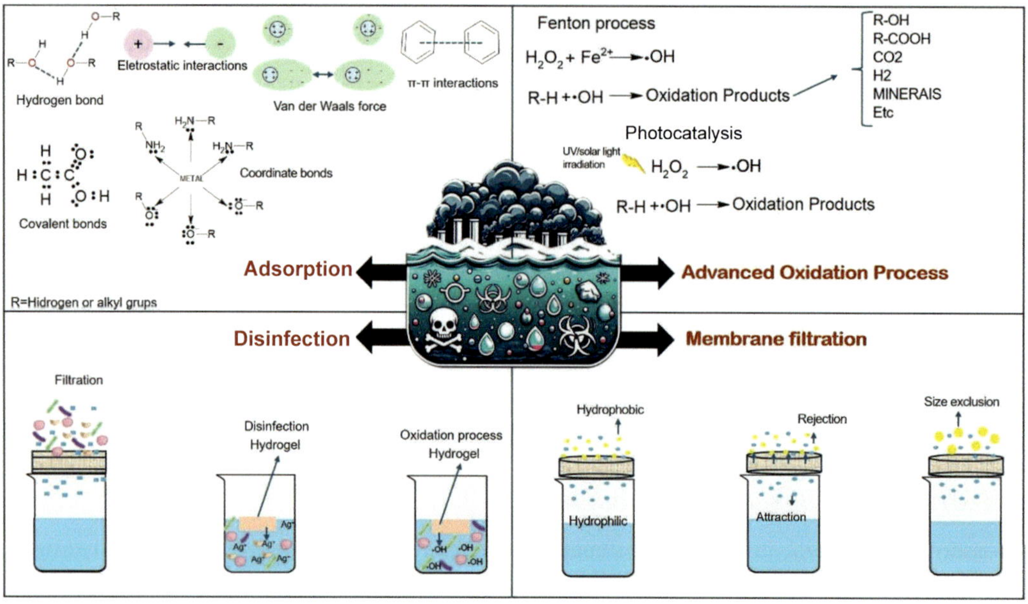

Figure 2: Functional mechanisms of hydrogel for pollution remediation.

interactions characterized by the exchange of electrons between the adsorbent and the adsorbate, such as covalent, ionic, and coordinate bonds (Sivakumar and Lee 2022). The type of interaction depends on the environmental pollutant and functional moieties present in the hydrogel.

4.1.1 Heavy Metal Pollutants

In the adsorption method, electrostatic interactions are fundamental to the removal of pollutants, thanks to the presence of organic groups with charges opposite to those of the pollutants (Khan and Lo 2016). For example, hydrogels with chitosan/sodium alginate and calcium ions effectively adsorb ions such as cadmium (Cd^{2+}) and lead (Pb^{2+}) through electrostatic interactions, due to the presence of COO^- groups on the hydrogel surface (Tang et al. 2020).

Additionally, heavy metals can be adsorbed through chemical interactions (ionic bonds), as shown by jute hydrogels with polyacrylic acid and alginate/polyaniline/polypyrrole hydrogels combined with graphene oxide. These hydrogels have been used to remove metals such as chromium (Cr^{6+}) and copper (Cu^{2+}) from water. The adsorption mechanisms in these systems include the formation of metal complexes, hydrogen bonds, and chemical interactions (Yu et al. 2021, Zhang et al. 2021). Hydrogels made from cellulose and chitosan have also proven effective in removing metals like copper (Cu), zinc (Zn^{2+}), and nickel (Ni^{2+}), adsorbing them onto the hydrogel surface via metal complex formation (Godiya et al. 2022). Additionally, hydrogels are useful for removing microplastics associated with metals such as iron, copper, or zinc (Guo et al. 2022).

4.1.2 Organic Pollutants

Acridine orange has been employed as a reference compound to evaluate the effectiveness of graphene hydrogels in removing cationic dyes through adsorption. It was observed that, in this adsorbent–adsorbate system, the primary interactions were electrostatic, indicating that the hydrogel is anionic. π-π interactions also played a significant role in the adsorption process (Jiang et al. 2017).

Hydrogels made of hydroxymethylacrylate and N-methyl maleic acid were employed to remove the pollutant sulfamethoxazole. The adsorption mechanism involved the formation of chemical bonds, likely covalent bonds, between parts of the molecule and the hydrogel, resulting in the degradation of the pollutant (Yang et al. 2017).

Silver nanoparticle (Ag NP) hydrogels functionalized with β-cyclodextrin and incorporated into alginate bubbles were also tested for sulfamethoxazole removal. These systems enable water treatment under real conditions, as cyclodextrins can form inclusion complexes that encapsulate nonpolar molecules. The host–guest interactions can be disrupted by physical interactions (Vassalini et al. 2020). β-Cyclodextrin–polyacrylamide hydrogels have been used to remove micropollutants such as phenolphthalein, bisphenol, propranolol hydrochloride, and 2-naphthol from water. These organic pollutants are removed primarily through the formation of inclusion complexes (Song et al. 2021).

Methylene blue, another common pollutant, is frequently targeted for removal using various hydrogels. For example, alginate-based hydrogels promote chemical adsorption (chemisorption) (Allangawi et al. 2023). Polyacrylamide- and polyacrylic acid-based hydrogels remove pollutants through electrostatic interactions between oppositely charged ions (Zamani-Babgohari et al. 2024). The chemical structure of these hydrogels also enables hydrogen bonding. Additionally, polyacrylic acid-based hydrogels can remove methylene blue through electrostatic interactions (Hu et al. 2018). While adsorption is a primary mechanism for pollutant removal, other processes, such as oxidation and membrane filtration, may also contribute.

4.2 Advanced Oxidation Processes (AOPs)

Advanced oxidation processes (AOPs) effectively degrade pollutant compounds, transforming them into more biodegradable intermediates or achieving complete mineralization. Commonly employed reactions to explain the oxidation mechanisms of pollutants include the Fenton process, photocatalysis, sonolytic oxidation, and ozonation (Nidheesh et al. 2022).

4.2.1 Fenton Process

The Fenton process encompasses a reaction between iron ions and hydrogen peroxide, leading to the production of hydroxyl radicals. These radicals react with organic pollutants via radical reactions, resulting in the production of carbon dioxide (CO_2), water, and acidic minerals such as sulfuric and nitric acids (Dong et al. 2018, Rodríguez et al. 2002).

For instance, hydrogels synthesized using Fe_3O_4/graphene oxide and polyacrylamide were studied for the elimination of L-Rhodamine B and heavy metals. Mass spectrometry analysis revealed that L-Rhodamine B degraded into smaller intermediates, including the propionate ion, which was further converted into carbon dioxide and water (Dong et al. 2018). Similarly, FeOOH-doped graphene hydrogels have been examined for pollutant removal from water via the Fenton process (Zhuang et al. 2020). Molecular dynamics simulations revealed that the adsorption efficiency was boosted by enhanced hydrogen bonding and the presence of active sites. This improvement increased the electron-donating capacity of the pollutant tetracycline for iron ions (Fe^{3+}) in the Fenton reaction. Additionally, these hydrogels exhibited remarkable stability over five consecutive cycles, highlighting their potential for effective water purification applications.

4.2.2 Photocatalysis

Photocatalysis represents another advanced oxidation method used to remove impurities. Photocatalytic materials accelerate reaction rates when exposed to light. These materials are typically semiconductors (such as TiO_2, V_2O_5, ZnO, Fe_2O_3, and CdO) and organic/inorganic composites (Moradeeya et al. 2022). The KI-PCN hydrogel, a carbon nitride material co-doped with potassium and iodine and confined in alginate, was used to remove pollutants such as methylene blue. It demonstrated high catalytic efficiency, primarily involving hydroxyl radical-driven reactions (Du et al. 2022). Additionally, nano-crystalline $MnFe_2O_4$ (MFO) hydrogels were applied to degrade sulfasalazine via photocatalysis, following degradation pathways that ultimately produced water, carbon dioxide, and ions such as sulfate and nitrate (Liu et al. 2022).

4.2.3 Sonolytic Oxidation and Ozonation

The sonolytic oxidation process uses ultrasound irradiation to form highly reactive hydroxyl radicals (OH) that effectively break down organic pollutants, with the presence of catalytic agents like nanostructured metal oxides (CuO, ZnO, and TiO_2) further enhancing the generation of these reactive species and increasing the oxidation potential (Adityosulindro et al. 2017, Ince 2018, Al-Bsoul et al. 2020, Ayare and Gogate 2020).

In this context, hydrogels can enhance oxidation power through the sonolysis mechanism by carrying catalyst agents. For instance, Yi et al. (2020) developed a 3D cobalt hydroxide/graphene oxide hydrogel as an accelerant using graft polymerization with an acrylamide monomer. The Co^{2+} ions inside the hydrogel promoted the activation of peroxymonosulfate, generating reactive species like SO_4^{-} and $^1O_2$, which are capable of oxidizing organic pollutants, including bisphenol A, tetracycline hydrochloride, p-chlorophenol, rhodamine B, sulfadiazine, phenol, and norfloxacin. Additionally, Zanias et al. (2020) tested the ability of a Co-Fe bimetallic carbon xerogel to degrade methylparaben using an ultrasound process (20 kHz). The catalyst's constructive collaboration for the degradation of particulate matter (PM) was confirmed, attributed to the xerogel's ability to catalyze the decomposition of hydrogen peroxide, formed by the sonolysis of water, into hydroxyl radicals.

Although promising for wastewater decontamination, the combination of sonolytic oxidation techniques and hydrogels faces challenges, including optimizing operating conditions such as the pH range, ultrasound frequency, and power density, as well as minimizing secondary pollution from metal ions used in catalytic processes (Adityosulindro et al. 2017, Yi et al. 2020, Ince 2018).

Similarly, the ozonation system can be optimized with catalysts such as metal oxides, metal-organic frameworks, and nanoparticles (Psaltou et al. 2021, Sun et al. 2021), which can be incorporated into hydrogels, although this is still underexplored. Despite providing complete mineralization of organic substances and rapid contaminant removal, the ozonation mechanism has drawbacks, such as unavoidable secondary pollution, low removal capacity for large quantities of contaminants, low selectivity of pollutants, and low solubility of O_3 (Baigorria et al. 2021). These factors need to be addressed, with the combination of ozonation and hydrogels offering a potential solution worth exploring.

4.3 Membrane Filtration

Hydrogels used in membrane filtration are extensively applied in water purification, effectively eliminating particles like sediments, dissolved solids, and turbidity (Ahmad et al. 2024). Several studies have highlighted their effectiveness in removing pollutants like heavy metals. Alginate hydrogels, for instance, retain metals and dyes through electrostatic interactions: their anionic surfaces repel cationic metals (Cai et al. 2023). Dyes, which can be either anionic or cationic, interact with membranes via adsorption (attraction) or electrostatic repulsion (rejection) (Park et al. 2024).

Alginate membranes have also been utilized for removing hydrophobic pollutants from water surfaces. These membranes allow hydrophilic water molecules to pass through while retaining oils or other nonpolar molecules (Ahmad et al. 2024). By incorporating cellulose nanofibrils into alginate and polyethylene oxide, hydrogels can remove pollutants through size exclusion. This property allows the specific retention of particles based on their scale and effectively separates oil from water due to polarity differences (Mokhena et al. 2018)

Calcium alginate hydrogels combined with polyester, polypropylene, or polyhydroxybutyrate have been tested for removing dyes such as methylene blue (799.80 Da), Rose Bengal (1017.64 Da), and Methyl Orange (327.33 Da), as well as sodium chloride (NaCl). These hydrogels exhibit a porous structure, and filtration is primarily governed by size exclusion. Materials with high molecular weights ($>$ 697 Da) were effectively retained (Zhang et al. 2021). When calcium alginate hydrogels were combined with polyethylene glycol, the rejection rates for Brilliant Blue G250 (854.02 Da) and Congo Red (696.68 Da) reached 100% and 99.5%, respectively (Zhao et al. 2015).

4.4 Disinfection

The disinfection process is essential for eliminating pathogenic microorganisms from water, which are responsible for countless deaths annually (Alvarez et al. 2018). Current water treatment methods involve chemical processes but can generate toxic waste and by-products associated with the development of cancer (Muellner et al. 2007). Therefore, the development of sustainable methodologies is crucial. Disinfection processes using hydrogels can involve oxidative reactions (Zhang et al. 2020), thermal degradation/thermal conductivity (Khan et al. 2024), filtration (molecular membranes), disinfectants, anti-fouling agents, among others (Guo et al. 2022).

The disinfection process through oxidative reactions involves the production of reactive oxygen species (ROS), such as peroxides and hydroxyl radicals (OH). These reactive species easily cross the membranes of microorganisms, accelerating the oxidation of glutathione (GSH). By oxidizing this antioxidant, cellular stress is induced, leading to cell death (Zhang et al. 2020). As an example, researchers have created antibacterial hydrogels (ABHs) that incorporate catechol-functionalized molecular hydrogen peroxide generators and quinone-bound activated carbon particles. These innovative hydrogels are designed to efficiently treat water by utilizing radical reactions, offering a promising solution for eliminating impurities and ensuring cleaner water (Guo et al. 2022).

Through filtration, hydrogels act like a sponge, forming a physical sieve that retains microorganisms present in water. For example, graphene oxide hydrogels containing molecules with five aromatic rings and two ammonium groups exhibit a synergistic effect between their constituents, resulting in antibacterial activity and retaining about 99.5% of gram-positive and gram-negative bacteria (Wang et al. 2019).

Nanosilver/diatomite hydrogels embedded in calcium alginate are used to disinfect water against Escherichia coli and Staphylococcus aureus bacteria. The disinfection process occurs via the controlled release of silver ions into the water. Ag^+ ions bind to the cytoplasmic membrane of bacteria, interfering with their biochemical processes and causing cell death (Zhao et al. 2021). Chitosan–PVA hydrogels containing silver nanoparticles have also demonstrated efficacy in eliminating four bacterial strains, achieving 100% elimination within a maximum of 4 hours, showing potential for disinfecting river, soil, or tap water samples (Dhiman and Agnihotri 2020). The potential of silver in hydrogels has also been observed in hydrogels formed of chitosan/graphene oxide/nano-Ag (CS/GO/Ag), tested on authentic water resources polluted with organic and biomedical impurities at high contamination levels (10^4–10^6 CFU mL^{-1}), where total disinfection was achieved (Kaur et al. 2024).

Polymeric hydrogels based on 2–hydroxyethyl methacrylate (HEMA) loaded with elemental iodine have been developed for drinking water disinfection through a controlled release system of various iodine species. Iodine has been studied for disinfecting drinking water and swimming pools, showing success against a wide range of microorganisms (Punyani et al. 2007).

5. Advanced Nanomaterial-Based Hydrogels for Environmental Pollution Remediation

The field of nanotechnology has captivated the scientific community, opening doors to a highly promising research area. Leveraging the unique and advantageous properties of nanostructured materials offers more efficient and sustainable solutions to current environmental challenges. Among the notable advancements in nanotechnology for pollution remediation are hydrogels derived from carbon-based materials, organic nanomaterials, metal oxide nanoparticles (NPs), hybrid composites, and advanced 2D materials (Figure 3). These hydrogels present innovative and effective solutions for contaminant mitigation, thereby promoting environmental sustainability and protecting ecosystems.

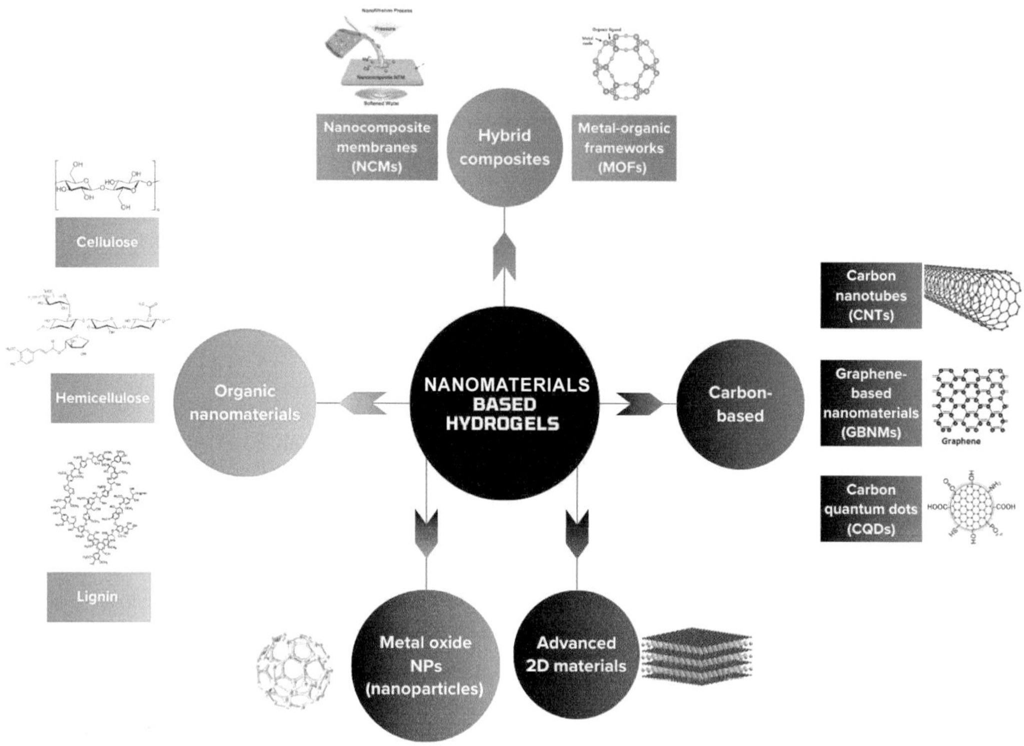

Figure 3: Nanomaterial-based hydrogels for environmental pollution remediation.

5.1 Carbon-based Materials

Carbon-based nanomaterials (CBNMs) stand out as exceptional adsorbents because of their distinctive chemical and physical attributes. These attributes allow them to effectively eliminate both organic and inorganic contaminants on a large scale, driven by their extensive surface area and high volatility (Asghar et al. 2024). Among CBNMs, carbon nanotubes (CNTs) and graphene-based nanomaterials (GBNMs) are significant. CNTs, noted for their large surface area, superior mechanical properties, and high void fraction, are effective in capturing pollutants, although their small particle size can pose challenges (Ozay et al. 2011). GBNMs, derived from graphene, offer outstanding mechanical strength and high electrical conductivity, rendering them appropriate for a wide range of applications, contributing to cleaner and safer ecosystems (Rittikulsittichai et al. 2016). Additionally, carbon quantum dots (CQDs), with their customizable fluorescence and biocompatibility, show great promise for detecting and removing organic pollutants (Zhang and Huang 2021, Asghar et al. 2024). Despite their advantages, the potential environmental risks associated with these nanomaterials call for further research to ensure their safe and effective use in pollution remediation (Sable et al. 2024, Karnwal and Malik 2024).

5.2 Organic Nanomaterials

Hydrogels sourced from natural polymers have been widely utilized in various research for their contaminant removal due to their multifunctional groups, high surface area, advanced retention, and chelation capacities (Mahinroosta et al. 2018). Natural polymers (lignin, cellulose, and hemicellulose), which are abundant in nature, offer positives such as availability, low cost, and effectiveness in capturing microparticles and metal nanoparticles (Moradeeya et al. 2017, Rathod et al. 2018). Chemical and physical modifications of these polymers and biomass-based adsorbents can further enhance their adsorption properties and interactions with environmental contaminants (Sahoo et al. 2024). Additionally, dendrimers, with their highly branched structure and functionalization capacity, show great potential for environmental remediation by purifying water and air through pollutant adsorption and acting as carriers for chelating agents (Rittikulsittichai et al. 2016).

5.2.1 Lignocellulosic and Cellulosic

Biopolymer-based hydrogels, like those derived from cellulose, provide numerous advantages and exhibit enhanced functionality, making them even more effective in their applications. This enhanced functionality could include increased absorbency, better mechanical strength, improved biocompatibility, or a higher degree of responsiveness to environmental stimuli, allowing them to perform more effectively in various practical scenarios. Moreover, the enhanced surface area, accessibility, high adsorption capacity, biodegradability, and recyclability further augment their effectiveness. The remarkable hydrophilicity of cellulose-based hydrogels specifically enhances their suitability as adsorbents for environmental remediation, ensuring they efficiently attract and retain contaminants (Akter et al. 2021). Employing cellulose, its offshoots, or cellulosic aggregates as efficient additives in polymeric mixtures for hydrogel formation effectively tackles the issues linked to the low solubility of pure cellulose. Different derivatives, including micro-nanocrystals of cellulose, micro-nano-bacterial cellulose, micro-nano-fibrillated cellulose, and micro-nano-cellulose, can be readily formulated using enzymatic or chemical techniques and utilized for these applications (Hassanisaadi et al. 2023).

5.3 Metal oxide NPs (Nanoparticles)

Nanometric-scale metal oxides such as cerium oxide, iron oxide, aluminum oxide, manganese oxide, and titanium oxide, among others, possess a range of unique attributes, including high reduction capacity and specific selectivity (Singh and Mishra 2020). Many of these oxides have the ability to participate in reduction reactions, transforming pollutants into less harmful substances, which is essential for the eradication of contaminants. Another important characteristic is their ability to selectively adsorb certain types of pollutants, enabling the efficient removal of specific contaminants and enhancing the overall efficacy of the remediation process. Finally, many metal oxides are chemically stable, resisting adverse environmental conditions without degrading easily, thereby ensuring long-term effectiveness (Sable et al. 2024). Therefore, these combined factors make nanometric metal oxides a promising option for environmental pollutant remediation.

5.4 Hybrid Composites

Nanocomposite materials, such as metal-organic frameworks (MOFs) and nanocomposite membranes (NCMs), highlight impressive properties for environmental remediation. MOFs, which are networked crystalline materials created by metal ions associated with organic bonds, are renowned for their surface area and selective adsorption ability for contaminants like heavy metals and organic pollutants (Férey et al. 2005, Kitagawa et al. 2004). These structures can be tailored to boost their stability and functionality through various modifications. Adjustments might involve altering the chemical composition, incorporating reinforcing agents, or optimizing the synthesis process. Such enhancements can lead to increased durability, improved performance, and expanded applications, making them even more effective for their intended purposes (Li et al. 2016). NCMs combine polymeric and non-polymeric materials, improving their thermal, mechanical, and chemical properties, as well as increasing hydrophobicity and pollutant removal capacity (Mansourpanah et al. 2011). The fabrication of NCMs involves incorporating nanoparticles into polymer matrices, thereby enhancing their adsorption and microbial inactivation capabilities (Goh et al. 2016). Moreover, nanosensors equipped with nanomaterials enable detailed detection and remediation of pollutants at the nanometric scale, offering high sensitivity and monitoring capabilities (Perumal et al. 2014). These advancements position nanocomposites as promising materials for various environmental remediation applications.

5.5 *Advanced 2D materials*

Two-dimensional (2D) nanomaterials, with their ultra-thin structures and minuscule sizes, are highly efficient and dependable for various applications. Their extensive specific surface area, resulting from their vast lateral scale and nanometric bulk, renders them particularly valuable for remediation tasks such as detection, adsorption, and catalysis. 2D nanomaterials have demonstrated remarkable catalytic, adsorptive, and separation capabilities in various water treatment systems (Sable et al. 2024). Of late, there has been growing attention to 2D materials owing to their unique framework and wide array of useful applications. Examples include graphene membranes with ultra-thin atomic layers, ultra-fast photocatalysts utilizing visible light to speed up processes, advanced oil-cleaning materials, and intelligent or self-repairing frameworks that can adjust the thickness and scale of their nanostructures (Fatima et al. 2022).

6. Conclusion and Future Remarks

Advanced nanomaterial-based hydrogels present a promising solution for environmental pollution remediation, overcoming the challenges associated with traditional methods, such as inefficiency and high costs. The outstanding characteristics of these nanomaterials, such as high surface area and reactivity, improve the hydrogels' ability to capture and neutralize pollutants. The integration of nanomaterials into hydrogels results in synergistic composites that significantly improve pollutant removal efficiency, applicable in containment barriers and high-efficiency filtration systems. The continuous development of smart nanomaterials expands remediation possibilities, promoting sustainability and environmental protection.

Therefore, nanomaterial-based hydrogels are essential mechanisms in the fight against pollution, offering effective and sustainable solutions. Future work should focus on developing new nanomaterials with enhanced properties, integrating smart nanomaterials that respond to environmental factors, scaling up production and testing under real-world conditions, promoting sustainability with regenerable and recyclable hydrogels, studying interactions with multiple contaminants, and evaluating the safety of nanomaterials. These promising directions will help refine nanomaterial-based hydrogels, contributing to a cleaner and more sustainable future.

Abbreviations

MOs	Microorganisms
WHO	World Health Organization
NMs	Nanomaterials
1D	One-dimensional
2D	Two-dimensional
3D	Three-dimensional
–COOH	Carboxyl
–OH	Hydroxyl
pH	Hydrogen ion potential
DMAEM	2-Dimethylaminoethyl methacrylate
Pb^{2+}	Lead ions
Ni^{2+}	Nickel ions
Co^{2+}	Cobalt ions
Cu^{2+}	Copper ions

–OH	Hydroxyl
$–NH_2$	Amino
$CONH_2$	Acylamino
COOH	Carboxyl
SO_3H	Sulfonic
Cd^{2+}	Cadmium
Cr^{6+}	Chromium
Zn^{2+}	Zinc
Ag NP	Silver nanoparticle
AOPs	Advanced oxidation processes
CO_2	Carbon dioxide
Fe_3O_4	Magnetite
FeOOH	Iron (III) oxyhydroxide (Goethite)
Fe (III)	Ferric ion (Iron (III) ion)
TiO_2	Titanium dioxide
V_2O_5	Vanadium pentoxide
ZnO	Zinc oxide
Fe_2O_3	Iron (III) oxide (Hematite)
CdO	Cadmium oxide
KI-PCN	Potassium iodide-phthalocyanine
$MnFe_2O_4$	Manganese ferrite
CuO	Copper (II) oxide
ZnO	Zinc oxide
TiO_2	Titanium dioxide
SO_4	Sulfate ion
O_2	Oxygen (molecular oxygen)
Co-Fe	Cobalt-iron (alloy or compound)
PM	Particle Material
O_3	Ozone
NaCl	Sodium chloride
ROS	Reactive oxygen species
GSH	Oxidation of glutathione
ABHs	Antibacterial hydrogels
PVA	Polyvinyl alcohol
Ag^+	Silver ion
CFU	Colony-Forming Unit
HEMA	2-Hydroxyethyl methacrylate
CBNMs	Carbon-based nanomaterials
CNTs	Carbon nanotubes
GBNMs	Graphene-based nanomaterials
CQDs	Carbon quantum dots
NCMs	Nanocomposite membranes
MOFs	Metal-organic frameworks

References

Adityosulindro, S., Barthe, L., González-Labrada, K., Jáuregui Haza, U. J., Delmas, H., and Julcour, C. (2017). Sonolysis and Sono-Fenton Oxidation for Removal of Ibuprofen in (Waste)Water. *Ultrasonics Sonochemistry,* 39: 889–896.

Ahmad, F., Nosheen, A., Atiq, M. S., Mushtaq, B., Ahmad, S., Azam, F. et al. (2024). An Eco-Friendly Hydroentangled Cotton Non-Woven Membrane with Alginate Hydrogel for Water Filtration. *International Journal of Biological Macromolecules,* 256: 128422.

Akter, M., Bhattacharjee, M., Dhar, A. K., Rahman, F. B., Haque, S., Rashid, T. U. et al. (2021). Cellulose-Based Hydrogels for Wastewater Treatment: A Concise Review. *Gels*, 7: 30.

Al-Bsoul, A., Al-Shannag, M., Tawalbeh, M., Al-Taani, A. A., Lafi, W. K., Al-Othman, A. et al. (2020). Optimal Conditions for Olive Mill Wastewater Treatment Using Ultrasound and Advanced Oxidation Processes. *Science of The Total Environment,* 700: 134576.

Allangawi, A., Aljar, M. A. A., Ayub, K., Abd El-Fattah, A., and Mahmood, T. (2023). Removal of Methylene Blue by Using Sodium Alginate-Based Hydrogel; Validation of Experimental Findings via DFT Calculations. *Journal of Molecular Graphics and Modelling,* 122: 108468.

Almomani, F., Bhosale, R., Khraisheh, M., Kumar, A., and Almomani, T. (2020). Heavy Metal Ions Removal from Industrial Wastewater Using Magnetic Nanoparticles (MNP). *Applied Surface Science,* 506: 144924.

Alvarez, P. J. J., Chan, C. K., Elimelech, M., Halas, N. J., and Villagrán, D. 2018. Emerging Opportunities for Nanotechnology to Enhance Water Security. *Nature Nanotechnology,* 13 (8): 634–641.

Aly, A. A., and El-Bisi, M. K. (2018). Grafting of Polysaccharides: Recent Advances. 469–519. *In:* Biopolym. Grafting; Elsevier: Amsterdam, The Netherlands.

Asghar, N., Hussain, A., Nguyen, D. A., Ali, S., Hussain, I., Junejo, A. et al. (2024). Advancement in Nanomaterials for Environmental Pollutants Remediation: A Systematic Review on Bibliometrics Analysis, Material Types, Synthesis Pathways, and Related Mechanisms. *Journal of Nanobiotechnology,* 22: 26.

Ayare, S. D., and Gogate, P. R. (2020). Sonochemical, Photocatalytic and Sonophotocatalytic Oxidation of Flonicamid Pesticide Solution Using Different Catalysts. *Chemical Engineering and Processing - Process Intensification* 154: 108040.

Badsha, M. A. H., Khan, M., Wu, B., Kumar, A., and Lo, I. M. C. (2021). Role of Surface Functional Groups of Hydrogels in Metal Adsorption: From Performance to Mechanism. *Journal of Hazardous Materials,* 408: 124463.

Baigorria, E., Galhardi, J. A., and Fraceto, L. F. (2021). Trends in Polymers Networks Applied to the Removal of Aqueous Pollutants: A Review. Journal of Cleaner Production. Elsevier Ltd.

Bashir, S., Hina, M., Iqbal, J., Rajpar, A., Mujtaba, M., Alghamdi, N. et al. (2020). Fundamental concepts of hydrogels: Synthesis, properties, and their applications. *Polymers*, 12: 2702.

Bilal, M., Rasheed, T., Zhao, Y., and Iqbal, H. M. N. (2019). Agarose-chitosan hydrogel-immobilized horseradish peroxidase with sustainable bio-catalytic and dye degradation properties. *International Journal of Biological Macromolecules*, 124: 742–749.

Cai, R., Chen, Y., Hu, J., Xiong, J., Lu, J., Liu, J. et al. (2023). A Self-Supported Sodium Alginate Composite Hydrogel Membrane and Its Performance in Filtering Heavy Metal Ions. *Carbohydrate Polymers,* 300: 120278.

Cruz-Acuna, R., Quiros, M., Huang, S., Siuda, D., Spence, J. R., Nusrat, A. et al. (2018). PEG-4MAL hydrogels for human organoid generation, culture, and *in vivo* delivery. *Nature Protocols*, 13: 2102–2119.

Cui, H., Webber, M. J., and Stupp, S. I. (2010). Self-assembly of peptide amphiphiles: From molecules to nanostructures to biomaterials. *Biopolymers*, 94: 1–18.

Dai, L., Cheng, T., Xi, X., Nie, S., Ke, H., Liu, Y. et al. (2020). A versatile TOCN/CGG self-assembling hydrogel for integrated wastewater treatment. *Cellulose*, 27: 915–925.

Darban, Z., Shahabuddin, S., Gaur, R., Ahmad, I., and Sridewi, N. (2022). Hydrogel-Based Adsorbent Material for the Effective Removal of Heavy Metals from Wastewater: A Comprehensive Review. *Gels*, 8: 263.

Delbecq, F., Kono, F., and Kawai, T. (2013). Preparation of PVP–PVA–exfoliated graphite cross-linked composite hydrogels for the incorporation of small tin nanoparticles. *European Polymer Journal*, 49: 2654–2659.

Dhiman, N. K., and Agnihotri, S. (2020). Hierarchically Aligned Nano Silver/Chitosan–PVA Hydrogel for Point-of-Use Water Disinfection: Contact-Active Mechanism Revealed. *Environmental Science: Nano,* 7(8): 2337–2350.

Dil, N. N., and Sadeghi, M. (2018). Free radical synthesis of nanosilver/gelatin-poly (acrylic acid) nanocomposite hydrogels employed for antibacterial activity and removal of Cu (II) metal ions. *Journal of Hazardous Materials*, 351: 38–53.

Dong, C., Lu, J., Qiu, B., Shen, B., Xing, M., and Zhang, J. (2018). Developing Stretchable and Graphene-Oxide-Based Hydrogel for the Removal of Organic Pollutants and Metal Ions. *Applied Catalysis B: Environmental,* 222: 146–156.

Du, Y., Che, H., Wang, P., Chen, J., and Ao, Y. (2022). Highly Efficient Removal of Organic Contaminant with Wide Concentration Range by a Novel Self-Cleaning Hydrogel: Mechanism, Degradation Pathway and DFT Calculation. *Journal of Hazardous Materials,* 440: 129738.

Dutta, D., Arya, S., and Kumar, S. (2021). Industrial Wastewater Treatment: Current Trends, Bottlenecks, and Best Practices. *Chemosphere*, 285: 131245.

Fatima, J., Shah, A. N., Tahir, M. B., Mehmood, T., Shah, A. A., Tanveer, M. et al. (2022). Tunable 2D nanomaterials, their key roles and mechanisms in water purification and monitoring. *Frontiers in Environmental Science*, 10: 766743.

Feng, Y., Gong, J.-L., Zeng, G.-M., Niu, Q.-Y., Zhang, H.-Y., Niu, C.-G. et al. (2010). Adsorption of Cd (II) and Zn (II) from aqueous solutions using magnetic hydroxyapatite nanoparticles as adsorbents. *Chemical Engineering Journal*, 162: 487–494.

Férey, G., Mellot-Draznieks, C., Serre, C., Millange, F., Dutour, J., Surblé, S. et al. (2005). A chromium terephthalate-based solid with unusually large pore volumes and surface area. *Science*, 309(5743): 2040–2042.

Gaharwar, A. K., Rivera, C., Wu, C. J., Chan, B. K., and Schmidt, G. (2013). Photocrosslinked nanocomposite hydrogels from PEG and silica nanospheres: Structural, mechanical and cell adhesion characteristics. *Materials Science and Engineering: C*, 33: 1800–1807.

Godiya, C. B., Cheng, X., Li, D., Chen, Z., and Lu, X. (2019). Carboxymethyl cellulose/polyacrylamide composite hydrogel for cascaded treatment/reuse of heavy metal ions in wastewater. *Journal of Hazardous Materials*, 364: 28–38.

Godiya, C. B., Revadekar, C., Kim, J., and Park, B. J. (2022). Amine-Bilayer-Functionalized Cellulose-Chitosan Composite Hydrogel for the Efficient Uptake of Hazardous Metal Cations and Catalysis in Polluted Water. *Journal of Hazardous Materials* 436: 129112.

Goh, K., Karahan, H. E., Wei, L., Bae, T. H., Fane, A. G., Wang, R. et al. (2016). Carbon nanomaterials for advancing separation membranes: A strategic perspective. *Carbon*, 19: 694–710.

González-Martín, J., Kraakman, N. J. R., Pérez, C., Lebrero, R., and Muñoz, R. (2021). A state of–the-art review on indoor air pollution and strategies for indoor air pollution control. *Chemosphere*, 262: 128376.

Guan, Y., Bian, J., Peng, F., Zhang, X.-M., and Sun, R.-C. (2014). High strength of hemicelluloses-based hydrogels by freeze/thaw technique. *Carbohydrate Polymers*, 101: 272–280.

Guerrero, R., Acibar, C., Alarde, C. M., Maslog, J., and Pacilan, C. J. (2020). Evaluation of Pb (II) Removal from Water Using Sodium Alginate/Hydroxypropyl Cellulose Beads. *E3S Web Conf.*, 148: 02002.

Guo, Q., Liu, Y., Liu, J., Wang, Y., Cui, Q., Song, P. et al. (2022). Hierarchically Structured Hydrogel Actuator for Microplastic Pollutant Detection and Removal. *Chemistry of Materials* 34 (11): 5165–5175.

Hao, T., Li, J., Yao, F., Dong, D., Wang, Y., Yang, B. et al. (2017). Injectable fullerenol/alginate hydrogel for suppression of oxidative stress damage in Brown adipose-derived stem cells and cardiac repair. *ACS Nano* 11: 5474–5488.

Hassanisaadi, M., Riseh, R. S., Rabiei, A., Varma, R. S., and Kennedy, J. F. (2023). Nano/Micro-Cellulose-Based Materials as Remarkable Sorbents for the Remediation of Agricultural Resources from Chemical Pollutants. *International Journal of Biological Macromolecules*, 246: 125763.

Hu, X., Cheng, W., Nie, W., and Shao, Z. (2015). Synthesis and characterization of a temperature-sensitive hydrogel based on sodium alginate and N-isopropylacrylamide. *Polymer Advanced Technologies*, 26: 1340–1345.

Hu, X.-S., Liang, R., and Sun, G. (2018). Super-Adsorbent Hydrogel for Removal of Methylene Blue Dye from Aqueous Solution. *Journal of Materials Chemistry*, A 6 (36): 17612–17624.

Ince, N. H. (2018). Ultrasound-Assisted Advanced Oxidation Processes for Water Decontamination. *Ultrasonics Sonochemistry*, 40: 97–103.

Isobe, N., Komamiya, T., Kimura, S., Kim, U. J., and Wada, M. (2018). Cellulose hydrogel with tunable shape and mechanical properties: from rigid cylinder to soft scaffold. *International Journal of Biological Macromolecules*, 117: 625–631.

Jiang, Y., Chowdhury, S., and Balasubramanian, R. (2017). Nitrogen-Doped Graphene Hydrogels as Potential Adsorbents and Photocatalysts for Environmental Remediation. *Chemical Engineering Journal*, 327: 751–763.

Ju, M. J., Oh, J., and Choi, Y. H. (2021). Changes in air pollution levels after COVID-19 outbreak in Korea. *Science of the Total Environment*, 750: 141521.

Karnwal, A., and Malik, T. (2024). Nano-revolution in heavy metal removal: engineered nanomaterials for cleaner water. *Frontiers in Environmental Science*, 12: 1393694.

Kasinski, A., Zielinska-Pisklak, M., Oledzka, E., and Sobczak, M. (2020). Smart hydrogels–synthetic stimuli-responsive antitumor drug release systems. *International Journal of Nanomedicine*, 15: 4541–4572.

Kaur Dhiman, N., Agnihotri, S., and Reddy, M. S. (2024). High Permeation, Biodegradable Hydrogel Based Flow-through Reactor for Water Disinfection with Controlled Silver Release. *Chemical Engineering Journal*, 489: 151335.

Kausar, A. (2020). Nanocarbon in Polymeric Nanocomposites Hydrogel-Design and Multi-Functional Tendencies. *Polymer-Plastics Technology and Materials*, 59: 1505–1521.

Khan, M., and Lo, I. M. (2016). A Holistic Review of Hydrogel Applications in the Adsorptive Removal of Aqueous Pollutants: Recent Progress, Challenges, and Perspectives. *Water Research*, 106: 259–271.

Khan, S. A., Malik, A. N., Velioglu, B., Gul, S., Kavakli, I. H., and Lazoglu, I. (2024). A Novel Smart Disinfection System Using 3D-Printed and Electrically Conductive Composite Hydrogel. Emergent Materials.

Kitagawa, S., Kitaura, R., and Noro, S. (2004). Functional porous coordination polymers. *Angewandte Chemie International Edition*, 43(18): 2334–2375.

Klinpituksa, P., and Kosaiyakanon, P. (2017). Superabsorbent polymer based on sodium carboxymethyl cellulose grafted polyacrylic acid by inverse suspension polymerization. International *Journal of Polymer Science*, 2017: 3476921.

Li, J. R., Kuppler, R. J., and Zhou, H. C. (2009). Selective gas adsorption and separation in metal–organic frameworks. *Chemical Society Reviews*, 38(5): 1477–1504.

Lin, J., Zheng, S. Y., Xiao, R., Yin, J., Wu, Z. L., Zheng, Q. et al. (2020). Constitutive behaviours of tough physical hydrogels with dynamic metal-coordinated bonds. *Journal of the Mechanics and Physics of Solids*, 139: 103935.

Liu, Y., Zhang, X., Zhou, Y., Ma, H., Cheng, X., Wei, L. et al. (2022). MFO@NZVI/Hydrogel for Sulfasalazine Degradation: Performance, Mechanism and Degradation Pathway. *Separation and Purification Technology*, 282: 120054.

Mahinroosta, M., Farsangi, Z. J., Allahverdi, A., and Shakoori, Z. (2018). Hydrogels as intelligent materials: A brief review of synthesis, properties and applications. *Materials Today Chemistry*, 8: 42–55.

Malik, D., Jain, C., and Yadav, A. K. (2017). Removal of heavy metals from emerging cellulosic low-cost adsorbents: A review. *Applied Water Science*, 7: 2113–2136.

Mansourpanah, Y., Alizadeh, K., Madaeni, S. S., Rahimpour, A., and Afarani, H. S. (2011). Using different surfactants for changing the properties of poly(piperazineamide) TFC nanofiltration membranes. *Desalination*, 271: 169–177.
Maziad, N., Mohsen, M., Gomaa, E., and Mohammed, R. (2015). Radiation copolymerization of hydrogels based in polyacrylic acid/polyvinyl alcohol applied in water treatment processes. *Journal of Materials Science and Engineering*: A, 5: 381–390.
Mokhena, T. C., Jacobs, N. V., and Luyt, A. S. (2018). Nanofibrous Alginate Membrane Coated with Cellulose Nanowhiskers for Water Purification. *Cellulose*, 25 (1): 417–427.
Mondal, S., Das, S., and Nandi, A. K. (2020). A review on recent advances in polymer and peptide hydrogels. *Soft Matter*, 16: 1404–1454.
Moradeeya, P. G., Kumar, M. A., Thorat, R. B., Rathod, M., Khambhaty, Y., and Basha, S. (2017). Nanocellulose for biosorption of chlorpyrifos from water: chemometric optimization, kinetics and equilibrium. *Cellulose*, 24(3): 1319–1332.
Moradeeya, P. G., Sharma, A., Kumar, M. A., and Basha, S. (2022). Titanium dioxide based nanocomposites – Current trends and emerging strategies for the photocatalytic degradation of ruinous environmental pollutants. *Environmental Research*, 204: 112384.
Mu, R., Liu, B., Chen, X., Wang, N., and Yang, J. (2020). Hydrogel Adsorbent in Industrial Wastewater Treatment and Ecological Environment Protection. *Environ. Technology and Innovation*, 20: 101107.
Muellner, M. G., Wagner, E. D., McCalla, K., Richardson, S. D., Woo, Y. T., and Plewa, M. J. (2007). Haloacetonitriles vs. Regulated Haloacetic Acids: Are Nitrogen-Containing DBPs More Toxic? *Environmental Science & Technology*, 41(2): 645–651.
Nguyen, N. T. P., Nguyen, L. V. H., Thanh, N. T., Van Toi, V., Quyen, T. N., Tran, P. A. et al. (2019). Stabilization of silver nanoparticles in chitosan and gelatin hydrogel and its applications. *Materials Letters*, 248: 241–245.
Nidheesh, P. V., Couras, C., Karim, A. V., and Nadais, H. (2022). A Review of Integrated Advanced Oxidation Processes and Biological Processes for Organic Pollutant Removal. *Chemical Engineering Communications*, 209(3): 390–432.
Ouyang, Q. Q., Hu, Z., Lin, Z. P., Quan, W. Y., Deng, Y. F., Li, S. D. et al. (2018). Chitosan hydrogel in combination with marine peptides from tilapia for burns healing. *International Journal of Biological Macromolecules*, 112: 1191–1198.
Ozay, O., Aktas, N., and Sahiner, N. (2011). Hydrogels as a potential chromatographic system: Absorption, speciation, and separation of chromium species from aqueous media. *Separation Science and Technology*, 46: 1450–1461.
Pakdel, P. M., and Peighambardoust, S. J. (2018). A review on acrylic based hydrogels and their applications in wastewater treatment. *Journal of Environmental Management*, 217: 123–143.
Park, K., Kandel, D. R., Radoor, S., Park, P.-K., and Lee, J. (2024). Comparative Evaluation of Adsorptive and Repulsive Separation Using Cellulose Nanofiber/Calcium Alginate Composite Hydrogel Filtration Membranes with Improved Degradability. *Journal of Environmental Chemical Engineering*, 12(6): 114351.
Pedroso-Santana, S., and Fleitas-Salazar, N. (2020). Ionotropic gelation method in the synthesis of nanoparticles/microparticles for biomedical purposes. *Polymer International*, 69: 443–447.
Pena, B., Laughter, M., Jett, S., Rowland, T. J., Taylor, M. R. G., Mestroni, L. et al. (2018). Injectable hydrogels for cardiac tissue engineering. *Macromolecular Bioscience*, 18: e1800079.
Perumal, V., and Uda, M. N. A. (2014). Advances in biosensors: Principle, architecture and applications. *Journal of Applied Biomedicine*, 12: 115.
Priya, A. S., Premanand, R., Ragupathi, I., Bhaviripudi, V. R., Aepuru, R., Kannan, K. et al. (2024). Comprehensive Review of Hydrogel Synthesis, Characterization, and Emerging Applications. *Journal of Composites Science*, 8: 457.
Psaltou, S., Sioumpoura, K., Kaprara, E., Mitrakas, M., and Zouboulis, A. (2021). Transition Metal Ions as Ozonation Catalysts: An Alternative Process of Heterogeneous Catalytic Ozonation. Catalysts 11 (9). MDPI.
Punyani, S., Narayanan, P., Vasudevan, P., and Singh, H. (2007). Sustained Release of Iodine from a Polymeric Hydrogel Device for Water Disinfection. *Journal of Applied Polymer Science*, 103 (5): 3334–3340.
Qi, X., Hu, X., Wei, W., Yu, H., Li, J., Zhang, J. et al. 2015. Investigation of Salecan/poly(vinyl alcohol) hydrogels prepared by freeze/thaw method. *Carbohydrate Polymers*, 118: 60–69.
Qi, X., Lin, L., Shen, L., Li, Z., Qin, T., Qian, Y. et al. (2019). Efficient decontamination of lead ions from wastewater by salecan polysaccharide-based hydrogels. *ACS Sustainable Chemistry & Engineering*, 7: 11014–11023.
Rajbhandary, A., and Nilsson, B. L. (2016). Self-Assembling Hydrogels. pp. 219–250. *In:* GELS HANDBOOK: Fundamentals, Properties and Applications, Volume 1: Fundamentals of Hydrogels; World Scientific: Singapore.
Rando, G., Scalone, E., Sfameni, S., and Plutino, M. R. (2024). Functional Bio-Based Polymeric Hydrogels for Wastewater Treatment: From Remediation to Sensing Applications. *Gels*, 10: 498.
Rathod, M., Moradeeya, P. G., Haldar, S., and Basha, S. (2018). Nanocellulose/TiO2 composites: preparation, characterization and application in the photocatalytic degradation of a potential endocrine disruptor, mefenamic acid, in aqueous media. *Photochemical & Photobiological Sciences*, 17(10): 1301–1309.
Rittikulsittichai, S., Kolhatkar, A. G., Sarangi, S., Vorontsova, M. A., Vekilov, P. G., Brazdeikis, A. et al. (2016). Multi-responsive hybrid particles: Thermo-, pH-, photo-, and magneto-responsive magnetic hydrogel cores with gold nanorod optical triggers. *Nanoscale*, 8: 11851–11861.
Rodríguez, M., Abderrazik, N. B., Contreras, S., Chamarro, E., Gimenez, J., and Esplugas, S. (2002). Iron(III) Photoxidation of Organic Compounds in Aqueous Solutions. *Applied Catalysis B: Environmental*, 37(2): 131–137.

Sable, R., Kumar, V., Singh, V., Rustagi, S., Ghahal, S., and Chaudhary, V. (2024). Rational Design of Smart Hydrogels for Biomedical Applications. *Frontiers in Chemistry*, 8: 615665.

Sahoo, T. P., Satasiya, G., Moradeeya, P. G., Saravaia, H. T., and Kumar, M. A. (2024). Removal of fluoroquinolone antibiotic and sulfonated dye by functionalized Persea americana seed powder: Appraisal on phase transfer kinetics, equilibrium, economics, and applications in rural settings. *Environmental Research*, 00139351.

Seida, Y., and Tokuyama, H. (2022). Hydrogel Adsorbents for the Removal of Hazardous Pollutants—Requirements and Available Functions as Adsorbent. *Gels*, 8(4): 220.

Shah, L. A., Khan, M., Javed, R., Sayed, M., Khan, M. S., Khan, A. et al. (2018). Superabsorbent polymer hydrogels with good thermal and mechanical properties for removal of selected heavy metal ions. *Journal of Cleaner Production*, 201: 78–87.

Siepmann, J., Siegel, R. A., and Rathbone, M. J. (2011). Fundamentals and Applications of Controlled Release Drug Delivery; Springer Science & Business Media: Berlin/Heidelberg, Germany.

Singh, B., and Singh, B. (2018). Modification of sterculia gum polysaccharide via network formation by radiation induced crosslinking polymerization for biomedical applications. *International Journal of Biological Macromolecules*, 116: 91–99.

Singh, I., and Mishra, P. K. (2020). Nano-membrane Filtration: A Novel Application of Nanotechnology for Waste Water Treatment. *In*: National Conference on Smart Materials: Energy and Environment for Smart Cities (NSES-2018), ScienceDirect. Review under responsibility of National Conference on Smart Materials: Energy and Environment for Smart Cities (NSES-2018).

Sivakumar, R., and Lee, N. Y. (2022). Adsorptive Removal of Organic Pollutant Methylene Blue Using Polysaccharide-Based Composite Hydrogels. *Chemosphere*, 286: 131890.

Song, X., Mensah, N. N., Wen, Y., Zhu, J., Zhang, Z., Tan, W. S. et al. (2021). β-Cyclodextrin-Polyacrylamide Hydrogel for Removal of Organic Micropollutants from Water. *Molecules*, 26(16): 5031.

Sun, Q., Zhu, G., Wu, J., Lu, J., and Zhang, Z. (2021). Simultaneous Catalytic Ozonation Degradation of Metronidazole and Removal of Heavy Metal from Aqueous Solution Using Nano-Magnesium Hydroxide. *Environmental Technology* (United Kingdom), 42 (6): 894–904.

Tang, S., Yang, J., Lin, L., Peng, K., Chen, Y., Jin, S. et al. (2020). Construction of Physically Crosslinked Chitosan/Sodium Alginate/Calcium Ion Double-Network Hydrogel and Its Application to Heavy Metal Ions Removal. *Chemical Engineering Journal*, 393: 124728.

Tran, T. H., Okabe, H., Hidaka, Y., and Hara, K. (2017). Removal of metal ions from aqueous solutions using carboxymethyl cellulose/sodium styrene sulfonate gels prepared by radiation grafting. *Carbohydrate Polymers*, 157: 335–343.

Vassalini, I., Ribaudo, G., Gianoncelli, A., Casula, M. F., and Alessandri, I. (2020). Plasmonic Hydrogels for Capture, Detection and Removal of Organic Pollutants. Environmental Science: *Nano*, 7(12): 3888–3900.

Wang, L., He, J., Zhu, L., Wang, Y., Feng, X., Chang, B. et al. (2019). Assembly of Pi-Functionalized Quaternary Ammonium Compounds with Graphene Hydrogel for Efficient Water Disinfection. *Journal of Colloid and Interface Science*, 535: 149–158.

Wang, L.-Y., and Wang, M.-J. (2016). Removal of heavy metal ions by poly (vinyl alcohol) and carboxymethyl cellulose composite hydrogels prepared by a freeze–thaw method. *ACS Sustainable Chemistry & Engineering*, 4: 2830–2837.

Yan, H., and Row, K. H. (2006). Characteristic and synthetic approach of molecularly imprinted polymer. *International Journal of Molecular Sciences*, 7: 155–178.

Yang, J., Dong, X., Gao, Y., and Zhang, W. (2015). One-step synthesis of methacrylated POSS cross-linked poly (N-isopropylacrylamide) hydrogels by-irradiation. *Materials Letters*, 157: 81–84.

Yang, J., Li, Z., and Zhu, H. (2017). Adsorption and Photocatalytic Degradation of Sulfamethoxazole by a Novel Composite Hydrogel with Visible Light Irradiation. *Applied Catalysis B: Environmental*, 217: 603–614.

Yang, Z., Liang, G., and Xu, B. (2007). Enzymatic control of the self-assembly of small molecules: A new way to generate supramolecular hydrogels. *Soft Matter*, 3: 515–520.

Yi, Q., Tan, J., Liu, W., Lu, H., Xing, M., and Zhang, J. (2020). Peroxymonosulfate Activation by Three-Dimensional Cobalt Hydroxide/Graphene Oxide Hydrogel for Wastewater Treatment through an Automated Process. *Chemical Engineering Journal*, 400: 125965.

Yu, F., Yang, P., Yang, Z., Zhang, X., and Ma, J. (2021). Double-Network Hydrogel Adsorbents for Environmental Applications. *Chemical Engineering Journal*, 426: 131900.

Zamani-Babgohari, F., Irannejad, A., Kalantari Pour, M., and Khayati, G. R. (2024). Synthesis of Carboxymethyl Starch Co (Polyacrylamide/ Polyacrylic Acid) Hydrogel for Removing Methylene Blue Dye from Aqueous Solution. International *Journal of Biological Macromolecules*, 269: 132053.

Zanias, A., Frontistis, Z., Vakros, J., Arvaniti, O. S., Ribeiro, R. S., Silva, A. M. T. et al. (2020). Degradation of Methylparaben by Sonocatalysis Using a Co–Fe Magnetic Carbon Xerogel. Ultrasonics Sonochemistry 64. Elsevier B.V.

Zhang, H., Zhang, F., and Wu, J. (2013). Physically crosslinked hydrogels from polysaccharides prepared by freeze–thaw technique. *Reactive & Functional Polymers*, 73: 923–928.

Zhang, W., Ou, J., Wang, B., Wang, H., He, Q., Song, J. et al. (2021). Efficient Heavy Metal Removal from Water by Alginate-Based Porous Nanocomposite Hydrogels: The Enhanced Removal Mechanism and Influencing Factor Insight. *Journal of Hazardous Materials*, 418: 126358.

Zhang, X., Zhang, G., Zhang, H., Liu, X., Shi, J., Shi, H. et al. (2020). A Bifunctional Hydrogel Incorporated with CuS@ MoS2 Microspheres for Disinfection and Improved Wound Healing. I, 382: 122849.

Zhang, Y., and Huang, Y. (2021). Rational Design of Smart Hydrogels for Biomedical Applications. *Frontiers in Chemistry*, 8: 615665.

Zhang, Y., Jiang, M., Zhang, Y., Cao, Q., Wang, X., Han, Y. et al. (2019). Novel Lignin-Chitosan-PVA Composite Hydrogel for Wound Dressing. *Materials Science and Engineering: C*, 104: 110002.

Zhao, D., Shen, Z., and Shen, X. (2021). Dual-Functional Calcium Alginate Hydrogel Beads for Disinfection Control and Removal of Dyes in Water. *International Journal of Biological Macromolecules*, 188: 253–262.

Zhao, K., Zhang, X., Wei, J., Li, J., Zhou, X., Liu, D. et al. (2015). Calcium Alginate Hydrogel Filtration Membrane with Excellent Anti-Fouling Property and Controlled Separation Performance. *Journal of Membrane Science*, 492: 536–546.

Zhou, Y., Fu, S., Zhang, L., Zhan, H., and Levit, M. V. (2014). Use of Carboxylated Cellulose Nanofibrils-Filled Magnetic Chitosan Hydrogel Beads as Adsorbents for Pb (II). *Carbohydrate Polymers*, 101: 75–82.

Zhuang, Y., Wang, X., Liu, Q., and Shi, B. (2020). N-Doped FeOOH/RGO Hydrogels with a Dual-Reaction-Center for Enhanced Catalytic Removal of Organic Pollutants. *Chemical Engineering Journal*, 379: 122310.

CHAPTER-5

The Scope of Smart Nanomaterials Against Other Conventional Methods in Environmental Remediation

Kusum Sharma,[1,*] *Lebanan Keisham,*[1] *Sachidanand Singh,*[2] *Rahul Datta,*[3] *Megha Barot*[4,5] and *Azamal Husen*[6]

1. Introduction

The entire world is currently dealing with a number of growing environmental issues that seriously jeopardize human health, global sustainability, and ecological integrity. Environmental quality has been severely deteriorated by rapid industrialization, urbanization, and intense agricultural methods, leading to enduring issues including soil contamination, air pollution, and water scarcity. The buildup of dangerous pollutants, such as greenhouse gases, particulate matter, heavy metals, and toxic organic compounds, make these problems worse. Numerous medical diseases, including cancer, neurological impairments, asthma, long-term and restrictive pulmonary diseases, and dermatological disorders, have been directly associated with environmental pollution. It has been determined that high-atomic-weight metals such as lead (Pb), oxide derivatives of sulfur and nitrogen, and persistent organic pollutants (POPs) are the main causes of threats to public health. A World Health Organization (WHO) report (Chan 2018) states that every year, environmental pollution, particularly a lack of access to clean water, causes the deaths of almost 270,000 newborns during their first month of life. Based-on exposure time and amount, lead is known to impart both acute and chronic health-consequences amongst diversified living beings (Singh et al. 2022),

[1] Department of Life Science, Rabindranath Tagore University, Raisen, Bhopal, M.P, India.
[2] Department of Biotechnology, School of Energy Technology, Pandit Deendayal Energy University, Gandhinagar, Gujarat, India.
[3] Department of Geology and Pedology, Faculty of Forestry and Wood echnology, Mendel University in Brno, Czech RepublicT.
[4] Research and Development Cell, Parul University, Vadodara, Gujarat, India.
[5] Department of Environmental Science, Parul Institute of Applied Sciences, Parul University, Gujarat, India.
[6] Wolaita Sodo University, P.O. Box:ʃ138, Wolaita, Ethiopia.
* Corresponding author: Kusum.sharma15@gmail.com, ORCID ID: 0000-0002-7135-3861

DOI: 10.1201/9781003558804-5

whereas high quantities of carbon monoxide (CO) can poison an organism immediately (Savioli et al. 2024). The need for efficient and long-lasting restoration techniques is thus highlighted by the fact that environmental deterioration is not only an ecological disaster but also a direct threat to human well-being. Conventional remediation techniques have long been used to reduce environmental pollution, including sand filtration, flocculation, condensation, and activated carbon (AC) adsorption (Lewoyehu 2021). Despite their proven usefulness, these methods frequently have serious drawbacks (Figure 1).

Methods such as flocculation and condensation require a lot of energy, are not very effective at eliminating some metal ions, and can produce secondary waste that is hard to recycle or manage (Saleh et al. 2022, Qasem et al. 2021). Similarly, activated carbon's limited adsorption capacity and poor effectiveness in treating complex pollutant mixes limit its application in water filtration, despite its widespread use. Nanotechnology is attracting greater interest in an effort to address the drawbacks of conventional remediation technologies and meet the expanding need for sustainable development. With their high efficiency, affordability, and environmental compatibility, smart nanomaterials (SNMs) have become a game-changer in environmental remediation in recent years (Khan et al. 2022). SNMs have remarkable sensitivity and reactivity to environmental pollutants because of their nanoscale size, vast surface area, and adjustable surface characteristics. Particularly, nanoparticles (NPs) can be designed to absorb, break down, or change contaminants into less dangerous forms (Thakur and Kumar 2022), which makes it possible to use them for soil remediation, air quality

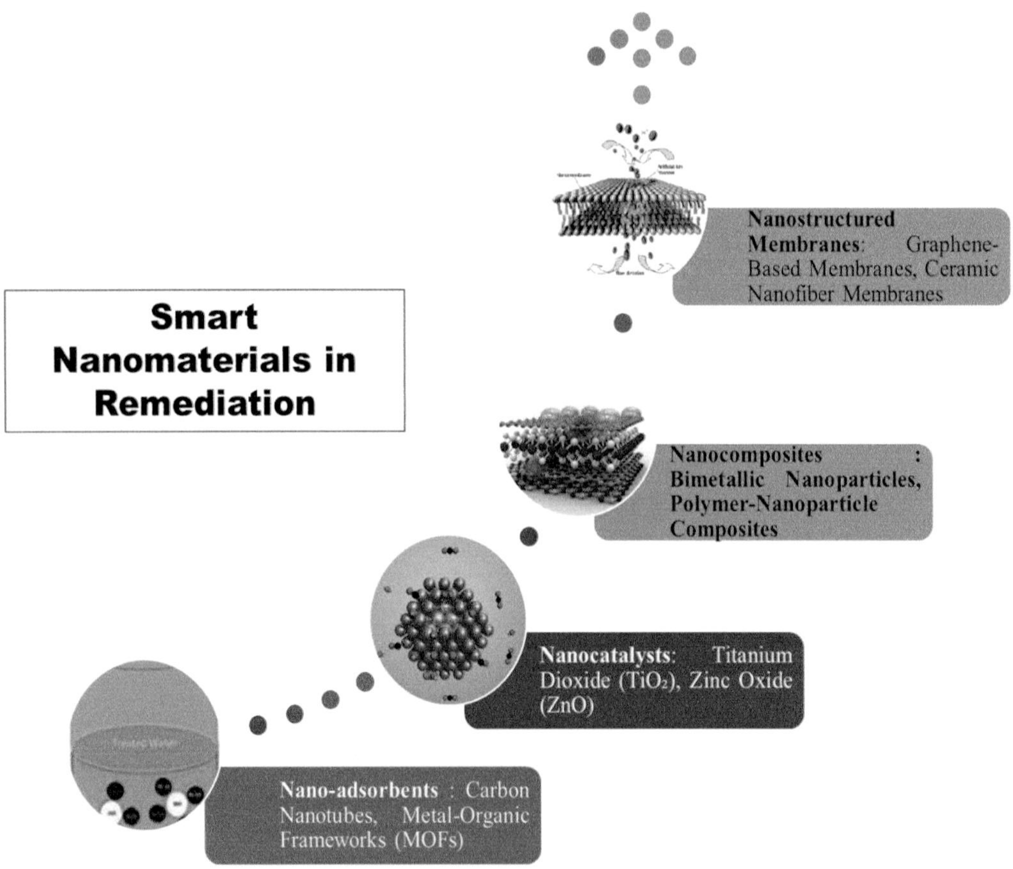

Figure 1: Types of smart nanomaterials in remediation.

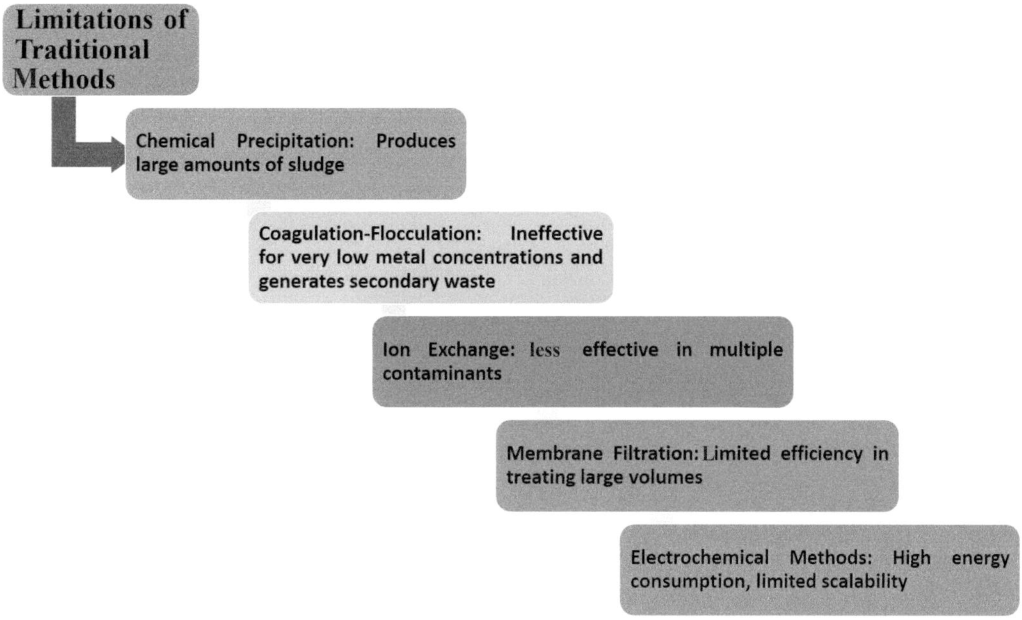

Figure 2: Limitations of traditional methods.

enhancement, and water purification. The use of nanomaterials into membrane-based filtration systems for water treatment is one of the most exciting developments in this field. Operating on low-energy principles, these membranes filter microbes, nanoparticles, and chemical contaminants more effectively than traditional systems (Saleem and Zaidi 2020, Zeng 2021, Joseph 2023). Furthermore, multifunctional remediation techniques that provide scalable, high-performance solutions with less infrastructure are made possible by nanotechnology. For instance, nanosensors are being created to track pollution in the air and water in real time, identifying pollutants at traces and allowing for prompt action.

Furthermore, through processes like reduction and photocatalysis, a variety of nanomaterials, including Titania and elemental iron, have demonstrated efficacy in the degradation of organic contaminants and hazardous metals (Hsieh 2010). They consistently proved effective in removing and neutralizing hazardous species when utilized as reactive media in PRBs for *in situ* remediation of groundwater (Obiri-Nyarko 2014) (Figure 2).

Nanoparticles' adaptable properties, such as their sensitivity to environmental factors like pH and temperature, further increase their suitability in a variety of contexts. Despite their enormous potential, issues including the high cost of producing nanoparticles, possible ecotoxicological hazards, and the absence of thorough regulatory frameworks make it difficult for nanotechnology to be widely used in environmental restoration (Hansen 2007). Furthermore, the type of pollutant, the environment, and the methods used all affect how well nanomaterials operate. This chapter will provide insights of SNMs and their benefits, drawbacks and potential application in remediation of environmental contaminants.

2. Smart Nanomaterials in Environmental Remediation

Nanomaterials are defined as materials with nanoscale building blocks (typically within 1–100 nm) which demonstrate distinct and novel physical or chemical characteristics as compared to their bulk counterpart. These properties make them highly effective in environmental remediation, offering several advantages over traditional methods (Figure 3a).

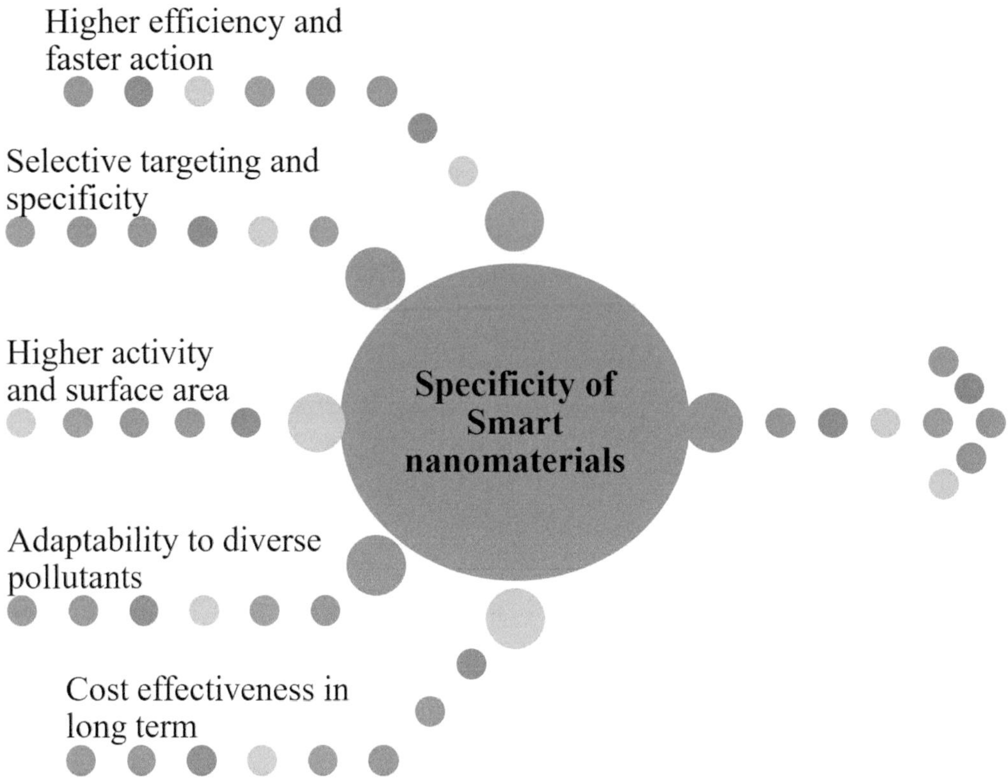

Figure 3a: Specificity of smart nanomaterials.

2.1 Smart Nanomaterials

Smart nanomaterials are engineered materials that exhibit the ability to respond dynamically to specific environmental factors. Nanomaterials are materials that are at least 100 nm in one dimension, while nanoparticles (NPs) are particles that are between 1 and 100 nm in size (Xu et al. 2013, Khan et al. 2022). Accordingly, SNMs are materials with nanoscale (1–100 nm) properties that can alter their thickness, density, texture, color, modulus, stiffness, and toughness in response to particular stimuli (Patadiya 2023). A key feature of smart nanomaterials is their high surface area per unit volume ratio, which significantly improves their interactivity with contaminants and results in more effective cleanup procedures. Due to their excellent mechanical, thermal and optical characteristics, nanoparticles further contribute to their effectiveness in pollution control, as they can be tailored to interact with a wide range of contaminants (Khan and Hossain 2022). These nanomaterials act as heterogeneous catalysts, and their small particle size maximizes the surface area exposed to reactants, resulting in more efficient chemical reactions. This can be illustrated by examples of cerium-, titanium- and zinc oxide nanoparticles in various catalytic applications such as methane combustion and photocatalysis (Mohamed et al. 2024). As per need or requirement, these SNMs can be designed to activate under specific environmental conditions, such as shifts in pH or the presence of pollutants (Ghanbari and Moradi 2017). pH-responsive nanomaterials, for instance, are especially helpful when contaminants show pH-dependent behavior, like when organic compounds degrade or heavy metals are removed (Chiappisi and Gradzielski 2015). These nanomaterials can change their structure, charge, or surface properties in response to changes in pH, allowing them to selectively target and neutralize contaminants in polluted environments (Ahmad et al. 2023). This ability to respond to environmental cues makes smart nanomaterials highly versatile and effective across diverse remediation practices, offering an advanced alternative to traditional methods for tackling pollution.

2.2 Types of Smart Nanomaterials in Remediation

These materials are modified with specific shapes and morphologies to carry out specific reactions and eliminate a variety of contaminants, toxic chemicals, organic pollutants, and heavy metals. The most investigated categories of smart nanomaterials for environmental remediation are listed below; each has a unique mode of action.

2.2.1 Magnetic Nanomaterials

Nanotechnology is concerned with developing materials at the nanoscale (1–100 nm). Magnetic nanoparticles (MNMs) provide advantageous characteristics such as superparamagnetism, low magnetic toughness, and high magnetic susceptibility (Wu et al. 2019). Magnetic nanoparticles (MNMs) (Table 1) are one of the most adaptable forms used in environmental restoration (Nnadozie and Ajibade 2020). These materials, which enable a variety of nanotechnology applications, are frequently made utilizing magnetic elements and compounds such as iron (Fe), nickel (Ni), cobalt (Co), manganese (Mn), magnetite, maghemite, and mixed ferrites. The magnetic nanoparticles (NPs) produced by different synthesis procedures are suitably surface-engineered or biofunctionalized with certain chemical species in order to establish an adequate structure and optimal ties (Comanescu 2023). They interact with the external magnetic field in a specific way, which gives them a variety of traits and uses. These nanoparticles are mostly made from the oxides of iron such as Fe_3O_4 and Fe_2O_3 or other related magnetic materials, allowing them to be easily manipulated with external magnetic fields (Kumar and Ansari 2024). Iron oxide nanoparticles are the most studied material to date due to their versatility, ease of production, biocompatibility, biostability, biodegradability, non-toxicity, and superparamagnetic nature (Ansari et al. 2024). According to reports, iron oxide nanoparticles (NPs) and nanocomposites have a wide range of applications, such as heavy metal

Table 1: Types of Metal Nanoparticles.

Metal nanoparticles	Nanosized particles of metals like silver, gold, iron, and zinc with high reactivity and surface area.	Removal of heavy metals, reduction of organic pollutants, and antimicrobial activity.	Fatehbasharzad 2021, Naseem and Durrani 2021
Metal oxide nanoparticles	Oxides of metals like titanium dioxide (TiO_2), zinc oxide (ZnO), and iron oxide (Fe_2O_3).	Photocatalysis, adsorption, and degradation of pollutants in water and air.	Pal 2024, Singh 2019
Carbon-based nanomaterials	Includes graphene, carbon nanotubes (CNTs), and fullerene-based nanomaterials.	Adsorption of organic pollutants, heavy metals, and oil spills; filtration.	Özgenç 2024, Rezania 2024, Singh and Upadhyay 2022
Polymeric nanomaterials	Nanomaterials based on polymers with tunable chemical and physical properties.	Controlled release of remediation agents and pollutant trapping.	Muthukumaran 2022, Alipoori 2021
Hybrid nanocomposites	Combinations of metal, metal oxide, and carbon-based nanomaterials for multifunctionality.	Synergistic removal of complex pollutants through combined mechanisms (adsorption, catalysis).	Rando 2022, Ribeiro 2016
Quantum dots	Semiconductor nanoparticles with unique optical and electronic properties.	Detection and degradation of pollutants through photocatalytic activity.	Sun 2023, Zalfani 2016
Nano-scale zero-valent iron (nzvi)	Nanosized particles of elemental iron with high reactivity.	Degradation of chlorinated compounds, heavy metal reduction, and soil remediation.	Ken and Sinha 2020, Jiang 2018
Magnetic nanomaterials	Nanoparticles with magnetic properties, such as magnetite (Fe_3O_4).	Recovery of pollutants and nanoparticles after remediation through magnetic separation.	Fernández 2020

removal (Natarajan et al. 2023), wastewater treatment (Jabbar et al. 2022), oil, organic pollutant, and dye removal from water (Yin et al. 2021), gene therapy and imaging (Yang et al. 2021), chemical nanosensors, and catalysis (Romanovski et al. 2023). Additionally, MNMs can effectively use magnetic separation to remove pollutants from water and other environmental media. Nanomaterials have unique characteristics, such as superparamagnetism, minuscule size, high recyclability, environmental friendliness, high stability, low cytotoxicity, and substantial adsorption capacity, which can be used to treat environmental pollution (Behera and Behera 2022). The environment is seriously threatened by groundwater contamination, which is a major concern for both the government and businesses. MNMs can improve their adsorption capability for particular chemical contaminants, ranging from metals to organics by adding functional groups like amino, carboxyl, or hydroxyl. A magnetic nanocomposite based on FeO_4 nanoparticles functionalized with mesoporous silica was created by Losito et al. (2021). This composite was shown to effectively remove toxic dyes from wastewater through adsorption, with the added benefit of being recoverable using a magnetic field, thus preventing secondary pollution. The ease of recovery of MNPs minimizes the risk of further environmental contamination, making them highly efficient and environmentally friendly for repeated use.

2.2.2 Nano-catalysts

Environmental remediation is one of the diverse disciplines that has seen a significant increase in interest in catalytically powered nanomotors (Table 1) in recent years. Orozco et al. (2013) reported on artificial enzymatic micromotors for water quality evaluation. A previous work by Naeem et al. (2021) found that the inner enzymatic catalase layer of the microtube produces oxygen bubbles as a result of the hydrogen peroxide fuel's biocatalytic breakdown. Catalytic micro/nanomotors can also be used as pH sensors by using their pH-taxis capabilities (He 2024). Guix et al. (2018) recently disclosed a catalytic, self-propelled polymeric microsphere with randomly dispersed Pd nanoparticles (NPs) that display pH-taxis behavior. A range of organic and inorganic pollutants can be efficiently degraded by nanoparticles made from noble metals like Au, Pt, and Pd (Mishra and Parida 2019). By reducing treatment time and converting waste into useful byproducts, there are some benefits of using a nano-catalyst in chemical oxidation as opposed to traditional treatment methods (Choudhary et al. 2024, Velusamy et al. 2021). Hildebrand et al. (2009) suggest that Pd may provide commercially applicable pathways for pollutants like chloro hydrocarbons to be selectively removed from wastewater.

Nano-catalysts are widely used in environmental remediation due to their exceptional catalytic properties (Sheerazi and Ahmed 2022). Nanomaterials, due to their high surface area and small size, offer more catalytic domain, which significantly enhances the degradation or transformation of pollutants. One of the most common applications of nanocatalysts is in photocatalysis, where nanoparticles, especially titanium dioxide (TiO_2), are used to target organics under UV light exposure (Moradeeya et al. 2022, Waghmode et al. 2019). A study by Zhang et al. (2019) demonstrated the use of TiO_2 nanoparticles to detoxify organic pollutants with high photocatalytic activities in wastewater (Zhang and Rhim 2022). Since it has high reactivity under UV light ($k < 390$ nm) and chemical stability, TiO_2 is one of the most commonly used materials in photocatalysis (Schneider et al. 2014). Similar to TiO_2, ZnO has also been the subject of much research due to its broad band gap and photocatalytic activity (Pasang et al. 2015). The photocatalytic activity of different synthesized catalysts has been demonstrated in numerous investigations. Under UV irradiation, TiO_2 nanoparticles generated reactive oxygen species (ROS), such as hydroxyl radicals, which efficiently broke down organic pollutants like phenol and methylene blue into non-toxic byproducts. Nano-scaled detoxifying agents include Fenton-based catalysts (Qi et al. 2023), photocatalysts (Jaleh et al. 2021), and electrocatalysts (Fawy et al. 2025), are preferred treatment for wastewater remediation to enhance the antimicrobial and chemical oxidation of organic pollutants (Ghazal et al. 2022). Photocatalysis has been shown to be a promising technique for treating and purifying various wastewater types (Shokri and Sanavi 2022). Additionally, it efficiently renders bacteria and other hazardous organisms in wastewater inert. Because nanomaterials have a greater surface area and ensure a uniform dispersion of the catalyst in the reaction solution, it has been claimed that utilizing them as electrocatalysts greatly increases fuel cell efficiency. Many studies have been conducted on the development of carbon-supported nanocatalysts for fuel cell applications (Chaturvedi et al. 2012). Among these, palladium nanoparticles (PdNPs) functionalized with single-walled carbon nanotubes have demonstrated potent electrocatalytic activity, making them suitable as supporting nanocarbon materials for the production of nanocomposites. Palladium can also be combined with six different carbon nanostructures to generate composites, each of which has unique characteristics and offers a variety of options for enhancing fuel cell performance.

2.2.3 Carbon Nanomaterials

Carbonaceous nanoparticles are distinguished by their wide surface area, high microporosity, superior sorption capabilities, and environmentally favorable nature. Environmental remediation uses carbon-based nanomaterials like fullerene C60, SWCNTs, MWCNTs, graphene, and activated carbon extensively (Gaganpreet and Pathania 2024). Because of their high surface area and functional groups like –COOH and –OH, carbon nanotubes (CNTs) have a superior adsorption capacity.

Functionalization with groups such as –NH2, –SH, and magnetic coatings further enhances their effectiveness. Stabilized by surfactants like poloxamer 407, CNTs remove heavy metals (e.g., Pb^{2+}, Cu^{2+}, Ni^{2+}, Zn^{2+}) and contaminants like petroleum hydrocarbons and pesticides, improving soil health and plant growth (Chen et al. 2015). Their applications include membrane-like filtration column-based separation, and aqueous dispersions. Carbonaceous nanostructures, including CNTs and GO, are becoming increasingly popular because of their exceptional adsorption capabilities, high surface area, and distinctive structural characteristics (Majumder and Gangopadhyay 2022). Heavy metals, chemical compounds, and even gases are among the many contaminants that these materials can absorb. Their effectiveness in environmental cleanup procedures can be increased by adding functional groups to their surface to improve their interaction with particular contaminants. Graphene oxide nanosheets were investigated for the adsorption of the dissolved heavy-metal species such as Cd^{2+} and Pb^{2+} (Chintalapudi and Pannem 2021). According to the study, GO's large surface area and the presence of functional groups like hydroxyl and carboxyl groups that promoted strong interactions with metal ions allowed it to adsorb heavy metals with a high capacity. Similarly, carbon nanotubes (CNTs) have been used to remove organic pollutants like dyes and medications. Several studies have demonstrated that CNTs are superior to conventional adsorbents in wastewater treatment because they can adsorb contaminants at concentrations significantly higher than those of conventional adsorbents.

2.2.4 Polymeric Nanomaterials

Polymeric nanomaterials are engineered to change their physical or chemical properties in response to environmental conditions like pH, temperature, or ionic strength. This responsiveness makes them highly suitable for targeted pollutant removal, particularly in scenarios where contaminants exhibit pH-dependent behavior or undergo chemical transformations in response to changes in their environment. Polymer nanotechnology offers solutions for water treatment through nanoparticles, nanocomposites, and NF membranes. Polymeric nanomembranes, such as chitosan-based NF membranes, immobilize pollutants via pore detouring and chemical interactions, effectively treating textile wastewater. Efficiency depends on the dye charge and size. Enhancements include incorporating materials like triethanolamine, metal oxide nanoparticles, and MWCNTs. For instance, carboxylated MWCNTs in polyamide membranes improve salt rejection, while polyethersulfone membranes with MWCNTs or graphene efficiently remove heavy metals and dyes from water. Yang et al. (2020) developed a pH-sensitive polymer nanocomposite that could selectively remove heavy metals from industrial wastewater. At acidic pH, the polymer nanoparticles exhibited enhanced adsorption of metal ions (Tan et al. 2017), while in neutral or alkaline conditions, the adsorption capacity was significantly reduced (Scrimieri et al. 2020). This ability to switch between different states depending on pH allows for efficient and controlled removal of contaminants, minimizing the need for extensive treatment processes. Polymeric nanomaterials can also be designed to change their morphology, such as swelling or shrinking, in response to temperature or ionic strength, providing an additional level of control over the remediation process (Khin et al. 2012, Alam et al. 2018). These materials can be used for more complex contaminant removal tasks, like elimination of organic contaminant or the recovery of rare earth metals from wastewater, where specific environmental conditions play a critical role in the material's efficiency.

3. Nanoremediation of Air Pollution

Air pollution is described as alterations in the atmosphere's composition brought on by physical, chemical, or biological pollutants that come from geogenic, biogenic, or human sources. Carbon monoxide, sulfur dioxide, and ozone are examples of gaseous pollutants, whereas particulate pollutants include aerosols from bacteria, viruses, and fungi (Figure 1). According to Aydın and İlkılıç (2017), the release of various pollutants, such as CO, NOx, SO_2, VOCs, and CO_2, is one of the detrimental ecological effects. Similarly, as per Li and Luo (2023), the main source of dangerous gases, including carbon monoxide and volatile organic compounds, is gas waste from industry.

Nanotechnology is regarded as a cutting-edge treatment method that uses the characteristics of nanomaterials as membranes/nanofilters, sensors, nanocatalysts, and nanoadsorbents to control and remediate air pollution in a number of ways (Amrane et al. 2020, Handojo et al. 2019, Nguyen-Tri et al. 2020). Nanoparticles are capable of absorbing a variety of airborne pollutants. To find and absorb pollutants like metal oxides, microbes, and volatile organic compounds (VOCs), a wide variety of nanoadsorbents have been created and categorized. To eliminate volatile organic compounds from industrial pollutants, nanocatalysts such as manganese oxide nanoparticles (NPs) are employed (Liu et al. 2024).

Gold NPs (Table 1) show promise in converting toxic carbon monoxide (CO) into carbon dioxide (CO_2) through a quantum size effect (Soleimani and Pourfath 2023). Additionally, nanostructured membranes, like those made from single-walled carbon nanotubes (SWCNTs), efficiently separate gases like methane or CO_2 (Salehi et al. 2020). These nanotubes, with their abundant pores and high surface area, are particularly effective in selective adsorption and desorption processes, making them ideal for air purification (Gulati et al. 2022, Liapun and Motola 2024). Silica–titania nanocomposites have been developed for efficient mercury vapor removal (Pitoniak et al. 2005). By combining the properties of nanosilica (high surface are) with titania (photocatalysis), these composites achieve superior mercury absorption and reduce the contact angle significantly, enhancing efficiency. Similarly, metal–organic frameworks (MOFs) have been utilized for cost-effective and selective CO_2 capture, offering 2–3 times greater efficiency than traditional sorbents. MOFs, can release CO_2 under vacuum for safe storage as carbonate minerals deep within the Earth. These materials are synthesized under mild conditions from inexpensive sources.

4. Nanoremediation (Wastewater and Industrial Effluent)

The excessive levels of inorganic and organic pollutants have severely impacted groundwater's quality, content, and usage. The increasing use of natural water bodies as disposal sites is further endangering groundwater quality. Furthermore, excessive use of fertilizers, weed killers, and pesticides poses a major threat to the quality and composition of groundwater (Nath et al. 2023). Nanoremediation has emerged as a promising alternative to conventional remediation techniques for groundwater decontamination. It has been successfully utilized to remove a variety of contaminants, including pesticides, hydrocarbons, heavy metals, chlorinated chemicals, organic compounds, and inorganic ions (Akhtar et al. 2023, Faisal 2024). *In situ* nanoremediation techniques for wastewater, groundwater, and water treatment employ nano-adsorbents, membrane systems based on nanocomposite technology, and nanocatalysts. Heavy metals, organic pollutants, and antibiotic-resistant microorganisms can be swiftly removed because of nanomaterials' excellent membrane permeability and efficient adsorption capabilities.

Nanomaterials have revolutionized wastewater treatment by offering high flexibility and efficiency for removing specific pollutants, including nitrogen, phosphorus, heavy metals, and chlorinated compounds. Nanoparticles (NPs) are versatile, enabling both *in situ* and *ex situ* purification. They can be utilized in slurry reactors for soil and sediment treatment or anchored on solid matrices like carbon, zeolite, or membranes for enhanced water purification. Zero-valent iron (Fe^0) NPs are prominent in environmental remediation, effectively breaking down organic pollutants like trichloroethylene (TCE) and dioxins into less toxic compounds (Nguyen et al. 2025). They reduce and immobilize heavy metals such as Cr (VI) and Pb (II) and remove nitrates and arsenic from water. Iron sulfide (FeS) NPs degrade organic pollutants like lindane, while core–shell iron NPs efficiently remove ions like Ni (II) and Ba^{2+} (Ahmad et al. 2024). Furthermore, Fe^0 NPs degrade chlorinated and polychlorinated organic compounds. Bimetallic NPs enhance the activity and stability of Fe^0 NPs, demonstrating superior performance in pollutant removal. These developments demonstrate how well nanoremediation methods can handle wastewater and groundwater contamination.

Photocatalytic nanomaterials additionally break down hazardous algal blooms, preserving the water's inherent cleanliness. Ni-doped $BiVO_4$ nanocomposites, for instance, have been demonstrated to effectively prevent the development and dissemination of algal blooms (Seema et al. 2025). The reduction of chlorinated chemicals is possible with nickel/iron nanoparticles (Wei et al. 2025, Venkateshaiah et al. 2025). Moreover, palladium NPs may remove chlorinated substances from water when supported on gold NPs (Mendes et al. 2021). By using a porous titanium silicate and alumina nanocomposite (Al_2O_3/TiO_2), heavy metals, especially Pb^{2+} and Cd^{2+}, can be eliminated (Adam 2024). These NPs, which are fixed to solid supports, demonstrate the revolutionary effect of nanoparticles in wastewater treatment by facilitating the effective *ex situ* treatment of polluted water and industrial effluents. Thus, the ability of TiO_2 NPs to filter water is currently the subject of extensive research. Through the photocatalytic activity of both the rutile and anatase forms of TiO_2 NPs, they are utilized in the wet oxidation process to remove harmful phenol pollution from wastewater (Nawaz et al. 2021). The creation of a composite reactor made of TiO_2 NPs and a UV light was a next-level breakthrough in this sector (Tawiah et al. 2024). Photocatalyst nano-TiO_2 molecules are excited by a UV lamp to create electron–hole pairs. Peng et al. are also developing mesoporous TiO_2 molecules for the elimination of rhodamine (Erdem et al. 2023). Organic pollutants like catechol is remediated with the combination of Titanium oxide-thin films and alumina (Li et al. 2025).

The elimination of benzene and toluene from carbonized bamboo (Phyllostachys pubescens) covered with TiO_2 NPs is another significant advancement in pollution control. Initially, Diallo et al. investigated the possibility of using polyamidoamine (PAMAM) dendrimers to extract copper from water (Yasir et al. 2023).

The removal and enrichment of harmful heavy metal ions could be possible (Ghosh and Sarkar 2022). For the removal of heavy metals including Pb, Cu, Co, Cd, Zn, Mn, and others, carbon nanotubes (CNTs) exhibit outstanding adsorption capabilities (Adam et al. 2021). Li et al. discovered that oxidized carbon nanotubes (CNTs) have a higher capacity for adsorbing cadmium (II) than regular CNTs (Li et al. 2024).

Aslam et al. (2021) reported sodium hypochlorite solutions to purify commercial SWCNTs and multi-walled carbon nanotubes (MWCNTs), which were then employed as adsorbents to remove zinc from water. In order to remove Cr (VI) from wastewater quickly, Hu et al. created surface-functionalized $MnFe_2O_4$ NPs as a new adsorbent (Gupta and Saxena 2024). Through the use of novel membranes, such as the reverse osmosis nanomembrane developed by the University of

California, Los Angeles (UCLA), nanotechnology is not only utilized to treat wastewater but also to improve the purification of drinking water. This membrane, which is composed of embedded nanoparticles and a cross-linked polymer matrix, detoxifies wastewater and desalinates saltwater by letting water molecules pass through while keeping bacteria and organics out. Compared to traditional RO membranes, its anti-clogging qualities provide a considerable cost advantage by prolonging the membrane's lifespan.

Nanoremediation of Soil: The transition from hunter-gatherer societies to farming and the subsequent rise of industrialization and urbanization have led to significant soil degradation, including erosion, contamination, and biodiversity loss. Nanoremediation has emerged as an effective solution to address soil contamination due to the unique properties of nanomaterials, such as high surface area, reactivity, and customizable physical and chemical traits. The use of nanoparticles, including metallic and magnetic nanoparticles, carbon nanotubes, and nanoscale zero-valent iron (nZVI), allows for selective pollutant removal, cost-effective *in situ* applications, and large-scale soil cleanup (Liang et al. 2022). These advanced materials enhance soil remediation efficiency while minimizing environmental impact. Twenty years ago, the first reported field-scale application of NPs for soil and groundwater remediation was placed. It was found that NPs could travel over 20 meters with groundwater and stay active in implanted soil for almost two months (Dibyanshu et al. 2023). In excess of 99 percent of trichloroethene (TCE) is capable of being eliminated from contaminated areas in a matter of days, according to Dutta et al. (2022). Because of its quick kinetics, affordability, non-toxicity, and minimal environmental impact, ZVI insertion is a viable option for soil restoration (Namakka et al. 2024). The main contaminants that ZVI was used to remove from water were halo-generated organic compounds, which usually contaminate soil and groundwater. Further, ferritin's ability to remediate chlorocarbons and groundwater contamination from nuclear waste leakage (Thakur and Kumar 2024) can indirectly address soil pollution by preventing the leaching of toxic substances from groundwater into surrounding soil, reducing long-term contamination risks. The nitrogen-rich branches of PAMAM dendrimers have been reported to allow chelation of metal ions such as copper, which can be applied to polluted soils for immobilizing or extracting heavy metals, thereby mitigating soil toxicity. Polymeric nanoparticles like UAA and PMUA are effective in breaking down hydrophobic organic compounds and polycyclic aromatic hydrocarbons (Kathalingam et al. 2023). These pollutants, often found in contaminated soils near industrial sites, can be remediated through *in situ* applications, reducing soil contamination levels. Carbon nanotubes (CNTs), single-walled or multi-walled, provide excellent adsorption capacities because of their affinity for pollutant functional groups. They have been studied in great detail for groundwater and soil remediation. Apul et al. assessed the effectiveness of CNTs with microwave assistance in eliminating total petroleum hydrocarbons (TPH) from soil (Alazaiza et al. 2021). The elimination efficiency of TPH was 82% after 60 seconds of microwave treatment, according to the results (Jagaba et al. 2022). Single-walled carbon nanotubes (SWCNTs) and multi-walled carbon nanotubes (MWCNTs) have been studied by Rudyak et al. (2022), for their ability to reduce dichlorobiphenyls-chloroethane (DDT) and hexachlorocyclohexane (HCH). The outcome shows that DDTs and HCHs might be successfully treated by CNTs. Doses of 0.29 weight percent were the most effective for the SWCNTs throughout a four-month period.

Evaluating metal nanoparticles for soil remediation has been the subject of several recent investigations. AG et al. (2025) made one such attempt, investigating the use of SiO_2 NPs to remediate Zn, Ni and Cd. The results demonstrated that the calcareous and non-calcareous soils experienced the greatest Cd reductions at 3% SiO_2 (56.1%) and 1% Al_2O_3 (38.3%), (NAFEES et al. 2025) respectively.

The effectiveness of iron phosphate (biochar-assisted) nanoparticles for the removal of Cd from polluted soil has been studied by Wang et al. (2022), in a previous work. Heavy metals and radioactive substances can be effectively removed by zeolites of both micron and nanosize

(Liu et al. 2021). When applied to contaminated soil, they can stabilize or extract these pollutants, preventing further leaching or environmental damage. Although primarily designed for self-cleaning surfaces, the photocatalytic properties of TiO_2 nanocrystals can be adapted for soil remediation by breaking down organic contaminants on the soil surface when exposed to light. Nanowire mats developed for oil spill remediation can also be applied to soils contaminated with hydrocarbons, such as those affected by industrial spills. Their high absorption capacity and reusability make them a sustainable option for soil decontamination.

5. Advantages of Smart Nanomaterials over Conventional Methods

Smart nanomaterials have a number of advantages over traditional remediation techniques, including increased efficiency and quicker action. By accelerating the destruction of contaminants through processes like nanocatalysis, they greatly increase efficacy and shorten remediation times. Because of their specificity and selective targeting, they can interact with particular contaminants—like organic pollutants or heavy metals—without upsetting the ecosystem. Furthermore, by converting hazardous elements into non-toxic chemicals like carbon dioxide and water and avoiding byproducts frequently connected with conventional techniques, nanoparticles reduce secondary contamination. Long-term cost-effectiveness and sustainability are facilitated by their reusability and multifunctionality, while their high surface area and reactivity improve interaction with contaminants. Economic viability is ensured by nanoparticles' capacity to treat higher pollutant volumes with fewer quantities, even in the face of higher initial production costs. Additionally, scalability has been enhanced by nanotechnology breakthroughs, which makes these materials appropriate for extensive use in soil, water, and air remediation. Smart nanomaterials often exhibit faster rates of contamination removal compared to conventional methods. Smart nanomaterials are categorized in different groups by means of the applied stimuli. By considering their various properties, many kinds of smart nanomaterials are known. Generally, these stimuli fall into one of three categories: chemical, biological, or physical. The time needed for cleanup procedures can be decreased by several orders of magnitude thanks to nanomaterials like nanocatalysts, which can speed up the breakdown of organic contaminants. On the other hand, it may take months or even years for biological techniques like bioremediation to produce comparable outcomes.

5.1 Selective Targeting and Specificity Against Traditional Methods

The efficiency of treating contaminated sites can be increased by designing smart nanomaterials that target particular contaminants. Chemical precipitation, coagulation-flocculation (Figure 3b), ion exchange, and membrane filtration are examples of traditional heavy metal cleanup techniques that have drawbacks that reduce their efficacy. These techniques frequently produce hazardous byproducts like sludge and remove heavy metals and critical ions with little selectivity (Zamora-Ledezma et al. 2021). Coagulation-flocculation operations, for example, might leave coagulant metal residues in the treated water and produce large amounts of chemical sludge (Ayach et al. 2024). Furthermore, a number of problems, such as elevated turbidity, decreased disinfection effectiveness, decreased hydraulic capacity, and possible negative health effects like Alzheimer's disease, are linked to residual coagulant metals like aluminum (Zinicovscaia et al. 2021). Furthermore, for large-scale water treatment applications, conventional heavy metal remediation techniques may be prohibitively expensive and difficult to scale up (Olawade et al. 2024).

Achieving selective adsorption, regeneration capability, scalability, and multifunctionality are the main goals of recent developments in nanomaterial-based techniques (Yang et al. 2019, Kolluru et al. 2021). By selectively adsorbing particular heavy metal ions, functionalized nanomaterials with customized surface characteristics reduce interference from other ions and improve treatment effectiveness (Mensah et al. 2021).

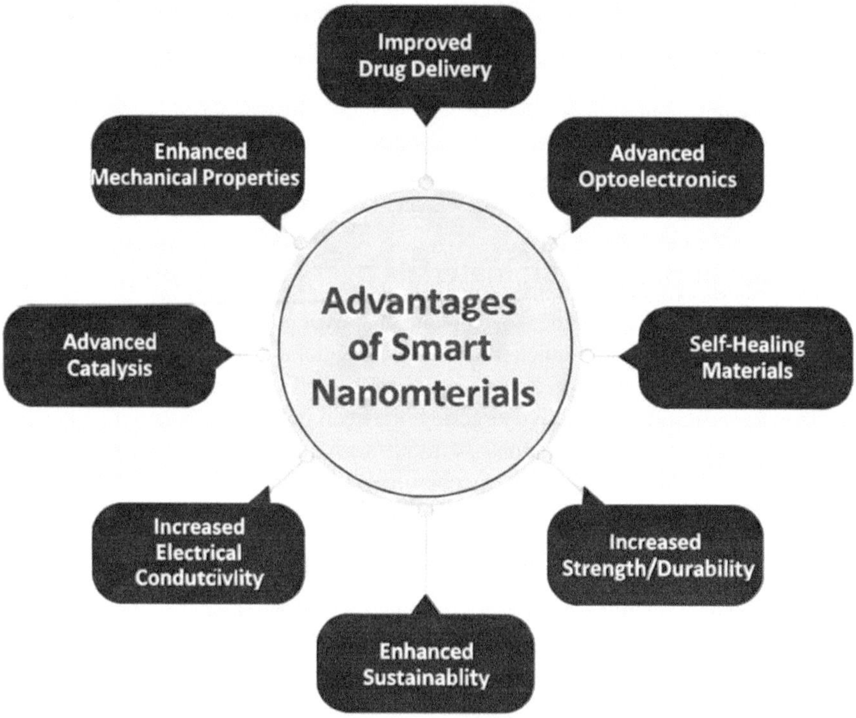

Figure 3b: Advantage of smart nanomaterials.

5.2 Minimization of Secondary Pollution

Unlike some conventional methods that produce harmful byproducts or generate waste during the remediation process, smart nanomaterials can often break down pollutants into non-toxic substances or transform them into harmless compounds. Air pollutants like NO(x), SO_2, highly reactive and toxic organic compounds, persistent organic pollutants (POPs) like dioxins, and PAHs (polycyclic aromatic hydrocarbons) are all hazardous for mankind (Akhtar et al. 2021). When inhaled in high quantities, carbon monoxide (CO) can cause immediate poisoning (Savioli et al. 2024). When certain heavy metals such as Pb are taken into a living organism, they can cause either immediate or chronic poisoning, depending on the different levels of exposure (Collin et al. 2022). Nanotechnology possesses a prospect to significantly contribute to the development of cleaner, greener technologies that have major environmental and health advantages (Chakrabartty et al. 2022). For the elimination of environmental pollutants, the environmental remediation process employs a variety of methods such as surface assimilation, absorption, synthetic reactions, catalysis using light, and filtering (Roy et al. 2021). Nanomaterials possess a high surface-area-to-volume ratio, frequently resulting in greater biogeological and chemical reactivity compared to other macroscale materials. Nanoscale materials are being used in a wide array of fields, including science, the environment, industry, and medicine.

5.3 Sustainability and Cost-Effectiveness

Persistent organic pollutants in the environment eventually enter the food chain and have a major detrimental effect on humans since they take a long time to decompose (Krithiga et al. 2022). Contamination in the environment is increasing daily due to the inability of present treatment procedures to fully break down resistant organic compounds. By creating more clever and environmentally friendly methods, researchers have been collaborating with governmental and non-governmental organizations to lower rising pollution levels (Hung et al. 2022, Rajabi et al. 2021). The photocatalytic process uses reactive hydroxyl and superoxide radicals to convert hazardous materials into environmentally benign products. Electron-hole pairs are created in the valence and conduction bands of a photocatalyst when photon absorption sets off a redox reaction on its surface. These combinations produce reactive species that can break down air pollutants including NO_2, CO, and NH_3, oxidize organic pollutants, sterilize aquatic microbes, and degrade plastics (Singh et al. 2021). It is an effective method for mineralizing hazardous substances, inorganic materials, and microbial disinfection due to the highly reactive radicals that are produced (Chen et al. 2021).

Cost control can not only reduce the cost of engineering construction but also minimize unnecessary waste. Smart nanomaterials can be reused multiple times, significantly reducing the costs associated with material disposal or replenishment (Figure 3b). Their high surface area and reactivity also allow for the treatment of large volumes of contaminated material with relatively small amounts of nanomaterial, making them more cost-effective in the long term compared to conventional methods.

Recommendation and Future Prospective: This chapter outlines how nanoremediation is being used in environmental cleanup, emphasizing both its promise and the challenges that still need to be addressed. Progress in smart nanomaterials is especially important for soil remediation, where nanoparticles (NPs) with multifunctional abilities are required—those that can interact with both hydrophilic and hydrophobic substances or catalyze different pollutant reactions at the same time. Continued research is essential to develop NPs capable of treating a broad range of contaminants and to enhance injection systems so they can be applied more effectively in real-world conditions.

Despite their promising potential, smart nanomaterials face several challenges that must be addressed before widespread adoption. These include ensuring the safety of nanomaterials in different environmental contexts and overcoming the limitations of current. To enable field-level application, it is crucial to develop standard protocols and dosing strategies while conducting extensive studies on the fate, transport behavior, and environmental impact of nanoparticles in soil, water, and sediments.

6. Conclusion

In conclusion, smart nanomaterials represent a substantial advancement over conventional methods for environmental remediation. Their unique properties—such as high surface area, reactivity, and environmental responsiveness—make them highly effective for addressing a wide range of contaminants. However, challenges related to safety, scalability, and regulation must be addressed to facilitate their practical use. As research progresses, the role of smart nanomaterials in environmental remediation is expected to expand, offering sustainable and efficient solutions to the growing environmental crisis.

References

Adam, A. M., Saad, H. A., Atta, A. A., Alsawat, M., Hegab, M. S., Altalhi, T. A. et al. (2021). An environmentally friendly method for removing Hg (II), Pb (II), Cd (II) and Sn (II) heavy metals from wastewater using novel metal–carbon-based composites. *Crystals*, 11(8): 882.

Adam, F. A. (2024). Purification of Water Contaminated with Heavy Metals, Exemplified by Lead Cations, Utilizing Fe2O3@ Al2O3 Nanocomposites. *Egyptian Journal of Chemistry*, 67(3): 457–466.

AG, M., Kaleeswari, R. K., and S, M. (2025). Innovative Nanoremediation Techniques for Soil Contamination: Exploring Metal Based Nanomaterial Approaches–a Review. *Soil and Sediment Contamination: An International Journal*, 1–69.

Ahmad, I., Aftab, M. A., Fatima, A., Mekkey, S. D., Melhi, S., and Ikram, S. (2024). A comprehensive review on the advancement of transition metals incorporated on functional magnetic nanocomposites for the catalytic reduction and photocatalytic degradation of organic pollutants. *Coordination Chemistry Reviews*, 514: 215904.

Ahmad, M. A., Adeel, M., Shakoor, N., Javed, R., Ishfaq, M., Peng, Y. et al. (2023). Modifying engineered nanomaterials to produce next generation agents for environmental remediation. *Science of the Total Environment*, 894: 164861.

Akhtar, A. B. T., Naseem, S., Yasar, A., and Naseem, Z. (2021). Persistent organic pollutants (POPs): sources, types, impacts, and their remediation. *Environmental Pollution and Remediation*, 213–246.

Akhtar, A. B. T., Naseem, S., Yasar, A., and Naseem, Z. (2021). Persistent organic pollutants (POPs): sources, types, impacts, and their remediation. *Environmental Pollution and Remediation*, 213–246.

Alam, A., Zhang, Y., Kuan, H. C., Lee, S. H., and Ma, J. (2018). Polymer composite hydrogels containing carbon nanomaterials—Morphology and mechanical and functional performance. *Progress in Polymer Science*, 77: 1–18.

Alazaiza, M. Y., Albahnasawi, A., Ali, G. A., Bashir, M. J., Copty, N. K., Amr, S. S. A. et al. (2021). Recent advances of nanoremediation technologies for soil and groundwater remediation: A review. *Water*, 13(16): 2186.

Ansari, K., Ahmad, R., Tanweer, M. S., and Azam, I. (2024). Magnetic iron oxide nanoparticles as a tool for the advancement of biomedical and environmental application: a review. *Biomedical Materials & Devices*, 2(1): 139–157.

Aslam, M. M. A., Kuo, H. W., Den, W., Usman, M., Sultan, M., and Ashraf, H. (2021). Functionalized carbon nanotubes (CNTs) for water and wastewater treatment: preparation to application. *Sustainability*, 13(10): 5717.

Ayach, J., El Malti, W., Duma, L., Lalevée, J., Al Ajami, M., Hamad, H. et al. (2024). Comparing conventional and advanced approaches for heavy metal removal in wastewater treatment: an in-depth review emphasizing filter-based strategies. *Polymers*, 16(14): 1959.

Aydın, H., and İlkılıç, C. (2017). Air pollution, pollutant emissions and harmfull effects. *Journal of Engineering and Technology*, 1(1): 8–15.

Behera, A., and Behera, A. (2022). Nanomaterials. *Advanced Materials: An Introduction To Modern Materials Science*, 77–125.

Chakrabartty, I., Hakeem, K. R., Mohanta, Y. K., and Varma, R. S. (2022). Greener nanomaterials and their diverse applications in the energy sector. *Clean Technologies and Environmental Policy*, 24(10): 3237–3252.

Chan, M. (2018). Ten years in public health 2007–2017: report by dr margaret chan director-general world health organization. World Health Organization.

Chen, G. Q., Wu, Y. H., Wang, Y. H., Chen, Z., Tong, X., Bai, Y. et al. (2021). Effects of microbial inactivation approaches on quantity and properties of extracellular polymeric substances in the process of wastewater treatment and reclamation: A review. *Journal of Hazardous Materials*, 413: 125283.

Chen, M., Xu, P., Zeng, G., Yang, C., Huang, D., and Zhang, J. (2015). Bioremediation of soils contaminated with polycyclic aromatic hydrocarbons, petroleum, pesticides, chlorophenols and heavy metals by composting: applications, microbes and future research needs. *Biotechnology Advances*, 33(6): 745–755.

Chiappisi, L., Simon, M., and Gradzielski, M. (2015). Toward bioderived intelligent nanocarriers for controlled pollutant recovery and pH-sensitive binding. *ACS Applied Materials & Interfaces*, 7(11): 6139–6145.

Chintalapudi, K., and Pannem, R. M. R. (2021). Enhanced Strength, Microstructure, and Thermal properties of Portland Pozzolana Fly ash-based cement composites by reinforcing Graphene Oxide nanosheets. *Journal of Building Engineering*, 42: 102521.

Choudhary, S. (2024). Role of Nano Catalysts In Green Chemistry. *Journal of Science Innovations and Nature of Earth*, 4(4): 08–11.

Collin, S., Baskar, A., Geevarghese, D. M., Ali, M. N. V. S., Bahubali, P., Choudhary, R. et al. (2022). Bioaccumulation of lead (Pb) and its effects in plants: A review. *Journal of Hazardous Materials Letters*, 3: 100064.

Comanescu, C. (2023). Recent advances in surface functionalization of magnetic nanoparticles. *Coatings*, 13(10): 1772.

Dibyanshu, K., Chhaya, T., and Raychoudhury, T. (2023). A review on the fate and transport behavior of engineered nanoparticles: Possibility of becoming an emerging contaminant in the groundwater. *International Journal of Environmental Science and Technology*, 20(4): 4649–4672.

Dutta, N., Usman, M., Ashraf, M. A., Luo, G., and Zhang, S. (2022). Efficacy of emerging technologies in addressing reductive dechlorination for environmental bioremediation: A review. *Journal of Hazardous Materials Letters*, 3: 100065.

Erdem, B., Sevinç, S., Erdem, S., and Öksüzoğlu, R. M. (2023). Comparative influence of adsorption assisted magnetic mesoporous TiO2 photocatalyst for the removal of methylene blue and rhodamine B. *Reaction Kinetics, Mechanisms and Catalysis*, 136(2): 1049–1065.

Faisal, t. (2024). A review of organic and inorganic chemicals, their sources, methods of mobilization and attenuation in groundwater. *Nigerian Journal of Water Resources*, 10(1): 67–80.

Fawy, K. F., Rodriguez-Ortiz, G., Ali, A., Jadeja, Y., Khan, H., Pathak, P. K. et al. (2025). Catalytic exploration metallic and nonmetallic nano-catalysts, properties, role in photoelectrochemistry for sustainable applications. *Reviews in Inorganic Chemistry*, (0).

Gaganpreet, and Pathania, Y. (2024). Carbon-Based Nanomaterials for Pollutants' Treatment. *Carbon-based Nanomaterials for Green Applications*, 355–381.

Ghanbari, F., and Moradi, M. (2017). Application of peroxymonosulfate and its activation methods for degradation of environmental organic pollutants. *Chemical Engineering Journal*, 310: 41–62.

Ghazal, H., Koumaki, E., Hoslett, J., Malamis, S., Katsou, E., Barcelo, D. et al. (2022). Insights into current physical, chemical and hybrid technologies used for the treatment of wastewater contaminated with pharmaceuticals. *Journal of Cleaner Production*, 361: 132079.

Ghosh, S., and Sarkar, B. (2022). Nanotechnology for advanced oxidation based water treatment processes. pp. 79–91. *In: Advanced Oxidation Processes for Wastewater Treatment*. CRC Press.

Guix, M., Weiz, S. M., Schmidt, O. G., and Medina-Sánchez, M. (2018). Self-propelled micro/nanoparticle motors. *Particle & Particle Systems Characterization*, 35(2): 1700382.

Gulati, S., Kumar, S., Goyal, K., Arora, A., and Varma, R. S. (2022). Improving the air quality with functionalized carbon nanotubes: sensing and remediation applications in the real world. *Chemosphere*, 299: 134468.

Gupta, K., and Saxena, R. (2024). Role of Nanomaterials in Remediating Environmental Pollutants. pp. 394–424. *In: Nanotechnology: A Quick Guide to Materials and Technologies*. Bentham Science Publishers.

Hansen, 2007Obiri-Nyarko, F., Grajales-Mesa, S. J., and Malina, G. (2014). An overview of permeable reactive barriers for in situ sustainable groundwater remediation. *Chemosphere*, 111: 243–259.

He, X., Zhang, H., Shi, H., Liu, W., and Sahle-Demessie, E. (2020). Fates of Au, Ag, ZnO, and CeO_2 nanoparticles in simulated gastric fluid studied using single-particle-inductively coupled plasma-mass spectrometry. *Journal of the American Society for Mass Spectrometry*, 31(10): 2180–2190.

Hildebrand, H., Mackenzie, K., and Kopinke, F. D. (2009). Pd/Fe3O4 nano-catalysts for selective dehalogenation in wastewater treatment processes—Influence of water constituents. *Applied Catalysis B: Environmental*, 91(1-2): 389–396.

Hung, L. Y., Wang, S. M., and Yeh, T. K. (2022). Collaboration between the government and environmental non-governmental organisations for marine debris policy development: The Taiwan experience. *Marine Policy*, 135: 104849.

Jabbar, K. Q., Barzinjy, A. A., and Hamad, S. M. (2022). Iron oxide nanoparticles: Preparation methods, functions, adsorption and coagulation/flocculation in wastewater treatment. *Environmental Nanotechnology, Monitoring & Management*, 17: 100661.

Jagaba, A. H., Kutty, S. R. M., Lawal, I. M., Aminu, N., Noor, A., Al-dhawi, B. N. S. et al. (2022). Diverse sustainable materials for the treatment of petroleum sludge and remediation of contaminated sites: A review. *Cleaner Waste Systems*, *2*: 100010.

Jaleh, B., Nasrollahzadeh, M., Nasri, A., Eslamipanah, M., Moradi, A., and Nezafat, Z. (2021). Biopolymer-derived (nano) catalysts for hydrogen evolution via hydrolysis of hydrides and electrochemical and photocatalytic techniques: A review. *International Journal of Biological Macromolecules*, 182: 1056–1090.

Kathalingam, A., Santhoshkumar, P., Ramesh, S., Sivanesan, I., and Kim, H. S. (2023). Biogenic polymer nanoparticles to remove hydrophobic organic contaminants from water. *Water Environment Research*, 95(10): e10935.

Khan, Y., Sadia, H., Ali Shah, S. Z., Khan, M. N., Shah, A. A., Ullah, N. et al. (2022). Classification, synthetic, and characterization approaches to nanoparticles, and their applications in various fields of nanotechnology: A review. *Catalysts*, 12(11): 1386.

Khin, M. M., Nair, A. S., Babu, V. J., Murugan, R., and Ramakrishna, S. (2012). A review on nanomaterials for environmental remediation. *Energy & Environmental Science*, 5(8): 8075–8109.

Krithiga, T., Sathish, S., Renita, A. A., Prabu, D., Lokesh, S., Geetha, R. et al. (2022). Persistent organic pollutants in water resources: Fate, occurrence, characterization and risk analysis. *Science of The Total Environment*, 831: 154808.

Kumar, M., Rany, M., and Ansari, J. R. (2024). Chemical Methods for Producing Iron Oxide Magnetic Nanoparticles: A Review. *Iran. J. Chem. Chem. Eng.(IJCCE) Review Article Vol*, 43(5).

Li, J., Zhou, L., Jin, J., Ouyang, J., Shuang, M., Liu, D. et al. (2025). Electrodeposition construction of template-synthesized hierarchical porous N, P-co-doped carbon/nano-TiO2/polyaniline composite electrodes for efficient U (VI) removal through hybrid capacitive deionization. *Journal of Environmental Chemical Engineering*, 116868.

Li, K., and Luo, X. (2023). Research progress on catalytic combustion of volatile organic compounds in industrial waste gas. *Catalysts*, 13(2): 268.

Liang, W., Wang, G., Peng, C., Tan, J., Wan, J., Sun, P. et al. (2022). Recent advances of carbon-based nano zero valent iron for heavy metals remediation in soil and water: A critical review. *Journal of Hazardous Materials*, 426: 127993.

Liapun, V., and Motola, M. (2024). Carbon Nanostructures Functionalization for Air Filtration and Purification. *Handbook of Functionalized Carbon Nanostructures: From Synthesis Methods to Applications*, 1821–1858.

Liu, W., Zhang, Z., Yuan, K., Dang, D., Jin, P., Han, X. et al. (2024). Catalytic oxidation degradation of volatile organic compounds (VOCs)–a review. *Reviews in Inorganic Chemistry*, 44(2): 209–229.

Liu, Y., Pang, H., Wang, X., Yu, S., Chen, Z., Zhang, P. et al. (2021). Zeolitic imidazolate framework-based nanomaterials for the capture of heavy metal ions and radionuclides: A review. *Chemical Engineering Journal*, 406: 127139.

Losito, D. W., de Araujo, D. R., Bezzon, V. D. N., Oseliero Filho, P. L., Fonseca, F. L. A., Chagas, C. D. S. et al. (2021). Mesoporous Silica–Fe3O4 nanoparticle composites as potential drug carriers. *ACS Applied Nano Materials*, 4(12): 13363–13378.

Majumder, P., and Gangopadhyay, R. (2022). Evolution of graphene oxide (GO)-based nanohybrid materials with diverse compositions: an overview. *RSC Advances*, 12(9): 5686–5719.

Mendes, L. D., Bernardi, G., Elias, W. C., de Oliveira, D. C., Domingos, J. B., and Carasek, E. (2021). A green approach to DDT degradation and metabolite monitoring in water comparing the hydrodechlorination efficiency of Pd, Au-on-Pd and Cu-on-Pd nanoparticle catalysis. *Science of the Total Environment*, 760: 143403.

Mensah, M. B., Lewis, D. J., Boadi, N. O., and Awudza, J. A. (2021). Heavy metal pollution and the role of inorganic nanomaterials in environmental remediation. *Royal Society open Science*, 8(10): 201485.

Mishra, P., Patnaik, S., and Parida, K. (2019). An overview of recent progress on noble metal modified magnetic Fe 3 O 4 for photocatalytic pollutant degradation and H 2 evolution. *Catalysis Science & Technology*, 9(4): 916–941.

Mohamed, M. V., Ahamed, A. J., Thakur, A., Ahamed, F. M., and Kumar, A. (2024). Facile Green Synthesis of Undoped ZnO and Copper-Doped Zinc Oxide Nanoparticles using Albizia lebbeck leaf extract: Electrochemical Insights and Anti-Corrosion Potential. *Journal of Bio-and Tribo-Corrosion*, 10(4): 113.

Moradeeya, P. G., Sharma, A., Kumar, M. A., and Basha, S. (2022). Titanium dioxide based nanocomposites–current trends and emerging strategies for the photocatalytic.

Naeem, S., Naeem, F., Mujtaba, J., Shukla, A. K., Mitra, S., Huang, G. et al. (2021). *Oxygen Generation Using Catalytic Nano/Micromotors. Micromachines,* 12: 1251.

Nafees, m., akbar, h., and rehman, a. (2024). Remediation technologies for heavy metals in agricultural soils nazish huma khan, fazli zuljalal, fazli wahid. *Restoration of Agriculture Ecosystem, Soil Nutrients and Carbon Dynamics*, 141.

Namakka, M., Rahman, M. R., Said, K. A. B. M., and Muhammad, A. (2024). Insights into micro-and nano-zero valent iron materials: synthesis methods and multifaceted applications. *RSC Advances*, 14(41): 30411–30439.

Natarajan, B., Kannan, P., Rather, J. A., and Sheikh, R. A. (2023). Recent developments in metal nanoparticles functionalized nanocomposite adsorbents for heavy metals removal from wastewaters. *Journal of the Taiwan Institute of Chemical Engineers*, 147: 104942.

Nath, A., Bhuyan, P., Gogoi, N., and Deka, P. (2023). Pesticides and chemical fertilizers: role in soil degradation, groundwater contamination, and human health. pp. 131–160. *In: Xenobiotics in Urban Ecosystems: Sources, Distribution and Health Impacts*. Cham: Springer International Publishing.

Nawaz, R., Kait, C. F., Chia, H. Y., Isa, M. H., Huei, L. W., and Sahrin, N. T. (2021). Synthesis of Black-TiO 2 and manganese-doped TiO 2 nanoparticles and their comparative performance evaluation for photocatalytic removal of phenolic compounds from agro-industrial effluent. *Journal of Nanoparticle Research*, 23: 1–18.

Nguyen, S. H., Vu, N. T., Van Nguyen, H., Nguyen, B., and Luong, T. T. (2025). Biologically synthesized Fe0-based nanoparticles and their application trends as catalysts in the treatment of chlorinated organic compounds: a review. *Environmental Science: Nano*.

Nnadozie, E. C., and Ajibade, P. A. (2020). Multifunctional magnetic oxide nanoparticle (MNP) core-shell: review of synthesis, structural studies and application for wastewater treatment. *Molecules*, (18): 4110.

Obiri-Nyarko Hsieh, W. P., Pan, J. R., Huang, C., Su, Y. C., and Juang, Y. J. (2010). Enhance the photocatalytic activity for the degradation of organic contaminants in water by incorporating TiO2 with zero-valent iron. *Science of the Total Environment*, 408(3): 672–679.

Olawade, D. B., Wada, O. Z., Egbewole, B. I., Fapohunda, O., Ige, A. O., Usman, S. O. et al. (2024). Metal and metal oxide nanomaterials for heavy metal remediation: novel approaches for selective, regenerative, and scalable water treatment. *Frontiers in Nanotechnology*, 6: 1466721.

Orozco, J., García-Gradilla, V., D'Agostino, M., Gao, W., Cortes, A., and Wang, J. (2013). Artificial enzyme-powered microfish for water-quality testing. *ACS nano*, 7(1): 818–824.

Pasang, T., Namratha, K., Parvin, T., Ranganathaiah, C., and Byrappa, K. (2015). Tuning of band gap in TiO2 and ZnO nanoparticles by selective doping for photocatalytic applications. *Materials Research Innovations*, 19(1): 73–80.

Patadiya, J., Naebe, M., Wang, X., Joshi, G., and Kanda Subramanian, B. (2023). Emerging 4D printing strategies for on-demand local actuation & micro printing of soft materials. *European Polymer Journal*, 184: 111778.

Pitoniak, E., Wu, C. Y., Mazyck, D. W., Powers, K. W., and Sigmund, W. (2005). Adsorption enhancement mechanisms of silica– titania nanocomposites for elemental mercury vapor removal. *Environmental Science & Technology*, 39(5): 1269–1274.

Qasem, N. A., Mohammed, R. H., and Lawal, D. U. (2021). Removal of heavy metal ions from wastewater: A comprehensive and critical review. *Npj Clean Water*, 4(1): 1–15.

Qi, J., Jiang, G., Wan, Y., Liu, J., and Pi, F. (2023). Nanomaterials-modulated Fenton reactions: Strategies, chemodynamic therapy and future trends. *Chemical Engineering Journal*, 466: 142960.

Rajabi, M., Ebrahimi, P., and Aryankhesal, A. (2021). Collaboration between the government and nongovernmental organizations in providing health-care services: A systematic review of challenges. *Journal of Education and Health Promotion*, 10(1): 242.

Romanovski, V., Matsukevich, I., Romanovskaia, E., and Periakaruppan, R. (2023). Nano metal oxide as nanosensors in agriculture and environment. pp. 321–352. *In: Nanometal Oxides in Horticulture and Agronomy*. Academic Press.

Roy, A., Sharma, A., Yadav, S., Jule, L. T., and Krishnaraj, R. (2021). Nanomaterials for remediation of environmental pollutants. *Bioinorganic Chemistry and Applications*, 2021(1): 1764647.

Rudyak, V. Y., Pryazhnikov, M. I., Minakov, A. V., and Shupik, A. A. (2023). Comparison of thermal conductivity of nanofluids with single-walled and multi-walled carbon nanotubes. *Diamond and Related Materials*, 139: 110376.

Saleh, T. A., Mustaqeem, M., and Khaled, M. (2022). Water treatment technologies in removing heavy metal ions from wastewater: A review. *Environmental Nanotechnology, Monitoring & Management*, 17: 100617.

Salehi, E., Heidary, F., Daraei, P., Keyhani, M., and Behjomanesh, M. (2020). Carbon nanostructures for advanced nanocomposite mixed matrix membranes: a comprehensive overview. *Reviews in Chemical Engineering*, 36(6): 723–748.

Savioli, G., Gri, N., Ceresa, I. F., Piccioni, A., Zanza, C., Longhitano, Y. et al. (2024). Carbon monoxide poisoning: From occupational health to emergency medicine. *Journal of Clinical Medicine*, 13(9): 2466.

Schneider, J., Matsuoka, M., Takeuchi, M., Zhang, J., Horiuchi, Y., Anpo, M. et al. (2014). Understanding TiO2 photocatalysis: mechanisms and materials. *Chemical Reviews*, 114(19): 9919–9986.

Scrimieri, L., Velardi, L., Serra, A., Manno, D., Ferrari, F., Cantarella, M. et al. (2020). Enhanced adsorption capacity of porous titanium dioxide nanoparticles synthetized in alkaline sol. *Applied Physics A*, 126: 1–8.

Seema, H., Zaman, S., Zahid, M., Zafar, Z., and Akitsu, T. (2025). Facile preparation and green synthesis of Ni/BiVO4 sensor for room temperature NH3 sensing. *Journal of the Indian Chemical Society*, 102(6): 101700.

Sheerazi, Z., and Ahmed, M. (2022). Nano-Adsorbents and Nano-Catalysts for Wastewater Treatment. pp. 517–539. *In: Emerging Nanomaterials for Advanced Technologies*. Cham: Springer International Publishing.

Shokri, A., and Sanavi Fard, M. (2022). A critical review in the features and application of photocatalysts in wastewater treatment. *Chemical Papers*, 76(9): 5309–5339.

Singh, A., Sharma, A., Verma, R. K., Chopade, R. L., Pandit, P. P., Nagar, V. et al. (2022). Heavy metal contamination of water and their toxic effect on living organisms. In *The Toxicity of Environmental Pollutants*. IntechOpen.

Singh, G., Singh, A., Singh, P., Shukla, R., Tripathi, S., and Mishra, V. K. (2021). The fate of organic pollutants and their microbial degradation in water bodies. *Pollutants and Water Management: Resources, Strategies and Scarcity*, 210–240.

Soleimani, M., and Pourfath, M. (2023). A comprehensive investigation of the plasmonic-photocatalytic properties of gold nanoparticles for CO 2 conversion to chemicals. *Nanoscale*, 15(15): 7051–7067.

Tan, Y. Z., Wu, D., Lee, H. T., Wang, H., Honciuc, A., and Chew, J. W. (2017). Synthesis of ligand-carrying polymeric nanoparticles for use in extraction and recovery of metal ions. *Colloids and Surfaces A: Physicochemical and Engineering Aspects*, 533: 179–186.

Tawiah, B., Abraham, A. M., Kappen, J. J., Raju, N. M., Rajan, C. S., and George, S. C. (2024). Nanotechnology and Water: Emerging Solutions for a Global Crisis. In *Nanotechnology in Societal Development* (pp. 333–374). Singapore: Springer Nature Singapore.

Thakur, A., and Kumar, A. (2024). Emerging paradigms into bioremediation approaches for nuclear contaminant removal: From challenge to solution. *Chemosphere*, 352: 141369.

Velusamy, K., Devanand, J., Kumar, P. S., Soundarajan, K., Sivasubramanian, V., Sindhu, J. et al. (2021). A review on nano-catalysts and biochar-based catalysts for biofuel production. *Fuel*, 306: 121632.

Venkateshaiah, A., Silvestri, D., Wacławek, S., Ramakrishnan, R. K., Krawczyk, K., Saravanan, P. et al. (2022). A comparative study of the degradation efficiency of chlorinated organic compounds by bimetallic zero-valent iron nanoparticles. Environmental Science: *Water Research & Technology*, 8(1): 162–172.

Waghmode, M. S., Gunjal, A. B., Mulla, J. A., Patil, N. N., and Nawani, N. N. (2019). Studies on the titanium dioxide nanoparticles: Biosynthesis, applications and remediation. *SN Applied Sciences*, 1(4): 310.

Wang, H., Chen, Q., Liu, R., Zhang, Y., and Zhang, Y. (2022). Synthesis and application of starch-stablized Fe–Mn/biochar composites for the removal of lead from water and soil. *Chemosphere*, 305: 135494.

Wei, C., Wu, W., Zhao, X., Sun, C., Shi, Z., Yang, J. et al. (2025). Influence of nickel loading on reactivity of Ni/Fe bimetallic nanoparticles toward trichloroethene and carbon tetrachloride. *Environmental Science: Nano*, 12(2): 1316–1327.

Wu, K., Su, D., Liu, J., Saha, R., and Wang, J. P. (2019). Magnetic nanoparticles in nanomedicine: a review of recent advances. *Nanotechnology*, 30(50): 502003.

Xu, L., Ma, W., Wang, L., Xu, C., Kuang, H., and Kotov, N. A. (2013). Nanoparticle assemblies: dimensional transformation of nanomaterials and scalability. *Chemical Society Reviews*, 42(7): 3114–3126.

Yang, Q., Zhou, Y., Chen, J., Huang, N., Wang, Z., and Cheng, Y. (2021). Gene therapy for drug-resistant glioblastoma via lipid-polymer hybrid nanoparticles combined with focused ultrasound. *International Journal of Nanomedicine*, 185–199.

Yasir, A. T., Benamor, A., Hawari, A. H., and Mahmoudi, E. (2023). Poly (amido amine) dendrimer based membranes for wastewater treatment–A critical review. *Chemical Engineering Science*, 273: 118665.

Yin, Z., Yuan, F., Li, M., Xue, M., Zhou, D., Chen, Y. et al. (2021). Self-cleaning, underwater writable, heat-insulated and photocatalytic cellulose membrane for high-efficient oil/water separation and removal of hazardous organic pollutants. *Progress in Organic Coatings*, 157: 106311.

Zeng, (2021). Recent advances on rapid detection and remediation of environmental pollutants utilizing nanomaterials-based (bio) sensors. *Science of The Total Environment*, 834: 155219.

Zhang, W., and Rhim, J. W. (2022). Titanium dioxide (TiO2) for the manufacture of multifunctional active food packaging films. *Food Packaging and Shelf Life*, 31: 100806.

Zhang, X., Li, X., Zhang, D., Su, N. Q., Yang, W., Everitt, H. O. et al. (2017). Product selectivity in plasmonic photocatalysis for carbon dioxide hydrogenation. *Nature Communications*, 8(1): 14542.

Zhang, X., Liu, Y., Yong, H., Qin, Y., Liu, J., and Liu, J. (2019). Development of multifunctional food packaging films based on chitosan, TiO2 nanoparticles and anthocyanin-rich black plum peel extract. *Food Hydrocolloids*, 94: 80–92.

Zinicovscaia, I., Ivlieva, A. L., Petritskaya, E. N., Rogatkin, D. A., Yushin, N., Grozdov, D. et al. (2021). Assessment of TiO2 Nanoparticles Accumulation in Organs and Their Effect on Cognitive Abilities of Mice. *Physics of Particles and Nuclei Letters*, 18(3): 378–384.

CHAPTER-6

Revolutionizing Pollution Control with Solid Waste-Derived Nanomaterials for a Greener Future

Pareshkumar G. Moradeeya,[1] *Janina Kanjee*[2] and *Geoffrey S. Simate*[1,*]

1. Introduction

Increased populations, urban expansion, and economic advancement result in elevated solid waste production in metropolitan areas, presenting critical challenges related to environmental degradation, public health hazards, and the exhaustion of natural resources (Chormare et al. 2023, Abdel-Shafy and Mansour 2018). According to the World Bank, nearly 30% of the 2.01 billion tonnes of solid waste produced by cities each year is not managed in a sustainable manner. It is predicted that this number will rise to 3.40 billion tonnes by the year 2050, outpacing population growth (Kaza et al. 2018). The effective and prompt treatment of solid waste in metropolitan regions is essential due to the intrinsic toxins it harbours, which present considerable threats to human health, ecosystems, soil, water, and the overall environment (Abdel-Shafy and Mansour 2018). Figure 1 has two parts: (a) showing major types of waste, and (b) illustrating their impacts on environmental components. Furthermore, the operations associated with the treatment, processing, and transportation of Solid Waste Management systems adversely affect the environment, atmosphere, and society (Mohsenizadeh et al. 2020). Figure 2 illustrates the collection and processing workflow of Municipal Solid Waste.

The issues arising from increasing populations, excessive waste generation, inadequate waste management, climate change, and other worldwide environmental disasters have made sustainable development an imperative in recent years. Presently, emerging nations are facing multiple issues in managing excessive waste (Kumari and Raghubanshi 2023). As a result, substantial quantities of waste, both liquid and solid, along with air emissions, are generated from various industries, discharging hazardous substances (Wang et al. 2023, Abdel-Shafy and Mansour 2018). Furthermore, in underdeveloped nations, inadequate infrastructure for sewage treatment and solid waste disposal significantly contributes to garbage accumulation (Kumari and Raghubanshi 2023).

[1] School of Chemical and Metallurgical Engineering, University of the Witwatersrand, Johannesburg, South Africa. https://orcid.org/0000-0003-2780-2532
[2] School of Civil and Environmental Engineering, University of the Witwatersrand, Johannesburg, South Africa. http://orcid.org/0000-0001-5322-7979
* Corresponding author: Geoffrey.Simate@wits.ac.za, https://orcid.org/0000-0001-9447-2663

DOI: 10.1201/9781003558804-6

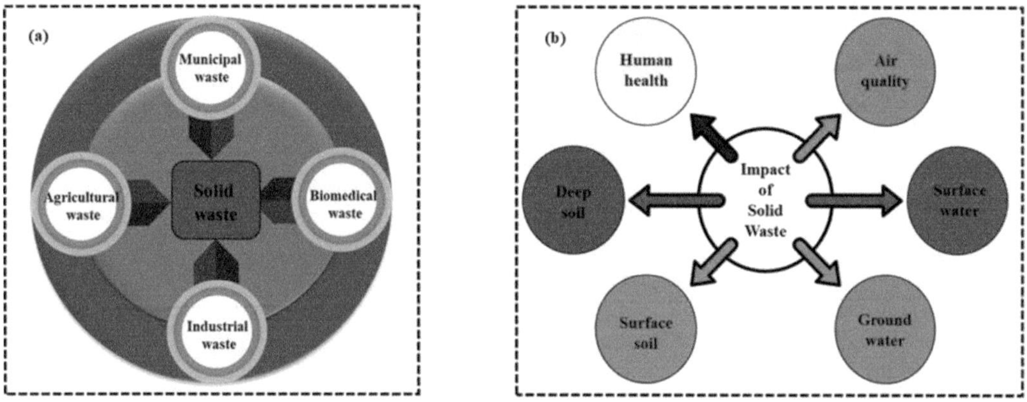

Figure 1: (a) Different types of waste, and (b) their corresponding impacts on key environmental components.

The necessity for effective management of these wastes is paramount during their disposal or storage to prevent environmental impact. Industries today generate a wide variety of waste, and effectively managing this waste has become a critical issue. Waste management practices emphasise collection, storage, and appropriate disposal, while regulating waste generation and maintenance is essential for a sanitary environment. This has led scientists and academics to suggest innovative approaches to waste management in all industries in an effort to lessen its negative impact on people and the ecosystem. There are serious concerns about the negative environmental impacts of using unsustainable practices in solid waste management, despite the fact that traditional approaches to disposing of solid waste are commonly used and frequently considered the most successful (Anil Kumar et al. 2024). Leveraging current scientific developments, researchers are exploring various strategies to recover and reutilise waste through the implementation of innovative treatment technologies (Sundramurthy et al. 2023).

According to a review by El-Kady et al. (2023), a plethora of synthetic nanomaterials have shown immense potential for innovative applications in fields such as agriculture, medicine, fuel and storage, and environmental protection. El-Kady et al. (2023) and Ray et al. (2009) note that sustainability is becoming more important in materials engineering as a means to reduce harmful consequences on the ecosystem and human health that are linked to nanomaterial manufacturing, utilisation, and disposal. Life cycle assessments indicate that the manufacturing of nanomaterials necessitates greater energy and natural resources compared to traditional technologies, even when created using ecologically benign methods (Samaddar et al. 2018, Nizam et al. 2021). Because they are inexpensive and easily accessible, waste materials are attractive potential feedstock, and synthesising nanomaterials is a new way to handle and recycle waste (Levchenko et al. 2022, Mubarik et al. 2021). Therefore, there has been a surge in interest in studying how to produce nanomaterials from waste in an environmentally responsible manner (Bhattacharya et al. 2021, Kumar et al. 2016). The need for waste recycling is on the rise, driven by stricter environmental restrictions. Under the present conditions, our community would gain from the efficient recovery and repurposing of certain waste materials for the production of nanomaterials. Utilising agricultural waste as a primary feedstock for nanomaterials is possibly economical (Kumar et al. 2016, Ikram et al. 2021). These wastes demonstrate harmful environmental impacts owing to their potential to be toxic, flammable,

Figure 2: Schematic representation of the collection and processing stages of municipal solid waste.

highly reactive, and radioactive (Kanagamani et al. 2020). Consequently, recycling industrial waste is vital for producing beneficial items that protect the environment.

We performed a comprehensive assessment of smart-material manufacturing utilising biobased and scrap from industry.

2. Waste Types and Environmental Impact

The waste sources, classified mainly as per Figure 1(a), have been employed as primary feedstock for the fabrication of smart nanomaterials. In recent years, researchers have synthesised several materials, including organic and metal-based nanoparticles, from industrial or biomass waste leftovers. The initial phase in the fabrication of nanoparticles from waste involves the processing of the waste, with further specifics delineated in Section 3. Various pretreatment procedures have been employed owing to the heterogeneous nature of wastes utilised by researchers (Samaddar et al. 2018).

2.1 Biological waste (Biowaste)

Extensive biomass resources are frequently incinerated, either deliberately or inadvertently. Due to the typically incomplete nature of combustion, this process generates pollutants that adversely affect human health and ecosystems. Jiang et al. (2024) examined the spatiotemporal characteristics of purposeful and unintentional biomass burning across several sectors, evaluated the effects of emissions on air quality, and assessed the subsequent repercussions on human health. Despite the fact that there have been several studies examining air pollutants caused by open biomass burning, this practice still adds a great deal of unpredictability to emission assessments. The incineration of agricultural residues in China necessitates careful scrutiny due to its substantial emissions of greenhouse gases as well as other gaseous pollutants and carcinogenic compounds (Chen et al. 2017). Levoglucosan, a dependable marker of biomass combustion, is generated from cellulose pyrolysis (Li et al. 2023c). The emission characteristics of certain toxic substances from agro-waste burning remain inadequately characterised (Jiang et al. 2024).

2.2 Industrial Waste

Industrial waste constitutes a major source of environmental contamination, comprising several extremely hazardous organic and inorganic contaminants that can induce severe toxicity in living beings. A significant cluster of industries, encompassing plastics, pesticides, pharmaceuticals, paints, petrochemicals, metals, leather, and textiles, serves as a primary source of hazardous wastes, mainly organic and inorganic, as examined by Misra and Pandey (2005). Industrial and hazardous waste pollution can adversely affect human health, leading to various ailments such as respiratory disorders, dermatological conditions, cancer, reproductive issues, and developmental problems in children. This occurs primarily through the contamination of air, water, and soil when inadequately managed, resulting in exposure to toxic chemicals via inhalation, ingestion, and dermal contact (Fazzo et al. 2017, Damalas and Koutroubas 2016). Cong (2018) performed a retrospective population-based study in Shanghai, China, examining the impact of industrial waste gas emissions on various cancer incidents. The primary pollutants in industrial waste gas emissions comprise fine particulate matter and gaseous pollutants. Factors related to industrial waste gas, along with emissions of soot, dust, and SO2, have garnered significant attention and represent a critical domain for monitoring and regulation (Manisalidis et al. 2020, Larki et al. 2023).

3. Methods for Nanomaterial Preparation

A graphical representation of the methodologies employed for synthesising smart nano sized materials from waste is depicted in Figure 3.

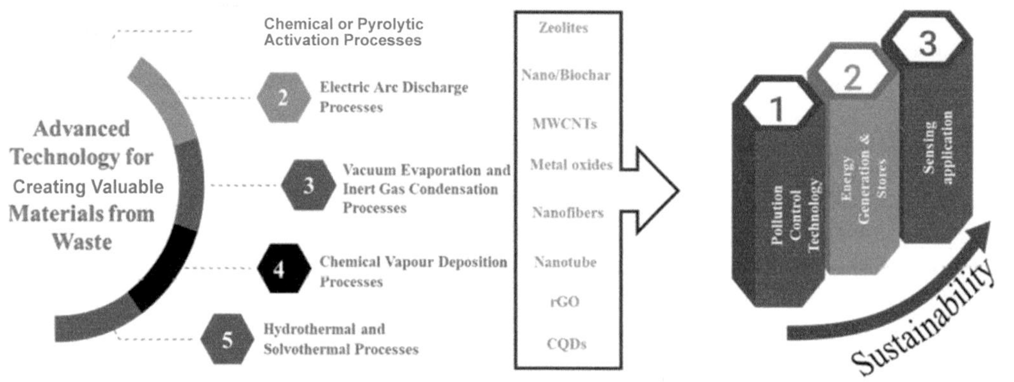

Figure 3: Methodologies for synthesizing smart nanosized materials from waste.

3.1 Chemical or Pyrolytic Activation Processes

Using chemical or thermal treatments to transform waste materials into usable nanomaterials is the chemical/thermal activation technique for synthesising nanomaterials from waste. Chemical and thermal treatments improve the material's properties, facilitating the development of important nanomaterials for a number of applications (Oleszczuk et al. 2016, Verma et al. 2023). During the activation stage, materials (including agricultural, industrial, or electronic waste) are subjected to chemical agents (such as acids, alkalis, or solvents) to decompose complex structures and eliminate undesirable impurities, resulting in the development of nanostructures by promoting the rearrangement of atoms/molecules into nanoscale materials (Wang et al. 2018). For electrochemical energy storage, Liu et al. (2014) synthesised functional carbon composites through rapidly pyrolysing sawdust. Hydrothermal carbonisation for the synthesis of graphite structures derived from lignocellulosic materials was conducted by Barin et al. (2014). Researchers Chen et al. (2016a) found a way to create carbon nanosheets by pyrolysing elm samaras that had been treated with an aqueous KOH solution, yielding materials sized 1–2 nm suitable for high-performance supercapacitors. According to Choi et al. (2017), thermal oxidation of linear Low-Density Polyethylene (LDPE) transformed it into a highly graphitic carbon, while graphitised polyacrylonitrile exhibited poorer electrical performance and less ordering.

Kim et al. (2018) synthesised porous sulfonated carbon from LDPE plastic for lithium-based battery applications. The effect of varying pyrolysis temperatures on ferromagnetic properties was studied by Saito (2009) in his study of carbon-based materials made from polyvinyl chloride powder. Yang et al. (2022) documented the subsequent activation of sulfonated polyethylene (PE) utilising KOH at an activation temperature of 900°C. In their study, Mhamane et al. (2016) collected rusted wires that were no longer needed, treated them with different chemicals, and then used heat to convert them into ferric oxide - Hematite (α-Fe2O3) for use as anodes. To fabricate a carbon-ferric oxide (Fe3O4) nanocomposite for supercapacitor applications, Vadiyar et al. (2018) employed waste materials (thermocol sheets and iron wires). Fu and Grant (2015) looked into the use of steel industry by-products in electrode fabrication.

3.2 Electric Arc Discharge (EAD)Processes

This approach to waste-to-nanomaterials synthesis is an eco-friendly and potentially game-changing strategy for making nanosized materials from waste and transforming them into versatile, high-performance nanoparticles (Yermagambet et al. 2021). This technique entails generating an electric arc between two electrodes within a gaseous or liquid medium, usually at elevated voltage. According to several studies (Larionov et al. 2023, Yermagambet et al. 2021, Al-Harbi and Abd-Elrahman

2024), nanomaterials are created when the vaporised material condenses due to the arc's extremely high heat. Yermagambet et al. (2021) synthesised and characterised graphene-based compounds derived from grade D coal.

Larionov et al. (2023) produced synthesis gas and nanosized graphite-like solid-phase products from various wastes. Berkmans et al. (2014) produced nanomaterials from Polyethylene Terephthalate (PET) waste. Zhang et al. (2022) transformed waste polymers into hydrogen gas and carbon nanotubes (CNTs). Singh et al. (2017) documented the synthesis of Multi-Walled Carbon Nanotubes (MWCNTs) using AC arcing of graphite rods, devoid of any catalyst. Pak et al. (2022a) produced ultrafine carbon materials from waste utilising electric arc plasma in an open-air environment. MWCNTs were synthesised on a cathode deposit utilising arc discharge plasma in hydrogen, argon, and air atmospheres, as performed by Roslan et al. (2017). Cui et al. (2004) synthesised MWCNTs and Single-Walled Carbon Nanotubes using DC arc discharge in a nitrogen atmosphere. Preparation of copper oxide from copper scrap and assessment of antibacterial activity were conducted by Tharchanaa et al. (2021). Mombelli et al. (2022) characterised cast iron and slag produced by reducing red muds using an arc-transferred plasma reactor under different smelting conditions. Using DC atmospheric arc plasma, Pak et al. (2022b) synthesised silicon carbide (SiC) from glass waste. Fullerene waste soot was utilised as the precursor for the synthesis of Double-Walled Carbon Nanotubes using arc discharge, employing an optimised mixture of Ar and H under pressure, as demonstrated by Qiu et al. (2010).

3.3 Vacuum Evaporation (VE) and Inert Gas Condensation (IGC) Processes

This approach is a versatile and efficient technology utilised for the creation of nanoparticles, especially in the context of waste material recycling. Vacuum evaporation entails heating a solid substance (namely, waste) within a vacuum chamber to elevated temperatures, resulting in the material's conversion into vapour. As the process transpires in a vacuum, the vapour will condense on a cooler surface, resulting in the formation of thin films or particles (Alvarado et al. 2015). Fujita et al. (2016) synthesised gold (Au) nanoparticles (NPs) in unsaturated fatty acids using the VE process. The production of thin films of indium tin oxide via this method was characterised by Jain et al. (2020). Zhan and Xu (2012) employed the vacuum evaporation method to separate metallic particles from electronic waste. Pathak et al. (2010) synthesised and characterised AgInSe2 films using the hot-wall vacuum evaporation technique. Zhan et al. (2016) synthesised spherical Pb NPs from electronic waste discarded printed circuit boards (DPCB) via this method. Highly distributed spherical Pb NPs were synthesised by the methods of vacuum evaporation and forced-flow IGC (Zhan et al. 2016).

IGC is a method wherein vaporised material condenses in the presence of argon or helium to inhibit oxidation during the low-pressure condensation process, which is crucial for managing reactive substances (Raffi et al. 2007). The synthesis of materials by IGC yields ultrafine-grained materials due to meticulous physical and chemical regulation of many parameters, resulting in grain size distribution (Suryanarayana and Prabhu 2007). At reduced evaporation temperature and inert gas pressure, smaller spherical Ag NPs exhibiting minimal agglomeration were produced, as shown by Raffi et al. (2007). Xiang et al. (2015) employed vacuum separation and IGC to synthesise Zn NPs from spent batteries.

3.4 Chemical Vapour Deposition (CVD) Processes

CVD is a prevalent approach for synthesising thin films, nanotubes, and nanocomposites (Mittal et al. 2021, Moradeeya et al. 2022). N-doped hollow carbon from common thermoplastic polymer waste polypropylene (PP) materials by the CVD method was performed by Tripathi et al. (2019). Tripathi et al. (2017) reported that MWCNTs were made from plastic waste employing this method. Aboul-Enein et al. (2024) prepared graphene sheets derived from PP waste in conjunction with metal oxide nanoparticles.

Waste synthetic polymers to functional graphene foil via thermal decomposition were shown by Cui et al. (2017b). Sharma et al. (2014) prepared graphene film from solid waste using the CVD method at atmospheric pressure. CNTs were synthesised via chemical vapour deposition (CVD) from carbonised iron-rich biomass, as performed by Jian-hua et al. (2012). Zhu et al. (2012) prepared CNTs using bamboo charcoal through the CVD method. A prior study utilised the CVD technique to produce MWCNTs from several forms of carbon precursors derived from coconut shells (Arga and Sharma 2017) and rice straw residues (Lotfi et al. 2018). The thermal decomposition of waste toner for synthesising MWCNTs using the CVD technique was conducted by Bhongade et al. (2019).

3.5 Hydrothermal (HT) and Solvothermal (ST) Processes

The hydro- and solvo-thermal method serves as an effective approach for synthesising nanomaterials, including those sourced from waste materials, by utilising a solvent at controlled temperatures to promote chemical reactions that yield nanostructures (Moradeeya et al. 2022). This process presents numerous benefits in waste management, including material recycling, reduced environmental impact, and the generation of value-added products from otherwise discarded resources. Coconut coir dust was converted to graphitic carbon nanostructures using HT carbonisation at 250°C followed by pyrolysis at 1500°C, as conducted by Barin et al. (2014). Biomass reed was utilised as a carbon source, subsequently transformed into porous carbon nanosheets using HT treatment, and then pyrolysed at 700°C to 900°C after being combined with a KOH solution for energy supercapacitors (Zhou et al. 2017). Chen et al. (2016b) subjected KOH-immersed wheat straw to HT treatment, followed by pyrolysis at 800°C and further treatment at 2600°C to synthesise graphene for use in Li-ion batteries. The preparation of amorphous silica nanowires from biomass (rice husk ash) was achieved through a Fe_2O_3-assisted HT method, with characterisations conducted using various techniques as reported by Bathla et al. (2018).

Chan and Zinchenko (2022) investigated the conversion of PET into luminous materials through the integration of HT treatment and several pretreatment processes. The synthesis of zeolite A from iron ore tailings using HT was conducted by Zhang and Li (2018). MCM-41 (Mesoporous Silica) was synthesised by an HT process utilising sodium silicate (Liu et al. 2021b). By employing the hydrothermal technique, Medina et al. (2020) created a geopolymer adsorbent from Blast Furnace Slag (BFS) and Fly Ash (FA), with the objective of removing Pb(II) from solutions in water.

Popova et al. (2020) performed a synthesis of Zeolite NaX utilising coal ash waste derived from a lignite coal-fired thermal power station; three distinct synthesis methods were employed. Pedrolo et al. (2017) synthesised zeolites from Brazilian coal ash using the hydrothermal technique with an alkaline KOH solution. Periyasamy et al. (2022) synthesised Fe_3O_4 from Fe Ore Tailings (FeT) using the HT reduction method.

4. Utilisation of Nanoparticles Produced from Waste

A group of researchers has utilised nanomaterials for adsorption removal, photodegradation, energy storage, fuel production, and sensing applications. To achieve this, they have engineered materials with enhanced stability, functionality, and reactivity. In recent years, many waste-derived nanomaterials and nanocomposites have been reported for water treatment. The adsorptive properties, characterised by low cost and an abundance of functional groups, have garnered significant interest from researchers for applications in adsorption and photocatalysis. Waste-derived nanomaterials are widely utilised in various energy storage applications, battery materials, hydrogen fuel production, and optical sensors. This segment discusses the current advancements of various nanomaterials across multiple new fields.

4.1 Applications in Wastewater and air Pollution Control Technology

Water pollution constitutes a major global concern, arising from various causes such as industrial effluents, agricultural runoff, sewage, and garbage disposal. Industrial wastewater comprises a distinct amalgamation of pollutants specific to the industries generating it. Mitigating water pollution necessitates collaborative initiatives, encompassing improved waste management, advanced treatment technology, stringent policy enforcement, and heightened public awareness to safeguard water supplies for future generations. Multiple research teams are exploring innovative nanotechnology-based solutions for industrial wastewater treatment utilising recycled nanoparticles, hence advancing circularity and sustainability objectives. Nonetheless, issues such as expense, ecological consequences, and scalability must be resolved for wider implementation in practical contexts, and the utilisation of waste-derived advanced materials may represent the optimal solution, as it can diminish solid waste and transform it into valuable resources (Anil Kumar et al. 2024).

Adsorption utilising nanoparticles is an applicable and promising technique for pollution abatement, owing to the active surface area and adjustable characteristics of the materials. This method, bolstered by ongoing material improvements, has the capacity to tackle various water pollutants, providing sustainable solutions for access to clean water (Moradeeya et al. 2017, Sahoo et al. 2024, Nama et al. 2024). Functionalised carbon adsorbents were synthesised from waste for carbon dioxide (CO2) adsorption by Kaur et al. (2019). Graphene produced from PET bottles was used to remove dyes (Methylene Blue (MB) and Acid Blue 25 (AB25)) by El Essawy et al. (2017). Ezzat et al. (2022) utilised PET waste bottles to synthesise graphene, which they then used to separate dyes (MB and Methyl Orange (MO)) from water. The porous carbon prepared from PET waste and utilised for the elimination of MB was conducted by Efimov et al. (2023). Janqamsari et al. (2021) developed a nanocomposite utilising polyvinyl alcohol and waste plastic (PET) for effective oil spill remediation. Surface-modified iron oxide, produced from mill scale, has been utilised for dye separation from water, performed by Arifin et al. (2017). Reported by Bai et al. (2020), dye removal from water was achieved using nanocomposite adsorbents synthesised from high-silicon FeT. Dong et al. (2020) synthesised Zn2SiO4 composites from FeT for dye removal.

Duan et al. (2016) prepared a geopolymer utilising FA and FeT for the separation of toxic metals from aqueous solution. In order to eliminate 17 alpha-ethinylestradiol, Silva et al. (2020) used FeT to produce CNTs in a fluidised bed. Magnetite powder and kaolinite sourced from leftover FeT were utilised for the removal of dye from water (Giri et al. 2011a). Giri et al. (2011b) synthesised Fe3O4 from waste FeT for the separation of MB and Congo Red (CR) from aqueous solution. Guerreiro et al. (2020) synthesised carbon nanomaterials for the elimination of Rhodamine B and Indigo-carmine utilising raw iron ore tailings as a catalyst. Porous silica was synthesised for effective adsorption of MB (Han et al. 2021). An aerogel composite was synthesised for dye removal (Hong et al. 2021). Kaewmee et al. (2020) created a geopolymer from coal FA waste and applied it for the separation of dye from aqueous solution. Kim et al. (2015) prepared mesoporous silica from mine tailings for the elimination of copper(II) ion (Cu(II)), cadmium(II) ion (Cd(II)), and lead(II) ion (Pb(II)) from water. Zeolite 13X was synthesised from palm oil mill FA to improve its CO_2 adsorption ability (Kongnoo et al. 2017). FeT and Mn Ore Tailings were utilised for the treatment of arsenic-contaminated soil, as reported by Lei et al. (2019).

The separation of toxic metals from water using the FA-based adsorbent material-supported Zero-Valent Iron (ZVI) was performed by Liu et al. (2017). A ZVI-supported FA-based adsorbent material for Crystal Violet (CV) and MB separation from aqueous solution was created by Liu et al. (2018). MCM-41 was synthesised for the adsorption-desorption study of organic vapours, as explored by Liu et al. (2021b). A geopolymer adsorbent for the removal of Pb(II) from water was created by Medina et al. (2020) by combining BFS with FA. Pedrolo et al. (2017) synthesised zeolites from Brazilian coal ash using the hydrothermal technique with an alkaline KOH solution.

Utilising agricultural, industrial, and electronic waste enables the development of effective photocatalysts that aid in water purification while mitigating the environmental effects of waste products. Periyasamy et al. (2022) synthesised visible-light photocatalyst (magnetite/carbon) nanocomposites from low-grade iron ore for dye removal. Using waste materials, Altın and Sökmen (2014) created a titanium dioxide (TiO2)/polystyrene photocatalyst and tested its effectiveness in degradatively removing MB and chromium(VI) (Cr(VI)). In order to degrade MB, Tanniratt et al. (2016) created a zinc oxide (ZnO)/diatomite composite using waste materials such as zinc dust and diatomite. To degrade antibiotic residues, Lu et al. (2013) synthesised TiO2/Fe3O4/fly-ash cenospheres derived from coal FA for waste treatment purposes. Photodegradation of MB dye in sunlight using mpg-C3N4-WO3 (derived from waste tungsten filament) was reported by Shinde et al. (2022). Investigation of a waste-derived Cu–Cu2O/SiO2 photocatalyst for the elimination of chromium(VI) was conducted by Zhu et al. (2023). Chen et al. (2014) prepared TiO2@Mn-Zn ferrite powders from waste for dye degradation. Nath et al. (2023) synthesised a ZnO–Fe2O3/polymer composite for the degradation of MB. Lumbaque et al. (2021) developed a solar light-activated photocatalyst for decomposing pharmaceutical residues in diverse water matrices. Dhmees et al. (2019) synthesised SiO2@CeO2 utilising BFS and uranium extraction waste for the purpose of dye degradation. Du et al. (2018) reported the utilisation of iron oxide-containing waste in the Fenton process to remove tri(2-chloroethyl)phosphate from water.

Nguyen et al. (2021) showed that Fe-containing tailings ore enriched with iron(iii) chloride can be successfully reutilized as a heterogeneous Fenton catalyst, enabling efficient decolorization of organic dyes from aqueous solutions. de Freitas et al. (2019) utilised FeT as a catalyst in a Fenton-treatment for the oxidation of MB. The Fe/MCM-41, synthesised using FeT, showed efficacy in the removal of MB (Deng et al. 2020). For the purpose of degrading CR dye, Behura et al. (2021) Cobalt ferrite nanoparticles were synthesised by a coprecipitation method using iron recovered from waste iron ore tailings and cobalt from spent lithium-ion batteries and were applied for photo- and sono-catalytic degradation of Congo red. Yang et al. (2019) synthesised SiC nanoparticles from DPCBs and investigated the degradation of MB. The MCM-48/TiO2, utilising e-waste as a silica source, was developed by Liou and Lai (2012). Calcium oxide (CaO)/Carbon was synthesised through thermal breakdown of eggshells and coffee waste, applied as photocatalysts for the elimination of hazardous dyes, as demonstrated by Mohamed et al. (2021). The breakdown of organic compounds was achieved by Babar et al. (2019) using a solar light-activated catalyst (g-C3N4-Fe2O3) that was produced from waste toner. Kamal et al. (2024) developed a mesoporous nanocomposite Ag/Carbon from medical X-ray sheets and potato peels for the purposes of CV elimination and antibacterial applications. CaO prepared from chicken eggshells for photodegradation of MB dye was performed by Jaiswal et al. (2021). For the purpose of degrading MB dye in water, Meena et al. (2024) employed a simple calcination method involving waste banana peel to create a green biochar-supported Zinc Ferrite composite catalyst that is extremely efficient when driven by visible light. Alberti et al. (2019) prepared a photocatalyst composite from industrial waste and applied it to eliminate dye. The synthesis of Co-doped TiO_2 was carried out by Li et al. (2023b) using biomass waste as a template for the decomposition of tetracycline. Nayak et al. (2019) created ZnO/Copper(II) Oxide (CuO) from electronic waste for the photodegradation of dye. Photodegradation of phenol utilising metal-doped (Fe, Mg, Ca, and Al) ZnO nanoparticles derived from zinc-bearing dust was performed by Wu et al. (2014).

4.2 Applications in Energy Storage Systems

Carbon nanomaterials generated from waste have been reported to significantly contribute to the production of Li-ion battery (Wang et al. 2018, da Silva et al. 2023, Li et al. 2023a). Waste-derived carbon NPs are emerging as a novel and sustainable alternative in the fabrication of battery materials (da Silva et al. 2023). The conversion of diverse waste materials into advanced functional materials provides significant scientific and environmental benefits, including cost efficiency, reduced ecological impact, and promotion of circular economy strategies.

Agricultural waste, specifically rice husk, was selected by Autthawong et al. (2020) as a cost-effective source for synthesising SiO_2/C and Si/C composite for the production of lithium-ion battery anodes. SiO_2/C composites, synthesised from agricultural waste (rice husks) using $ZnCl_2$ as an activating agent in an Ar environment, were reported by Cui et al. (2017a) as anode materials for Li-ion batteries. Autthawong et al. (2021) produced Sn–SiO_2@TiO_2 anode materials, utilising SiO_2 obtained from rice husk. Ratsameetammajak et al. (2022) created PANI-SiO2@rGO composites that demonstrate improved electrochemical performance, utilising nano-silica derived from rice husks. The recycling of mixed-waste plastics (PP/PE/Polystyrene (PS)) to carbon was prepared by Li et al. (2023a) and applied for battery materials.

Kumar et al. (2019) synthesised an anode for Sodium-Ion Batteries (SIBs) with activated carbon derived from electronic waste plastic. Sun et al. (2015) reported the straightforward creation of carbon material from biomass for SIBs. Li et al. (2016) documented the generation of HC microtubes obtained from cotton and their application as an anode. HC material produced from eucalyptus wood for SIBs, demonstrating notable electrochemical performance, was emphasised by Liu et al. (2022). The C/Fe3O4 nanocomposite made of waste toner was used as a supercapacitor, as described by Kaipannan et al. (2019). Deka et al. (2020) developed N-doped carbon utilising discarded PS foam for energy storage and deionisation. Efficient carbonisation of waste PP into carbon nanosheets was accomplished utilising a combination catalyst of ferrocene and sulphur, facilitating enhanced capacitive energy storage, as indicated by Liu et al. (2021a).

4.3 Surveillance of Contaminants and Sensing Applications

Nanosized sensors utilising smart materials offer considerable benefits for the detection and monitoring of hazardous compounds in the ecosystem (Kapoor et al. 2024). Laboratory tests employed to detect water pollutants are financially prohibitive in certain places due to the necessity of qualified workers and costly equipment. Researchers are creating cost-effective sensors utilising nanoparticles to monitor contaminants in order to address this issue. Certain researchers advance the field by employing recycled nanoparticles (Bhattacharya et al. 2021). ZnO was synthesised from discarded zinc–manganese batteries, as proven by Zhan et al. (2018), for possible sensing and catalytic applications. Deshmukh et al. (2020) illustrated the utility of red-mud-rGO nanocomposites for detection of arsenic (III). Biomass waste-derived aerogels utilised for electrochemical sensing were developed by Sha et al. (2018). Kasinathan et al. (2022) synthesised multicoloured fluorescent Carbon Quantum Dots (CQDs) from agro waste, exploring their possibilities in mercury (II) ion detection and bioimaging. Hu et al. (2019) prepared fluorescent Carbon Dots (CDs) from waste PET for detecting ferric and pyrophosphate ions in water. Kumari and Chaudhary (2020) synthesised highly photoluminescent and biocompatible CQDs from plastic waste and applied them for the selective detection of *Escherichia coli*. Chen et al. (2021) developed CuCo2O4@biomass-derived carbon for detection of tryptophan. Liu et al. (2019) developed fluorescent nanosensors from biomass for detection of toxic metals. Sachdev and Gopinath (2015) proposed multifunctional CDs obtained from biomass, emphasising their prospective uses in sensing and bioimaging. Shi et al. (2017) synthesised fluorescent CDs for detection of Fe(II) and H2O2. Hossain et al. (2022) developed an ammonia sensor utilising NiO-doped CuO nanoflakes derived from electronic trash. Bressi et al. (2024) synthesised CDs generated from waste for detection of toxic metals (Hg(II) and Pb(II)). Ngu et al. (2016) synthesised carbon NPs from agricultural waste, which were examined as sensing materials for Sn(II) ions.

5. Conclusions

Nanomaterials derived from waste exhibit considerable potential for diverse applications, and these materials can be synthesised by straightforward and cost-effective techniques. Additionally, energy storage and pollution removal applications and characteristics are often markedly improved relative to pure nanomaterials. The benefits of improvements in electrode, photocatalytic efficiency, and sensing performance in their individual functions will be amplified by selecting viable production methods for nanomaterials obtained from waste. In conclusion, to actualise the potential advantages of utilising waste-derived nanomaterials, a comprehensive understanding of the myriad contributing elements is essential, and decisions should prioritise cautious and ethical considerations. The use and extent of nanomaterials, as well as strategies for controlling engineered nanomaterials, must be defined locally, with management rules informed by health and environmental science.

References

Abdel-Shafy, H. I., and Mansour, M. S. (2018). Solid waste issue: Sources, composition, disposal, recycling, and valorization. *Egyptian Journal of Petroleum*, 27(4): 1275–1290.

Aboul-Enein, A. A., Haggar, A. M., Awadallah, A. E., and Azab, M. A. (2024). Synthesis of polypropylene waste-derived graphene sheets in the presence of metal oxide nanoparticles. *Materials Chemistry and Physics*, 315: 129021.

Alberti, S., Caratto, V., Peddis, D., Belviso, C., and Ferretti, M. (2019). Synthesis and characterization of a new photocatalyst based on TiO2 nanoparticles supported on a magnetic zeolite obtained from iron and steel industrial waste. *Journal of Alloys and Compounds*, 797: 820–825.

Al-Harbi, N., and Abd-Elrahman, N. K. (2024). Physical methods for preparation of nanomaterials, their characterization and applications: a review. *Journal of Umm Al-Qura University for Applied Sciences*, 1–22.

Altın, İ., and Sökmen, M. (2014). Preparation of TiO2-polystyrene photocatalyst from waste material and its usability for removal of various pollutants. *Applied Catalysis B: Environmental*, 144: 694–701.

Alvarado, J. A., Maldonado, A., Juarez, H., Pacio, M., and Perez, R. (2015). Characterization of nanostructured ZnO thin films deposited through vacuum evaporation. *Beilstein Journal of Nanotechnology*, 6(1): 971–975.

Anil Kumar, M., Moradeeya, P. G., Manikanda Bharath, K., Jakulin Divya Mary, P., and Giridharan, K. S. (2024). Pyrolytic Conversion of Heterogenic Natural Waste Biomass from Rural Communities with Concomitant Valorization. pp. 259–287. *In: Circular Bioeconomy Perspectives in Sustainable Bioenergy Production*. Singapore: Springer Nature Singapore.

Araga, R., and Sharma, C. S. (2017). One step direct synthesis of multiwalled carbon nanotubes from coconut shell derived charcoal. *Materials Letters*, 188: 205–207.

Autthawong, T., Namsar, O., Yu, A., and Sarakonsri, T. (2020). Cost-effective production of SiO 2/C and Si/C composites derived from rice husk for advanced lithium-ion battery anodes. *Journal of Materials Science: Materials in Electronics*, 31: 9126–9132.

Autthawong, T., Yodbunork, C., Yodying, W., Boonprachai, R., Namsar, O., Yu, A. S. et al. (2021). Fast-charging anode materials and novel nanocomposite design of rice husk-derived SiO2 and Sn nanoparticles self-assembled on TiO2 (B) nanorods for lithium-ion storage applications. *ACS omega*, 7(1): 1357–1367.

Babar, S., Gavade, N., Shinde, H., Gore, A., Mahajan, P., Lee, K. H. et al. (2019). An innovative transformation of waste toner powder into magnetic g-C3N4-Fe2O3 photocatalyst: sustainable e-waste management. *Journal of Environmental Chemical Engineering*, 7(2): 103041.

Bai, S., Tian, G., Gong, L., Tang, Q., Meng, J., Duan, X. et al. (2020). Mesoporous manganese silicate composite adsorbents synthesized from high-silicon iron ore tailing. *Chemical Engineering Research and Design*, 159: 543–554.

Barin, G. B., de Fátima Gimenez, I., da Costa, L. P., Souza Filho, A. G., and Barreto, L. S. (2014). Influence of hydrothermal carbonization on formation of curved graphite structures obtained from a lignocellulosic precursor. *Carbon*, 78: 609–612.

Bathla, A., Narula, C., and Chauhan, R. P. (2018). Hydrothermal synthesis and characterization of silica nanowires using rice husk ash: an agricultural waste. *Journal of materials science: Materials in electronics*, 29: 6225–6231.

Behura, R., Sakthivel, R., and Das, N. (2021). Synthesis of cobalt ferrite nanoparticles from waste iron ore tailings and spent lithium ion batteries for photo/sono-catalytic degradation of Congo red. *Powder technology*, 386: 519–527.

Berkmans, A. J., Jagannatham, M., Priyanka, S., and Haridoss, P. (2014). Synthesis of branched, nano channeled, ultrafine and nano carbon tubes from PET wastes using the arc discharge method. *Waste management*, 34(11): 2139–2145.

Bhattacharya, G., Fishlock, S. J., McLaughlin, J. A., and Roy, S. S. (2021). Metal-oxide nanomaterials recycled from E-waste and metal industries: A concise review of applications in energy storage, catalysis, and sensing. *International Journal of Energy Research*, 45(5): 8091–8102.

Bhongade, T., Gogaram, Gautam, D. M., and Vijayakumar, R. P. (2019). Synthesis of MWCNTs using waste toner powder as carbon source by chemical vapor deposition method. *Fullerenes, Nanotubes and Carbon Nanostructures*, 27(11): 864–872.

Bressi, V., Celesti, C., Ferlazzo, A., Len, T., Moulaee, K., Neri, G. et al. (2024). Waste-derived carbon nanodots for fluorimetric and simultaneous electrochemical detection of heavy metals in water. *Environmental Science: Nano*, 11(3): 1245–1258.

Chan, K., and Zinchenko, A. (2022). Aminolysis-assisted hydrothermal conversion of waste PET plastic to N-doped carbon dots with markedly enhanced fluorescence. *Journal of Environmental Chemical Engineering*, 10(3): 107749.

Chen, B., Xie, Q., Zhang, S., Lin, L., Zhang, Y., Zhang, L. et al. (2021). A novel electrochemical molecularly imprinted senor based on CuCo2O4@ biomass derived carbon for sensitive detection of tryptophan. *Journal of Electroanalytical Chemistry*, 901: 115680.
Chen, C. C., Butler, E., Al Ahmad, M., Hung, Y. T., and Fu, Y. P. (2014). Characterizations of TiO2@ Mn-Zn ferrite powders for magnetic photocatalyst prepared from used alkaline batteries and waste steel pickling liquor. *Materials Research Bulletin*, 50: 178–182.
Chen, C., Yu, D., Zhao, G., Du, B., Tang, W., Sun, L. et al. (2016a). Three-dimensional scaffolding framework of porous carbon nanosheets derived from plant wastes for high-performance supercapacitors. *Nano Energy*, 27: 377–389.
Chen, F., Yang, J., Bai, T., Long, B., and Zhou, X. (2016b). Facile synthesis of few-layer graphene from biomass waste and its application in lithium ion batteries. *Journal of Electroanalytical Chemistry*, 768: 18–26.
Chen, J., Li, C., Ristovski, Z., Milic, A., Gu, Y., Islam, M. S. et al. (2017). A review of biomass burning: Emissions and impacts on air quality, health and climate in China. *Science of the Total Environment*, 579: 1000–1034.
Choi, D., Jang, D., Joh, H. I., Reichmanis, E., and Lee, S. (2017). High performance graphitic carbon from waste polyethylene: thermal oxidation as a stabilization pathway revisited. *Chemistry of Materials*, 29(21): 9518–9527.
Chormare, R., Moradeeya, P. G., Sahoo, T. P., Seenuvasan, M., Baskar, G., Saravaia, H. T. et al. (2023). Conversion of solid wastes and natural biomass for deciphering the valorization of biochar in pollution abatement: A review on the thermo-chemical processes. *Chemosphere*, 139760.
Cong, X. (2018). Air pollution from industrial waste gas emissions is associated with cancer incidences in Shanghai, China. *Environmental Science and Pollution Research*, *25*: 13067–13078.
Cui, J., Cheng, F., Lin, J., Yang, J., Jiang, K., Wen, Z. et al. (2017a). High surface area C/SiO2 composites from rice husks as a high-performance anode for lithium ion batteries. *Powder Technology*, 311: 1–8.
Cui, L., Wang, X., Chen, N., Ji, B., and Qu, L. (2017b). Trash to treasure: converting plastic waste into a useful graphene foil. *Nanoscale*, 9(26): 9089–9094.
Cui, S., Scharff, P., Siegmund, C., Schneider, D., Risch, K., Klötzer, S. et al. (2004). Investigation on preparation of multiwalled carbon nanotubes by DC arc discharge under N2 atmosphere. *Carbon*, 42(5–6): 931–939.
da Silva, E. P., Fragal, V. H., Fragal, E. H., Sequinel, T., Gorup, L. F., Silva, R. et al. (2023). Sustainable energy and waste management: how to transform plastic waste into carbon nanostructures for electrochemical supercapacitors. *Waste Management*, 171: 71–85.
Damalas, C. A., and Koutroubas, S. D. (2016). Farmers' exposure to pesticides: toxicity types and ways of prevention. *Toxics*, 4(1): 1.
de Freitas, V. A. A., Breder, S. M., Silvas, F. P. C., Rouse, P. R., and de Oliveira, L. C. A. (2019). Use of iron ore tailing from tailing dam as catalyst in a fenton-like process for methylene blue oxidation in continuous flow mode. *Chemosphere*, 219: 328–334.
Deka, N., Barman, J., Kasthuri, S., Nutalapati, V., and Dutta, G. K. (2020). Transforming waste polystyrene foam into N-doped porous carbon for capacitive energy storage and deionization applications. *Applied Surface Science*, 511: 145576.
Deng, Y., Xu, X., Wang, R., and Zhao, Y. (2020). Characterization and photocatalytic evaluation of Fe-loaded mesoporous MCM-41 prepared using iron and silicon sources extracted from iron ore tailing. *Waste and Biomass Valorization*, 11: 1491–1498.
Deshmukh, S., Banerjee, D., Bhattacharya, G., Fishlock, S. J., Barman, A., McLaughlin, J. et al. (2020). Red mud-reduced graphene oxide nanocomposites for the electrochemical sensing of arsenic. *ACS Applied Nano Materials*, 3(5): 4084–4090.
Dhmees, A. S., Rashad, A. M., Eliwa, A. A., and Zawrah, M. F. (2019). Preparation and characterization of nano SiO2@ CeO2 extracted from blast furnace slag and uranium extraction waste for wastewater treatment. *Ceramics International*, 45(6): 7309–7317.
Dong, G., Tian, G., Gong, L., Tang, Q., Li, M., Meng, J., and Liang, J. (2020). Mesoporous zinc silicate composites derived from iron ore tailings for highly efficient dye removal: Structure and morphology evolution. *Microporous and Mesoporous Materials*, 305: 110352.
Du, L., Wang, X., and Wu, J. (2018). Degradation of tri (2-chloroethyl) phosphate by a microwave enhanced heterogeneous Fenton process using iron oxide containing waste. *RSC advances*, 8(32): 18139–18145.
Duan, P., Yan, C., Zhou, W., and Ren, D. (2016). Development of fly ash and iron ore tailing based porous geopolymer for removal of Cu (II) from wastewater. *Ceramics International*, 42(12): 13507–13518.
Efimov, M. N., Vasilev, A. A., Muratov, D. G., Kostev, A. I., Kolesnikov, E. A., Kiseleva, S. G. et al. (2023). Conversion of polyethylene terephthalate waste into high-yield porous carbon adsorbent via pyrolysis of dipotassium terephthalate. *Waste Management*, 162: 113–122.
El Essawy, N. A., Ali, S. M., Farag, H. A., Konsowa, A. H., Elnouby, M., and Hamad, H. A. (2017). Green synthesis of graphene from recycled PET bottle wastes for use in the adsorption of dyes in aqueous solution. *Ecotoxicology and environmental safety*, 145: 57–68.
El-Kady, M. M., Ansari, I., Arora, C., Rai, N., Soni, S., Verma, D. K. et al. (2023). Nanomaterials: A comprehensive review of applications, toxicity, impact, and fate to environment. *Journal of Molecular Liquids*, 370: 121046.

Ezzat, M. N., and Abd Ali, Z. T. (2022). Green approach for fabrication of graphene from polyethylene terephthalate (PET) bottle waste as reactive material for dyes removal from aqueous solution: Batch and continuous study. *Sustainable Materials and Technologies*, *32*, e00404.

Fazzo, L., Minichilli, F., Santoro, M., Ceccarini, A., Della Seta, M., Bianchi, F. et al. (2017). Hazardous waste and health impact: a systematic review of the scientific literature. *Environmental Health*, 16: 1–11.

Fu, C., and Grant, P. S. (2015). Toward low-cost grid scale energy storage: supercapacitors based on up-cycled industrial mill scale waste. *ACS Sustainable Chemistry & Engineering*, 3(11): 2831–2838.

Fujita, A., Matsumoto, Y., Takeuchi, M., Ryuto, H., and Takaoka, G. H. (2016). Growth behavior of gold nanoparticles synthesized in unsaturated fatty acids by vacuum evaporation methods. *Physical Chemistry Chemical Physics*, 18(7): 5464–5470.

Giri, S. K., Das, N. N., and Pradhan, G. C. (2011a). Magnetite powder and kaolinite derived from waste iron ore tailings for environmental applications. *Powder Technology*, 214(3): 513–518.

Giri, S. K., Das, N. N., and Pradhan, G. C. (2011b). Synthesis and characterization of magnetite nanoparticles using waste iron ore tailings for adsorptive removal of dyes from aqueous solution. *Colloids and Surfaces A: Physicochemical and Engineering Aspects*, 389(1–3): 43–49.

Guerreiro, G. G., de Andrade, F. V., and de Freitas, M. R. (2020). Carbon nanostructures based-adsorbent obtained from iron ore tailings. *Ceramics International*, 46(18): 29271–29281.

Han, X., Wang, Y., Zhang, N., Meng, J., Li, Y., and Liang, J. (2021). Facile synthesis of mesoporous silica derived from iron ore tailings for efficient adsorption of methylene blue. *Colloids and Surfaces A: Physicochemical and Engineering Aspects*, 617: 126391.

Hong, W., Liu, X., Srinivasakannan, C., Duan, X., and Wang, X. (2021). Novel aerogel absorbent derived from iron tailings via atmospheric drying. *Arabian Journal for Science and Engineering*, 1–14.

Hossain, R., Hassan, K., and Sahajwalla, V. (2022). Utilising problematic waste to detect toxic gas release in the environment: fabricating a NiO doped CuO nanoflake based ammonia sensor from e-waste. *Nanoscale Advances*, 4(19): 4066–4079.

Hu, Y., Gao, Z., Yang, J., Chen, H., and Han, L. (2019). Environmentally benign conversion of waste polyethylene terephthalate to fluorescent carbon dots for "on-off-on" sensing of ferric and pyrophosphate ions. *Journal of Colloid and Interface Science*, 538: 481–488.

Ikram, R., Mohamed Jan, B., Sidek, A., and Kenanakis, G. (2021). Utilization of eco-friendly waste generated nanomaterials in water-based drilling fluids; state of the art review. *Materials*, 14(15): 4171.

Jain, P., Singh, S., and Pundir, S. K. (2020). Synthesis of highly transparent indium tin oxide thin films using vacuum evaporation technique. *Journal of Nanoscience and Nanotechnology*, 20(6): 3845–3853.

Jaiswal, K. K., Dutta, S., Pohrmen, C. B., Verma, R., Kumar, A., and Ramaswamy, A. P. (2021). Bio-waste chicken eggshell-derived calcium oxide for photocatalytic application in methylene blue dye degradation under natural sunlight irradiation. *Inorganic and Nano-Metal Chemistry*, 51(7): 995–1004.

Janqamsari, Y., Ashjari, M., and Niazi, Z. (2021). Carbon nanotube promoted porous nanocomposite based on PVA and recycled PET fibers for efficient oil spills cleanup applications. *Chemical Papers*, 75: 3443–3456.

Jiang, K., Xing, R., Luo, Z., Huang, W., Yi, F., Men, Y. et al. (2024). Pollutant emissions from biomass burning: A review on emission characteristics, environmental impacts, and research perspectives. *Particuology*, 85: 296–309.

Jian-hua, S. H. I., Jian-guo, Z. H. A. O., Bao-yan, X. I. N. G., and Ren, Z. H. A. O. (2012). Carbon nanotubes produced by chemical vapor depositionusing natural biomass-based catalysts. *New Carbon Materials*, 27(3): 175–180.

Kaewmee, P., Song, M., Iwanami, M., Tsutsumi, H., and Takahashi, F. (2020). Porous and reusable potassium-activated geopolymer adsorbent with high compressive strength fabricated from coal fly ash wastes. *Journal of Cleaner Production*, 272: 122617.

Kaipannan, S., Govindarajan, K., Sundaramoorthy, S., and Marappan, S. (2019). Waste toner-derived carbon/Fe3O4 nanocomposite for high-performance supercapacitor. *ACS Omega*, 4(14): 15798–15805.

Kamal, A. B., Hassane, A. M., An, C., Deng, Q., Hu, N., Abolibda, T. Z. et al. (2024). Developing a cost-effective and eco-friendly adsorbent/photocatalyst using biomass and urban waste for crystal violet removal and antimicrobial applications. *Biomass Conversion and Biorefinery*, 1–19.

Kanagamani, K., Geethamani, P., and Narmatha, M. (2020). Hazardous waste management. In *Environmental Issues and Sustainable Development*. IntechOpen.

Kapoor, A., Raghunathan, M., Lal, B., Kumar, P., Srivastava, N., Devnani, G. L. et al (2024). Sustainable valorization of waste plastic into nanostructured materials for environmental, energy, catalytic and biomedical applications: A review. *Chemosphere*, 143279.

Kasinathan, K., Samayanan, S., Marimuthu, K., and Yim, J. H. (2022). Green synthesis of multicolour fluorescence carbon quantum dots from sugarcane waste: Investigation of mercury (II) ion sensing, and bio-imaging applications. *Applied Surface Science*, 601: 154266.

Kaur, B., Gupta, R. K., and Bhunia, H. (2019). Chemically activated nanoporous carbon adsorbents from waste plastic for CO2 capture: Breakthrough adsorption study. *Microporous and Mesoporous Materials*, 282: 146–158.

Kaza, S., Yao, L., Bhada-Tata, P., and Van Woerden, F. (2018). *What a Waste 2.0: A Global Snapshot of Solid Waste Management to 2050*. World Bank Publications.

Kim, P. J., Fontecha, H. D., Kim, K., and Pol, V. G. (2018). Toward high-performance lithium–sulfur batteries: upcycling of LDPE plastic into sulfonated carbon scaffold via microwave-promoted sulfonation. *ACS Applied Materials & Interfaces*, 10(17): 14827–14834.

Kim, S., Han, Y., Park, J., and Park, J. (2015). Adsorption characteristics of mesoporous silica SBA-15 synthesized from mine tailing. *International Journal of Mineral Processing*, 140: 88–94.

Kongnoo, A., Tontisirin, S., Worathanakul, P., and Phalakornkule, C. (2017). Surface characteristics and CO2 adsorption capacities of acid-activated zeolite 13X prepared from palm oil mill fly ash. *Fuel*, 193: 385–394.

Kumar, R., Singh, R. K., and Singh, D. P. (2016). Natural and waste hydrocarbon precursors for the synthesis of carbon based nanomaterials: Graphene and CNTs. *Renewable and Sustainable Energy Reviews*, 58: 976–1006.

Kumar, U., Goonetilleke, D., Gaikwad, V., Pramudita, J. C., Joshi, R. K., Sharma, N. et al. (2019). Activated carbon from e-waste plastics as a promising anode for sodium-ion batteries. *ACS Sustainable Chemistry & Engineering*, 7(12): 10310–10322.

Kumari, M., and Chaudhary, S. (2020). Modulating the physicochemical and biological properties of carbon dots synthesised from plastic waste for effective sensing of E. coli. *Colloids and Surfaces B: Biointerfaces*, 196: 111333.

Kumari, T., and Raghubanshi, A. S. (2023). Waste management practices in the developing nations: challenges and opportunities. *Waste Management and Resource Recycling in the Developing World*, 773–797.

Larionov, K. B., Povalyaev, P. V., Kaltaev, A. Z., Slysarsky, K. V., Gorshkov, A. S., Gubin, V. E. et al. (2023). Electric arc gasification of pyrolysis oil with the production of hydrogen-enriched synthesis gas and carbon nanomaterial. *Fuel Processing Technology*, 245: 107746.

Larki, I., Zahedi, A., Asadi, M., Forootan, M. M., Farajollahi, M., Ahmadi, R. et al. (2023). Mitigation approaches and techniques for combustion power plants flue gas emissions: A comprehensive review. *Science of The Total Environment*, 166108.

Lei, M., Tang, L., Du, H., Peng, L., Tie, B., Williams, P. N. et al. (2019). Safety assessment and application of iron and manganese ore tailings for the remediation of As-contaminated soil. *Process Safety and Environmental Protection*, 125: 334–341.

Levchenko, I., Mandhakini, M., Prasad, K., Bazaka, O., Ivanova, E. P., Jacob, M. V. et al. (2022). Functional Nanomaterials from Waste and Low-Value Natural Products: A Technological Approach Level. *Advanced Materials Technologies*, 7(11): 2101471.

Li, J., Dou, F., Gong, J., Gao, Y., Hua, Y., Sielicki, K. et al. (2023a). Recycling of plastic wastes for the mass production of yolk–shell-nanostructured Co3O4@ C for lithium-ion batteries. *ACS Applied Nano Materials*, 6(2): 1171–1180.

Li, Q., Jiang, L., Li, Y., Wang, X., Zhao, L., Huang, P. et al. (2023b). Enhancement of visible-light photocatalytic degradation of tetracycline by Co-doped TiO2 templated by waste tobacco stem silk. *Molecules*, 28(1): 386.

Li, Y., Chen, M., Wang, Y., Huang, T., Wang, G., Li, Z. et al. (2023c). Seasonal characteristics and provenance of organic aerosols in the urban atmosphere of Liaocheng in the North China Plain: Significant effect of biomass burning. *Particuology*, 75: 185–198.

Li, Y., Hu, Y. S., Titirici, M. M., Chen, L., and Huang, X. (2016). Hard carbon microtubes made from renewable cotton as high-performance anode material for sodium-ion batteries. *Advanced Energy Materials*, 6(18): 1600659.

Liou, T. H., and Lai, B. C. (2012). Utilization of e-waste as a silica source for the synthesis of the catalyst support MCM-48 and highly enhanced photocatalytic activity of supported titania nanoparticles. *Applied Catalysis B: Environmental*, 115: 138–148.

Liu, J., Mwamulima, T., Wang, Y., Fang, Y., Song, S., and Peng, C. (2017). Removal of Pb (II) and Cr (VI) from aqueous solutions using the fly ash-based adsorbent material-supported zero-valent iron. *Journal of Molecular Liquids*, 243: 205-211.

Liu, J., Wang, L., Huang, Z., Fan, F., Jiao, L., and Li, F. (2022). Facile synthesis of high quality hard carbon anode from Eucalyptus wood for sodium-ion batteries. *Chemical Papers*, 76(12): 7465–7473.

Liu, J., Wang, Y., Fang, Y., Mwamulima, T., Song, S., and Peng, C. (2018). Removal of crystal violet and methylene blue from aqueous solutions using the fly ash-based adsorbent material-supported zero-valent iron. *Journal of Molecular Liquids*, 250: 468–476.

Liu, W. J., Tian, K., He, Y. R., Jiang, H., and Yu, H. Q. (2014). High-yield harvest of nanofibers/mesoporous carbon composite by pyrolysis of waste biomass and its application for high durability electrochemical energy storage. *Environmental Science & Technology*, 48(23): 13951–13959.

Liu, X., Ma, C., Wen, Y., Chen, X., Zhao, X., Tang, T. et al. (2021a). Highly efficient conversion of waste plastic into thin carbon nanosheets for superior capacitive energy storage. *Carbon*, 171: 819–828.

Liu, Y., Li, C., Peyravi, A., Sun, Z., Zhang, G., Rahmani, K. et al. (2021b). Mesoporous MCM-41 derived from natural Opoka and its application for organic vapors removal. *Journal of Hazardous Materials*, 408: 124911.

Liu, Z., Jin, W., Wang, F., Li, T., Nie, J., Xiao, W. et al. (2019). Ratiometric fluorescent sensing of Pb2+ and Hg2+ with two types of carbon dot nanohybrids synthesized from the same biomass. *Sensors and Actuators B: Chemical*, 296: 126698.

Lotfy, V. F., Fathy, N. A., and Basta, A. H. (2018). Novel approach for synthesizing different shapes of carbon nanotubes from rice straw residue. *Journal of Environmental Chemical Engineering*, 6(5): 6263–6274.

Lu, Z., Zhou, W., Huo, P., Luo, Y., He, M., Pan, J. et al. (2013). Performance of a novel TiO2 photocatalyst based on the magnetic floating fly-ash cenospheres for the purpose of treating waste by waste. *Chemical Engineering Journal*, 225: 34–42.

Lumbaque, E. C., Schweigert, C. E., Dos Santos, L. O., Lavayen, V., Dos Santos, J. H. Z., and Sirtori, C. (2021). Degradation of pharmaceuticals in wastewater matrices through solar light-driven photocatalyst prepared from petrochemical waste. *Environmental Science and Pollution Research*, 28: 24124–24137.

Manisalidis, I., Stavropoulou, E., Stavropoulos, A., and Bezirtzoglou, E. (2020). Environmental and health impacts of air pollution: a review. *Frontiers in Public Health*, 8: 14.

Medina, T. J., Arredondo, S. P., Corral, R., Jacobo, A., Zárraga, R. A., Rosas, C. A. et al. (2020). Microstructure and Pb2+ adsorption properties of blast furnace slag and fly ash based geopolymers. *Minerals*, 10(9): 808.

Meena, R., Abdullah, M. M., Vasanthakumar, V., Ravichandran, D., and Murugesan, S. (2024). Green biochar-supported ZnFe2O4 composite photocatalyst derived from waste banana peel: a sustainable approach for highly efficient visible light-driven degradation of organic pollutants in wastewater. *Ionics*, 30(9): 5639–5650.

Mehta, A., Mishra, A., Kainth, S., and Basu, S. (2018). Carbon quantum dots/TiO2 nanocomposite for sensing of toxic metals and photodetoxification of dyes with kill waste by waste concept. *Materials & Design*, 155: 485–493.

Mhamane, D., Kim, H. K., Aravindan, V., Roh, K. C., Srinivasan, M., and Kim, K. B. (2016). Rusted iron wire waste into high performance anode (α-Fe 2 O 3) for Li-ion batteries: an efficient waste management approach. *Green Chemistry*, 18(5): 1395–1404.

Misra, V., and Pandey, S. D. (2005). Hazardous waste, impact on health and environment for development of better waste management strategies in future in India. *Environment International*, 31(3): 417–431.

Mittal, M., Sardar, S., and Jana, A. (2021). Nanofabrication techniques for semiconductor chemical sensors. pp. 119–137. *In: Handbook of nanomaterials for sensing applications*. Elsevier.

Mohamed, F., Shaban, M., Aljohani, G., and Ahmed, A. M. (2021). Synthesis of novel eco-friendly CaO/C photocatalyst from coffee and eggshell wastes for dye degradation. *Journal of Materials Research and Technology*, 14: 3140–3149.

Mohsenizadeh, M., Tural, M. K., and Kentel, E. (2020). Municipal solid waste management with cost minimization and emission control objectives: A case study of Ankara. *Sustainable Cities and Society*, 52: 101807.

Mombelli, D., Mapelli, C., Barella, S., Gruttadauria, A., Ragona, M., Pisu, M. et al. (2020). Characterization of cast iron and slag produced by red muds reduction via Arc Transferred Plasma (ATP) reactor under different smelting conditions. *Journal of Environmental Chemical Engineering*, 8(5): 104293.

Moradeeya, P. G., Kumar, M. A., Thorat, R. B., Rathod, M., Khambhaty, Y., and Basha, S. (2017). Nanocellulose for biosorption of chlorpyrifos from water: chemometric optimization, kinetics and equilibrium. *Cellulose*, 24: 1319–1332.

Moradeeya, P. G., Sharma, A., Kumar, M. A., and Basha, S. (2022). Titanium dioxide based nanocomposites–current trends and emerging strategies for the photocatalytic degradation of ruinous environmental pollutants. *Environmental Research*, 204: 112384.

Mubarik, S., Qureshi, N., Sattar, Z., Shaheen, A., Kalsoom, A., Imran, M. et al. (2021). Synthetic approach to rice waste-derived carbon-based nanomaterials and their applications. *Nanomanufacturing*, 1(3): 109–159.

Nama, M., Satasiya, G., Sahoo, T. P., Moradeeya, P. G., Sadukha, S., Singhal, K. et al. (2024). Thermo-chemical behaviour of Dunaliella salina biomass and valorising their biochar for naphthalene removal from aqueous rural environment. *Chemosphere*, 353: 141639.

Nath, A., Mahato, D. K., Singh, L. R., Kuiri, P. K., and Mahato, M. (2023). Waste-derived green ZnO–Fe2O3 photocatalyst for degradation of MB dye pollutant in water. *Inorganic and Nano-Metal Chemistry*, 1–14.

Nayak, P., Kumar, S., Sinha, I., and Singh, K. K. (2019). ZnO/CuO nanocomposites from recycled printed circuit board: preparation and photocatalytic properties. *Environmental Science and Pollution Research*, 26: 16279–16288.

Ngu, P. Z. Z., Chia, S. P. P., Fong, J. F. Y., and Ng, S. M. (2016). Synthesis of carbon nanoparticles from waste rice husk used for the optical sensing of metal ions. *New Carbon Materials*, 31(2): 135–143.

Nguyen, L. H., Ngo, Q. N., Van, H. T., Nguyen, T. P., and Thi, K. O. P. (2021). Reutilization of Fe-containing tailings ore enriched by iron (iii) chloride as a heterogeneous Fenton catalyst for decolorization of organic dyes. *RSC Advances*, 11(26): 15871–15884.

Nizam, N. U. M., Hanafiah, M. M., and Woon, K. S. (2021). A content review of life cycle assessment of nanomaterials: current practices, challenges, and future prospects. *Nanomaterials*, 11(12): 3324.

Oleszczuk, P., Ćwikła-Bundyra, W., Bogusz, A., Skwarek, E., and Ok, Y. S. (2016). Characterization of nanoparticles of biochars from different biomass. *Journal of Analytical and Applied Pyrolysis*, 121: 165–172.

Pak, A. Y., Larionov, K. B., Kolobova, E. N., Slyusarskiy, K. V., Bolatova, J., Yankovsky, S. A. et al. (2022a). A novel approach of waste tires rubber utilization via ambient air direct current arc discharge plasma. *Fuel Processing Technology*, 227: 107111.

Pak, A. Y., Bolatova, Z., Nikitin, D. S., Korchagina, A. P., Kalinina, N. A., and Ivashutenko, A. S. (2022b). Glass waste derived silicon carbide synthesis via direct current atmospheric arc plasma. *Waste Management*, 144: 263–271.

Pathak, D., Bedi, R. K., and Kaur, D. (2010). Characterization of AgInSe2 films deposited by hot-wall vacuum evaporation method. *Materials and Manufacturing Processes*, 25(9): 1012–1017.

Pedrolo, D. R. S., de Menezes Quines, L. K., de Souza, G., and Marcilio, N. R. (2017). Synthesis of zeolites from Brazilian coal ash and its application in SO2 adsorption. *Journal of Environmental Chemical Engineering*, 5(5): 4788–4794.

Periyasamy, M., Sain, S., Ghosh, E., Jenkinson, K. J., Wheatley, A. E., Mukhopadhyay, S. et al. (2022). Visible light photocatalysts from low-grade iron ore: the environmentally benign production of magnetite/carbon (Fe3O4/C) nanocomposites. *Environmental Science and Pollution Research*, 29(5): 6698–6709.

Popova, M., Boycheva, S., Lazarova, H., Zgureva, D., Lázár, K., and Szegedi, Á. (2020). VOC oxidation and CO2 adsorption on dual adsorption/catalytic system based on fly ash zeolites. *Catalysis Today*, 357: 518–525.

Qiu, J., Chen, G., Li, Z., and Zhao, Z. (2010). Preparation of double-walled carbon nanotubes from fullerene waste soot by arc-discharge. *Carbon*, 48(4): 1312–1315.

Raffi, M., Rumaiz, A. K., Hasan, M. M., and Shah, S. I. (2007). Studies of the growth parameters for silver nanoparticle synthesis by inert gas condensation. *Journal of Materials Research*, 22(12): 3378–3384.

Ratsameetammajak, N., Autthawong, T., Chairuangsri, T., Kurata, H., Yu, A. S., and Sarakonsri, T. (2022). Rice husk-derived nano-SiO 2 assembled on reduced graphene oxide distributed on conductive flexible polyaniline frameworks towards high-performance lithium-ion batteries. *RSC Advances*, 12(23): 14621–14630.

Ray, P. C., Yu, H., and Fu, P. P. (2009). Toxicity and environmental risks of nanomaterials: challenges and future needs. *Journal of Environmental Science and Health Part C*, 27(1): 1–35.

Roslan, M. S., Chaudary, K. T., Haider, Z., Aziz, M. S., and Ali, J. (2017). Multi-walled carbon nanotubes grow under low pressure hydrogen, air, and argon ambient by arc discharge plasma. *Fullerenes, Nanotubes and Carbon Nanostructures*, 25(4): 269–272.

Sachdev, A., and Gopinath, P. (2015). Green synthesis of multifunctional carbon dots from coriander leaves and their potential application as antioxidants, sensors and bioimaging agents. *Analyst*, 140(12): 260–4269.

Sahoo, T. P., Satasiya, G., Moradeeya, P. G., Saravaia, H. T., and Kumar, M. A. (2024). Removal of fluoroquinolone antibiotic and sulfonated dye by functionalized Persea americana seed powder: Appraisal on phase transfer kinetics, equilibrium, economics, and applications in rural settings. *Environmental Research*, 261: 119727.

Saito, T. (2009). Magnetic properties of carbon materials prepared from polyvinyl chloride. *Journal of Applied Physics*, 105(1).

Samaddar, P., Ok, Y. S., Kim, K. H., Kwon, E. E., and Tsang, D. C. (2018). Synthesis of nanomaterials from various wastes and their new age applications. *Journal of Cleaner Production*, 197: 1190–1209.

Sha, T., Li, X., Liu, J., Sun, M., Wang, N., Bo, X. et al. (2018). Biomass waste derived carbon nanoballs aggregation networks-based aerogels as electrode material for electrochemical sensing. *Sensors and Actuators B: Chemical*, 277: 195–204.

Sharma, S., Kalita, G., Hirano, R., Shinde, S. M., Papon, R., Ohtani, H. et al. (2014). Synthesis of graphene crystals from solid waste plastic by chemical vapor deposition. *Carbon*, 72: 66–73.

Shi, J., Ni, G., Tu, J., Jin, X., and Peng, J. (2017). Green synthesis of fluorescent carbon dots for sensitive detection of Fe 2+ and hydrogen peroxide. *Journal of Nanoparticle Research*, 19: 1–10.

Shinde, S. B., Nille, O. S., Gore, A. H., Birajdar, N. B., Kolekar, G. B., and Anbhule, P. V. (2022). Valorization of Waste Tungsten Filament into mpg-C3N4–WO3 Photocatalyst: A Sustainable e-Waste Management and Wastewater Treatment. *Langmuir*, 38(44): 13543–13557.

Silva, R. C. F., Ardisson, J. D., Cotta, A. A. C., Araujo, M. H., and de Carvalho Teixeira, A. P. (2020). Use of iron mining tailings from dams for carbon nanotubes synthesis in fluidized bed for 17α-ethinylestradiol removal. *Environmental Pollution*, 260: 114099.

Singh, K. K., Chaudhary, S. K., Venugopal, R., and Gaurav, A. (2017). Bulk synthesis of multi-walled carbon nanotubes by AC arc discharge method. *Proceedings of the Institution of Mechanical Engineers, Part N: Journal of Nanomaterials, Nanoengineering and Nanosystems*, 231(3): 141–151.

Sun, N., Liu, H., and Xu, B. (2015). Facile synthesis of high performance hard carbon anode materials for sodium ion batteries. *Journal of materials chemistry a*, 3(41): 20560–20566.

Sundramurthy, V. P., Nithya, T. G., Masi, C., Gomadurai, C., and Abda, E. M. (2023). Recent advances and prospects for industrial waste management and product recovery for environmental appliances: a review. *Physical Sciences Reviews*, 8(9): 2341–2359.

Suryanarayana, C., Prabhu, B., Raffi, M., Rumaiz, A. K., Hasan, M. M., and Shah, S. I. (2007). Studies of the growth parameters for silver nanoparticle synthesis by inert gas condensation. *Journal of Materials Research*, 22(12): 3378–3384.

Tanniratt, P., Wasanapiarnpong, T., Mongkolkachit, C., and Sujaridworakun, P. (2016). Utilization of industrial wastes for preparation of high performance ZnO/diatomite hybrid photocatalyst. *Ceramics International*, 42(15): 17605–17609.

Tharchanaa, S. B., Priyanka, K., Preethi, K., and Shanmugavelayutham, G. (2021). Facile synthesis of Cu and CuO nanoparticles from copper scrap using plasma arc discharge method and evaluation of antibacterial activity. *Materials Technology*, 36(2): 97–104.

Tripathi, P. K., Durbach, S., and Coville, N. J. (2017). Synthesis of multi-walled carbon nanotubes from plastic waste using a stainless-steel CVD reactor as catalyst. *Nanomaterials*, 7(10): 284.

Tripathi, P. K., Durbach, S., and Coville, N. J. (2019). CVD Synthesis of solid, hollow, and nitrogen-doped hollow carbon spheres from polypropylene waste materials. *Applied Sciences*, 9(12): 2451.

Vadiyar, M. M., Liu, X., and Ye, Z. (2018). Utilizing Waste Thermocol Sheets and Rusted Iron Wires to Fabricate Carbon–Fe3O4 Nanocomposite-Based Supercapacitors: Turning Wastes into Value-Added Materials. *ChemSusChem*, 11(14): 2410–2420.

Verma, R., Dutta, S., Kumar, A., Dabodiya, T. S., Kumar, N., Karuppasamy, K. S. K. et al. (2023). Biochar-based nanocomposite materials: types, characteristics, physical activation, and diverse application scenarios. pp. 3–18. *In: Biochar-Based Nanocomposites for Contaminant Management: Synthesis, Contaminants Removal, and Environmental Sustainability*. Cham: Springer International Publishing.

Wang, X., Firouzkouhi, H., Chow, J. C., Watson, J. G., Ho, S. S. H., Carter, W. et al. (2023). Chemically speciated air pollutant emissions from open burning of household solid waste from South Africa. *Atmospheric Chemistry and Physics*, 23(24): 15375–15393.

Wang, Z., Shen, D., Wu, C., and Gu, S. (2018). State-of-the-art on the production and application of carbon nanomaterials from biomass. *Green Chemistry*, 20(22): 5031–5057.

Wu, Z. J., Huang, W., Cui, K. K., Gao, Z. F., and Wang, P. (2014). Sustainable synthesis of metals-doped ZnO nanoparticles from zinc-bearing dust for photodegradation of phenol. *Journal of Hazardous Materials*, 278: 91–99.

Xiang, X., Xia, F., Zhan, L., and Xie, B. (2015). Preparation of zinc nano structured particles from spent zinc manganese batteries by vacuum separation and inert gas condensation. *Separation and Purification Technology*, 142: 227–233.

Yang, J., Feng, J., Li, W., Chen, X., Liu, X., Ruan, J. et al. (2019). A resource-utilization way of the waste printed circuit boards to prepare silicon carbide nanoparticles and their photocatalytic application. *Journal of Hazardous Materials*, 373: 640–648.

Yermagambet, B. T., Kazankapova, M. K., Kasenov, B. K., Aitmagambetova, A. Z., and Kuanyshbekov, E. E. (2021). Synthesis of graphene-containing nanomaterials based on a carbon product using electric arc discharge. *Solid Fuel Chemistry*, 55: 380–390.

Zhan, L., and Xu, Z. (2012). Separating criterion of Pb, Cd, Bi and Zn from metallic particles of crushed electronic wastes by vacuum evaporation. *Separation Science and Technology*, 47(6): 913–919.

Zhan, L., Li, O., Wang, Z., and Xie, B. (2018). Recycling zinc and preparing high-value-added nanozinc oxide from waste zinc–manganese batteries by high-temperature evaporation-separation and oxygen control oxidation. *ACS Sustainable Chemistry & Engineering*, 6(9): 12104–12109.

Zhan, L., Xiang, X., Xie, B., and Sun, J. (2016). A novel method of preparing highly dispersed spherical lead nanoparticles from solders of waste printed circuit boards. *Chemical Engineering Journal*, 303: 261–267.

Zhang, C., and Li, S. (2018). Utilization of iron ore tailing for the synthesis of zeolite A by hydrothermal method. *Journal of Material Cycles and Waste Management*, 20: 1605–1614.

Zhang, P., Liang, C., Wu, M., Chen, X., Liu, D., and Ma, J. (2022). High-efficient microwave plasma discharging initiated conversion of waste plastics into hydrogen and carbon nanotubes. *Energy Conversion and Management*, 268: 116017.

Zhou, D., Wang, H., Mao, N., Chen, Y., Zhou, Y., Yin, T. et al. (2017). High energy supercapacitors based on interconnected porous carbon nanosheets with ionic liquid electrolyte. *Microporous and Mesoporous Materials*, 241: 202–209.

Zhu, J., Jia, J., Kwong, F. L., Ng, D. H. L., and Tjong, S. C. (2012). Synthesis of multiwalled carbon nanotubes from bamboo charcoal and the roles of minerals on their growth. *biomass and bioenergy*, 36: 12–19.

Zhu, P., Li, Y., Ma, Y., Ruan, X., and Zhang, Q. (2023). Preparation of waste-based Cu–Cu2O/SiO2 photocatalyst from serpentine tailings and waste printed circuit boards and photoreduction of Cr (VI). *Ceramics International*, 49(8): 12518–12528.

CHAPTER-7

Conventional and Advanced Techniques of Environmental Waste Management and Remediation

*Keisham Lebanan** and *Kusum Sharma*

1. Introduction

Environmental degradation brought on by humans, such as pollution and resource exploitation, has become a major global concern. The massive increase in waste generation caused by the continuously growing human population is one of the main causes of the ecosystem's ongoing degradation. Waste is defined as any material that has lost its intended value or use and is discarded by its owner, manufacturer, or user (Koshta et al. 2022). The types of waste vary depending on their origin. Municipal solid waste, agricultural waste, biological waste, and industrial waste are examples of common waste categories. Food scraps, paper, plastics, metals, glass, and other debris are commonly found in municipal solid waste, which comes from homes and businesses (Valavanidis 2023). Livestock operations, crop residues, agro-industrial byproducts, and even agricultural operations connected to tourism are sources of agricultural waste (Gupta et al. 2023). However, a variety of manufacturing processes, mining operations, and factory-related activities generate industrial waste.

Global waste generation is increasing at an unprecedented rate and is expected to reach approximately 2.2 billion tons annually by 2025 (Sharma et al. 2021). With the report from Akhtar et al. (2022), around 242 million tons of plastic waste were produced globally in 2016 alone. In India, small towns generate about 0.41 kg of waste per capita daily, while rural areas produce between 0.3 to 0.4 million metric tons of solid waste per day (Pal and Bhatia 2022), most of which is organic in nature, including livestock manure and crop residues. The surge in waste generation is strongly associated with unsustainable economic development, industrial expansion, urban growth, and population increase, with low-income countries predicted to experience the highest growth in waste production (Voukkali et al. 2024).

Waste management can be accomplished with a variety of methods, such as recycling and waste-to-energy conversion as highlighted in Figure 1. Reducing the quantity of waste that pollutes the environment or ends up in landfills is crucial to ensuring a sustainable future. This can be achieved by using effective waste management practices. Presently, more than 75% of waste management

Department of Life Science, Rabindranath Tagore University, Raisen, Madhya Pradesh-464993, India.
ORCID ID: 0000-0002-6535-071X
* Corresponding author: keisham.lebanan@rntu.ac.in, keisham.lebanan@aisectuniversity.ac.in

DOI: 10.1201/9781003558804-7

Figure 1: Sources and impacts of municipal solid waste.

procedures globally involve landfilling (Maturi et al. 2022). However, leachate, which is produced as a result of poor landfill management, contaminates surface and groundwater (Parvin and Tareq 2021). Furthermore, solid waste degradation exacerbates climate change by contributing around 5% of global greenhouse gas emissions, especially methane (Gupta et al. 2022). Despite these obstacles, every solid waste component has the potential to be recycled and transformed into products with additional value with the right scientific and technological advancements (Ashokkumar et al. 2022). The growing global population, industrialization, and urbanization contribute to the production of enormous amounts of waste—making this study necessary. The sustainability of the earth is threatened by improperly disposed waste, which also causes major environmental problems. Waste management practices including landfilling, incineration, and negligent disposal have several drawbacks, including health risks to people, groundwater and soil contamination, and environmental harm. As a result, this chapter examines both traditional and cutting-edge methods for managing and cleaning up environmental pollution. It highlights the pressing need for effective, science-based, and sustainable waste treatment methods that reduce environmental impact and support resource recovery and climate resilience (Agarwal 2015, Kumar 2015). The given Table 1 presents a comparative analysis of traditional and modern techniques of environmental waste management based on various parameters such as technology, environmental impact, cost, efficiency, and sustainability.

2. Conventional Techniques of Waste Management

Traditional methods for managing hazardous and non-hazardous waste streams include disposal and diversion strategies. Non-hazardous solid waste can be disposed of in two ways: landfilling and thermal treatments like incineration and solid recovered fuel (SRF) systems. In nations like Canada, landfilling is still the most common practice despite the fact that it causes major environmental issues due to the emission of landfill gases and leachate (Ghosh et al. 2023). However, thermal treatments are frequently linked to increased operating costs as well as the production of fly ash, bottom ash, and air pollutants, even if they provide significant volume reduction and energy recovery (heat and electricity). Physical, chemical, biological, and thermal techniques are employed to reduce the amount and toxicity of hazardous waste during disposal. On the other hand, diversion methods aim to reduce the quantity of waste that ends up in landfills by promoting recycling, reuse, and waste reduction (Lai 2014). Material Recovery Facilities (MRFs), which separate and prepare materials for remanufacturing, can be used to recycle glass, metals, plastics, and paper (Nahman 2021). Composting, which is primarily used for organic wastes including crop residues, wood, manure, and food scraps, is another significant diversion strategy. It converts biodegradable materials into nutrient-rich compost for use in agriculture.

2.1 Landfilling

One of the most popular ways to dispose of solid waste worldwide, especially in industrialized and developing countries, is still landfilling. Conventional landfills are designed waste disposal sites in which waste is systematically covered with earth filling after being compacted into certain cells. Environmental control systems are incorporated into the design of these facilities to reduce possible risks to the environment and public health (Fjeld et al. 2023). Environmental organizations like the Ministry of the Environment (MOE) tightly regulate the construction and management of such landfills in areas like Ontario, Canada (Wolters et al. 2025). To reduce odor, litter, and pest infestation, incoming debris is dispersed, compacted, and covered with a thin layer of soil every day in a typical traditional (dry) landfill. A final cover layer, typically composed of clay or synthetic materials, is applied once the landfill has reached its intended capacity. But there are long-term environmental issues with this approach. It might take anything from 30 to 200 years for waste to decompose in dry landfills. Additionally, the possibility of liner failure over time could result in leachate leakage and significant groundwater contamination (Zhang et al. 2021).

Bioreactor landfills have emerged as a state-of-the-art remedy for these issues. By actively controlling moisture and boosting microbial activity, a bioreactor landfill speeds up the degradation of garbage (Srivastava et al. 2022). The controlled intake of liquids significantly accelerates the breakdown of organic compounds that are easily and moderately decomposable. According to Srivastava et al. (2022), bioreactor landfills can stabilize waste in 5–10 years, which is a substantial improvement over typical methods. Bioreactor landfills enable better landfill volume use while also improving organic waste decomposition, increasing methane recovery possibilities, and shortening post-closure care periods. Importantly, even in the event of partial containment system failure, they achieve environmental stabilization—as seen by steady levels of landfill gas production and leachate contamination—in a shorter period of time. In contrast, a bioreactor landfill has a significantly smaller stabilized waste volume after ten years than a regular landfill, indicating increased waste processing and site utilization efficiency. Despite these benefits, extensive monitoring and leachate/gas management systems are required for the successful deployment of bioreactor landfills. The transition to bioreactor landfill technology provides a more efficient and sustainable method for modern waste management and remediation operations, even though conventional landfilling is still widely employed.

2.2 Incineration

Incineration is a common method for handling solid waste, especially plastics, because it significantly reduces the volume and weight of waste materials, hence lowering the amount of land required for landfill disposal. This high-temperature thermal treatment method converts waste into heat, flue gas, and ash, which can be utilized to recover energy. Despite its effectiveness in lowering volume and creating energy, incineration has some serious drawbacks that limit its environmental sustainability.

Incineration is mostly used in waste management to reduce the load on landfills by burning non-recyclable waste materials, particularly plastics (Ferdous et al. 2021). This approach is very expensive and environmentally damaging, even though it successfully lowers the amount of physical trash. One of the primary problems with incineration is the potential release of dangerous chemicals into the atmosphere, such as dioxins, furans, and polychlorinated biphenyls (PCBs). These substances are recognized as highly persistent organic pollutants that negatively impact human and environmental health (Okoye et al. 2021). Strict environmental laws, such as the Clean Air Act (Currie et al. 2023), the Toxic Substances Control Act (Adebiyi 2022), and the National Pollutant Discharge Elimination System (NPDES) (Theodore and Theodore 2021), govern how modern incinerators work in order to mitigate these dangers. Modern incinerators use high-temperature thermal cracking (~850°C) to degrade hazardous pollutants before disposal in the environment (Cui et al. 2021).

Despite these regulatory safeguards, incineration nevertheless contributes significantly to greenhouse gas (GHG) emissions. Burning plastic waste emits methane (CH_4) and carbon dioxide (CO_2), which contribute significantly to global warming (Wu 2024). According to studies, incineration releases far more CO_2 (between 343 and 4605 g CO_2 per kg of plastic) than landfilling (254 g CO_2 per kg of plastic) (Wu 2024), which makes it a less desirable alternative when considering climate change.

Despite its effectiveness in reducing waste volume and recovering energy, incineration is not thought to be the most sustainable way to manage plastic waste because of these environmental effects. Minimizing the ecological impact of incineration requires ongoing advancements in pollution control technologies as well as research into other waste management techniques, including recycling, composting, and advanced bioremediation. Even though incineration is essential for minimizing waste volume and may provide energy advantages, its high operating costs and significant environmental impact—especially when considering toxic emissions and greenhouse gas emissions—make careful implementation under strict environmental supervision necessary.

2.3 Composting

The controlled aerobic decomposition of organic waste materials by microorganisms under particular environmental circumstances is known as composting, and it is an efficient biological waste management strategy. It has several advantages for the environment, agriculture, and economy, making it a sustainable substitute for traditional waste disposal techniques like landfilling. Composting considerably reduces the chance of groundwater contamination, in contrast to landfilling (Mor and Ravindra 2023). This is because, through microbial absorption and modification, composting lowers the concentration of chemical pollutants and pathogenic microorganisms in the waste, such as endocrine-disrupting residues (EDRs) (Sharma et al. 2023) and persistent organic pollutants (POPs) (Kumar et al. 2023).

Composting differs from natural decomposition in that it occurs under carefully controlled environmental conditions, such as optimal moisture, temperature, aeration, and microbial activity. Beneficial bacteria that assist in composting not only decompose organic matter but also absorb harmful chemicals, which aids in soil fertility and detoxification (Michael 2021). Furthermore, composted waste contains a high concentration of organic matter, which enhances soil structure and contains organisms that encourage plant growth. It is also high in essential nutrients. This enhances food security and aligns with sustainable farming practices.

In addition to being used as fertilizer, compost is also utilized in bioremediation, wetland restoration, weed control, pollution prevention, erosion management, and plant disease suppression. By reducing the need for synthetic fertilizers and boosting soil biodiversity, it helps to minimize environmental hazards, including chemical contamination and nutrient runoff (Pahalvi et al. 2021).

Composting does, however, present some real-world challenges. Long maturation times, the likelihood of thermotolerant diseases, disagreeable odors from anaerobic microbial activity (especially the emission of hydrogen sulfide) (Andraskar et al. 2021), and occasional nutritional imbalances are only a few examples. These issues often hinder farmers from adopting composting widely in favor of readily available chemical fertilizers (Zheng et al. 2022). Innovations are required to address odor control, rapid pathogen and heavy metal detection, and improvements in the rate of mineralization in order to advance composting as a common waste management technique. Compost may be made more useful and effective by adding nutrient-rich activators, antibacterial, antifungal, antiviral, and anti-nematode chemicals.

The effects of composting on the environment are a serious concern as well. Despite being more environmentally benign than landfilling or incineration, it nevertheless releases greenhouse gases (GHGs), including methane (CH_4) and carbon dioxide (CO_2), especially when anaerobic conditions are present (Filonchyk et al. 2024). As a result, laws and policies have been put in place in a number of nations to control composting activities and reduce health hazards. Temperature control and oxygen flow management are two key areas for technological advancement since they are essential for sustaining efficient aerobic microbial activity. Monitoring temperatures and optimizing oxygen levels suppress anaerobic microbial processes, which lowers toxic gas emissions and enhances compost quality overall. In summary, composting has a lot to offer in terms of agricultural production and environmental waste management, but it also has drawbacks that need to be resolved through technological advancement and government regulation. Composting can be a key component of sustainable waste management and remediation by emphasizing microbial efficiency, pathogen control, nutrient enrichment, and emissions reduction.

2.4 Recycling

With rapid population growth, particularly in low-income and middle-income countries, the generation of waste is increasing at an unprecedented rate. For example, the global yearly waste from electrical and electronic equipment increased from 33.8 million tons in 2010 to 49.8 million tons in 2018 (Hussain 2021). Despite the availability of waste treatment technologies such as incineration, landfill disposal remains the most common approach in many nations due to a lack of finance, infrastructure, and limited adoption of modern waste management systems. Poorly managed landfills pose substantial health and environmental concerns, as demonstrated by the fatal landslide in Shenzhen, China on December 20, 2015 (Skiveren 2022). Furthermore, around 15 million people globally participate in informal garbage recycling (Ogwueleka and Naveen 2021), primarily processing plastics, metals, glass, and paper. Although they contribute substantially to reducing landfill burdens and promoting resource recovery, the primitive techniques used by informal recyclers—such as open burning, acid leaching, and unprotected handling—lead to severe environmental pollution and public health hazards. These workers are frequently exposed to harmful chemicals, physical injuries, respiratory and skin ailments, infections, and long-term health issues, all of which contribute to reduced life expectancy.

To address these challenges, various recycling and remediation techniques are employed to manage environmental waste. Mechanical recycling is routinely employed to sift, clean, and reprocess plastics, metals, and paper into raw materials, albeit this frequently results in downcycling (Schyns and Shaver 2021). Chemical recycling degrades waste materials into their chemical components, allowing the recovery of monomers or fuels from complex waste such as mixed polymers; nevertheless, this approach demands a significant amount of energy and cost (Chanda 2021). Organic waste, such as food scraps and sewage sludge, is managed using biological treatments such as composting and anaerobic digestion, which produce valuable compost and biogas (Liew 2022). Electronic waste recycling entails disassembly and processing to recover valuable metals using either environmentally benign processes in official settings or dangerous operations in informal settings. Furthermore, contaminated land and waste sites are remediated utilizing techniques such as phytoremediation with plants, bioremediation with microbes, washing soil with chemical agents, and thermal desorption to remove pollutants via heating (Aparicio et al. 2022). In addition, landfill mining is gaining popularity as a way to recover rich resources from abandoned waste sites while also cleaning up legacy pollutants. Integrating the informal sector into formal waste management systems has demonstrated potential for improving both environmental and socioeconomic outcomes. Several Latin American countries have shown that recognizing and supporting informal waste workers through training, protective equipment, legal inclusion, and improved infrastructure can result in increased waste recycling efficiency and lower poverty. Strengthening cooperation between governments, private sectors, and local communities can significantly enhance waste management efforts, particularly in resource-constrained regions. Adopting sustainable recycling techniques and ensuring the inclusion of all stakeholders are crucial steps toward addressing the mounting environmental and public health challenges associated with waste.

3. Advanced Techniques of Waste Management and Remediation

3.1 Bioremediation and Phytoremediation

The main sources of municipal solid waste (MSW) include homes, eateries, marketplaces, retail establishments, public areas, and commercial structures (Yakah et al. 2023). Leachate, which seeps into the environment and accumulates in significant amounts at disposal sites, is a common type of contamination seen in these wastes. The inappropriate processing and disposal of MSW cause serious societal concerns by polluting soil and water supplies, disrupting ecosystems, and having a significant detrimental impact on human health (Choudhary et al. 2024). The sources of MSW and its adverse effects on human health as well as the environment are illustrated in Figure 1. Therefore, proper segregation, recycling, and sustainable waste management practices are essential to effectively cope with municipal solid waste. In order to filter and interact with these pollutants through the air and water channels, soil—a vital natural resource—is essential.

MSW adds to the buildup of heavy metals and other organic contaminants in soil, which causes extensive soil contamination. An estimated 500 million acres worldwide are covered by 5 million polluted soil sites, where levels of pollutants like arsenic, copper, cadmium, chromium, mercury, lead, and nickel surpass baseline or regulatory limitations (Dongre 2021). Because of their high atomic mass and specific gravity, these heavy metals and metalloids are persistent in the environment and present long-term hazards to ecological and human health. As a result, creating efficient soil cleanup methods and putting safe disposal plans into action have become top priorities worldwide.

In order to make contaminated soil healthy and usable again, the technique is known as soil remediation. Various techniques, such as chemical, biological, and physical methods, are employed based on the type and degree of contamination. Because they are economical and environmentally beneficial, biological processes like bioremediation and phytoremediation have drawn a lot of interest. Bioremediation uses microorganisms including fungi, bacteria, and actinomycetes to break down, change, or purify dangerous contaminants in soil into less toxic or dangerous forms. These microorganisms efficiently lower pollution concentrations (Bala et al. 2022) by metabolizing the pollutants as a source of nutrition and energy. Depending on site conditions and contaminant types, bioremediation techniques might include *ex-situ* methods like compost bioremediation and land farming, as well as *in-situ* methods like bioventing and bioaugmentation. On the other hand, phytoremediation employs green plants to remove, trap, degrade, or stabilize pollutants from contaminated soils. This method employs specific plant species that can accumulate heavy metals or degrade organic contaminants through a range of processes such as rhizofiltration, phytoextraction, phytostabilization, and phytodegradation (Bakshe and Jugade 2023). Phytoremediation is a sustainable and visually appealing approach that often improves landscape beauty while restoring ecosystem function (Rocha et al. 2023). It also increases soil health and biodiversity by facilitating interactions between microorganisms and roots. Even though it takes longer than chemical treatments, it is less intrusive and provides the possibility of long-term healing, especially in large areas with low to moderate contamination. Both phytoremediation and bioremediation, taken together, are viable approaches to soil remediation and environmental waste management. By using them, the toxic load of municipal solid waste can be lessened, environmental and human health hazards can be reduced, and contaminated land can be restored for development, recreation, or farming. In order to achieve long-term ecological resilience and public health protection, it will be crucial to integrate this green technology into larger waste management systems as the demand for sustainable solutions increases.

3.2 Nanotechnology-Based Remediation

The increasing volume and complexity of contemporary waste streams, many of which contain heavy metals, microplastics, and persistent organic pollutants that pose long-term threats to ecosystems and human health, make traditional methods like landfilling and burning insufficient. Through the use of nanoparticles with exceptional physicochemical properties, nanotechnology offers creative and practical waste management solutions in this area (Khan et al. 2022). These materials provide improved performance in pollution detection, removal, and degradation because of their high surface area, reactivity, and adjustable surface functions. Techniques using various nanomaterials, including carbon nanotubes, graphene, metal oxide nanoparticles such as TiO_2 and ZnO (Thakre et al. 2021), silver nanoparticles (Dawadi et al. 2021), and modified zeolites, are all included in the category of remediation approaches based on nanotechnology. These substances are strong adsorbents that effectively bind and extract a variety of contaminants from waste streams, such as organic chemicals, dyes, and heavy metals. By using pores at the nanoscale size to selectively trap impurities while permitting cleaned water to flow through, nano-filters—when combined with nanomaterials—improve the effectiveness of wastewater treatment systems (Rajpoot 2025). Another cutting-edge technique based on nanotechnology is photocatalysis, which breaks down organic pollutants into innocuous byproducts like carbon dioxide and water by using light-activated nanoparticles like titanium dioxide. This method has been used to remediate leachates and industrial effluents because it works especially well in UV or visible light. Additionally, real-time, extremely sensitive detection of pollutants in soil, water, and air is made possible by nanosensors, allowing for proactive environmental pollution management and monitoring. These sensors facilitate prompt decision-making and intervention by detecting traces of dangerous pollutants like arsenic, mercury, and volatile organic compounds. The usefulness of nanosensors in waste management is illustrated by case studies from various geographical areas, which range from tracking emissions from landfills to evaluating the quality of wastewater in treatment facilities.

In addition to enhancing the effectiveness and specificity of waste treatment procedures, nanotechnology-based remediation technologies contribute to the transition to circular and sustainable waste management practices. Nanotechnology represents a paradigm shift in the sector by improving waste monitoring systems, lowering environmental toxicity, and allowing resource recovery (Malik et al. 2023). The safe and widespread use of nanotechnology in waste management, which eventually leads to a cleaner and more resilient environment, is based on continued research and development, regulatory support, and environmental risk assessment.

3.3 Plasma Gasification

A state-of-the-art waste-to-energy method that supports the circular economy and sustainable waste management is plasma gasification. In contrast to conventional disposal techniques like landfilling and incineration, plasma gasification uses the extraordinarily high temperatures produced by plasma torches to transform waste into vitrified slag and synthesis gas (syngas), which can both be recycled—slag for building materials and syngas for energy production. This approach has been thoroughly examined to compare its environmental performance across multiple impact categories with traditional waste management techniques using an environmental Life Cycle Assessment (LCA) lens. According to a survey of various LCA studies, plasma gasification has a lot to offer the environment, particularly in terms of pollution reduction (Zhang et al. 2025), resource recovery, and a reduced risk of global warming. With 67% of the assessed studies demonstrating a lower environmental burden compared to landfilling and 61% compared to incineration, LCA results specifically show that plasma gasification outperforms landfilling and incineration in terms of global warming impact (Sanjaya and Abbas 2023). This is because of its ability to destroy waste while also recovering energy, which significantly reduces greenhouse gas emissions and the requirement for fossil fuels. Furthermore, plasma gasification continues to have greater environmental performance than classical (non-plasma) gasification. The extra value of plasma-based systems in attaining a lower environmental footprint is highlighted by the fact that, despite 89% of LCAs favoring conventional gasification over landfilling in decreasing global warming potential (Nagar and Kaushal 2024), only 56% find it more advantageous than incineration.

Remediation using plasma gasification has benefits beyond energy recovery. It reduces the production of toxic residues because the high-temperature plasma arc stabilizes inorganic materials into an inert, non-leachable slag and completely decomposes dangerous organic molecules. In contrast to landfilling procedures, this process eliminates the requirement for post-treatment and long-term environmental monitoring. Systems for plasma gasification are also very effective and versatile since they can handle a variety of waste kinds, such as hazardous, medical, and electronic waste. Overall, plasma gasification is a sustainable alternative to waste management that lessens its negative effects on the environment while recovering valuable resources. Its exceptional performance in a number of LCA impact categories indicates that it has the potential to be a crucial part of integrated waste management systems in the future. With additional advancements, refinement, and incorporation into regulatory structures, plasma gasification has the potential to greatly lessen waste's ecological impact and promote international initiatives for environmental sustainability.

3.4 Electrokinetic Remediation Technology (EKRT)

A novel and environmentally friendly method for treating multi-metal contaminated soils *in-situ*, electrokinetic remediation technology has a lot of promise for managing waste in the environment. It is a promising developing technique because of its high efficiency for fine-grained soils. The use of an electric field in this method to remove contaminants from soil is effective on a wide range of matrices irrespective of heterogeneity (Taneja et al. 2024). The LCA approach provides a comprehensive evaluation of the environmental impacts of renewable energy sources, such as wind, photovoltaic (PV), and biogas, by considering the entire life cycle from production to decommissioning (Elzoghby 2024). Researchers applied LCA methods to compare electrokinetic devices to traditional cleanup procedures (Safwat et al. 2025). The findings reveal that, when built appropriately, EKRT significantly reduces global warming potential (GWP) over a 100-year period (Eckert-Ross 2021). System modifications that minimize energy consumption and mitigate the consequences of system installation and operation are primarily responsible for the GWP reduction. EKRT reduces emissions associated with soil removal and landfill disposal by eliminating the need for transportation and off-site processing (Vocciante et al. 2024). Furthermore, by mobilizing and extracting various heavy metals from soil using a low-intensity electric field, including lead, cadmium, arsenic, and chromium, the technology enables targeted, on-site treatment without creating considerable landscape disruption.

One of the most significant benefits of this remediation strategy is its compatibility with renewable energy sources. EKRT systems, which dramatically reduce reliance on fossil fuels and the overall carbon footprint of cleanup operations, can be combined with solar or wind power to further boost environmental benefits. Thus, by supporting green energy projects and the principles of the circular economy, electrokinetic remediation not only addresses the pressing issue of soil contamination but also advances broader environmental objectives. To sum up, electrokinetic remediation—once thought of as a specialized technology—is now demonstrating itself to be a practical and environmentally preferable substitute for traditional cleanup techniques, particularly when system configurations are properly adjusted. Its function in sustainable waste management techniques has been validated and its advantages quantified with the use of LCA tools like GaBi. Thus, this technology is a progressive approach to environmentally and energy-efficiently managing hazardous sites.

3.5 Circular Economy Practices/ 11 R Approach to Sustainable Waste Management

Modern systems focus on designing products and systems that prioritize reuse, recycling, and minimal waste generation, ensuring resources remain in use for as long as possible. The improper disposal and buildup of organic waste pose significant risks to both human health and the environment, leading to issues like unpleasant odors, contamination of groundwater, and soil pollution (Sharholy et al. 2008, 2005, Rathi 2007, Mishra et al. 2016). To counter these harmful effects, it is crucial to adopt safe and eco-friendly disposal practices. One such strategy is the "11 R formula," according to Hajam et al. 2023 (Figure 2), which encourages rethinking consumption habits, avoiding unnecessary products, minimizing waste production, reusing and repurposing items, recycling materials, repairing damaged items, refilling containers, composting organic waste, and continuously repeating this cycle to foster sustainability (Kaza et al. 2018).

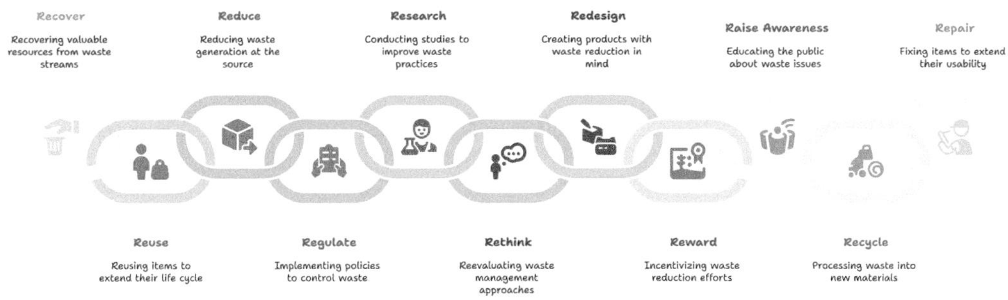

Figure 2: 11 R Strategy for Waste Management.

3.6 Remote Sensing & GIS based Management

Remote Sensing and Geographic Information System (GIS)-based waste management is an innovative approach that utilizes satellite imagery, aerial data, and spatial analysis tools to monitor, manage, and optimize waste management practices. Through remote sensing, real-time data can be collected on waste accumulation, landfill sites, and environmental impacts from a distance. In Coimbatore landfill sites, 75% of municipal solid waste was dumped without treatment, and it was identified using vector data and remote sensing (RS) (Gautam et al. 2020). GIS plays a crucial role by providing spatial visualization and analysis of this data, helping decision-makers identify high-waste areas, plan efficient waste collection routes, and assess the environmental consequences of waste disposal. This approach also integrates data on population density, land use, and infrastructure, allowing for better planning of recycling programs, waste treatment, and disposal facilities.

The datasets and spots join the spatial examination devices given by GIS to arrange and survey in order to choose possible waste areas (Vishnuvardhan and Elangovan 2020). Overall, GIS and remote sensing-based waste management enhances efficiency, reduces costs, and minimizes the environmental impact of waste, contributing to more sustainable urban and rural development.

4. Challenges and Future Prospects

Even though both traditional and cutting-edge methods have been developed, environmental waste management and cleanup still confront many obstacles. Despite their widespread use, traditional techniques like landfilling, incineration, and simple chemical treatments can present long-term health and environmental risks because they produce leachate, emit harmful chemicals, and release greenhouse gases. In low- and middle-income nations, where insufficient infrastructure, a lack

Table 1: Comparative Analysis of Traditional and Modern Techniques of Environmental Waste Management.

Parameter	Traditional Techniques	Modern Techniques
Definition	Conventional, age-old methods used by communities for waste disposal and management.	Advanced, technology-based methods for efficient and sustainable waste management.
Technology Involvement	Minimal or none	High; involves sophisticated machinery and software systems
Environmental Impact	High — leads to soil, water, and air pollution	Low to moderate — controlled emissions and safe disposal
Health Hazards	Higher due to unregulated disposal and open burning	Lower due to strict health and safety norms
Waste Segregation	Rarely practiced	Systematic and mandatory in most modern systems
Recycling Rate	Low	High — formal recycling plants and organized systems
Energy Recovery	Not possible or very limited	Possible via waste-to-energy (WTE) plants and biogas units
Cost Involved	Low initial cost but higher long-term environmental cost	High initial investment, lower long-term environmental and health costs
Public Participation	High — community managed	Moderate — managed by municipal bodies and private operators
Waste Volume Reduction Efficiency	Low (no volume control methods)	High (mechanical, chemical, and biological processes involved)
Sustainability	Less sustainable due to accumulation of waste and pollution	More sustainable with recycling, reuse, and recovery strategies
Use of Information Technology	None	Extensive — for monitoring, GIS mapping, data analytics, smart waste collection

of money, and insufficient legal frameworks impede effective waste disposal and management, these approaches also struggle with the increasing amount and complexity of trash. Furthermore, many areas are dominated by the informal trash sector, which puts workers' health in serious danger and contributes to environmental degradation. However, by increasing waste conversion efficiency, encouraging resource recovery, and reducing ecological footprints, cutting-edge methods like bioremediation, phytoremediation, plasma gasification, electrokinetic remediation, and nanotechnology-based approaches present viable substitutes. These approaches may not be widely adopted, however, due to their high upfront costs, technological complexity, and requirement for specialized workers. Furthermore, in order to reach their full potential, the integration of renewable energy sources with remediation technology is still in its infancy and needs funding and supportive regulations.

Prospects for the future rest on the comprehensive implementation of integrated waste management plans that blend conventional and contemporary techniques, bolstered by research and development expenditures, public awareness campaigns, and legislative changes. The circular economy's principles for optimizing resource recovery, life cycle assessment for assessing environmental impact, and technology transfer to poorer nations to increase capacity should all be prioritized. To overcome these obstacles and guarantee the sustainable development of waste management and remediation techniques worldwide, cooperation between governments, businesses, and academic institutions will be crucial.

5. Conclusion

Effective environmental waste management and remediation require a harmonious blend of conventional and advanced technologies and are vital to combating the growing environmental pollution crisis. Traditional methods such as landfilling, incineration, and composting continue to serve as foundational waste management strategies, while innovative technologies have emerged to enhance their effectiveness and sustainability. The integration of advanced techniques, including bioremediation, phytoremediation, and nanotechnology-based solutions, has revolutionized our ability to treat contaminated environments and manage waste more efficiently. These modern approaches offer more environmentally friendly alternatives that minimize secondary pollution and promote resource recovery. Digital technologies and smart waste management systems have transformed the sector, enabling real-time monitoring, optimized collection routes, and improved sorting efficiency. The emergence of circular economy principles has shifted focus toward waste prevention, reuse, and recycling, challenging the traditional linear consumption model. However, successful environmental waste management requires a holistic approach that combines both conventional and advanced techniques. The selection of appropriate methods depends on various factors, including waste type, local infrastructure, economic feasibility, and environmental conditions. Looking forward, continued research and development in this field will be crucial for addressing emerging waste challenges and environmental concerns. The future of waste management lies in sustainable, integrated solutions that balance technological innovation with practical implementation, ultimately working toward a cleaner, healthier planet for future generations. The evolution of waste management techniques demonstrates our growing commitment to environmental stewardship and highlights the importance of continued innovation in this critical field.

References

Abu-Rayash, A., and Dincer, I. (2023). Development and assessment of an integrated wind-solar based energy system for sustainable communities. *Energy Conversion and Management*, 277: 116680.

Adebiyi, F. M. (2022). Air quality and management in petroleum refining industry: A review. *Environmental Chemistry and Ecotoxicology*, 4: 89–96.

Akhtar, R., Sirwal, M. Y., Hussain, K., Dar, M. A., Shahnawaz, M., and Daochen, Z. (2022). Impact of plastic waste on the coral reefs: an overview. *Impact of Plastic Waste on the Marine Biota*, 239–256.

Andraskar, J., Yadav, S., and Kapley, A. (2021). Challenges and control strategies of odor emission from composting operation. *Applied Biochemistry and Biotechnology*, 193: 2331–2356.

Aparicio, J. D., Raimondo, E. E., Saez, J. M., Costa-Gutierrez, S. B., Álvarez, A., Benimeli, C. S. et al. (2022). The current approach to soil remediation: A review of physicochemical and biological technologies, and the potential of their strategic combination. *Journal of Environmental Chemical Engineering*, 10(2): 107141.

Ashokkumar, V., Flora, G., Venkatkarthick, R., SenthilKannan, K., Kuppam, C., Stephy, G. M. et al. (2022). Advanced technologies on the sustainable approaches for conversion of organic waste to valuable bioproducts: Emerging circular bioeconomy perspective. *Fuel*, 324: 124313.

Bakshe, P., and Jugade, R. (2023). Phytostabilization and rhizofiltration of toxic heavy metals by heavy metal accumulator plants for sustainable management of contaminated industrial sites: A comprehensive review. *Journal of Hazardous Materials Advances*, 10: 100293.

Bala, S., Garg, D., Thirumalesh, B. V., Sharma, M., Sridhar, K., Inbaraj, B. S. et al. (2022). Recent strategies for bioremediation of emerging pollutants: a review for a green and sustainable environment. *Toxics*, 10(8): 484.

Chanda, M. (2021). Chemical aspects of polymer recycling. *Advanced Industrial and Engineering Polymer Research*, 4(3): 133–150.

Choudhary, M., Singh, D., Parihar, M., Choudhary, K. B., Nogia, M., Samal, S. K. et al. (2024). Impact of municipal solid waste on the environment, soil, and human health. pp. 33–58. *In: Waste Management for Sustainable and Restored Agricultural Soil*. Academic Press.

Cui, X., Zhang, J., Wang, X., Pan, M., Lin, Q., Khan, K. Y. et al. (2021). A review on the thermal treatment of heavy metal hyperaccumulator: Fates of heavy metals and generation of products. *Journal of Hazardous Materials*, 405: 123832.

Currie, J., Voorheis, J., and Walker, R. (2023). What caused racial disparities in particulate exposure to fall? New evidence from the Clean Air Act and satellite-based measures of air quality. *American Economic Review*, 113(1): 71–97.

Dawadi, S., Katuwal, S., Gupta, A., Lamichhane, U., Thapa, R., Jaisi, S. et al. (2021). Current research on silver nanoparticles: synthesis, characterization, and applications. *Journal of Nanomaterials*, 1: 6687290.

Dongre, R. S. (2021). Chromium & lead as soil pollutants: insights on toxicity profiles and their remediation. *Journal of Advanced Biotechnology and Bioengineering*, 9: 1–16.

Eckert-Ross, A. (2021). An analysis of the global warming impact of Humboldt State University's refrigerant inventory.

Elzoghby, A. (2024). *An Elaborated Environmental Life Cycle Assessment on Renewable Energy in Germany* (Doctoral dissertation, Hochschule Rhein-Waal).

Ferdous, W., Manalo, A., Siddique, R., Mendis, P., Zhuge, Y., Wong, H. S. et al. (2021). Recycling of landfill wastes (tyres, plastics and glass) in construction–A review on global waste generation, performance, application and future opportunities. *Resources, Conservation and Recycling*, 173: 105745.

Filonchyk, M., Peterson, M. P., Zhang, L., Hurynovich, V., and He, Y. (2024). Greenhouse gases emissions and global climate change: Examining the influence of CO2, CH4, and N2O. *Science of The Total Environment*, 173359.

Fjeld, R. A., DeVol, T. A., and Martinez, N. E. (2023). *Quantitative environmental risk analysis for human health.* John Wiley & Sons.

Gautam, S., Brema, J., and Dhasarathan, R. (2020). Spatio-temporal estimates of solid waste disposal in an urban city of India: A remote sensing and GIS approach. Environmental Technology & Innovation, 18: 100650.

Ghosh, A., Kumar, S., and Das, J. (2023). Impact of leachate and landfill gas on the ecosystem and health: Research trends and the way forward towards sustainability. *Journal of Environmental Management*, 336: 117708.

Gupta, A. P., Upadhyay, P., Sen, T., and Dutta, J. (2023). Agricultural waste as a resource: the lesser travelled road to sustainability. *Agriculture Waste Management and Bioresource: The Circular Economy Perspective*, 1–20.

Gupta, J., Ghosh, P., Kumari, M., and Thakur, I. S. (2022). Solid waste landfill sites for the mitigation of greenhouse gases. pp. 315-340. *In: Biomass, biofuels, biochemicals*. Elsevier.

Hajam, Y. A., Kumar, R., and Kumar, A. (2023). Environmental waste management strategies and vermi transformation for sustainable development. *Environmental Challenges*, 100747.

Hussain, C. M. (Ed.). (2021). *Environmental Management of Waste Electrical and Electronic Equipment*. Elsevier.

Kaza, S., Yao, L., Bhada-Tata, P., and Van Woerden, F., (2018). What a Waste 2.0: a Global Snapshot of Solid Waste Management to 2050. World Bank Publications.

Khan, Y., Sadia, H., Ali Shah, S. Z., Khan, M. N., Shah, A. A., Ullah, N. et al. (2022). Classification, synthetic, and characterization approaches to nanoparticles, and their applications in various fields of nanotechnology: a review. *Catalysts*, 12(11): 1386.

Koshta, N., Patra, S., and Singh, S. P. (2022). Sharing economic responsibility: Assessing end user's willingness to support E-waste reverse logistics for circular economy. *Journal of Cleaner Production*, 332: 130057.

Kumar, M., Saggu, S. K., Pratibha, P., Singh, S. K., and Kumar, S. (2023). Exploring the role of microbes for the management of persistent organic pollutants. *Journal of Environmental Management*, 344: 118492.

Lai, K., Li, L., Mutti, S., Staring, R., Taylor, M., Umali, J. et al. (2014). Evaluation of waste reduction and diversion as alternatives to landfill disposal. pp. 183–187. *In: 2014 Systems and Information Engineering Design Symposium (SIEDS)*. IEEE.

Liew, C. S., Yunus, N. M., Chidi, B. S., Lam, M. K., Goh, P. S., Mohamad, M. et al. (2022). A review on recent disposal of hazardous sewage sludge via anaerobic digestion and novel composting. *Journal of Hazardous Materials*, 423: 126995.

Malik, S., Muhammad, K., and Waheed, Y. (2023). Nanotechnology: a revolution in modern industry. *Molecules*, 28(2): 661.

Maturi, K. C., Gupta, A., Haq, I., and Kalamdhad, A. S. (2022). A glance over current status of waste management and landfills across the globe: A review. *Biodegradation and Detoxification of Micropollutants in Industrial Wastewater*, 131–144.

Michael, P. S. (2021). Role of organic fertilizers in the management of nutrient deficiency, acidity, and toxicity in acid soils–A review. *Journal of Global Agriculture and Ecology*, 12(3): 19–30.

Mishra, R. K., Mohammad, N., and Roychoudhury, N. (2016). Soil pollution: Causes, effects and control. *Van Sangyan*, 3(1): 1–14.

Mor, S., and Ravindra, K. (2023). Municipal solid waste landfills in lower-and middle-income countries: Environmental impacts, challenges and sustainable management practices. *Process Safety and Environmental Protection*, 174: 510–530.

Nagar, V., and Kaushal, R. (2024). A review of recent advancement in plasma gasification: A promising solution for waste management and energy production. *International Journal of Hydrogen Energy*, 77: 405–419.

Nahman, A. (2021). Incentives for municipalities to divert waste from landfill in South Africa. *Waste research development and innovation roadmap research report. Pretoria: Council for Scientific and Industrial Research.*

Ogwueleka, T. C., and Naveen, B. P. (2021). Activities of informal recycling sector in North-Central, Nigeria. *Energy Nexus*, 1: 100003.

Okoye, C. O., Addey, C. I., Oderinde, O., Okoro, J. O., Uwamungu, J. Y., Ikechukwu, C. K. et al. (2022). Toxic chemicals and persistent organic pollutants associated with micro-and nanoplastics pollution. *Chemical Engineering Journal Advances*, 11: 100310.

Padhan, D., Rout, P. P., Kundu, R., Adhikary, S., and Padhi, P. P. (2021). Bioremediation of heavy metals and other toxic substances by microorganisms. *Soil Bioremediation: An Approach Towards Sustainable Technology*, 285-329.

Pahalvi, H. N., Rafiya, L., Rashid, S., Nisar, B., and Kamili, A. N. (2021). Chemical fertilizers and their impact on soil health. *Microbiota and Biofertilizers, Vol 2: Ecofriendly Tools For Reclamation of Degraded Soil Environs*: 1–20.

Pal, M. S., and Bhatia, M. (2022). Current status, topographical constraints, and implementation strategy of municipal solid waste in India: a review. *Arabian Journal of Geosciences*, 15(12): 1176.

Parvin, F., and Tareq, S. M. (2021). Impact of landfill leachate contamination on surface and groundwater of Bangladesh: a systematic review and possible public health risks assessment. *Applied Water Science*, 11(6): 100.

Rajpoot, S. (2025). Smart and Innovative Nanotechnology Applications for Water Purification: Exploring Technologies for a Cleaner Future. In *Emerging Trends and Technologies in Water Management and Conservation* (pp. 87-120). IGI Global Scientific Publishing.

Rathi, S. (2007). Optimization model for integrated municipal solid waste management in Mumbai, India. *Environment and development economics*, 12(1): 105–121.

Rocha, C. S., Rocha, D. C., Kochi, L. Y., Carneiro, D. N. M., Dos Reis, M. V., and Gomes, M. P. (2022). Phytoremediation by ornamental plants: a beautiful and ecological alternative. *Environmental Science and Pollution Research*, 1–19.

Safwat, S. M., Duraid, S., and Galal, M. M. (2025). Performance evaluation and life cycle assessment of electrokinetic technology with various types of electrodes for the treatment of Co-contaminated soil in brick industry site. *Separation Science and Technology*, 1–17.

Sanjaya, E., and Abbas, A. (2023). Plasma gasification as an alternative energy-from-waste (EFW) technology for the circular economy: An environmental review. *Resources, Conservation and Recycling*, 189: 106730.

Schyns, Z. O., and Shaver, M. P. 2021. Mechanical recycling of packaging plastics: a review. *Macromolecular Rapid Communications*, 42(3): 2000415.

Sharholy, M., Ahmad, K., Mahmood, G., and Trivedi, R. C. (2005). Analysis of municipal solid waste management systems in Delhi–a review. In *Book of proceedings for the second International Congress of Chemistry and Environment, Indore, India* (pp. 773–777).

Sharholy, M., Ahmad, K., Mahmood, G., and Trivedi, R. C. (2008). Municipal solid waste management in Indian cities–A review. *Waste management*, 28(2): 459–467.

Sharma, H. B., Vanapalli, K. R., Samal, B., Cheela, V. S., Dubey, B. K., and Bhattacharya, J. (2021). Circular economy approach in solid waste management system to achieve UN-SDGs: Solutions for post-COVID recovery. *Science of the Total Environment*, 800: 149605.

Sharma, P., Nadda, A. K., and Kumar, S. (2023). Microbial community profiling in bio-stimulated municipal solid waste for effective removal of organic pollutants containing endocrine disrupting chemicals. *Microbiological Research*, 267: 127273.

Skiveren, N. (2022). Cinematic Waesthetics: Wasted Worlds, Wasted Lives and Becoming-Waste in Contemporary Science Fiction Film. *Revenant Journal*.

Srivastava, A. N., Singh, R., Chakma, S., and Birke, V. (2022). Advancements in operations of bioreactor landfills for enhanced biodegradation of municipal solid waste. pp. 153–166. *In: Circular Economy in Municipal Solid Waste Landfilling: Biomining & Leachate Treatment: Sustainable Solid Waste Management: Waste to Wealth*. Cham: Springer International Publishing.

Suresh, K., Anbalagan, S., Ponnambalam Ragini, Y., and Agaram Sundaram, V. (2024). An In-Depth Analysis of Sustainable Approaches, Conventional Techniques, and Policy Frameworks for Managing Municipal Solid Waste in India. *Journal of Renewable Energy and Environment*, 11(2): 102–117.

Taneja, S., Karaca, O., and Haritash, A. K. (2024). Electrokinetic remediation: Past experiences and future roadmap for sustainable remediation of metal-contaminated soils. *Journal of Geochemical Exploration*, 259: 107437.

Thakre, K. G., Barai, D. P., and Bhanvase, B. A. (2021). A review of graphene-TiO2 and graphene-ZnO nanocomposite photocatalysts for wastewater treatment. *Water Environment Research*, 93(11): 2414–2460.

Theodore, M. K., and Theodore, L. (2021). Environmental Regulations. In *Introduction to Environmental Management* (pp. 7-24). CRC Press.

Valavanidis, A. (2023). Global municipal solid waste (MSW) in crisis. *Two Billion Tonnes of MSW Every Year, a Worrying Worldwide Environmental Problem. Researchgate*.

Vishnuvardhan, K., and Elangovan, K. (2020). Application of remote sensing and GIS for identifying suitable sites for solid waste disposal in Erode Corporation, Tamil Nadu, India.

Vocciante, M., Alessandra de Folly, D., Franchi, E., Petruzzelli, G., and Ferro, S. (2021). CO2 footprint analysis of consolidated and innovative technologies in remediation activities. *Journal of Cleaner Production*, 297: 126723.

Voukkali, I., Papamichael, I., Loizia, P., and Zorpas, A. A. (2024). Urbanization and solid waste production: Prospects and challenges. *Environmental Science and Pollution Research*, 31(12): 17678–17689.

Wolters, E. A., Steel, B. S., Krawchenko, T., and Farooq, S. (2025). Small and Rural Local Government Environmental Sustainability Plans, Programs and Policies in Cascadia: A Comparative Analysis. *Society & Natural Resources*, 1–27.

Wu, Y. (2024). Waste-to-energy technology towards energy and carbon neutrality: exploring different waste treatment scenarios in China.

Yakah, N., Samavati, M., Akuoko Kwarteng, A., Martin, A., and Simons, A. (2023). Prospects of waste incineration for improved municipal solid waste (MSW) management in Ghana—a review. *Clean Technologies*, 5(3): 997–1011.

Zhang, J., Zhang, J. M., Xing, B., Liu, G. D., and Liang, Y. (2021). Study on the effect of municipal solid landfills on groundwater by combining the models of variable leakage rate, leachate concentration, and contaminant solute transport. *Journal of Environmental Management*, 292: 112815.

Zhang, R., Yin, K., Wei, R., Ruan, J., Yang, J., and Wang, Y. (2025). Comprehensive analysis of life cycle energy consumption and environmental impact of hydrogen production process via plasma co-gasification of coal and biomass. *Energy*, 135976.

Zheng, S., Yin, K., and Yu, L. (2022). Factors influencing the farmer's chemical fertilizer reduction behavior from the perspective of farmer differentiation. *Heliyon*, 8(12).

CHAPTER-8

Role of Nanomaterials in Soil Remediation

Ankita Ghosh[1,*] and *Pritom Biswas*[2]

1. Introduction

Soil is a crucial life-supporting system and a fundamental component of terrestrial ecosystems (Sol-Magdaleno et al. 2025). Soil not only serves as a fundamental medium for plant growth and food production but also plays a crucial role in regulating vital planetary processes that are essential for sustaining life on Earth (Abhilash et al. 2013). These processes encompass biodiversity conservation, biogenic gas regulation, biogeochemical cycles, climate regulation, ecosystem restoration, pollutant detoxification, and the water cycle (Gil-Díaz et al. 2017). Proper soil management is critical for ensuring agricultural productivity and safeguarding the equilibrium of Earth's ecological functions (Lal 2015). However, soil contamination and degradation have emerged as major environmental concerns that jeopardize agricultural output and food security (Lal 2009, New et al. 2023). Key factors driving soil degradation include declining fertility, deforestation, erosion, overgrazing, and urbanization (Lal 2015). Furthermore, chemical pollutants, including heavy metals, pesticides, persistent organic pollutants (POPs), and polycyclic aromatic hydrocarbons (PAHs), have exacerbated the issue of soil degradation (Ghosh et al. 2023). These environmental risks are considered major threats to ecosystems. Due to these factors and ongoing human activities, it is estimated that around 30% of the land is either contaminated or degraded, with the extent of degradation continuing to rise over time (Abhilash et al. 2013, Ghosh et al. 2024). Contaminated soils are unsuitable for food production and pose significant risks of contaminating the food chain, as contaminants like heavy metals, POPs, and pesticides can bioaccumulate and biomagnify (Ghosh et al. 2023). Consuming crops cultivated in contaminated soils can lead to significant health risks, with pollutants potentially entering the human body through the food chain. The widespread degradation of land and soil has sparked global concerns, emphasizing the pressing need for the restoration of these damaged resources (Sah et al. 2022).

Nanotechnology is recognized as a rapidly advancing field, expanding its application across various sectors. It encompasses the design, measurement, and manipulation of matter at the atomic, molecular, macromolecular, and micromolecular levels, resulting in materials with unique

[1] Department of Environmental Sciences, Govind Ballabh Pant University of Agriculture & Technology, Pantnagar-263145, Uttarakhand, India. ORCID iD- 0000-0002-1158-4836

[2] Department of Biotechnology, Central University of South Bihar, Gaya- 824236, Bihar, India. ORCID iD-0000-0001-8246-2506

* Corresponding author: ankitaghoshdav@gmail.com

DOI: 10.1201/9781003558804-8

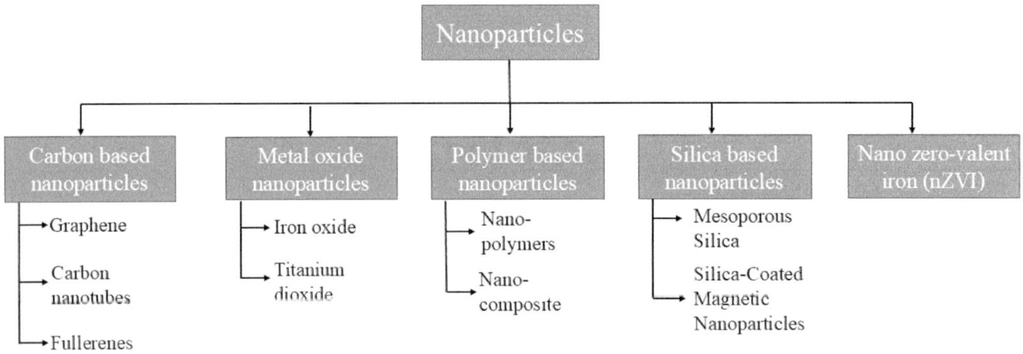

Figure 1: Classification of nanoparticles used in soil remediation (Thangavelu et al. 2022).

properties distinct from their bulk counterparts (Khan et al. 2017). The entities produced via this method are categorized as nanoparticles, characterized as particles exhibiting at least one external dimension within the range of 1 to 100 nanometers (Biswas and Kumar 2021). Among the diverse applications of nanotechnology, its potential in environmental restoration, particularly in treating wastewater, groundwater, and soil contamination, has gained considerable attention (Chakraborty et al. 2023). The increasing adoption of nanotechnology for environmental cleanup is driven by the need for cleaner, cost-effective, readily available, and efficient technologies that minimize residue and ecological persistence (Cai et al. 2019).

The application of nanoparticles in the remediation of chemically contaminated soils, especially those polluted with heavy metals, pesticides, and POPs, is a subject of significant research. Frequently utilized nanoparticles for soil remediation encompass bimetallic nanoparticles, fullerenes, multiwalled carbon nanotubes (MWCNTs), nanoscale zerovalent iron (nZVI), stabilized nanoparticles, titanium dioxide (TiO_2), and zinc oxide (ZnO) (Cai et al. 2019, Ng et al. 2025) nanomaterials are also being increasingly utilized for the remediation of contaminated sediments due to their excellent capacity of immobilizing/adsorbing metal(loid. Figure 1 illustrates the several types of nanoparticles utilized in soil remediation. Nanoremediation refers to the use of reactive nanomaterials to treat contaminated soils by transforming or neutralizing pollutants (Yan et al. 2012, Dhanapal et al. 2024). Owing to their small size and distinctive properties, nanomaterials possess larger surface areas and enhanced sorption capacities, which make them highly efficient adsorbents (Gong et al. 2018). Other advantageous characteristics include the ability to modify properties at lower temperatures, reduced interparticle diffusion distances, adjustable pore sizes, and unique surface chemistries (Tang et al. 2014). These properties allow nanomaterials to function as highly effective catalysts for the chemical reduction and catalysis required to address pollutants. As a result, nanomaterials have attracted considerable interest for their potential in remediating soil environments contaminated by hazardous substances, including heavy metals, chlorinated organic solvents, organochlorine pesticides, PAHs, and polychlorinated biphenyls (PCBs) (Chakraborty et al. 2023). This chapter explores the distinct types of nanomaterials employed in soil remediation, their mechanisms of action, and the efficacy of these materials in improving soil health. Additionally, it addresses the challenges and limitations associated with the use of nanotechnology in soil remediation.

2. Types of Nanomaterials Used in Soil Remediation

2.1 Carbon-Based Nanomaterials (CBNs)

Carbon-based nanomaterials (CBNs) have garnered significant scientific attention due to their unique attributes, including exceptional thermal and electrical conduction, remarkable mechanical robustness, and extraordinary optical properties. These materials encompass various carbon allotropes with different geometries, with the predominant hybridization state of carbon-carbon bonds influencing their chemical and electronic properties. The flexibility in hybridization states enables the formation of distinct structural configurations (Rahman et al. 2011).

Graphene, a two-dimensional sheet comprising a solitary atomic layer of sp^2-hybridized carbon atoms organized in a hexagonal lattice, has a surface area approximately twice that of single-walled carbon nanotubes. It is essentially a single layer extracted from graphite. The robust carbon-carbon bonds in graphene, combined with weak van der Waals forces between layers, create a stable, durable, and flexible material. Furthermore, the high electron mobility facilitated by delocalized π-electrons renders graphene an excellent conductor (Brownson and Banks 2010). Graphene oxide (GO), an oxidized form of graphene, incorporates oxygen-containing functional groups, including carboxylic acids, epoxides, and hydroxyl groups. These functional groups enhance GO's utility as a cost-effective precursor for producing graphene-based materials (Dreyer et al. 2009). The acidic layered structure of GO facilitates interactions with basic contaminants such as ammonia and cationic metals, including lead (Pb^{2+}) and chromium (Cr^{6+}) (Wang et al. 2017).

Carbon nanotubes (CNTs), which are essentially graphene sheets curled into cylindrical shapes, are classified as single-walled (SWCNTs) or multi-walled (MWCNTs) depending on the number of graphene layers. SWCNTs can have diameters as narrow as 0.4 nm (Zhao et al. 2016). Similar to graphene, CNTs possess outstanding mechanical properties due to the robust covalent carbon bonds within their structure, making them some of the most durable and inflexible materials in terms of tensile strength and elasticity (Kumar and Harsha 2016). Their extensive specific surface area allows for efficient adsorption and capture of heavy metals, including cadmium (Cd^{2+}), copper (Cu^{2+}), Pb^{2+}, mercury (Hg), and zinc (Zn^{2+}), among others (Anitha et al. 2015). However, the significant surface area of CNTs also results in van der Waals forces that promote aggregation, reducing their adsorption sites. Mechanical methods, such as ultrasonication, or chemical modifications, like the addition of surfactants, are used to disperse CNTs and maintain their long-term effectiveness (Rossell et al. 2013).

Fullerenes, characterized by their spherical, hollow structures derived from graphene, are another notable carbon-based nanomaterial. Buckminsterfullerene (C_{60}), the most stable form of fullerene, consists of 60 carbon atoms arranged in a spherical configuration resembling a soccer ball. It has a diameter of ~0.7 nm and features a truncated icosahedron structure composed of 20 hexagons and 12 pentagons (Liang et al. 2015). SWCNTs, MWCNTs, and graphene have been widely investigated for environmental purposes because of their vast surface areas and low propensity to clump together (Jha et al. 2016). These materials are especially efficient in eliminating both inorganic and organic pollutants from soil, water, and air. Furthermore, CBNs can break down contaminants through photocatalytic reactions. Upon exposure to photons of appropriate energies, electrons within CBNs undergo excitation from the valence band to the conduction band, leading to the generation of electron-hole pairs (e^-/h^+) (Tang et al. 2012). In the valence band, h^+ interacts with water molecules to produce hydroxyl radicals (•OH), which are powerful oxidants ($E^o = 2.7$ eV) that break down persistent chlorinated organic compounds. Simultaneously, in the conduction band, e^- reacts with oxygen (O_2) to generate superoxide radicals (O_2^-•), which contribute to the reduction of metallic pollutants (Guerra et al. 2018).

The integration of TiO_2 nanoparticles onto graphene surfaces enhances photocatalytic activity by improving light absorption, expanding the effective wavelength range, and reducing the recombination of e^-/h^+ pairs (He et al. 2018). Several hybrid composites of graphene combined with semiconductor materials have been developed for photocatalytic applications. The photocatalytic performance typically improves with higher graphene concentrations; however, surpassing the optimal graphene proportions may diminish efficiency. For instance, Zhang et al. (2010) achieved optimal gas-phase benzene removal with TiO_2-graphene composites containing 0.5% graphene. Similarly, Liu et al. (2011) reported high photocatalytic performance using ZnO-graphene composites with 1.0% graphene for Cr^{6+} reduction under UV irradiation. Additionally, CNT-based photocatalysts incorporating TiO_2 nanoparticles have been successfully employed in the photooxidation of organic pollutants (Xu et al. 2005).

2.2 Metal and Metal Oxide-Based Nanomaterials

Magnetite (Fe_3O_4) is among the primary iron ores and exhibits the highest magnetism of any natural mineral on Earth (Cornell and Schwertmann 2003). When reduced to nanoscale dimensions, Fe_3O_4 nanoparticles exhibit superparamagnetic properties, enabling simple separation and recovery from the medium through the application of an external magnetic field. These nanoparticles are also biocompatible and nontoxic (Wei et al. 2012). They have been utilized in various environmental applications, including the adsorption-based removal of heavy metals and organic pollutants, as well as in chemical methods (Gutierrez et al. 2017). For instance, Dong et al. (2018) synthesized Fe_3O_4 nanoparticles supported on wood biochar, successfully utilizing them in the remediation of PAH-contaminated sediments through chemical oxidation with sodium persulfate as the source of SO_4^- radicals. Konate et al. (2017) revealed that Fe_3O_4 nanoparticles within soil environments diminished the concentration of heavy metals, exerting a beneficial impact on wheat seed growth and mitigating oxidative stress responses. This advantageous outcome was attributed not only to the electrostatic attraction and retention of metals but also to the stimulation of defense mechanisms within wheat seedlings.

Fe_3O_4 nanoparticles have garnered significant attention as heterogeneous Fenton catalysts for the degradation of organic pollutants (Gamallo et al. 2020). The Fenton reaction initiates with electron transfer, wherein ferrous iron (Fe^{2+}) reacts with hydrogen peroxide (H_2O_2), producing •OH, which are highly reactive species that oxidize organic matter to CO_2 and H_2O. Subsequently, ferric iron (Fe^{3+}) is reduced back to Fe^{2+}, forming hydroperoxyl radicals (•OOH), another oxidative agent for organic compounds (Barzegar et al. 2009). Natural Fe_3O_4, containing impurities, has also been shown to catalyze H_2O_2 decomposition, leading to •OH generation for degrading pollutants like p-nitrophenol (He et al. 2018). Fenton-like reactions in natural soils have similarly been effective in degrading compounds such as pentachlorophenol (Konate et al. 2017). Luo et al. (2018) investigated the elimination of phenanthrene from polluted soils by employing a photo-Fenton system comprising mesoporous magnetite and carboxylate-rich carbon composites, operating under visible light irradiation at neutral pH conditions.

TiO_2 is another widely applied nanomaterial in soil remediation (Gu et al. 2012). Upon exposure to light, TiO_2 undergoes electron excitation from the valence to the conduction band, leading to the generation of photoinduced holes (h^+). These h^+ can interact with water or hydroxyl ions to generate •OH, while the conduction band electrons (e^-) reduce oxygen to superoxide ions (O_2^-). Reactive species such as •OH, O_2^-, and h^+ exhibit high oxidative potential, enabling the degradation of organic pollutants and microbial inactivation (Elmolla and Chaudhuri 2010). Wang et al. (2016) developed a UV-driven TiO_2 nanoparticle-based photocatalytic approach for the remediation of soils contaminated with diphenylarsinic acid, achieving optimal results by fine-tuning parameters such as soil-to-water ratio, TiO_2 dosage, irradiation duration, and light intensity. Similarly, TiO_2 in micro- and nanoscale forms has been effective in degrading pyrene in various soil types under solar or dark conditions (Chang Chien et al. 2011). However, TiO_2 encounters challenges such as the swift recombination of photogenerated electron-hole pairs and restricted absorption within the visible-light spectrum. Enhancements in its photocatalytic performance have been achieved through modifications such as doping with noble metals (e.g., silver (Ag), gold (Au), platinum (Pt), and palladium (Pd)) to improve visible-light absorption and inhibit e^-/h^+ recombination by acting as electron traps (Pan and Xu 2013). Nolan et al. (2010) attributed the visible-light activity of TiO_2 doped with silver to surface plasmon resonance and the oxidized surface of Ag. Similarly, Zhang et al. (2014) synthesized Pt/CeTiO_2 nanoparticles with 0.5–5 wt% Pt, demonstrating enhanced visible-light activity and reduced charge recombination in degrading toluene gas and rhodamine B in water. Salami et al. (2012) reported high photocatalytic efficiency for Ag-doped TiO_2 in degrading methylene blue, 4-chlorophenol, and benzene under visible light. Recently, titanates such as bismuth titanate ($Bi_{12}TiO_{20}$) have garnered attention for their visible red-region photosensitivity and their potential use in environmental applications (He et al. 2013). $Bi_{12}TiO_{20}$ has proven effective under visible light exposure (wavelengths under 500 nm) in breaking down contaminants like acid orange 7, methanol, methyl orange, and rhodamine B (Skiker et al. 2018).

2.3 Polymer-Based Nanomaterials

Nanomaterials offer significant benefits due to their high surface-to-volume ratio, which boosts their catalytic efficiency and adsorption capacity. However, this property can also cause nanoparticle aggregation and decreased stability, limiting their practical use. To overcome these challenges, the integration of polymers as a matrix or support has been widely investigated. Polymers help address stability concerns and impart valuable physicochemical properties, including enhanced durability, mechanical strength, recyclability, and thermal and moisture stability (Hussain 2018). Polymeric nanocomposites can be synthesized through diverse methodologies. These approaches encompass melt intercalation, wherein the polymer is heated beyond its softening temperature and amalgamated with the host material in the absence of solvents; *in situ* polymerization, which facilitates superior nanoparticle dispersion and robust interfacial interactions between the polymer and the nanomaterial; and direct blending, a simpler process that combines the polymer and nanoparticles directly (Kalaitzidou et al. 2007). Polymer-based nanomaterials have demonstrated efficacy in treating diverse environmental contaminants, including heavy metals, organic pollutants, and dyes, across various media such as gases, groundwater, industrial effluents, and soils (He et al. 2009). Inert organic polymers like carboxymethyl cellulose (CMC), polyaspartate, or polyacrylic acid are often used as nanoparticle coatings. These polymers act as stabilizers by introducing negatively charged surfaces, which reduce particle aggregation and enhance nanoparticle migration through soil systems (Kumari and Singh 2016). Radoičić et al. (2017) synthesized and prepared an innovative photocatalytically active nanocomposite material, the polymerization product of polyaniline (PANI)/TiO_2, via the oxidative polymerization of PANI in the presence of TiO_2 nanoparticles. This nanocomposite exhibited superior photocatalytic performance in degrading rhodamine B and methylene blue compared to pure TiO_2 nanoparticles (Cameselle et al. 2013).

A notable advancement in photocatalysis involves polymeric graphitic carbon nitride (g-C_3N_4), a metal-free and readily accessible photocatalyst. g-C_3N_4 exhibits a remarkable ability to harness solar energy for the production of electricity and fuels, alongside demonstrating significant potential in environmental remediation applications (Iqbal et al. 2018). However, the utility of g-C_3N_4 under visible light is limited due to its relatively wide bandgap (~2.7 eV) and the rapid recombination of electron-hole pairs. To overcome these challenges, Wang et al. (2019) designed a nanocomposite of g-C_3N_4 and Fe_3O_4 nanoparticles. This composite demonstrated improved photocatalytic performance in the visible light range, a reduced rate of electron-hole pair recombination, and minimal toxicity, making it highly effective for remediating phenanthrene-contaminated soil. The substantial electrical conductivity of Fe_3O_4 ($\sim 1.9 \times 10^6$ S/m) enables its function as an efficient electron acceptor, thereby mitigating charge recombination within the g-C_3N_4 matrix. Additionally, a diverse array of polymers, including cellulose, polypyrrole, poly(methyl methacrylate), and polythiophene, have demonstrated efficacy as adsorbents for the sequestration of heavy metal ions (Anjum et al. 2019). A magnetic nanocomposite, produced through inverse microemulsion polymerization of acrylamide in the presence of a magnetic liquid and N,N'-methylenebisacrylamide as a cross-linking agent, has been investigated for its capacity to remove Cd(II), Co(II), Ni(II), and Pb(II) ions (Zhao et al. 2009). Amphiphilic polyurethane (APU) nanoparticles have also shown promise in remediating hydrophobic organic pollutants like 2-methylnaphthalene (2-MNPT) in soils. The hydrophilic surface of APU particles enhances soil mobility, while their hydrophobic core exhibits a strong affinity for hydrophobic contaminants. Once these particles immobilize the pollutants, they can be biodegraded by microorganisms such as *Acinetobacter* sp., with most nanopolymeric particles recovered through minimal washing steps after soil treatment (Kim et al. 2009).

2.4 Silica-Based Nanomaterials

Silica-derived nanomaterials, such as fumed silica nanoparticles, mesoporous silica materials, and silica-coated magnetic nanoparticles, exhibit surface hydroxyl groups that facilitate their functionalization for applications such as adsorption and immobilization of active moieties. Among these materials, mesoporous silica stands out due to its exceptionally high surface area and tunable pore size, endowing it with remarkable versatility. As an illustrative example, mesoporous silica nanoparticles functionalized with polyethylenimine (PEI), a polymer abundant in primary, secondary, and tertiary amine groups, have demonstrated significant efficacy in adsorbing organic pollutants (Tsai et al. 2016). These materials also serve as supports for immobilizing catalytic enzymes in biological remediation. Wang et al. (2015) developed magnetic mesoporous silica particles with large pores as supports for laccase, which exhibited improved catalytic efficiency and stability over several reuse cycles. Similarly, Mekatel et al. (2015) functionalized silica particles with sulfonate groups to anchor iron oxide nanoparticles, preventing their aggregation and enabling their use in the oxidative degradation of trichloroethylene (TCE).

Silica nanomaterials are extensively used in environmental cleanup because of their versatility, ease of surface modification, high pore volume, and large surface area relative to their volume. Functionalization of silica surfaces and pore walls with various groups has expanded their utility in adsorption and catalysis. For example, hydroxyl groups enhance gas adsorption, surface modification, and wetting (Guerra et al. 2018). Amine-modified aluminosilicates have been employed for CO_2 and carbonyl compound adsorption (Nomura and Jones 2013). Functionalization with carboxylic groups enables the removal of organic dyes, metal ions, and pollutants from wastewater, while modifications with amino, aminopropyl, and thiol groups enhance the remediation of metal ions such as Cd^{2+}, Co^{2+}, Cu^{2+}, Zn^{2+}, Ni^{2+}, Al^{3+}, Cr^{3+}, Pb^{2+}, Hg^{2+}, and U^{6+} (Guerra et al. 2018).

2.5 Nano Zero-Valent Iron (nZVI)

Nano zero-valent iron (nZVI) is widely used in soil remediation, with applications ranging from laboratory experiments to large-scale field implementations. Its popularity stems from its high efficiency in removing pollutants and its cost-effectiveness (O'Carroll et al. 2013). It has proven effective in degrading various contaminants, including biphenyls, chlorinated pesticides, metals, nitrates, and polychlorinated nitroaromatic compounds (Stefaniuk et al. 2016). nZVI, typically measuring less than 100 nm in diameter, can be directly injected into soil for *in situ* remediation. When exposed to water and dissolved oxygen in an aqueous environment, nZVI undergoes oxidation, resulting in the formation of a core–shell structure. This structure consists of a zero-valent iron core encased by an oxide or hydroxide shell (Martin et al. 2008). The thin shell promotes electron transfer from the core, enhancing the strong reducing capacity of nZVI ($Fe^{2+} + 2e^- \rightarrow Fe(s)$, $E^0 = -0.44$ V) for degrading contaminants. Additionally, it acts as an effective adsorbent for metalloids and heavy metals (Fu et al. 2014).

nZVI is commonly synthesized through two primary approaches: top-down and bottom-up methods. Top-down methods, such as lithography, involve reducing the size of bulk iron to the nanoscale. Conversely, bottom-up approaches involve building nZVI from smaller units. These bottom-up techniques encompass the reduction of Fe^{2+}/Fe^{3+} ions in aqueous environments and the reduction of iron oxides in the gas phase utilizing hydrogen gas (Zhao et al. 2016). However, conventional synthesis methods often pose environmental concerns, prompting the development of green chemistry-based techniques. These involve utilizing natural resources such as tree leaf extracts, green mango husks, and food industry residues, leveraging antioxidant extracts to reduce Fe^{3+} and Fe^{2+} ions into nZVI (Machado et al. 2015). Desalegn et al. (2018) demonstrated the use of nZVI combined with sodium persulfate ($Na_2S_2O_8$) for the oxidation of 16 PAHs in contaminated soils. Their study optimized variables such as nZVI doses (0.01–1 g/L), $Na_2S_2O_8$ concentrations (0.017–170 g/L), and temperatures (25–70°C). This method demonstrated a substantial reduction in total organic carbon, achieving approximately 60% removal at ambient temperature. These results indicate that this approach may offer a suitable alternative for the *ex situ* treatment of persistent pollutants, such as PAHs, obviating the need for elevated temperatures (Chen et al. 2015).

Despite its efficacy, nZVI has practical limitations, primarily due to high interparticle interactions that cause agglomeration into larger aggregates, diminishing its reactivity and mobility in soils. Stabilizing agents can mitigate these issues by adsorbing onto nZVI surfaces, enhancing stability and diffusion within porous media. Common stabilizing agents include synthetic polymers such as polyelectrolytes and triblock polymers (Hydutsky et al. 2007, He et al. 2007), or naturally occurring organic matter (Johnson et al. 2009), which prevent aggregation through electrostatic repulsion. Encapsulation in emulsified vegetable oil droplets is another stabilization strategy (Quinn et al. 2005). Quinn et al. (2005) evaluated emulsified nZVI for remediating dense non-aqueous phase liquids (DNAPLs) containing TCE. The hydrophobic emulsion facilitated DNAPL migration into the emulsion's aqueous phase, enhancing TCE degradation. Incorporating secondary metals such as Ag, Cu, Ni, or Pd into nZVI to create bimetallic particles (Fe–Ag, Fe–Cu, Fe–Ni, Fe–Pd) has been investigated to enhance pollutant degradation capacity. Significantly, Fe–Pd exhibits exceptional efficacy in converting chlorinated hydrocarbons into dehalogenated products (Lien and Zhang 2001). These bimetallic particles facilitate electron transfer from the Fe(0) core and catalyze degradation reactions when chlorinated pollutants are present (Schrick et al. 2002).

A further advancement in nZVI stabilization is the development of Carbo-Iron, a nanostructured hybrid material incorporating iron clusters (approximately 50 nm) embedded within activated carbon colloid (ACC). With an iron content of 20–30% Fe(0) by weight, Carbo-Iron demonstrates enhanced stability attributed to electrostatic repulsion within the negatively charged colloidal environment. Furthermore, the addition of low concentrations of CMC significantly improves its mobility (Mackenzie et al. 2012). Carbo-Iron has been utilized at a pilot scale for the remediation of tetrachloroethene (PCE) contamination in Germany, demonstrating efficient transport and rapid PCE degradation over several meters. Its potential applicability also encompasses hybrid treatment systems designed for soils contaminated with organochlorine compounds and heavy metals (Galdames et al. 2017). The initial stage involves nZVI application for heavy metal and chlorine atom removal, followed by bioremediation. Biostimulation, which enriches native microbial activity by adding nutrients and oxygen, and bioaugmentation, which introduces specific degrading bacteria, are two common bioremediation strategies. This combined approach leverages nZVI's abiotic dechlorination and microbial-mediated reduction for efficient soil decontamination. As an emerging *in situ* remediation technology, it holds promise for significantly reducing decontamination timelines compared to standalone bioremediation (Němeček et al. 2016).

3. Mechanisms of Action of Nanomaterials

3.1 Immobilization

The mechanism of *in situ* immobilization has gained considerable global attention due to its cost-effectiveness and environmentally sustainable approach for remediating contaminated soils (Usman et al. 2020). The selection of nanomaterials for immobilization remediation is largely governed by the properties of the contaminants and the existing soil conditions. A wide range of nanoparticles has been widely employed to immobilize both organic and inorganic pollutants in the soil environment (Qian et al. 2020). These nanomaterials include metal oxide nanoparticles and carbon-based nanomaterials. In the case of carbon nanomaterials, organic contaminants are mainly adsorbed via van der Waals forces and π-π interactions. Carbon nanomaterials, such as fullerene, carbon nanotubes, and graphene, are effective adsorbents for immobilization remediation due to their hydrophobic surfaces and high adsorption capacities, which significantly improve their efficiency in removing organic contaminants from soils (Guerra et al. 2018). Among these, CNTs are especially notable for their enhanced adsorption properties for organic compounds, surpassing those of soil organic matter (Zhang et al. 2019).

CNTs demonstrate distinct adsorption abilities for ionizable organic compounds, such as pesticides, via π-π and cation-π interactions. Additionally, their low-barrier surfaces and ability to form hydrogen bonds with charged electrons further enhance their adsorption efficiency (Zhang et al. 2019). The use of CNTs in the reduction of PAHs has been widely investigated under different conditions (Qian et al. 2020). For example, CNTs can restrict the mobility of PAHs in soils and reduce their bioavailability to crops and microorganisms, thereby mitigating their adverse environmental impact (Guerra et al. 2018). The oxygen content in CNTs also plays a critical role in their adsorption capacity, as functional groups like –OH strengthen π-π and –OH interactions, thereby increasing adsorption efficiency (Guerra et al. 2018). In the case of metal oxide nanomaterials, contaminants are immobilized through surface complexation mechanisms. Commonly used metal oxides in soil remediation include Fe(III) oxide and TiO_2. Fe(III) oxide, with its high adsorption capacity, substantially reduces arsenic and cadmium. For instance, it can adsorb up to 37.03 mg/g of cadmium under a pH of 6.0 (Qian et al. 2020). Similarly, TiO_2 demonstrates a high sorption rate, achieving a 94% removal efficiency for thorium (Th(IV)) under acidic conditions at a pH of 4.0. Acidic soil conditions promote the sorption process by enhancing surface complexation, which becomes the primary adsorption mechanism (Tan et al. 2007).

3.2 Photocatalytic Degradation

The photocatalytic degradation process leverages nano-photocatalysts activated by sunlight or ultraviolet (UV) radiation to decompose organic pollutants. This method has been extensively applied to address soil contamination. *Ex situ* approaches involve extracting contaminants from the soil using nonpolar solvents, followed by treating the resulting leachate with photocatalysts to neutralize harmful substances (Fenoll et al. 2014). The integration of nanomaterials in photocatalysis significantly improves the efficiency of conventional processes, addressing the prolonged duration typically associated with traditional photocatalytic methods (Zhu et al. 2019). In contrast, the *in situ* method involves the direct application of nano-photocatalysts into contaminated soil, followed by exposure to sunlight. However, this approach faces significant challenges, including limited light penetration into the soil and the requirement for continuous land plowing (Gu et al. 2012). The efficiency of photocatalysis is influenced by various factors such as soil properties, pH levels, and organic matter content. Wang et al. (2016) highlighted that these parameters significantly influence the production of OH· by nanomaterials and the competitive adsorption of pollutants onto soil particles.

3.3 Fenton-Like Reactions

Fenton reagents, comprising H_2O_2 and Fe^{2+}, are widely employed for the treatment of organic wastewater. These reagents achieve pollutant degradation by oxidizing organic molecules through OH· generated from H_2O_2, with ferrous ions serving as catalysts (Hou et al. 2016). Recently, the use of Fenton reagents for soil remediation has attracted growing attention (Usman et al. 2016). Various catalysts have been investigated for subsurface Fenton-like oxidation processes, with iron-based nanomaterials (such as Fe_3O_4) and iron-containing nanocomposites being the most extensively researched (Garrido-Ramírez et al. 2010). Iron oxide, especially Fe_3O_4, is highly efficient for *in situ* soil remediation because of its widespread availability, cost-effectiveness, ease of synthesis, reusability, and environmentally friendly characteristics within the soil matrix (Garrido-Ramírez et al. 2010). Fe_3O_4-catalyzed Fenton-like reactions have demonstrated significant potential for remediating PAH-contaminated soils, facilitating degradation without the production of harmful byproducts or the need for pH adjustments (Usman et al. 2016). The efficiency of these reactions in degrading PAHs is closely influenced by the concentration of PAHs, organic matter, and mineral composition of the soil matrix. Introducing a pretreatment step using availability-enhancing agents, such as cyclodextrin, ethanol, or nitrilotriacetic acid, can significantly improve treatment outcomes. Despite its advantages, Fe_3O_4-catalyzed Fenton-like reactions can exhibit slow reaction rates. Enhancing the dispersion of solid catalysts within the soil and increasing degradation efficiency may require supplementary techniques, such as sonication. Sonication facilitates the cleaning of catalyst surfaces, ensuring sustained reactivity and efficiency throughout the process (Hou et al. 2016).

3.4 Reduction Reaction

Reduction reactions have shown considerable potential in mitigating heavy metals and organic contaminants in polluted soils, as well as in addressing water and groundwater contamination, especially through the use of nZVI nanoparticles (Qian et al. 2020). nZVI particles have been widely applied on a large field scale owing to their nanoscale size and high surface area, which improve remediation efficiency by enabling direct interaction with contaminants. Typically, nZVI is introduced into contaminated soil via injection walls. These nanoparticles possess strong reduction and adsorption capacities, enabling the transformation of toxic substances into less harmful compounds. For instance, Cr(VI) is reduced to Cr(III), forming ferrous chromite as a byproduct (Rabbani et al. 2015).

The integration of biochar with nZVI has been shown to improve the reduction reaction capacity and removal efficiency of nZVI in addressing Cr(VI) contamination. This enhancement is attributed to the prevention of particle agglomeration and reduced mobility of the iron-biochar mixture within the soil (Qian et al. 2020). For example, a combination of biochar and nZVI successfully removed 66% of Cr(VI) content from the soil (Qian et al. 2020). Additionally, a study reported that 1 gram of nZVI injected into contaminated soil reduced 28% of 1 kilogram of Cr(VI), and 98% of Cr(VI) was eliminated within 24 hours under treatment conditions with a pH of 5 (Chrysochoou et al. 2012).

The use of CMC as a stabilizer in conjunction with nZVI further enhances its capacity to convert Cr(VI) into carbonate- and iron-manganese oxide-bound forms, significantly reducing its bioavailability and leachability by approximately 50% under a soil-to-solution ratio of 1 gram to 10 milliliters (Zhang et al. 2018). The application of CMC-stabilized nZVI has shown promise in the removal of DDT, TCE, and pesticides within soil column systems. Notably, a 30-hour treatment period in a soil matrix containing 9.2% organic matter demonstrated a 44% dechlorination of TCE following the injection of CMC-stabilized nZVI (El-Temsah et al. 2016). In a similar study, 25% of DDT was removed from 1 kilogram of soil, which contained 24 milligrams of DDT per kilogram, within 72 hours using a 20% aqueous nZVI solution (Tian et al. 2020). To enhance the reactivity of nZVI, higher doses of nanoparticles may be required, especially for soils with long-term contamination (Diao et al. 2021).

4. Efficacy of Nanomaterials in Soil Remediation

Soil contamination and degradation remain significant environmental concerns, with remediation posing a global challenge. The degradation of soil reservoirs worldwide greatly affects agricultural productivity and food safety, necessitating immediate action (Roy et al. 2021). The issue is exacerbated by soil pollution with heavy metals (toxic residues), herbicides, and POPs. Due to biomagnification, soils contaminated with these substances increase the risk of contamination within the food chain. The urgency of boosting food production to support a growing population, alongside the challenge of preventing further land degradation, heavily impacts agricultural output (Roy et al. 2021). Adsorption or immobilization is among the most employed remediation methods. The advantages of this method include its effectiveness, low cost, and environmentally friendly nature, making it a popular choice for the removal of metal pollutants from soils (Zhang et al. 2020). Engineered nanomaterials, including carbon nanotubes, metal oxides (such as ferric oxide and TiO_2), and a diverse array of nanocomposites, have been effectively employed to immobilize contaminants within soil matrices (Habib et al. 2023). Table 1 provides a comprehensive overview of the

Table 1: Overview of the efficacy of nanomaterials in *in situ* soil remediation.

Pollutants	Pollutants Conc.	Nanomaterials	Nanomaterials Conc.	Results	References
Arsenic	70.2 mg/g	nZVI	5 wt%	70% removal of As	(Tang et al. 2014, Sol-Magdaleno et al. 2025)
Chromium (VI)	50 mg/L	nZVI	45.7 m^2/g	62% removal within 60 min	(Cai et al. 2019
Cobalt	185 mg/kg	Iron (III) oxide	0.5 wt%	31% removal within 60 days	(Abhilash et al. 2013, New et al. 2023)
Copper	547 µg/g	Nano hydroxyapatite (nHAP)	5 wt%	Fully removal of Cu	(Hussain et al. 2016)
Dichloro-diphenyl-trichloroethane (DDT)	24 mg/kg	nZVI	20% aqueous	25% of DDT removal within 3 days	(Gaur et al. 2014, Ng et al. 2025)
Diphenylarsinic acid (DPAA)	20 µg/g	Titanium dioxide (TiO_2)	5 wt%	82% removal within 3 hours	(Zheng et al. 2020)
Polycyclic aromatic hydrocarbons (PAHs)	4000 µg/g	Carbon nanotubes (CNTs)	0.25 wt%	66.1% removal within 24 hours	(Roy et al. 2021)
Pb	Soil-to-solution ($Pb(NO_3)_2$) ratio of 1:5	nHAP@biochar	8 wt%	Immobilizing 71.9% of Pb within 28d	(Gil-Díaz et al. 2017, Dhanpal et al. 2024)
Tri- chloroethylene (TCE)	727 µg/g	PVP-nZVI	0.4 g	84.73% dechlorination of TCE	(Zhang et al. 2018)
Zn	70.4 µg/g	nHAP	5 wt%	Complete reduction of Zn	(Gil-Díaz et al. 2017)

effectiveness of various nanomaterials utilized in *in situ* soil remediation processes. For instance, ferric oxide nanoparticles have demonstrated remarkable capability in adsorbing and immobilizing heavy metals like Cd and As from different media samples (Gaur et al. 2014). Oil-based wastewater poses a significant threat to aquatic ecosystems (Wei et al. 2024). Iron nanoparticles have proven effective in removing total petroleum hydrocarbons (TPHs) from water, achieving a removal efficiency of 88.34% (Murgueitio et al. 2018). Thus, nanotechnology-based solutions offer high-performance treatments that not only reduce impurities and toxic substances in water but also facilitate heavy metal detoxification. Additionally, nanoscience can be integrated with biological treatments to mitigate environmental risks. Industrial processes driven by nanoparticles have also been shown to efficiently decolorize organic dyes (Habib et al. 2023). Gold and silver nanoparticles, as catalysts, are particularly effective in dye degradation. This process proceeds through two sequential steps: first, electrons accumulate on the particle surface, subsequently facilitating dye diffusion and enabling the reduction reaction to occur (Chen et al. 2020).

Magnetic nanoparticles have shown widespread applications in the adsorption and catalytic remediation of pollutants (Ajmal et al. 2018). Polyacrylamide-modified magnetite nanomaterials have shown potential in controlling soil erosion and reducing arsenate leaching (Zheng et al. 2020). On-site immobilization of contaminants has gained significant attention as a cost-effective and practical approach for remediating polluted soils. Numerous modifications of nanomaterials have been explored to identify optimal materials that balance cost-efficiency, performance stability, minimal environmental impact, and high reusability. For instance, nanohydroxyapatite particles have successfully immobilized metal concentrations in pore water by decreasing the exchangeable fraction of metals in contaminated sediments and soils (Zhang et al. 2018). Research also indicates that sodium carboxymethyl cellulose-stabilized nZVI can achieve an 80% removal rate of soil-bound Cr(VI) (Jeevanandam et al. 2018).

Redox-sensitive pollutants can be effectively addressed through abiotic reductive processes, which directly influence their mobility and toxicity within soil environments. These processes can be implemented independently or in combination with immobilization strategies. For instance, the use of nZVI in pyrene-contaminated soil created strongly reducing conditions, promoting the reduction and removal of pyrene. Additionally, nanomaterials have been incorporated into advanced oxidation processes (AOPs), which utilize oxidants to degrade organic pollutants in soil (Usman et al. 2020). The generation of reactive radicals, such as hydroxyl and sulfate, through the interaction of oxidants like hydrogen peroxide or persulfate with soluble Fe(II), effectively drives the oxidative degradation of organic contaminants. Despite this, the need for an acidic pH to prevent Fe(II) precipitation, along with the excessive costs of acidification and its negative effects on soil quality and microbial communities, limits the broader use of this method. To overcome these challenges, solid iron phases have been suggested as alternatives to soluble Fe(II), allowing chemical oxidation without the need for pH adjustment. The use of iron particles combined with chelating agents has further enhanced these processes (Fu et al. 2018).

Green nanoparticles offer considerable promise in mitigating environmental pollution and are particularly suited for localized treatment of toxic metal ion contamination. These materials are useful for remediating polluted environments and cleaning waste sites. Ferrous nanoparticles, for example, are effective for both disinfecting heavy metal-contaminated water and remediating soils containing heavy metals. Additionally, they can serve as eco-friendly nano fertilizers, enhancing nutrient uptake by plants while reducing groundwater contamination. These fertilizers are immobilized and prevent transformation into non-bioavailable chemical or gaseous forms (Lu et al. 2016). Fe_3O_4 and greigite (Fe_3S_4), synthesized with siliceous materials by bacteria, have also been utilized in optical coatings for solar energy technologies (Hussain et al. 2016).

5. Challenges Associated With Nanotechnology in Soil Remediation

Nanomaterials can provide benefits to soil organisms under specific conditions. However, when their concentrations exceed acceptable limits, they may pose significant risks to microorganisms, terrestrial plants, and the overall soil ecosystem. Extensive research has documented the adverse effects of nanomaterials on plants, showing that they can disrupt various growth stages, including roots, shoots, and seedlings, as well as impair photosynthetic activity, reproduction, and yield parameters (Zhang et al. 2018). The absorption of nanomaterials into plant cells is a critical factor in their application for remediation purposes. Lin et al. (2009) examined the pathways of carbon-based nanomaterials within plant tissues and cells, finding that their performance improved with the presence of natural organic matter. Despite these advancements, metal and metal oxide nanomaterials, such as copper nanoparticles, have been shown to negatively affect plant growth. For example, copper nanoparticles inhibited wheat root growth by nearly 60% while promoting lateral root development, primarily due to induced oxidative stress (Zhang et al. 2018).

Jośko et al. (2017) highlighted that factors such as particle size, solubility, bioavailability, dosage, test duration, and plant physicochemical properties significantly influence the harmful effects of nanomaterials on organisms. These materials not only threaten terrestrial plants but also pose risks to soil microbes and fauna. According to Asadishad et al. (2018), nanomaterials can induce genotoxicity in microbial cells, suppress soil enzyme activity, and alter microbial community structures. Ge et al. (2011) reported that TiO_2 and ZnO nanoparticles reduced microbial abundance and biomass, illustrating their detrimental effects on soil microbiota. In addition to ecological concerns, nanomaterials can pose health risks to site workers who are exposed to them over extended periods in various scenarios. Human exposure through ingestion, inhalation, or dermal contact can result in illnesses, cellular and genetic damage, organ dysfunction, or biochemical changes, as they enter the bloodstream and circulate within the body (Iavicoli et al. 2017). Current methods for detecting and tracking nanoparticles in the environment often lack efficiency and accuracy, underscoring the need for advanced analytical tools capable of elucidating the mechanisms and distribution of nanoparticles across diverse environmental matrices, including soil, sediments, and water (Khanna et al. 2021).

A promising approach involves using biomarkers in biological monitoring systems to trace nanoparticles in environmental settings. Another strategy includes designing permeable iron barriers (PIBs) for shallow aquifers to capture nZVI during post-remediation (Patil et al. 2016). These barriers, constructed from granular iron or zero-valent metals, function as single or modular units, allowing for contamination mitigation through immobilization and transformation processes. The barriers can be permanent, semi-permanent, or interchangeable within polluted flow pathways, enabling real-time monitoring of nZVI levels post-decontamination. Captured nanoparticles can subsequently be removed to prevent their migration into deeper aquifers. In recent advancements, a geophysical method incorporating detailed electric conductivity visualization has been employed to monitor the high-pressure injection of microscale zero-valent iron (mZVI) into groundwater (Wang et al. 2022). While initial results are promising, further research is required. Moreover, developing advanced engineered nanomaterials with novel coatings or functional groups presents a strategic avenue for enhancing environmental remediation applications.

6. Conclusion and Future Perspective

Soil is among the most static environmental compartments, making it particularly vulnerable to persistent pollution, especially from inorganic contaminants like heavy metals. These pollutants are non-degradable, and their accumulation poses significant risks to ecosystems and human health. Polluted soils exhibit altered natural characteristics, including diminished self-purification capacity, reduced microbial activity, lower agricultural productivity, contamination of water resources, and impaired functionality for housing and other purposes. To address these challenges, various soil remediation techniques have been developed, with nanotechnology emerging as a promising and innovative approach. Nanotechnology has revolutionized environmental remediation by providing sustainable solutions for pollution mitigation, both at small and large scales, with minimal toxic effects during production. Specific nanoparticles have proven effective in localized applications, such as sewage treatment and landmine remediation, and offer potential for further refinement to reduce production costs and enhance compatibility. Nanomaterials, including carbon-based, metal-based, and metal oxide nanomaterials, along with nZVI, have demonstrated significant efficacy in reducing soil contamination through processes like chemical degradation, Fenton-type oxidation, photocatalysis, immobilization, and bioremediation integration. nZVI is particularly noteworthy due to its cost-effectiveness and catalytic efficiency against a wide range of contaminants. However, its application comes with potential environmental risks, necessitating careful assessment of its impact on terrestrial plants, soil organisms, and human health. Balancing the benefits of nanotechnology with its potential hazards is critical to advancing sustainable soil remediation technologies. Future research should prioritize the development of multifunctional nanocomposites, integrating them with complementary remediation technologies. Real-time monitoring and control of nanomaterial applications, as well as evaluations of their effectiveness in pollutant removal, are essential to ensure their safe and efficient use. An integrated approach to understanding the environmental implications of nanotechnology is vital for fostering innovation while mitigating risks, ultimately contributing to more effective and sustainable soil remediation practices.

References

Abhilash, P. C., Dubey, R. K., Tripathi, V., Srivastava, P., Verma, J. P., and Singh, H.B. (2013). Remediation and management of POPs-contaminated soils in a warming climate: challenges and perspectives. *Environ. Sci. Pollut. Res. Int.*, 20(8): 5879–5885.

Ajmal, Z., Muhmood, A., Usman, M., Kizito, S., Lu, J., Dong, R. et al. (2018). Phosphate removal from aqueous solution using iron oxides: Adsorption, desorption and regeneration characteristics. *J. Colloid Interface Sci.*, 528: 145–155.

Anitha, K., Namsani, S., and Singh, J. K. (2015). Removal of Heavy Metal Ions Using a Functionalized Single-Walled Carbon Nanotube: A Molecular Dynamics Study. *J. Phys. Chem.*, A 119(30): 8349–8358.

Anjum, M., Miandad, R., Waqas, M., Gehany, F., and Barakat, M.A. (2019). Remediation of wastewater using various nanomaterials. *Arab. J. Chem.*, 12(8): 4897–4919.

Asadishad, B., Chahal, S., Akbari, A., Cianciarelli, V., Azodi, M., Ghoshal, S. et al. (2018). Amendment of Agricultural Soil with Metal Nanoparticles: Effects on Soil Enzyme Activity and Microbial Community Composition. *Environ. Sci. Technol.*, 52(4): 1908–1918.

Barzegar, G., Jorfi, S., Soltani, R. D. C., Ahmadi, M., Saeedi, R., Abtahi, M. et al. (2009). Enhanced Sono-Fenton-Like Oxidation of PAH-Contaminated Soil Using Nano-Sized Magnetite as Catalyst: Optimization with Response Surface Methodology. *Soil Sediment Contam.*, 26(5): 538–557.

Biswas, P., and Kumar. N. (2021). Application of Nanotechnology in Crop Improvement: An Overview. *Crop Improv.*, 211–224.

Brownson, D. A. C., and Banks, C. E. (2010). Graphene electrochemistry: an overview of potential applications. *Analyst*, 135(11): 2768–2778.

Cai, C., Zhao, M., Yu, Z., Rong, H., and Zhang,V. (2019). Utilization of nanomaterials for *in-situ* remediation of heavy metal(loid) contaminated sediments: A review. Sci. *Total Environ.*, 662: 205–217.

Cameselle, C., Reddy, K. R., Darko-Kagya, K., and Khodadoust,V. (2013). Effect of Dispersant on Transport of Nanoscale Iron Particles in Soils: Zeta Potential Measurements and Column Experiments. *J. Environ. Eng.*, 139(1): 23–33.

Chakraborty, M., Sharma, B., Ghosh, A., Sah, D., and Rai, J. P. N. (2023). Elicitation of E-waste (acrylonitrile-butadiene styrene) enriched soil bioremediation and detoxification using *Priestia aryabhattai* MGP1. *Environ. Res.* 238(December): 1–26.

Chang Chien, S. W., Chang, C. H., Chen, S. H., Wang, M. C., Madhava Rao, M., and Satya Veni,V. (2011). Effect of sunlight irradiation on photocatalytic pyrene degradation in contaminated soils by micro-nano size TiO2. *Sci. Total Environ.*, 409(19): 4101–4108.

Chen, C. F., Binh, N. T., Chen, C.W., and Dong, C.D. (2015). Removal of polycyclic aromatic hydrocarbons from sediments using sodium persulfate activated by temperature and nanoscale zero-valent iron. *J. Air Waste Manag. Assoc.*, 65(4): 375–383.

Chen, S., Drehmel, J. R., and Penn, R. L. (2020). Facile Synthesis of Monodispersed Ag NPs in Ethylene Glycol Using Mixed Capping Agents. *ACS Omega,* 5(11): 6069–6073.

Chrysochoou, M., Johnston, C.P., and Dahal, V. (2012). A comparative evaluation of hexavalent chromium treatment in contaminated soil by calcium polysulfide and green-tea nanoscale zero-valent iron. *J. Hazard. Mater.*, 201–202: 33–42.

Cornell, R.M., and Schwertmann, U. (2003). The Iron Oxides. The Iron Oxides.

Desalegn, B., Megharaj, M., Chen, Z., and Naidu, R. (2018). Green mango peel-nanozerovalent iron activated persulfate oxidation of petroleum hydrocarbons in oil sludge contaminated soil. *Environ. Technol. Innov.*, 11: 142–152.

Dhanapal, A. R., Thiruvengadam, M., Vairavanathan, J., Venkidasamy, B., Easwaran, M., and Ghorbanpour, M. (2024). Nanotechnology approaches for the remediation of agricultural polluted soils. *ACS Omega,* 9(12): 13522–13533.

Diao, Z.H., Yan, L., Dong, F.X., Chen, Z.L., Guo, P.R., Qian, W. et al. (2021). Ultrasound-assisted catalytic reduction of Cr(VI) by an acid mine drainage based nZVI coupling with FeS2 system from aqueous solutions: Performance and mechanism. *J. Environ. Manage.*, 278(Pt 1).

Dong, C. D., Chen, C. W., Kao, C. M., Chien, C. C., and Hung, C. M. (2018). Wood-Biochar-Supported Magnetite Nanoparticles for Remediation of PAH-Contaminated Estuary Sediment. *Catalysts*, 8(2): 73.

Dreyer, D. R., Park, S., Bielawski, C.W., and Ruoff, R.S. (2009). The chemistry of graphene oxide. *Chem. Soc. Rev.*, 39(1): 228–240.

Elmolla, E. S., and Chaudhuri, M. (2010). Degradation of amoxicillin, ampicillin and cloxacillin antibiotics in aqueous solution by the UV/ZnO photocatalytic process. *J. Hazard. Mater.*, 173(1–3): 445–449.

El-Temsah, Y. S., Sevcu, A., Bobcikova, K., Cernik, M., and Joner, E. J. (2016). DDT degradation efficiency and ecotoxicological effects of two types of nano-sized zero-valent iron (nZVI) in water and soil. *Chemosphere*, 144: 2221–2228.

Fenoll, J., Flores, P., Hellín, P., Hernández J., and Navarro, S. (2014). Minimization of methabenzthiazuron residues in leaching water using amended soils and photocatalytic treatment with TiO2 and ZnO. *J. Environ. Sci.*, 26(4): 757–764.

Fu, F., Dionysiou, D. D., and Liu, H. (2014). The use of zero-valent iron for groundwater remediation and wastewater treatment: A review. *J. Hazard. Mater.*, 267: 194–205.

Fu, H., Yang, Y., Zhu, R., Liu, J., Usman, M., Chen, Q. et al. (2018). Superior adsorption of phosphate by ferrihydrite-coated and lanthanum-decorated magnetite. *J. Colloid Interface Sci.*, 530: 704–713.

Galdames, A., Mendoza, A., Orueta, M., de Soto García, I.S., Sánchez, M., Virto, I. et al. (2017). Development of new remediation technologies for contaminated soils based on the application of zero-valent iron nanoparticles and bioremediation with compost. *Resour.-Effic. Technol.*, 3(2): 166–176.

Gamallo, M., Fernández, L., Feijoo, G., and Moreira, M. T. (2020). Nano-based technologies for environmental soil remediation. In Nanomaterials for Sustainable Energy and Environmental Remediation. Elsevier Inc.

Garrido-Ramírez, E.G., Theng, B. K. G., and Mora, M. L. (2010). Clays and oxide minerals as catalysts and nanocatalysts in Fenton-like reactions- A review. *Appl. Clay Sci.*, 47(3–4): 182–192.

Gaur, N., Flora, G., Yadav, M., and Tiwari, A. (2014). A review with recent advancements on bioremediation-based abolition of heavy metals. Environ. Sci. Process. *Impacts*, 16(2): 180–193.

Ge, Y., Schimel, J. P., and Holden, P. A. (2011). Evidence for negative effects of TiO2 and ZnO nanoparticles on soil bacterial communities. *Environ. Sci. Technol.*, 45(4): 1659–1664.

Ghosh, A., Sah, D., Chakraborty, M., and Rai, J. P. N. (2023). Bio-mediated detoxification of heavy metal contaminated soil and phytotoxicity reduction using novel strain of *Brevundimonas vancanneytii* SMA3. *Heliyon*, 9(11): e22344.

Ghosh, A., Sah, D., Chakraborty, M., and Rai, J. P. N. (2024). Mechanism and application of bacterial exopolysaccharides: An advanced approach for sustainable heavy metal abolition from soil. *Carbohydr. Res.*, 544(May).

Gil-Díaz, M., Pinilla, P., Alonso, J., and Lobo, M. C. (2017). Viability of a nanoremediation process in single or multi-metal(loid) contaminated soils. *J. Hazard. Mater.*, 321: 812–819.

Gong, X., Huang, D., Liu, Y., Peng, Z., Zeng, G., Xu, P. et al. (2018). Remediation of contaminated soils by biotechnology with nanomaterials: bio-behavior, applications, and perspectives. *Crit. Rev. Biotechnol.*, 38(3): 455–468.

Gu, J., Dong, D., Kong, L., Zheng, Y., and Li, X. (2012). Photocatalytic degradation of phenanthrene on soil surfaces in the presence of nanometer anatase TiO2 under UV-light. *J. Environ. Sci.*, 24(12): 2122–2126.

Guerra, F. D., Attia, M. F., Whitehead, D. C., and Alexis, F. (2018). Nanotechnology for environmental remediation: Materials and applications. *Molecules*, 23(7): 1–23.

Gutierrez, A. M., Dziubla, T. D., and Hilt J. Z. (2017). Recent advances on iron oxide magnetic nanoparticles as sorbents of organic pollutants in water and wastewater treatment. Rev. Environ. *Health*, 32(1–2): 111–117.

Habib, A., Ikram, M., Haider, A., Ul-Hamid, A., Shahzadi, I., Haider, J. et al. (2023). Experimental and theoretical study of catalytic dye degradation and bactericidal potential of multiple phase Bi and MoS2 doped SnO2 quantum dots. *RSC Adv.*, 13(16): 10861–10872.

He, F., Zhao, D., Liu, J., and Roberts C. B. (2007). Stabilization of Fe - Pd nanoparticles with sodium carboxymethyl cellulose for enhanced transport and dechlorination of trichloroethylene in soil and groundwater. *Ind. Eng. Chem. Res.*, 46(1): 29–34.

He, F., Li, Z., Shi, S., Xu, W., Sheng, H., Gu, Y. et al. (2018). Dechlorination of Excess Trichloroethene by Bimetallic and Sulfidated Nanoscale Zero-Valent Iron. *Environ. Sci. Technol.*, 52(15): 8627–8637.

He, T., Ma, H., Zhou, Z., Xu, W., Ren, F., Shi, Z., and Wang, J. (2009). Preparation of ZnS-Fluoropolymer nanocomposites and its photocatalytic degradation of methylene blue. *Polym. Degrad. Stab.*, 94(12): 2251–2256.

He, Y., Basnet, P., Murph, S. E. H., and Zhao, Y. (2013). Ag nanoparticle embedded TiO(2) composite nanorod arrays fabricated by oblique angle deposition: toward plasmonic photocatalysis. *ACS Appl. Mater. Interfaces*, 5(22): 11818–11827.

Hou, L., Wang, L., Royer, S., and Zhang, H. (2016). Ultrasound-assisted heterogeneous Fenton-like degradation of tetracycline over a magnetite catalyst. *J. Hazard. Mater.*, 302: 458–467.

Hussain, C. M. (2018). Polymer nanocomposites- An intro. New Polymer Nanocomposites for Environmental Remediation, xxi–xxv.

Hussain, I., Singh, N. B., Singh, A., Singh, H., and Singh, S. C. (2016). Green synthesis of nanoparticles and its potential application. *Biotechnol. Lett.*, 38(4): 545–560.

Hydutsky, B.W., Mack, E. J., Beckerman, B. B., Skluzacek, J. M., and Mallouk, T. E. (2007). Optimization of nano- and microiron transport through sand columns using polyelectrolyte mixtures. *Environ. Sci. Technol.*, 41(18): 6418–6424.

Iavicoli, I., Leso, V., Beezhold, D. H., and Shvedova, A. A. (2017). Nanotechnology in agriculture: Opportunities, toxicological implications, and occupational risks. *Toxicol. Appl. Pharmacol.*, 329: 96–111.

Iqbal, W., Yang, B., Zhao, X., Rauf, M., Waqas, M., Gong, Y. et al. (2018). Controllable synthesis of graphitic carbon nitride nanomaterials for solar energy conversion and environmental remediation: The road travelled and the way forward. Catal. *Sci. Technol.*, 8(18): 4576–4599.

Jeevanandam, J., Barhoum, A., Chan, Y.S., Dufresne, A., and Danquah, M.K. (2018). Review on nanoparticles and nanostructured materials: History, sources, toxicity and regulations. *Beilstein J. Nanotechnol.*, 9(1): 1050–1074.

Jha, K.C., Liu, Z., Vijwani, H., Nadagouda, M., Mukhopadhyay, S.M., and Tsige, M. (2016). Carbon Nanotube Based Groundwater Remediation: The Case of Trichloroethylene. *Molecules*, 21(7): 953.

Jiemvarangkul, P., Zhang, W. X., and Lien, H.L. (2011). Enhanced transport of polyelectrolyte stabilized nanoscale zero-valent iron (nZVI) in porous media. *Chem. Eng. J.*, 170(2–3): 482–491.

Johnson, R. L., Johnson, G. O. B., Nurmi, J.T., and Tratnyek, P. G. (2009). Natural organic matter enhanced mobility of nano zerovalent iron. *Environ. Sci. Technol.*, 43(14): 5455–5460.

Jośko, I., Oleszczuk, P., and Skwarek, E. (2017). Toxicity of combined mixtures of nanoparticles to plants. *J. Hazard. Mater.*, 331: 200–209.

Kalaitzidou, K., Fukushima, H., and Drzal, L.T. (2007). A new compounding method for exfoliated graphite-polypropylene nanocomposites with enhanced flexural properties and lower percolation threshold. *Compos. Sci. Technol.*, 67(10): 2045–2051.

Khan, M. N., Mobin, M., Abbas, Z. K., AlMutairi, K. A., and Siddiqui, Z. H. (2017). Role of nanomaterials in plants under challenging environments. *Plant Physiol. Biochem.*, 110: 194–209.

Khanna, K., Kohli, S. K., Handa, N., Kaur, H., Ohri, P., Bhardwaj, R. et al. (2021). Enthralling the impact of engineered nanoparticles on soil microbiome: A concentric approach towards environmental risks and cogitation. *Ecotoxicol. Environ. Saf.*, 222: 112459.

Kim, Y. B., Kim, J. Y., and Kim, E. K. (2009). Mobilization and biodegradation of 2-methylnaphthalene by amphiphilic polyurethane nanoparticle. *Appl. Biochem. Biotechnol.*, 159(1): 1–10.

Konate, A., He, X., Zhang, Z., Ma, Y., Zhang, P., Alugongo, G.M. et al. (2017). Magnetic (Fe3O4) Nanoparticles Reduce Heavy Metals Uptake and Mitigate Their Toxicity in Wheat Seedling. *Sustainability,* 9(5): 790.

Kuiken, T. (2010). Cleaning up contaminated waste sites: Is nanotechnology the answer? *Nano Today*, 5(1): 6–8.

Kumar, L., and Harsha, S. P. (2016). Effect of Carbon Nanotubes on CNT Reinforced FGM Nano Plate under Thermo Mechanical Loading. *Procedia Technol.*, 23: 130–137.

Kumari, B., and Singh, D.P. (2016). A review on multifaceted application of nanoparticles in the field of bioremediation of petroleum hydrocarbons. *Ecol. Eng.*, 97: 98–105.

Lal, R. (2009). Soil degradation as a reason for inadequate human nutrition. *Food Secur.*, 1(1): 45–57.

Lal, R. 2015. Restoring Soil Quality to Mitigate Soil Degradation. *Sustainability*, 7(5): 5875–5895.

Liang, J., Liu, J., Yuan, X., Dong, H., Zeng, G., Wu, H. et al. (2015). Facile synthesis of alumina-decorated multi-walled carbon nanotubes for simultaneous adsorption of cadmium ion and trichloroethylene. *Chem. Eng. J.*, 273: 101–110.

Lien, H. L., and Zhang, W. X. (2001). Nanoscale iron particles for complete reduction of chlorinated ethenes. *Colloids Surf. A Physicochem. Eng. Asp.*, 191(1–2): 97–105.

Lin, S., Reppert, J., Hu, Q., Hudson, J. S., Reid, M. L., Ratnikova, T. A. et al. (2009). Uptake, translocation, and transmission of carbon nanomaterials in rice plants. *Small*, 5(10): 1128–1132.

Liu, X., Pan, L., Lv, T., Lu, T., Zhu, G., Sun, Z. et al. (2011). Microwave-assisted synthesis of ZnO--graphene composite for photocatalytic reduction of Cr(VI). *Catal. Sci. Technol.*, 1(7): 1189–1193.

Lu, H., Wang, J., Stoller, M., Wang, T., Bao, Y., and Hao, H. (2016). An Overview of Nanomaterials for Water and Wastewater Treatment. *Adv. Mater. Sci. Eng.* 2016.

Lu, Y., Song, S., Wang, R., Liu, Z., Meng, J., Sweetman, A.J. et al. (2015). Impacts of soil and water pollution on food safety and health risks in China. *Environ. Int.*, 77: 5–15.

Luo, Z., Wang, J., Song, Y., Zheng, X., Qu, L., Qian, J. et al. (2019). Remediation of phenanthrene contaminated soil by g-C3N4/Fe3O4 composites and its phytotoxicity evaluation. *Chemosphere*, 221: 554–562.

Machado, S., Pacheco, J. G., Nouws, H. P. A., Albergaria, J. T., and Delerue-Matos, C. (2015). Characterization of green zero-valent iron nanoparticles produced with tree leaf extracts. *Sci. Total Environ.*, 533: 76–81.

Mackenzie, K., Bleyl, S., Georgi, A., and Kopinke, F. D. (2012). Carbo-Iron - An Fe/AC composite - As alternative to nano-iron for groundwater treatment. *Water Res.*, 46(12): 3817–3826.

Martin, J. E., Herzing, A. A., Yan, W., Li, X. Q., Koel, B. E., Kiely, C. J., and Zhang, W. X. (2008). Determination of the oxide layer thickness in core-shell zerovalent iron nanoparticles. *Langmuir*, 24(8): 4329–4334.

Mekatel, E. H., Amokrane, S., Aid, A., Nibou, D., and Trari, M. (2015). Adsorption of methyl orange on nanoparticles of a synthetic zeolite NaA/CuO. C. R. Chim. 18(3): 336–344.

Murgueitio, E., Cumbal, L., Abril, M., Izquierdo, A., Debut, A., and Tinoco, O. (2018). Green Synthesis of Iron Nanoparticles: Application on the Removal of Petroleum Oil from Contaminated Water and Soils. *J. Nanotechnol.*, 2018(1): 4184769.

Němeček, J., Pokorný, P., Lhotský, O., Knytl, V., Najmanová, P., Steinová, J. et al. (2016). Combined nano-biotechnology for in-situ remediation of mixed contamination of groundwater by hexavalent chromium and chlorinated solvents. *Sci. Total Environ.*, 563–564: 822–834.

New, W. X., Ogbezode, J. E., and Gani, P. (2023). Nanoparticles in soil remediation: Challenges and opportunities. Ind. Domest. *Waste Manag.*, 3(2): 127–140.

Ng, W. M., Chong, W. H., Abdullah, A. Z., and Lim, J. (2025). Magnetic Capture of Functionalized Nanoscale Zerovalent Iron for Soil Remediation: A Feasibility Study on Transport Control. Langmuir.

Nolan, N. T., Seery, M. K., Hinder, S. J., Healy, L. F., and Pillai, S. C. (2010). A Systematic Study of the Effect of Silver on the Chelation of Formic Acid to a Titanium Precursor and the Resulting Effect on the Anatase to Rutile Transformation of TiO2. *J. Phys. Chem. C,* 114(30): 13026–13034.

Nomura, A., and Jones, C. W. (2013). Amine-functionalized porous silicas as adsorbents for aldehyde abatement. *ACS Appl. Mater. Interfaces*, 5(12): 5569–5577.

O'carroll, D., Sleep, B., Krol, M., Boparai, H., and Kocur, C. (2013). Nanoscale zero valent iron and bimetallic particles for contaminated site remediation. *Adv. Water Resour.*, 51: 104–122.

Pan, X., and Xu, Y. J. (2013). Defect-mediated growth of noble-metal (Ag, Pt, and Pd) nanoparticles on TiO2 with oxygen vacancies for photocatalytic redox reactions under visible light. *J. Phys. Chem. C*, 117(35): 17996–18005. https://doi.org/10.1021/JP4064802

Patil, S. S., Shedbalkar, U. U., Truskewycz, A., Chopade, B. A., and Ball, A. S. (2016). Nanoparticles for environmental clean-up: A review of potential risks and emerging solutions. *Environ. Technol. Innov.*, 5: 10–21.

Qian, Y., Qin, C., Chen, M., and Lin, S. (2020). Nanotechnology in soil remediation – applications vs. implications. *Ecotoxicol. Environ. Saf.*, 201: 110815.

Quinn, J., Geiger, C., Clausen, C., Brooks, K., Coon, C., O'Hara, S. et al. 2005. Field Demonstration of DNAPL Dehalogenation Using Emulsified Zero-Valent Iron. *Environ. Sci. Technol.*, 39(5): 1309–1318.

Rabbani, M. M., Ahmed, I., and Park, S. J. (2015). Application of nanotechnology to remediate contaminated soils. *Environ. Remediation Technologies for Metal-Contaminated Soils*, 219–229.

Radoičić, M., Ćirić-Marjanović, G., Spasojević, V., Ahrenkiel, P., Mitrić, M., Novaković, T. et al. (2017). Title: Superior photocatalytic properties of carbonized PANI/TiO 2 nanocomposites.

Rahman, A., Ali, I., Al Zahrani, S. M., and Eleithy, R. H. (2011). A review of the applications of nanocarbon polymer composites. *Nano*, 6(3): 185–203.

Rossell, M. D., Kuebel, C., Ilari, G., Rechberger, F., Heiligtag, F. J., Niederberger, M. et al. (2013). Impact of sonication pretreatment on carbon nanotubes: A transmission electron microscopy study. *Carbon*, 61: 404–411.

Roy, A., Sharma, A., Yadav, S., Jule, L. T., and Krishnaraj, R. (2021). Nanomaterials for Remediation of Environmental Pollutants. *Bioinorg. Chem. Appl.*, 2021: 1764647.

Sah, D., Rai, J. P. N., Ghosh, A., and Chakraborty, M. (2022). A review on biosurfactant producing bacteria for remediation of petroleum contaminated soils. 3 Biotech, 12(9).

Salami, N., Bayati, M. R., Golestani-Fard, F., and Zargar, H. R. (2012). UV and visible photodecomposition of organic pollutants over micro arc oxidized Ag-activated TiO2 nanocrystalline layers. *Mater. Res. Bull.*, 47(4): 1080–1088.

Schrick, B., Blough, J. L., Jones, A. D., and Mallouk, T. E. (2002). Hydrodechlorination of trichloroethylene to hydrocarbons using bimetallic nickel-iron nanoparticles. *Chem. Mater.*, 14(12): 5140–5147.

Skiker, R., Zouraibi, M., Saidi, M., and Ziat, K. (2018). Facile coprecipitation synthesis of novel Bi 12 TiO 20 /BiFeO 3 heterostructure serie with enhanced photocatalytic activity for removal of methyl orange from water. *J. Phys. Chem. Solids*, 119: 265–275.

Sol-Magdaleno, M., Aguilar-Aguilar, J. I., Beltrán-Naturi, E., Valencia-Ordóñez, L. D., Díaz-González, A., Trejo-Flores, P. et al. 2025. Carbon nanomaterials as an environmental technology in the remediation of agricultural soils contaminated with heavy metals: a review. *Discov. Soil*, 2(1): 26.

Stefaniuk, M., Oleszczuk, P., and Ok, Y. S. (2016). Review on nano zerovalent iron (nZVI): From synthesis to environmental applications. *Chem. Eng. J.*, 287: 618–632.

Tan, X., Wang, X., Chen, C., and Sun, A. (2007). Effect of soil humic and fulvic acids, pH and ionic strength on Th(IV) sorption to TiO2 nanoparticles. *Appl. Radiat. Isot.*, 65(4): 375–381.

Tang, W. W., Zeng, G. M., Gong, J. L., Liang, J., Xu, P., Zhang, C. et al. (2014). Impact of humic/fulvic acid on the removal of heavy metals from aqueous solutions using nanomaterials: a review. *Sci. Total Environ.*, 468–469: 1014–1027.

Tang, W. W., Zeng, G. M., Gong, J. L., Liu, Y., Wang, X. Y., Liu, Y. Y. et al. (2012). Simultaneous adsorption of atrazine and Cu (II) from wastewater by magnetic multi-walled carbon nanotube. *Chem. Eng. J.*, 211–212: 470–478.

Thangavelu, L., Veeraragavan, G. R., Mallineni, S. K., Devaraj, E., Parameswari, R. P., Syed, N. H. et al. (2022). Role of Nanoparticles in Environmental Remediation: An Insight into Heavy Metal Pollution from Dentistry. *Bioinorg. Chem. Appl.* 2022.

Tian, H., Liang, Y., Yang, D., and Sun, Y. (2020). Characteristics of PVP-stabilised NZVI and application to dechlorination of soil-sorbed TCE with ionic surfactant. *Chemosphere* 239.

Tsai, C. H., Chang, W. C., Saikia, D., Wu, C. E., and Kao, H. M. (2016). Functionalization of cubic mesoporous silica SBA-16 with carboxylic acid via one-pot synthesis route for effective removal of cationic dyes. *J. Hazard. Mater.*, 309: 236–248.

Usman, M., Hanna, K., and Haderlein, S. (2016). Fenton oxidation to remediate PAHs in contaminated soils: A critical review of major limitations and counter-strategies. *Sci. Total Environ.*, 569–570: 179–190.

Usman, M., Farooq, M., Wakeel, A., Nawaz, A., Cheema, S. A., ur Rehman, H. et al. (2020). Nanotechnology in agriculture: Current status, challenges and future opportunities. Sci. Total Environ. 721.

Wang, A. N., Teng, Y., Hu, X. F., Wu, L. H., Huang, Y. J., Luo, Y. M. et al. (2016). Diphenylarsinic acid contaminated soil remediation by titanium dioxide (P25) photocatalysis: Degradation pathway, optimization of operating parameters and effects of soil properties. *Sci. Total Environ.*, 541: 348–355.

Wang, B., Gao, C., Li, X., Zhang, Y., Qu, T., Du, X. et al. (2022). Remediation of groundwater pollution by in situ reactive zone: A review. *Process Saf. Environ. Prot.*, 168: 858–871.

Wang, D., Zhang, G., Zhou, L., Wang, M., Cai, D., and Wu, Z. (2017). Synthesis of a Multifunctional Graphene Oxide-Based Magnetic Nanocomposite for Efficient Removal of Cr(VI). *Langmuir*, 33(28): 7007–7014.

Wang, J., Luo, Z., Song, Y., Zheng, X., Qu, L., Qian, J., et al. (2019). Remediation of phenanthrene contaminated soil by g-C3N4/Fe3O4 composites and its phytotoxicity evaluation. *Chemosphere*, 221: 554–562.

Wang, S., Wang, K., Dai, C., Shi, H., and Li, J. (2015). Adsorption of Pb2+ on amino-functionalized core-shell magnetic mesoporous SBA-15 silica composite. *Chem. Eng. J.*, 262: 897–903.

Wei, Y., Han, B., Hu, X., Lin, Y., Wang, X., and Deng, X. (2012). Synthesis of Fe3O4 nanoparticles and their magnetic properties. *Procedia Eng.*, 27: 632–637.

Wei, Z., Long, W., Li, S., Zhao, Y., Yu, S., and Zhou, F. (2024). Preparation of Cationic Polyacrylamide Suspension and Its Application in Oilfield Wastewater Treatment. *Polymers*, 16(1): 151.

Xu, J. C., Lu, M., Guo, X. Y., and Li, H. L. (2005). Zinc ions surface-doped titanium dioxide nanotubes and its photocatalysis activity for degradation of methyl orange in water. *J. Mol. Catal. A Chem.*, 226(1): 123–127.

Yan, W., Lien, H. L., Koel, B.E., and Zhang, W. X. (2012). Iron nanoparticles for environmental clean-up: recent developments and future outlook. Environ. Sci. Process. *Impacts*, 15(1): 63–77.

Zhang, D., Ma, X. L., Gu, Y., Huang, H., and Zhang, G. W. (2020). Green Synthesis of Metallic Nanoparticles and Their Potential Applications to Treat Cancer. *Front. Chem.*, 8: 799.

Zhang, L., Tse, M. S., and Tan, O. K. (2014). Facile in situ synthesis of visible light-active Pt/C-TiO2 nanoparticles for environmental remediation. *J. Environ. Chem. Eng.*, 2(2): 1214–1220.

Zhang, R., Zhang, N., and Fang, Z. (2018). *In situ* remediation of hexavalent chromium contaminated soil by CMC-stabilized nanoscale zero-valent iron composited with biochar. *Water Sci. Technol.*, 77(5-6): 1622–1631.

Zhang, T., Lowry, G. V., Capiro, N. L., Chen, J., Chen, W., Chen, Y. et al. (2019). In situ remediation of subsurface contamination: opportunities and challenges for nanotechnology and advanced materials. *Environ. Sci. Nano*, 6(5): 1283–1302.

Zhang, T. T., Xue, Q., and Wei, M. L. (2018). Leachability and Stability of Hexavalent-Chromium-Contaminated Soil Stabilized by Ferrous Sulfate and Calcium Polysulfide. *Appl. Sci.*, 8(9): 1431.

Zhang, Y., Tang, Z. R., Fu, X., and Xu, Y. J. (2010). TiO2-graphene nanocomposites for gas-phase photocatalytic degradation of volatile aromatic pollutant: Is TiO2-graphene truly different from other TiO2-carbon composite materials? *ACS Nano*, 4(12): 7303–7314.

Zhang, Z., Ke, M., Qu, Q., Peijnenburg, W. J. G. M., Lu, T., Zhang, Q. et al. (2018). Impact of copper nanoparticles and ionic copper exposure on wheat (*Triticum aestivum* L.) root morphology and antioxidant response. *Environ. Pollut.*, 239: 689–697.

Zhao, D., Gao, X., Wu, C., Xie, R., Feng, S., and Chen, C. (2016). Facile preparation of amino functionalized graphene oxide decorated with Fe3O4 nanoparticles for the adsorption of Cr(VI). *Appl. Surf. Sci.*, 384: 1–9.

Zhao, F., Tang, W.Z., Zhao, D., Meng, Y., Yin, D., and Sillanpää, M. (2009). Adsorption kinetics, isotherms and mechanisms of Cd(II), Pb(II), Co(II) and Ni(II) by a modified magnetic polyacrylamide microcomposite adsorbent. *J. Water Process Eng.*, 4(C): 47–57.

Zhao, X., Liu, W., Cai, Z., Han, B., Qian, T., and Zhao, D. (2016). An overview of preparation and applications of stabilized zero-valent iron nanoparticles for soil and groundwater remediation. *Water Res.* 100: 245–266.

Zheng, M., Huang, Z., Ji, H., Qiu, F., Zhao, D., Bredar, A. R. C. et al. (2020). Simultaneous control of soil erosion and arsenic leaching at disturbed land using polyacrylamide modified magnetite nanoparticles. *Sci. Total Environ.*, 702: 134997.

Zhu, Y., Xu, F., Liu, Q., Chen, M., Liu, X., Wang, Y. et al. (2019). Nanomaterials and plants: Positive effects, toxicity and the remediation of metal and metalloid pollution in soil. Sci. *Total Environ.*, 662: 414–421.

CHAPTER-9

Role of Nanomaterials in Water Remediation

Margi Patel, Rita Mahapatra and *Poonam Mishra**

1. Introduction

Environmental pollution is a major concern across the globe, particularly in developing countries, due to increasing pollutant levels caused by human activities. The prominent occurrence of both organic and inorganic contaminants, which are harmful to living organisms, has grown significantly in recent years. Human actions are exerting immense pressure on oceans, rivers, and other inland water systems, leading to considerable deterioration of water quality. Contaminants due to agricultural waste, chemicals such as herbicides and fertilizers, industrial contaminants like untreated industrial effluents, heavy metals, pharmaceutical residues, polycyclic aromatic hydrocarbons, and household effluent containing pathogens and other pollutants enter water bodies, where they remain dissolved, suspended, or settle in sediments, causing water contamination (Mishra et al. 2023, Venkatraman et al. 2024). Several strategies have been implemented to treat natural water sources and make them safe for human use. From pretreatment to the treatment of waste, at each step of water remediation, nanomaterials offer advanced solutions to demanding environmental challenges, such as energy consumption and hazardous chemicals (Aghalari et al. 2020). Recently, nanotechnology has played an important role in combating these environmental issues in developing nations. In this regard, incorporating cutting-edge nanotechnology into conventional engineering opens new possibilities for advanced water decontamination technologies.

Contamination of the open water bodies, such as rivers, lakes, oceans, and groundwater, is known as water pollution. Frequent human activities, including insufficient waste management from housing societies, partial or untreated industrial wastewater, and release from hospitals, significantly contribute to the contamination of natural water resources. This contamination reduces water quality, rendering it unsafe and harmful to human health and wildlife. This may have an undesirable effect on the different trophic levels of ecological pyramids. Hence, it is imperative to improve the quality of the water to make it consumable for all living entities present on the earth. The fundamental objective of the conventional methods is to remove the organic matter from the polluted water under controlled conditions. This involves three steps: primary, secondary, and tertiary treatment of contaminated water (Epelle et al. 2022). Primary treatment comprises the

Krishna School of Science, Drs. Kiran & Pallavi Patel Global University, Varnama, Vadodara- 391243, India.
0000-0001-6424-5435; 0000-0003-1545-9730; 0000-0002-5410-5217
* Corresponding author: poonammishra.kss@kpgu.ac.in

DOI: 10.1201/9781003558804-9

removal of larger debris, for example, twigs, tires, branches, plastics, etc., through the physical separation of these materials by passing them onto moving metal screens that filter them out from the wastewater. This remaining stream is then propelled into a sedimentation tank where organic solids settle down and are known as primary sludge. At this point, flotation is also used to remove the larger particles. This sludge contains dissolved chemical impurities, such as organic and inorganic chemical compounds, as well as suspended solids that require further treatment, which is carried out as secondary treatment (Radetic 2024). For this, the use of microorganisms for the degradation of the suspended organic solids is well known. Here, the biological treatment is carried out in bio-trickling filters and aeration tanks, preceded by a disinfection step. Currently, these trickling filters are composed of plastic compared to earlier stone beds onto which the wastewater dips. These trickling filters, when pumped with wastewater, provide a bed for the formation of biofilms known as zooleal film (Abyar and Nowrouzi 2023). Further, this sludge proceeds to the tertiary treatment where the removal of turbidity, inorganic substances, especially nitrogen and phosphorus, and metallic impurities are taken care of. This involves flocculation, coagulation, membrane filtration, ozonation, and chlorination (Omarova et al. 2018, Zahmatkesh et al. 2022, Qi and He 2025). Apart from these methods, there are several low-cost alternatives for water treatment plants available, such as aerated lagoons or ponds, and aerobic as well as anaerobic ponds (Rana et al. 2024). Although all these conventional treatment processes have been instrumental for a long period, using them on a large scale presents significant challenges. Hence, we must discuss innovative and newer emerging techniques to counteract these challenges. There are many wastewater treatment (WWT) methods that can be used; smart nanomaterials offer an alternative for their effectiveness in pollutant removal through absorption, adsorption, chemical reactions, photocatalysis, and filtration (Guerra et al. 2018, Sharma et al. 2024). Currently, nanotechnology research is being conducted to synthesize more efficient designs, with alteration of material properties ranging from roughly 1 nm to 100 nm in one dimension. Recently, nanotechnology has become a general-purpose technology, helpful to various sectors and industries. Nanotechnology is quickly acquiring importance in several areas such as health services, cosmetics, food and environment remediation, biomedical sciences, chemical industries, and drug-gene delivery (Malik et al. 2023). Nanotechnology is a promising alternative for the remediation of almost all waste from effluents. Because of its enhanced properties, such as a high surface-area-to-volume ratio and high reactivity, nanotechnology has emerged as the most effective approach for remediation (Mathur et al. 2022). Further, even a small volume of nanoparticles/nanomaterials (NPs/NMs) offers a multifold larger surface area for interaction with contaminants present in water bodies. Moreover, the surface of these NPs/NMs can be modified by conjugating different small chemical molecules based on requirements, enhancing the properties of NPs/NMs, such as their reactivity and efficiency. The present chapter discusses the application of various NPs/NMs in water remediation.

2. Characteristics of Nanoparticles

Nanoparticles (NPs) have unique characteristics like a greater surface-area-to-volume ratio, enhanced activity, and the capability for functionalization, making these small particles suitable for a range of applications, particularly in fields like water remediation, medicine, and catalysis. In general, nanoparticles typically range from 1 to 100 nm in size, having a greater surface-area-to-volume ratio, which affects their reactivity, solubility, and cross-reactivity to biological barriers. Their shape can also influence how nanoparticles interact with cells and tissues, affecting their uptake and distribution (Lakshmi et al. 2024). As already mentioned earlier, a high surface-area-to-volume ratio, which significantly increases the number of surface atoms compared to the bulk material, leads to enhanced reactivity. This high aspect ratio is crucial for catalytic and adsorption processes, making these particles effective in removing contaminants from water or air and enhancing the efficiency of catalytic reactions in industrial processes (Mekuye and Abera 2023). Also, the increased surface energy due to the greater surface-area-to-volume ratio makes NPs/NMs more chemically active than their bulk counterparts, enhancing their reactivity. This is particularly useful in environmental remediation, where nanoparticles can interact more readily with pollutants, breaking them down or binding to them more effectively (Eker et al. 2024). Moreover, these can be modified or "functionalized" with various chemical groups, enhancing their interactions with specific targets. For example, functionalizing nanoparticles with certain molecules can make them more selective towards particular pollutants or biological targets, improving their efficiency in applications such as targeted drug delivery or biosensing (Lakshmi et al. 2024). Furthermore, nanoparticles possess distinctive physicochemical properties, including a large surface area, high mechanical strength, optical activity, and reactivity with different chemicals, making them suitable for a variety of applications (Joudeh and Linke 2022). These properties have been characterized by the famous Beer-Lambert law based on the absorption properties of the material (Swinehart 1962), which provides information about their characteristic absorption maxima, reflection, and phosphorescence. Metallic and semiconductor nanoparticles possess distinct colors, which makes them suitable for photonics and optoelectronics (Ogochukwu et al. 2024). Lastly, nanoparticles exhibit a high adsorption ability due to their unique physicochemical properties, allowing them to attract and hold substances on their surface. This high adsorption capability is essential for applications in environmental clean-up, drug delivery, and catalysis (Joudeh and Linke 2022).

These aforesaid properties of the nanoparticles are crucial and play a very significant role in using them as an alternative for water remediation.

3. Types of Nanoparticles used in Water Remediation

Since ancient times, silver, known for its low toxicity and effective antimicrobial properties, has been a widely used chemical in various applications. For the synthesis of AgNPs, there are mainly three approaches: physical, chemical, and biological. Laser ablation, thermal evaporation, and microwave-assisted synthesis are widely used physical methods for producing silver nanoparticles (AgNPs). Research indicates that microwave-assisted synthesis yields high-purity AgNPs with a uniform size distribution in significantly less time compared to other techniques (Sati et al. 2025). On the other hand, chemical methods include the use of reducing agents or irradiation techniques such as gamma rays, lasers, or microwaves. The stabilization technique is also a type of chemical method that utilizes stabilizers to prevent aggregation (Lakshani et al. 2025). The selection of different stabilizers can lead to the formation of nanoparticles with different physical characteristics, such as varying sizes and shapes (Algebaly et al. 2020). Currently, many researchers are focusing on green chemistry methods to make metal nanoparticles. These methods use natural materials like plants, fungi, bacteria, and yeast as reducing, capping, and stabilizing agents. They help produce nanoparticles with the desired shape and size (Sharma et al. 2024). The major advantage of using herbal extracts for the production of metal nanoparticles is their ease of availability, as well as their being safe to use without toxicity. Silver nanoparticles (AgNPs), synthesized from silver salts like silver nitrate and silver chloride, have been shown to exhibit strong biocidal activity (Amin et al. 2014). According to a report, 97% of chromium (Cr VI) from a solution was removed by silver nanoparticles made using leaf extract from *Moringa stenopetala*, based on the Freundlich isotherm analysis (Kakavandi et al. 2014).

Gold nanoparticles (AuNPs) are mainly used to remove mercury. During this process, they can form compounds like AuHg, $AuHg_3$, and Au_3Hg. Citrate-coated gold nanoparticles have been attached to aluminium oxide (Al_2O_3), and their ability to remove mercury ions was studied. Other materials, such as sodium borohydride, have also been added to gold nanoparticles to improve their mercury removal performance (Hussain et al. 2016).

Titanium dioxide (TiO_2) is a popular white powder and a commonly used inorganic compound due to its chemically stable nature, being inexpensive, and having photocatalytic properties. TiO_2 is mainly produced using sol-gel methods (Mathur et al. 2022), but it can also be made through hydrothermal techniques and chemical vapor deposition (Hsu et al. 2024). It exists in different crystal forms, and each form has unique photocatalytic abilities (Yan and Chen 2015).

Metal oxide nanoparticles like TiO_2 help to reduce the accumulation of harmful metals and metalloids such as arsenic (As), cadmium (Cd), and copper (Cu) (Mathur et al. 2022). Nowadays, herbal materials such as flowers, leaves, fruit, and root extracts have been broadly utilized for the reduction of zero-valent iron (ZVI) nanoparticles. In the reduction of metal ions, herbal biomolecules, secondary metabolites, and enzymes play a key role. The main herbal biomolecules used for the reduction of iron ions are polyphenols, flavonoids, amino acids, terpenoids, reducing sugars, steroids, saponins, vitamins, alkaloids, and organic acids (Puthukkara et al. 2021). Nanoscale zero-valent iron (nZVI) nanoparticles effectively remove metal ions from water. Their short lifespan and high reactivity make them suitable for treating a broad spectrum of metalloid contaminants. Magnetic separation, which utilizes an external magnetic field, offers a convenient method for recovering nZVI after treatment, enhancing the efficiency and sustainability of the water purification process. nZVI holds even more advantages: (1) as compared to mZVI (macroscale ZVI), it can successfully remove some aqueous contaminant species due to its large surface area, (2) it also reduces some contaminants more rapidly, and (3) it avoids the formation of some undesirable by-products due to more complete degradation (Guan et al. 2015). Xu et al. (2018) reported that oxides of iron nanoparticles have usually been used to eliminate aqueous arsenic, with an adsorptive ability for diverse Fe_2O_3 and Fe_3O_4 of 5.99 mg g^{-1}. According to Xu et al. (2018), iron oxide nanoparticles are often used to remove arsenic from water due to their good adsorption abilities. For example, Fe_2O_3 and Fe_3O_4 can adsorb up to 5.99 mg of arsenic per gram. Iron oxides combined with carbon nanotubes showed even better performance, adsorbing up to 24.05 mg/g of arsenic (As^{3+}) and 47.41 mg/g of arsenic (As^{5+}). According to a report, Zn-doped Al_2O_3 NPs with high surface area and hydrophilicity efficiently removed 87% As and 98% Pb (Sherugar et al. 2021).

For the synthesis of multi-walled carbon nanotubes (MWCNTs), a DC arc in a helium-filled chamber at sub-atmospheric pressure can be utilized. Arc discharge using a composite anode in a hydrogen or argon environment can be utilized to synthesize single-walled carbon nanotubes (SWCNTs). Maintaining a constant electrode gap is crucial for achieving optimal current density and maximizing anode utilization, but it can lead to the production of unwanted products like multi-walled carbon nanotubes (MWCNTs) or fullerenes (Prasek et al. 2011). Another technique, laser ablation, which is a key part of pulsed laser deposition, is considered more effective for producing high-quality and pure single-walled carbon nanotubes (SWCNTs). The most commonly used method for making carbon nanotubes is catalytic chemical vapor deposition (CCVD). For large-scale production of pure carbon nanotubes, CCVD is a more cost-effective method compared to laser ablation. CCVD has several benefits, such as easy control of the process and the ability to produce very pure materials (Steiner et al. 2009). Palashuddin and his team have developed an advanced innovation in water quality monitoring: color-changing carbon and sulfur dots. These synthetic nanomaterials, akin to tiny, light-emitting semiconductors, exhibit unique color shifts in the presence of specific contaminants like lead, cobalt, and chromium. This remarkable characteristic allows for the rapid and accurate identification of pollutants without introducing any additional metallic contaminants into the water sample. Furthermore, these carbon dots possess the added benefit of degrading harmful substances such as pesticides and dyes, making them a promising dual-function tool for water purification and environmental remediation. Studies have shown the effectiveness of single-walled carbon nanotube (SWCNT) nanocomposites in heavy metal removal. For instance, SWCNTs incorporated with magnetite, cobalt, and sulfide exhibited excellent mercury adsorption, above 99.56% within 7 minutes (Alijani & Shariatinia, 2018). Furthermore, research by Alguacil and Lopez (2019) highlighted the potential of multi-walled carbon nanotubes (MWCNTs) in removing gold ions. Additionally, the incorporation of magnetic iron oxide nanoparticles (Fe_3O_4 NPs) into CNTs enhances their functionality (Alguacil and López 2019). These magnetic CNTs demonstrate promising applications in the removal of various metals (loids) and offer the advantage of easy recovery and reuse after the extraction process (Iqbal et al. 2021).

Various methods can be used to synthesize graphene, including exfoliation, arc discharge, chemical oxidation-reduction, heating (thermal pyrolysis), layer-by-layer growth (epitaxial growth), chemical reactions, chemical vapor deposition, and laser-based techniques (Lakshani et al. 2025). Exfoliation methods for graphene synthesis encompass three primary approaches: mechanical, chemical, and electrochemical. Mechanical exfoliation, the pioneering technique for isolating graphene sheets from graphite, involves applying mechanical stress to the graphite surface. Liquid-assisted mechanical exfoliation includes various techniques like ball milling, sonication, high-shear mixing, and high-pressure homogenization. Chemical exfoliation is a method used to make single-layer graphene by the addition of special chemicals known as intercalating agents to a graphite mixture. These chemicals attach to the graphite and push the layers apart, making it easier to separate them. Commonly utilized reducing agents in this process consist of ionic liquids, hydrazine hydrate, N-methyl-2-pyrrolidone, and methane sulfonic acid (Adetayo and Runsewe 2019, Farjadian et al. 2020, Kumar et al. 2021). Electrochemical exfoliation is used to obtain few-layered graphene from graphite and produce functional graphene. The arc discharge method is commonly used to produce carbon-based nanomaterials including CNT, fullerene, and few-layered graphene. Graphene production often involves the use of two high-purity graphite electrodes, one as the cathode and the other as the anode within a water-cooled stainless-steel chamber. A specific gas is used inside the chamber, and the electric current is usually kept between 100 and 150 amperes. Common gases used in this process include helium (He), ammonia (NH_3), hydrogen (H_2), nitrogen (N_2), argon (Ar), air, and carbon dioxide (CO_2) (Lakshani et al. 2025). Recently, redox reaction has been considered the most viable procedure for maximum graphene production. Oxidative exfoliation of graphite involves introducing oxygen-containing groups onto the graphite sheets, which effectively increases the interlayer spacing and results in the formation of graphene oxide (GO). Chemical vapor deposition (CVD) is a well-known method for graphene synthesis. In this process, a gas that contains carbon undergoes decomposition on a metal catalyst, typically copper or nickel, at elevated temperatures, leading to the growth of graphene layers on the metal surface. A diverse range of CVD methods exists, each distinguished by specific process parameters such as temperature, reactor wall or substrate material, pressure, precursor type, deposition time, activation method, and gas flow conditions (Saeed et al. 2020). The functionalized derivatives of carbon nanotubes and graphene with high surface area and adsorption sites are proposed to remove heavy metals via adsorption (Xu et al. 2018). Because graphene has a high capability to adsorb substances, it can be used to remove fluoride from water. Samad et al. (2021) reported that an altered graphene oxide material called GO-Zr-P (graphene oxide-zirconium phosphate) was very effective. Using 150 mg of this material, about 99% of 50 ppm of heavy metals like lead (Pb^{2+}), copper (Cu^{2+}), cadmium (Cd^{2+}), and zinc (Zn^{2+}) was removed in just 20 minutes. Also, using 50 mg of GO-Zr-P, around 98% of 50 ppb of these metals was removed within 30 minutes. Some commonly used inorganic nanoparticles for the elimination of contaminants from water are shown in Figure 1.

3.1 Organic Nanoparticles

Several types of organic nanoparticles, some of which are micelles, dendrimers, liposomes, nanogels, and ferritin, are utilized for water remediation. Dendrimers, with their large, branched structures, exhibit a high capacity for binding and trapping various contaminants, as shown in Figure 2. This unique characteristic enables them to effectively prevent the passage of these pollutants through membranes (Abbasi et al. 2014). Furthermore, dendrimers offer significant potential for environmental applications, including water purification and the development of eco-friendly industrial processes. A diverse range of dendrimers can be utilized for water treatment, including those designed to bind anions, organic compounds, cations, biological compounds, and even viruses. Moreover, dendrimers can be engineered to possess multiple binding functionalities, offering enhanced versatility in water treatment applications (Patanjali et al. 2019). Allegritti and team demonstrated that $>$ 80% alkylphenol ethoxylates (APEO), common pollutants in industrial

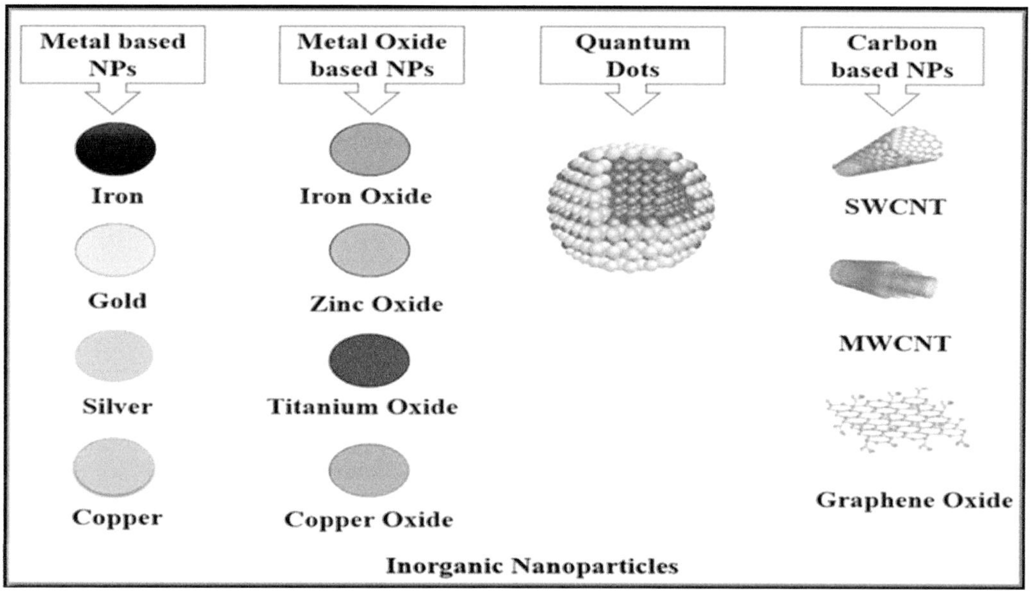

Figure 1: Inorganic nanoparticles used in water remediation.

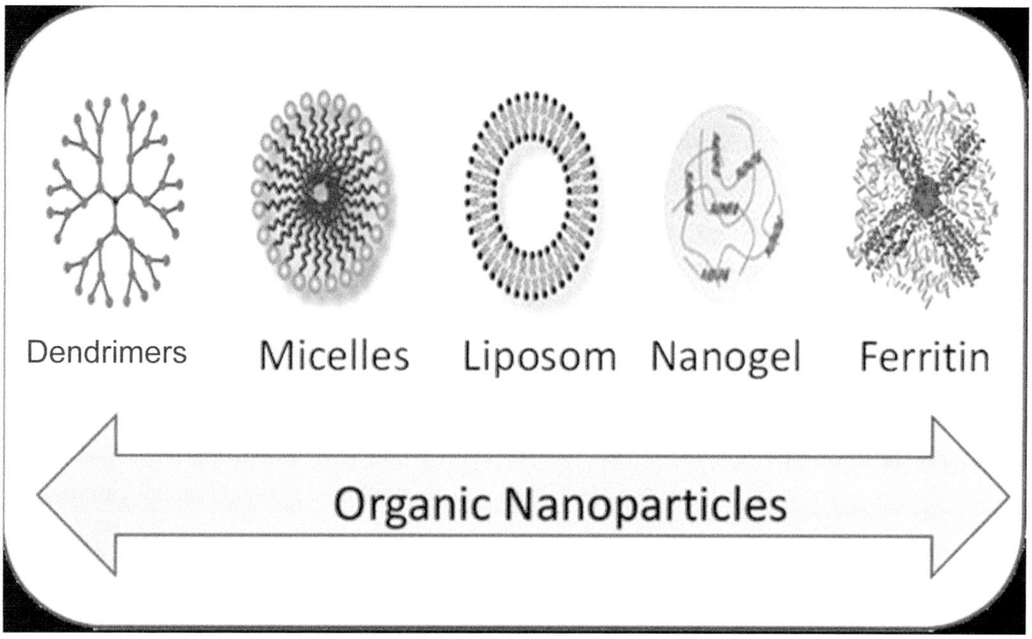

Figure 2: Different types of organic nanoparticles.

wastewater, were removed by utilizing solid magnetic liposomes (Allegritti et al. 2025). Recently, it was reported that ferritin bio-nanomaterial treatment in water was very effective in reducing pollutants. Fluoride and sulfate levels in water from metro rail underground tunnels were lowered using the Geographical Information System (GIS) (Lilly et al. 2022).

Figure 2 displays several types of organic nanoparticles, some of which are micelles, dendrimers, liposomes, nanogels, and ferritin, utilized for water remediation. Dendrimers are usually made using either a divergent or convergent method. In the divergent method, the dendrimer grows outward starting from a central core molecule with multiple branches. The main molecule undergoes a reaction with monomer units that have a single reactive functional group and two unreactive groups, resulting

Table 1: Application of nanoparticles in the removal of pollutants.

Sr. No.	Type of Nanoparticle	Type of pollutants removed	Reference
1	Carbon nanotubes	Organic contaminant	(Jamil et al. 2024)
2	Nano-scale metal oxide	Heavy metals, Radionuclides	(Salwadi and Kurapati 2023)
3	Nanocatalyst	PCB, Azodyes, Pesticides, etc	(Karimi et al. 2023, Lakhani et al. 2024)
4	Nano-structured catalytic material.	Decomposition of organic pollutants, inactivation of microorganisms	(Lakhani et al. 2024)
5	Bioactive nanoparticle	Removal of bacteria, fungi	(Mondal et al. 2024)
6	Biomimetic membranes	Removing salts	(Sharma et al. 2024)

in the formation of the first-generation dendrimer. Dendrimers, with their large, branched structures, exhibit a high capacity for binding and trapping various contaminants. This unique characteristic enables them to effectively prevent the passage of these pollutants through membranes (Abbasi et al. 2014). Furthermore, dendrimers offer significant potential for environmental applications, including water purification and the development of eco-friendly industrial processes. A diverse range of dendrimers can be utilized for water treatment, including those designed to bind anions, organic compounds, cations, biological compounds, and even viruses. Moreover, dendrimers can be engineered to possess multiple binding functionalities, offering enhanced versatility in water treatment applications (Patanjali et al. 2019). Liposomes are utilized for the removal of common pollutants. Liposomes can be synthesized using three main methods: mechanical dispersion, solvent dispersion, and detergent removal. Allegritti et al. (2025) demonstrated that > 80% alkylphenol ethoxylates (APEO), common pollutants in industrial wastewater, were removed by utilizing solid magnetic liposomes. Earlier, it was reported that the ferritin bio-nanomaterial treatment in water was very effective in reducing the pollutant level of fluoride and sulfate salts from water in the underground passages of the metro rail using the Geographical Information System (Lilly et al. 2022). Several research groups have extensively reviewed the performance of diverse hydrogel materials for water purification. These materials are cost-effective, environmentally friendly, and efficient in removing various contaminants from wastewater, including dyes, heavy metals, oils, pharmaceuticals, and pesticides. These hydrogels can break down naturally and are safe for living things, which makes them an eco-friendly choice for cleaning water. Carbon dot-hydrogel composites, concentrating on their synthesis and applications in detecting and eliminating water pollutants (Jiang et al. 2024). Also, another report briefly explained the efficiency of cellulose-based hydrogels for water treatment and purification (Le et al. 2023). The usage of polymer-based hydrogels to clean wastewater has been studied, and gum acacia-based hydrogels and their mixtures have also been tested for treating wastewater (Ahmaruzzaman et al. 2023). Also, in another review, the usefulness of alginate- and cellulose-based hydrogels for water and wastewater treatment was highlighted (Radoor et al. 2024). A summary is shown in Table 1 describing the application of nanotechnology in water treatment.

For the removal of water pollutants from water, nanoremediation utilizes nanotubes, carbon, and carbon-based NMs, NPs, nanofibers, nanoclusters, and nanocomposites. A wide range of nanoparticles possess unique properties and modes of action, which are discussed below.

4. Mechanisms of Action

Each nanoparticle type has distinct mechanisms that make it effective for specific water remediation purposes. These mechanisms, which include adsorption, reduction, photocatalysis, magnetic separation, and antibacterial action, provide diverse solutions for removing various contaminants from water. The breakdown of the mechanisms of action for different nanoparticles used in water remediation is discussed here in detail.

4.1 Metal Oxide Nanoparticles (e.g., TiO_2, ZnO)

4.1.1 Photocatalytic Degradation

Metal oxide nanoparticles such as titanium dioxide (TiO_2) and zinc oxide (ZnO) can absorb ultraviolet (UV) light. This absorption leads to the excitation of electrons from the valence band to the conduction band in the nanoparticle, leading to the formation of electron-hole pairs. These electron-hole pairs can interact with water and oxygen in the surroundings to create highly reactive oxygen species (ROS) such as hydroxyl radicals (OH•) and superoxide anions (O_2•−). These reactive oxygen species are extremely reactive and can break down organic contaminants (e.g., pesticides, dyes, and pharmaceutical waste) and pathogens (e.g., bacteria and viruses) (Yang et al. 2024). This is the primary way metal oxides remove harmful substances from water via oxidation and degradation, as shown in Figure 3.

Many studies have focused on the expansion of the solar absorption band by combining TiO_2 with new metals, intermediates, and anions, which maximizes the performance of the photocatalyst (GATTO 2014). Yadav et al. (2020) used a boron-doped TiO_2 (B-TiO_2) nanocatalyst to remove 4-nitrophenol from water. The highest removal rates were 79% with regular TiO_2 and 90% with B-TiO_2, showing that adding boron improved its effectiveness (Yadav et al. 2017). Also, carbon-TiO_2 nanocatalysts were able to remove methyl ethyl ketone under UV light with a high efficiency of 94% (Shayegan et al. 2020).

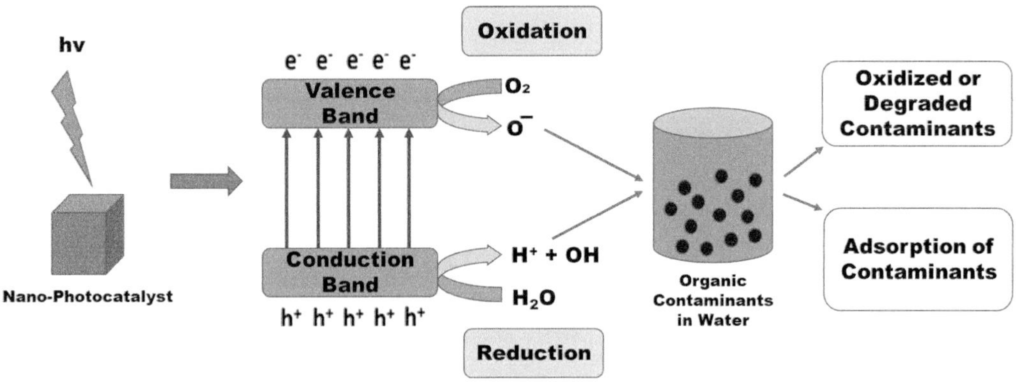

Figure 3: Nano-Photocatalytic Degradation of Organic Contaminants in Water.

4.1.2 Surface Adsorption

Metal oxide nanoparticles have a high surface area-to-volume ratio, making them excellent for adsorbing pollutants. These particles can capture heavy metals (e.g., lead, mercury, and arsenic) and organic pollutants (e.g., dyes and oil) onto their surfaces (Figure 4). The contaminants adhere due to electrostatic forces and hydrogen bonding, effectively concentrating pollutants on the particle surface. Once adsorbed, contaminants can be removed physically by filtering out the nanoparticles. One study highlighted the remarkable effectiveness of NiO and Co_3O_4 nanoparticles as effective adsorption agents for the simultaneous elimination of contaminants in water (Iqbal et al. 2024). Sarah et al. reported that chiral mesoporous silica nanoparticles (MSNPs) are highly effective for the removal of sulfoxaflor from water. The adsorption capacity of functionalized MSNPs was significantly higher (435.45 mg g^{-1}) owing to the highly porous nature of the silica particles and the presence of chiral functional groups. The magnetic nature of the adsorbent helped to remove it from the aqueous medium using a magnet. Both the removal efficiency and adsorption capacity were reasonable (Alharthi et al. 2025).

The specially treated MSNPs could hold a lot more sulfoxaflor (435.45 mg per gram) because they have many tiny pores and special chiral groups. Their magnetic properties made it easy to take

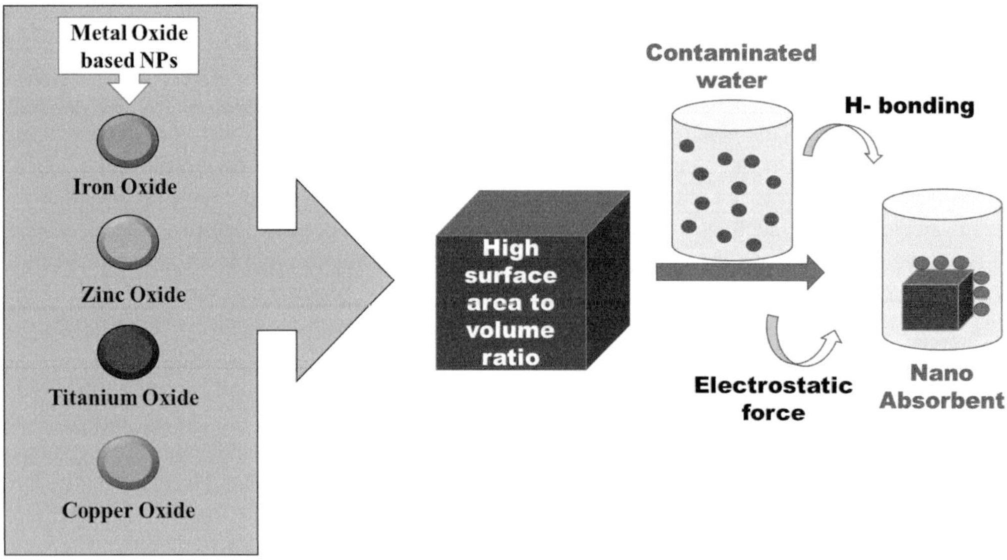

Figure 4: Metal Oxide-based Nanoparticles Used for Wastewater Treatment.

them out of the water using a magnet. Overall, they removed sulfoxaflor efficiently and could hold a good amount (Alharthi et al. 2025).

4.2 Zero-Valent Iron Nanoparticles (Fe°)

The Fenton reaction using nanoscale zero-valent iron (nZVI) is a useful catalytic process for the degradation of organic and inorganic pollutants in wastewater (Prashanth et al. 2024). Due to high efficiency in the treatment of industrial wastewater, the Fenton-like methods are being widely studied. Zero-valent iron (ZVI) nanoparticles are highly reactive due to their electron-donating nature. ZVI donates electrons to toxic contaminants in the water, primarily heavy metals (e.g., Cr^{6+}, Pb^{2+}) and chlorinated organic compounds (e.g., pesticides, solvents). For instance, Chromium VI (Cr^{6+}) is highly toxic, but when it interacts with Fe^0 nanoparticles, the iron reduces the chromium to Chromium III (Cr^{3+}), which is less toxic and less mobile. This process is called reductive dechlorination or chemical reduction.

4.2.1 Fenton-like Reactions

ZVI nanoparticles can catalyze Fenton-like reactions by reacting with hydrogen peroxide (H_2O_2) present in the water. The reaction generates hydroxyl radicals (OH•), which are highly reactive and capable of oxidizing organic contaminants (Galdames et al. 2020, Keenan and Sedlak 2008). This oxidation process breaks down complex organic pollutants (such as industrial chemicals) into simpler, less toxic compounds. This mechanism also contributes to decontamination by degrading pathogens or organic toxins into non-toxic compounds.

4.3 Carbon-Based Nanoparticles (e.g., Activated Carbon, Graphene Oxide)

Carbon nanotubes (CNTs) and their composites can absorb dyes from water. This includes both multi-walled nanotubes (MWCNTs) and single-walled nanotubes (SWCNTs) (Ijaz et al. 2023).

4.3.1 Adsorption

Carbon-based nanoparticles like activated carbon and graphene oxide have a highly porous structure and a large surface area, making them great for adsorbing different types of pollutants in water. These nanoparticles are effective at adsorbing heavy metals (such as mercury, cadmium, and lead) and organic pollutants (such as pesticides, hydrocarbons, and pharmaceuticals). The mechanism relies on both electrostatic and van der Waals forces to attract contaminants to the surface of the carbon particles. Activated carbon is widely used in water filtration systems because it can physically trap contaminants within its pores.

4.3.2 Catalytic Properties (Graphene Oxide)

Graphene oxide (GO) nanoparticles not only adsorb contaminants but also have catalytic properties. These particles can enhance the breakdown of organic pollutants through oxidation reactions, often in the presence of other agents (such as H_2O_2 or UV light). GO's surface chemistry (presence of hydroxyl, epoxide, and carboxyl groups) enables catalytic oxidation to break down complex organic contaminants into less harmful or simpler compounds (Dayana Priyadharshini et al. 2022).

4.4 Polymeric Nanoparticles

4.4.1 Ion Exchange

Polymeric nanoparticles are often functionalized with groups that can exchange ions. These materials can be tailored to eliminate harmful ions such as heavy metals (e.g., Cd^{2+}, Cu^{2+}, Hg^{2+}) from water by exchanging these harmful ions with less toxic ions (e.g., Na^+, K^+, Ca^{2+}). The high surface area and specific functionalization of these particles allow them to selectively remove certain contaminants through ion exchange processes (Yu et al. 2022).

4.4.2 Adsorption and Encapsulation

These nanoparticles can also adsorb organic contaminants (e.g., pesticides, dyes) through hydrophobic interactions or hydrogen bonding. Additionally, certain types of polymeric nanoparticles are designed to encapsulate toxic materials, preventing their release back into the environment and reducing contamination in the water. The hydrophilic or hydrophobic properties of these materials allow for the targeted removal of pollutants, depending on the water composition and contaminant type.

4.5 Magnetic Nanoparticles (e.g., Fe_3O_4, Fe_3O_4-TiO_2)

4.5.1 Magnetic Separation

Magnetic nanoparticles (e.g., Fe_3O_4) can be quickly removed from water using an external magnetic field, making the entire remediation process more efficient and cost-effective (Kumar et al. 2021). After the nanoparticles have adsorbed the contaminants, the magnetic field can pull the particles (along with adsorbed contaminants) out of the water for further treatment or disposal. One study utilized magnetic mesoporous silica nanospheres to remove lead (Pb^{2+}), mercury (Hg^{2+}), and palladium (Pd^{2+}) ions from a solution (Farahbakhsh et al. 2021)

4.5.2 Adsorption

Like other nanoparticles, magnetic nanoparticles can adsorb contaminants (e.g., heavy metals, dyes, organic compounds) onto their surface. The magnetic properties allow for easy removal after adsorption.

4.5.3 Catalysis

Magnetic nanoparticles like Fe_3O_4-TiO_2 combine the properties of magnetic nanoparticles and photocatalytic nanoparticles (like TiO_2). When exposed to UV light, the TiO_2 component generates reactive oxygen species (ROS) that degrade organic pollutants. These combined properties allow for pollutant removal through both adsorption and catalysis, providing a dual approach to water purification.

4.6 Metal Nanoparticles (e.g., Ag, Au, Pd)

4.6.1 Antibacterial Properties (Silver Nanoparticles - AgNPs)

Silver nanoparticles (AgNPs) have strong antibacterial properties and can effectively disinfect water. The small size of AgNPs allows them to interact with microbial cells, causing the release of silver ions (Ag^+), which interfere with cell function by damaging the cell membrane, proteins, and DNA. This mechanism kills bacteria and viruses in the water, making it safer for consumption (Figure 5).

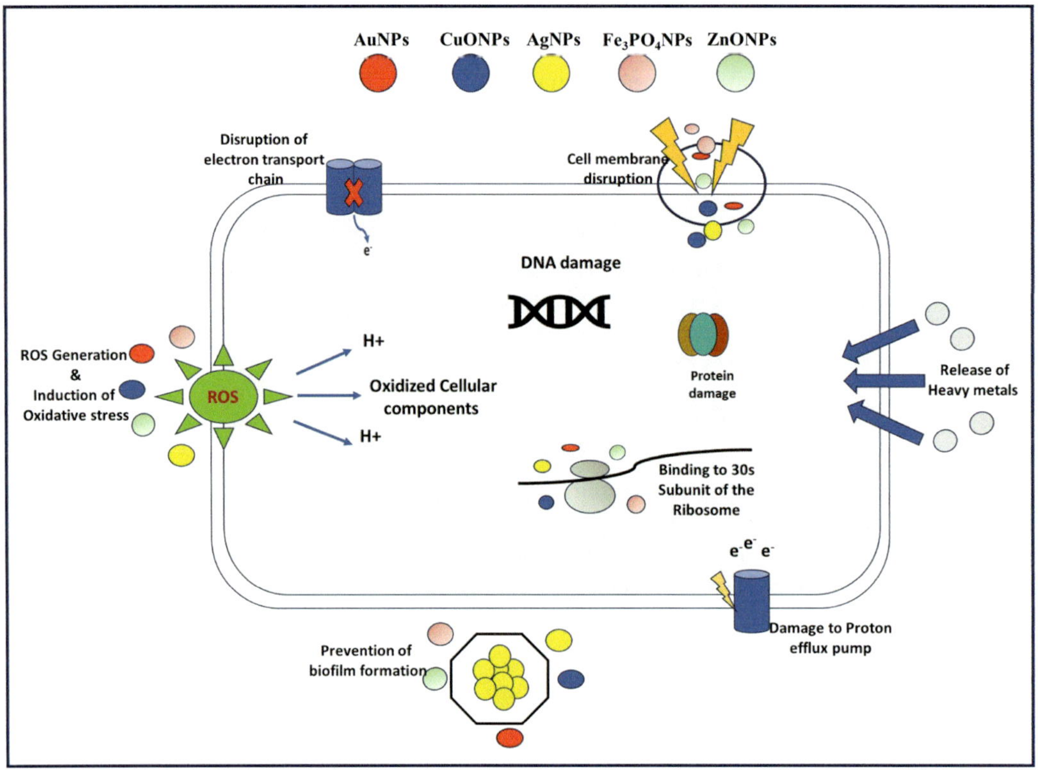

Figure 5: Mechanism of Action of Different Metal Nanoparticles for Wastewater Treatment.

4.6.2 Catalysis (Gold and Palladium Nanoparticles - AuNPs, PdNPs)

Gold (AuNPs) and palladium (PdNPs) nanoparticles have catalytic properties that can facilitate the breakdown of organic contaminants in water, such as chlorinated compounds, pesticides, and solvents. These nanoparticles can catalyze reduction reactions when in the presence of hydrogen or other reducing agents. The catalytic properties of PdNPs are particularly useful for breaking down pollutants in wastewater treatment systems.

5. Applications of Nanoparticles in Water Remediation

The hydrosphere, being 71% of the total Earth, is important to preserve for the survival of human populations in the future. With the arrival of the Industrial Revolution in the 19th century, water bodies have been at ultimate risk of becoming sinks for different types of pollutants, such as organic contaminants, dyes, inorganic heavy metals, and biological pathogens, in the last few decades (Chandran et al. 2023, Rybarczyk et al. 2024). This contamination of water bodies comes from different diffuse sources such as industries, agricultural waste, and other natural phenomena, like the littering of organic waste into the water. All these factors are responsible for water pollution, making it unsuitable for human use. To make the water suitable for human use, various physicochemical and biological methods have been used, such as ion exchange, adsorption, filtration, bioremediation, and phytoremediation. Further, another technique known as nanoremediation employs small-scale particles or nanoparticles (NPs) to eliminate contaminants from water sources. These nanoparticles can be synthesized chemically as well as biologically, as extracted from bacteria, plants, and fungi (Kulkarni et al. 2023, Singh et al. 2021). This section of the review will discuss nanoparticles' role in removing various contaminants from contaminated water.

5.1 Nanoremediation of Heavy Metals

Small metallic particles, also known as transition elements, with a higher molecular weight of 63.5–200 g/mol and a density greater than 5 g/cm^3 (Shadman et al. 2019). Contamination of water bodies with toxic heavy metals such as mercury (Hg), copper (Cu), arsenic (As), chromium (Cr), nickel (Ni), and others is caused mainly by the release of industrial effluent, energy plants, and environmental factors such as flooding, discharge from landfills, etc. These polluted water bodies can be a significant menace to human health due to persistent accumulation and magnification after channeling into food chains (Moukadiri et al. 2024). Due to this, researchers are keenly looking for certain solutions for these pollutants.

Among the advanced technologies utilized for solving this issue, using different kinds of nanoparticles to remove heavy metals from contaminated water is one of the favored choices. Large specific surface area, surface charge, and porosity make these small-sized particles of 1–100 nm, known as nanoparticles, one of the most sought-after and important methods to remove heavy metals from water. Due to high adsorption capacity and large specific area, nanoparticles behave as mini magnets to pull out the heavy metals from wastewater and hold them for a longer time, which later on can be removed. Among these, carbon, chitosan, ferrites, metal oxides, and magnetic particles have been used in several approaches, like nanosorption, to remove Hg, Cd, and Pb. This is highly efficient and cost-effective (Baby et al. 2022). NPs/NMs such as ZnO, TiO_2, carbon nanotubes, and metal oxides are available tools for removing heavy metal contaminants. For example, biodegradable ZnO nanoparticles can absorb heavy metals such as Pb(II), Ni(II), and Cu(II) ions efficiently (Garg et al. 2022, Samad et al. 2021). The application of silica-based nanomaterials has also been implicated as an effective strategy to remove heavy metals like lead, nickel, and cadmium due to their controlled pore size, regulated large surface area, and non-toxic properties (Sankareswaran et al. 2022). Recently, the use of nanomaterials like graphene oxides and carbon quantum dots in membrane filters has been shown to have higher efficiency of absorption of heavy metals such as Cu^{2+}, Cr^{2+}, Pb^{2+}, Co^{2+}, and Mn^{2+} (Mahmoud and Mostafa 2023). The efficiency of such nanomembrane filters can be explored by further investigation.

5.2 Removal of Organic Pollutants

These are hazardous chemicals generated in various chemical and biochemical industries as by-products of several chemical reactions. They are categorized as Persistent Organic Compounds (POCs) and Volatile Organic Compounds (VOCs). Some well-known POPs are polychlorinated biphenyls, chlorinated hydrocarbon derivatives known as organochlorine pesticides, polycyclic aromatic hydrocarbons, dioxins, dibenzofurans, etc. (Ashraf 2017). There are various other ways by which organic pollutants are released into the water, as summarized in Figure 6.

According to a report from the United Nations Environment Programme, Persistent Organic Pollutants accumulate in the fatty tissues and magnify in living organisms, remaining intact for a longer period. These POPs weaken the defense mechanism of the body, making it prone to various infections, congenital and reproductive organ malformation, as well as hormonal imbalance (Fei et al. 2022). Therefore, they have a toxic impact on humans and other living organisms. These organic pollutants include pesticides, pharmaceuticals, industrial chemicals, organic wastes from mining, and pollutants from household debris. Removal of these contaminants from the water is essential before further usage through various technologies. Nanoparticle-mediated methods offer a suitable

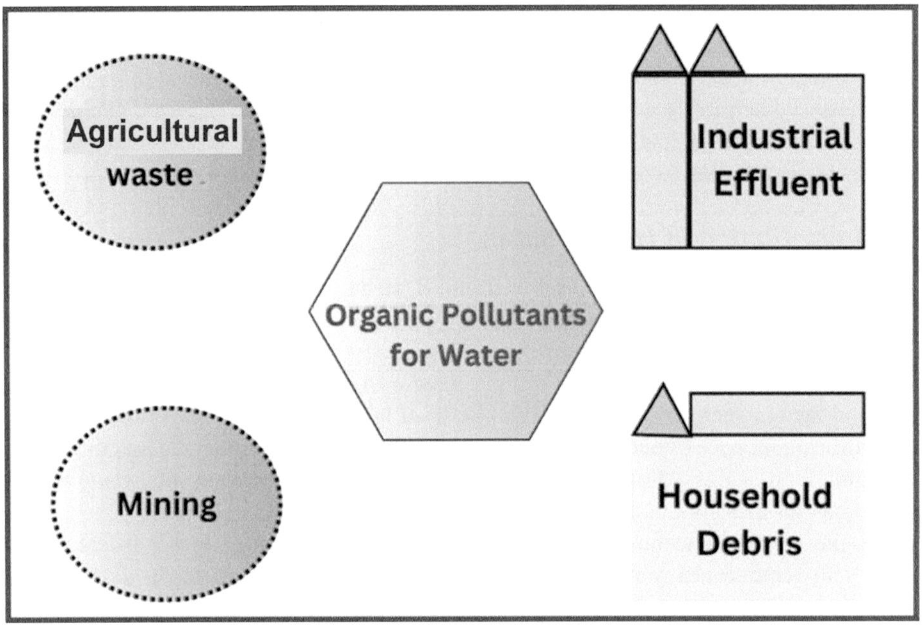

Figure 6: Multiple ways for the release of waste into the water bodies.

approach for water remediation to eliminate organic pollutants. Nano-adsorption can be used with a removal efficiency of more than 70% of the POPs. This high removal is attributed to the stronger interaction of pollutants and nanomaterials via hydrogen, electrostatic, and hydrophobic bonding (Ighalo et al. 2022). Another method is the usage of nanoparticles with catalytic activity, known as nanoenzymes, for the treatment and removal of impurities from the water. For example, laccase, peroxidases, and oxidases coupled on different nanoparticles like copper, iron, and PdNPs/PCNF have been shown to effectively remove POPs such as phenolic compounds, dyes (e.g., methylene blue, malachite green), and the pesticide simazine via catalytic oxidation, producing reactive oxygen intermediates to degrade organic pollutants (Dadigala et al. 2022, Singh et al. 2023). Thus, nanoparticles conjugated with biocatalysts provide a cost-effective, efficient method for removing organic particulate matter from water.

5.3 Nanoparticles in Disinfection and Pathogen Removal

Water-borne diseases such as acute diarrhea, cholera, typhoid, giardiasis, and hepatitis are some of the most common causes of fatality among children in developing countries. Open defecation, improper hygienic conditions, and contamination of water bodies by mismanaged biomedical waste and human feces are the major reasons for such prominent medical situations by adding one or another pathogenic microbial entity to the water bodies. Hence, disinfection and decontamination of these water bodies are essential for making them suitable for drinking and day-to-day activities. Conventional treatment involves UV irradiation, halogen treatments, ozonation, and filtration (Lanrewaju et al. 2022). Although this treatment helps to enhance water quality, the production of undesired by-products as a result of halogen treatment, such as haloacetic acids (HAA) and trihalomethanes (THM), can lead to various defects related to mutations causing cancer and cell death. These by-products can also be linked to prenatal deformities in the human population (Srivastav et al. 2020). The limited efficacy of UV radiation in eliminating UV-resistant microbes is another drawback of these conventional methods. Silver nanoparticles with antimicrobial properties offer another option for water disinfection. Also, metallic nanoparticles such as titanium oxide (TiO2) and zinc oxide (ZnO) are used for producing Reactive Oxygen Species (ROS) that help in the eradication of microorganisms to make the water more suitable for drinking purposes (Dimapilis et al. 2018, Mhadhbi et al. 2023, Slavin et al. 2017). Apart from this, carbon nanotubes have also been used for the control of microbial cell populations via cell wall rupture (Asaftei et al. 2023). Thus, nanoparticles offer a great range of antimicrobial activities, for example, against *Escherichia coli, Salmonella typhi, Enterobacter ludwigii, Staphylococcus aureus*, etc. (Olawade et al. 2024). Further, nanoparticles are also used as coatings on the instruments and equipment used for water treatment, as they can prevent biofilm formation on the surface of this equipment. For example, graphene oxide has been proven to be bacteriostatic by inhibiting bacterial growth on the surface (Yadav et al. 2017). Thus, nanoparticles offer substantial possibilities for wastewater treatment, removing pollutants and microbial contaminants to make it suitable for daily usage.

6. Challenges and Future Directions

To date, synthesizing NPs/NMs with desired and optimized properties for wastewater treatment remains a key challenge. These properties include enhanced stability within water environments, high efficacy in removing contaminants, and minimal environmental impact (Zhao et al. 2023). By using the green chemistry route and the utilization of renewable resources, future research activities should focus on the development of sustainable nanoparticle synthesis methodologies (Sachin et al. 2023). Articulation of nanoparticles with conventional water treatment methods is also needed in order to enhance efficacy and counteract their limitations in water treatment (Rikta 2019). By integrating nanoparticles with techniques like filtration, coagulation, and membrane separation, several advantages can be achieved, such as enhanced treatment efficiency, improved pathogen removal, and reduced treatment costs. For example, the enhancement of the antibacterial properties of NPs/NMs and prevention of biofouling can increase their immobilization onto filtration membranes (Chakachaka et al. 2023). This integrated approach holds significant promise for developing more efficient, sustainable, and cost-effective water treatment solutions for the future.

7. Conclusion

Water pollution has a significant and harmful impact on both humans and the ecosystem. The development of innovative and effective water treatment technologies is crucial to address this global challenge. Nanotechnology offers a promising approach with its exceptional characteristics, like greater surface area, enhanced reactivity, and tunable surface chemistry. This chapter highlights the diverse range of nanoparticles, including metal oxides, zero-valent iron, carbon-based materials, polymeric nanoparticles, magnetic nanoparticles, and noble metal nanoparticles, and their mechanisms of action in water remediation. These mechanisms include adsorption, photocatalysis, reduction, ion exchange, magnetic separation, and antibacterial activity. While nanotechnology presents significant potential for water remediation, it is important to address the potential risks linked with the use of nanoparticles. These risks include potential toxicity to aquatic organisms, environmental persistence, and potential for human exposure. Despite these challenges, nanotechnology holds immense promises for revolutionizing water treatment and guaranteeing access to clean and safe water for all. This requires continuous research to be carried out in this field to address the growing global water crisis and create a more sustainable future.

References

Abbasi, E., Aval, S. F., Akbarzadeh, A., Milani, M., Nasrabadi, H. T., Joo, S. W. et al. (2014). Dendrimers: Synthesis, applications, and properties. *Nanoscale Research Letters*, 9(1): 247.

Abyar, H., and Nowrouzi, M. (2023). Trickling filter systems for sustainable water supply: An evaluation of eco-environmental burdens and greenhouse gas emissions. *Environmental Research*, 237: 117011.

Adetayo, A., and Runsewe, D. (2019). Synthesis and Fabrication of Graphene and Graphene Oxide: A Review. *Open Journal of Composite Materials*, 09(02): 207–229.

Aghalari, Z., Dahms, H.-U., Sillanpää, M., Sosa-Hernandez, J. E., and Parra-Saldívar, R. (2020). Effectiveness of wastewater treatment systems in removing microbial agents: A systematic review. *Globalization and Health*, 16(1): 13.

Ahmaruzzaman, Md., Roy, P., Bonilla-Petriciolet, A., Badawi, M., Ganachari, S. V., Shetti, N. P. et al. (2023). Polymeric hydrogels-based materials for wastewater treatment. *Chemosphere*, 331: 138743.

Algebaly, A. S., Mohammed, A. E., Abutaha, N., and Elobeid, M. M. (2020). Biogenic synthesis of silver nanoparticles: Antibacterial and cytotoxic potential. *Saudi Journal of Biological Sciences*, 27(5): 1340–1351.

Alguacil, F. J., and López, F. A. (2019). On the Active Adsorption of Chromium(III) from Alkaline Solutions Using Multiwalled Carbon Nanotubes. *Applied Sciences*, 10(1): 36.

Alharthi, S., Ali, A., and Santali, E. Y. (2025). Synthesis of chiral mesoporous silica nanoparticles for the adsorptive removal of the chiral insecticide sulfoxaflor from water. *Nanoscale Advances*, 10.

Alijani, H., and Shariatinia, Z. (2018). Synthesis of high growth rate SWCNTs and their magnetite cobalt sulfide nanohybrid as super-adsorbent for mercury removal. *Chemical Engineering Research and Design*, 129: 132–149.

Allegritti, E., Palumbo, C., Battista, S., D'Orazio, F., Aschi, M., Mattera, M. et al. (2025). Exploring Solid Magnetic Liposomes for Organic Pollutant Removal from Wastewater: The Role of Lipid Composition. *ACS Applied Materials & Interfaces*, 17(2): 3489–3502.

Amin, M. T., Alazba, A. A., and Manzoor, U. (2014). A Review of Removal of Pollutants from Water/Wastewater Using Different Types of Nanomaterials. *Advances in Materials Science and Engineering*, 2014: 1–24.

Asaftei, M., Lucidi, M., Cirtoaje, C., Holban, A.-M., Charitidis, C. A., Yang, F. et al. (2023). Fighting bacterial pathogens with carbon nanotubes: Focused review of recent progress. *RSC Advances*, 13(29): 19682–19694.

Ashraf, M. A. (2017). Persistent organic pollutants (POPs): A global issue, a global challenge. *Environmental Science and Pollution Research*, 24(5): 4223–4227.

Baby, R., Hussein, M. Z., Abdullah, A. H., and Zainal, Z. (2022). Nanomaterials for the Treatment of Heavy Metal Contaminated Water. *Polymers*, 14(3): 583.

Chakachaka, V. M., Mahlangu, O. T., Tshangana, C. S., Mamba, B. B., and Muleja, A. A. (2023). Highly adhesive CoFe2O4 nanoengineered PES membranes for salts and Naproxen removal and antimicrobial activities. *Journal of Membrane Science*, 676: 121612.

Chandran, D. G., Muruganandam, L., and Biswas, R. (2023). A review on adsorption of heavy metals from wastewater using carbon nanotube and graphene-based nanomaterials. *Environmental Science and Pollution Research*, 30(51): 110010–110046.

Dadigala, R., Bandi, R., Alle, M., Park, C.-W., Han, S.-Y., Kwon, G.-J. et al. (2022). Effective fabrication of cellulose nanofibrils supported Pd nanoparticles as a novel nanozyme with peroxidase and oxidase-like activities for efficient dye degradation. *Journal of Hazardous Materials*, 436: 129165.

Dayana Priyadharshini, S., Manikandan, S., Kiruthiga, R., Rednam, U., Babu, P. S., Subbaiya, R. et al. (2022). Graphene oxide-based nanomaterials for the treatment of pollutants in the aquatic environment: Recent trends and perspectives – A review. *Environmental Pollution, 306*, 119377.

Dimapilis, E. A. S., Hsu, C.-S., Mendoza, R. M. O., and Lu, M.-C. (2018). Zinc oxide nanoparticles for water disinfection. *Sustainable Environment Research*, 28(2): 47–56.

Eker, F., Duman, H., Akdaşçi, E., Bolat, E., Sarıtaş, S., Karav, S. et al. (2024). A Comprehensive Review of Nanoparticles: From Classification to Application and Toxicity. *Molecules*, 29(15): 3482.

Epelle, E. I., Okoye, P. U., Roddy, S., Gunes, B., and Okolie, J. A. (2022). Advances in the Applications of Nanomaterials for Wastewater Treatment. *Environments*, 9(11): 141.

Farahbakhsh, J., Vatanpour, V., Ganjali, M. R., and Saeb, M. R. (2021). Magnetic nanoparticles in wastewater treatment. In *Magnetic Nanoparticle-Based Hybrid Materials* (pp. 547–589). Elsevier.

Farjadian, F., Abbaspour, S., Sadatlu, M. A. A., Mirkiani, S., Ghasemi, A., Hoseini-Ghahfarokhi, M. et al. Recent Developments in Graphene and Graphene Oxide: Properties, Synthesis, and Modifications: A Review. *ChemistrySelect*, 5(33): 10200–10219.

Fei, L., Bilal, M., Qamar, S. A., Imran, H. M., Riasat, A., Jahangeer, M. et al. (2022). Nano-remediation technologies for the sustainable mitigation of persistent organic pollutants. *Environmental Research, 211*, 113060. https://doi.org/10.1016/j.envres.2022.113060.

Galdames, A., Ruiz-Rubio, L., Orueta, M., Sánchez-Arzalluz, M., and Vilas-Vilela, J. L. (2020). Zero-Valent Iron Nanoparticles for Soil and Groundwater Remediation. *International Journal of Environmental Research and Public Health*, 17(16): 5817. https://doi.org/10.3390/ijerph17165817.

Garg, R., Garg, R., Okon Eddy, N., Ibrahim Almohana, A., Fahad Almojil, S., Amir Khan, M. et al. (2022). Biosynthesized silica-based zinc oxide nanocomposites for the sequestration of heavy metal ions from aqueous solutions. *Journal of King Saud University - Science*, 34(4): 101996.

GATTO, S. (2014). Photocatalytic activity assessment of micro-sized tio2 used as powders and as starting material for porcelain gres tiles production.

Guan, X., Sun, Y., Qin, H., Li, J., Lo, I. M. C., He, D. et al. (2015). The limitations of applying zero-valent iron technology in contaminants sequestration and the corresponding countermeasures: The development in zero-valent iron technology in the last two decades (1994–2014). *Water Research*, 75: 224–248.

Guerra, F. D., Attia, M. F., Whitehead, D. C., and Alexis, F. (2018). Nanotechnology for Environmental Remediation: Materials and Applications. *Molecules*, 23(7): 1760.

Hsu, C.-Y., Mahmoud, Z. H., Abdullaev, S., Ali, F. K., Ali Naeem, Y., Mzahim Mizher, R. et al. (2024). Nano titanium oxide (nano-TiO2): A review of synthesis methods, properties, and applications. *Case Studies in Chemical and Environmental Engineering*, 9: 100626.

Hussain, I., Singh, N. B., Singh, A., Singh, H., and Singh, S. C. (2016). Green synthesis of nanoparticles and its potential application. *Biotechnology Letters*, 38(4): 545–560.

Ighalo, J. O., Yap, P.-S., Iwuozor, K. O., Aniagor, C. O., Liu, T., Dulta, K. et al. (2022). Adsorption of persistent organic pollutants (POPs) from the aqueous environment by nano-adsorbents: A review. *Environmental Research*, 212: 113123.

Ijaz, H., Mahmood, A., Abdel-Daim, M. M., Sarfraz, R. M., Zaman, M., Zafar, N. et al. (2023). Review on carbon nanotubes (CNTs) and their chemical and physical characteristics, with particular emphasis on potential applications in biomedicine. *Inorganic Chemistry Communications*, 155: 111020.

Iqbal, A., Jalees, M. I., Farooq, M. U., Cevik, E., and Bozkurt, A. (2024). Effectual adsorptive performance of metal-based engineered nanoparticles for surface water remediation: Systematic optimization by box-behnken design. *International Journal of Environmental Science and Technology*.

Iqbal, M. J., Javed, Z., Sadia, H., Qureshi, I. A., Irshad, A., Ahmed, R. et al. (2021). Clinical applications of artificial intelligence and machine learning in cancer diagnosis: Looking into the future. *Cancer Cell International*, 21(1): 270.

Jamil, A., Anwar-ul-Haq, M., Ahmed, R., Rizwan, A., Umair, M., and Zia-ur-Rehman, M. (2024). Potential applications of carbon nanotubes for the removal of organic and inorganic pollutants from wastewater. In *Water Treatment Using Engineered Carbon Nanotubes* (pp. 307–329). Elsevier.

Jiang, M., Wang, Y., Li, J., and Gao, X. (2024). Review of carbon dot–hydrogel composite material as a future water-environmental regulator. *International Journal of Biological Macromolecules*, 269: 131850.

Joudeh, N., and Linke, D. (2022). Nanoparticle classification, physicochemical properties, characterization, and applications: A comprehensive review for biologists. *Journal of Nanobiotechnology*, 20(1): 262.

Kakavandi, B., Kalantary, R. R., Farzadkia, M., Mahvi, A. H., Esrafili, A., Azari, A. et al. (2014). Enhanced chromium (VI) removal using activated carbon modified by zero valent iron and silver bimetallic nanoparticles. *Journal of Environmental Health Science and Engineering*, 12(1): 115.

Karimi, F., Zare, N., Jahanshahi, R., Arabpoor, Z., Ayati, A., Krivoshapkin, P. et al. (2023). Natural waste-derived nano photocatalysts for azo dye degradation. *Environmental Research*, 238: 117202.

Keenan, C. R., and Sedlak, D. L. (2008). Factors Affecting the Yield of Oxidants from the Reaction of Nanoparticulate Zero-Valent Iron and Oxygen. *Environmental Science & Technology*, 42(4): 1262–1267.

Kulkarni, D., Sherkar, R., Shirsathe, C., Sonwane, R., Varpe, N., Shelke, S. et al. (2023). Biofabrication of nanoparticles: Sources, synthesis, and biomedical applications. *Frontiers in Bioengineering and Biotechnology*, 11: 1159193.

Kumar, R., Rauwel, P., and Rauwel, E. (2021). Nanoadsorbants for the Removal of Heavy Metals from Contaminated Water: Current Scenario and Future Directions. *Processes*, 9(8): 1379.

Lakhani, P., Bhanderi, D., and Modi, C. K. (2024). Nanocatalysis: Recent progress, mechanistic insights, and diverse applications. *Journal of Nanoparticle Research*, 26(7): 148.

Lakshani, N., Rathnayaka, R. M. D. S., and Sandaruwan, C. (2025). Advanced Nanomaterials for Water Treatment: Synthesis Methods and Application. pp. 257–277. *In:* Alam, M. W. (Ed.), *Breaking Boundaries: Pioneering Sustainable Solutions Through Materials and Technology*. Springer Nature Singapore.

Lakshmi, D., Diana, M. I., Helen, P. A., and Selvin, P. C. (2024). Functionalized Nanomaterials, Classification, Properties, and Functionalization Techniques. pp. 65–92. *In:* C. M. Hussain & M. B. Ahamed (Eds.), *Functionalized Nanomaterials Based Supercapacitor*. Springer Nature Singapore.

Lanrewaju, A. A., Enitan-Folami, A. M., Sabiu, S., and Swalaha, F. M. (2022). A review on disinfection methods for inactivation of waterborne viruses. *Frontiers in Microbiology*, 13: 991856.

Le, V. T., Joo, S.-W., Berkani, M., Mashifana, T., Kamyab, H., Wang, C. et al. (2023). Sustainable cellulose-based hydrogels for water treatment and purification. *Industrial Crops and Products*, 205: 117525.

Lilly, R., Prabhakaran, S., Giridharan, K., Sambandam, P., Stalin, B., Subhashini, S. J. et al. (2022). Efficiency of Ferritin bio-nanomaterial in reducing the pollutants level of water in the underground corridors of metro rail using GIS. *Scientific Reports*, 12(1): 20301.

Mahmoud, A. E. D., and Mostafa, E. (2023). Nanofiltration Membranes for the Removal of Heavy Metals from Aqueous Solutions: Preparations and Applications. *Membranes*, 13(9): 789.

Malik, S., Muhammad, K., and Waheed, Y. (2023). Nanotechnology: A Revolution in Modern Industry. *Molecules*, 28(2): 661.

Mathur, J., Goswami, P., Gupta, A., Srivastava, S., Minkina, T., Shan, S. et al. (2022). Nanomaterials for Water Remediation: An Efficient Strategy for Prevention of Metal(loid) Hazard. *Water*, 14(24): 3998.

Mekuye, B., and Abera, B. (2023). Nanomaterials: An overview of synthesis, classification, characterization, and applications. *Nano Select*, 4(8): 486–501.

Mhadhbi, M., Abderazzak, H., and Avar, B. (2023). Synthesis and Properties of Titanium Dioxide Nanoparticles. In Bejaoui ,B. (Ed.), *Updates on Titanium Dioxide*. IntechOpen.

Mishra, R. K., Mentha, S. S., Misra, Y., and Dwivedi, N. (2023). Emerging pollutants of severe environmental concern in water and wastewater: A comprehensive review on current developments and future research. *Water-Energy Nexus*, 6: 74–95.

Mondal, S. K., Chakraborty, S., Manna, S., and Mandal, S. M. (2024). Antimicrobial nanoparticles: Current landscape and future challenges. *RSC Pharmaceutics*, 1(3): 388–402.

Moukadiri, H., Noukrati, H., Ben Youcef, H., Iraola, I., Trabadelo, V., Oukarroum, A. et al. (2024). Impact and toxicity of heavy metals on human health and latest trends in removal process from aquatic media. *International Journal of Environmental Science and Technology*, 21(3): 3407–3444.

Ogochukwu, O. O., Fabiyi, M. B., Aworunse, O. S., Oyewole, O. A., and Isibor, P. O. (2024). Nanoparticle Properties and Characterization. pp. 23–40. *In:* Isibor, P. O., Devi, G., and Enuneku, A. A. (Eds.), *Environmental Nanotoxicology*.

Olawade, D. B., Wada, O. Z., Fapohunda, O., Egbewole, B. I., Ajisafe, O., and Ige, A. O. (2024). Nanoparticles for microbial control in water: Mechanisms, applications, and ecological implications. *Frontiers in Nanotechnology*, 6: 1427843.

Omarova, A., Tussupova, K., Berndtsson, R., Kalishev, M., and Sharapatova, K. (2018). Protozoan Parasites in Drinking Water: A System Approach for Improved Water, Sanitation and Hygiene in Developing Countries. *International Journal of Environmental Research and Public Health*, 15(3): 495.

Patanjali, P., Singh, R., Kumar, A., and Chaudhary, P. (2019). Nanotechnology for water treatment: A green approach. pp. 485–512. *In: Green Synthesis, Characterization and Applications of Nanoparticles*. Elsevier.

Prasek, J., Drbohlavova, J., Chomoucka, J., Hubalek, J., Jasek, O., Adam, V. et al. (2011). Methods for carbon nanotubes synthesis—Review. *Journal of Materials Chemistry*, 21(40): 15872.

Prashanth, C. B., Geetha, K. M., Wilson, B., and Banu, S. (2024). Traditional uses, bioactive composition and pharmacological activities of Lens culinaris. *South African Journal of Botany*, 168: 542–561.

Puthukkara P, A. R., Jose T, S., and S, D. L. (2021). Plant-mediated synthesis of zero-valent iron nanoparticles and its application in water treatment. *Journal of Environmental Chemical Engineering*, 9(1): 104569.

Qi, Y., and He, K. (2025). Science and Technology for Water Purification: Achievements and Strategies. *Water*, 17(1): 91.

Radetic, B. (2024). Sludge Treatment Technologies and Systems, an Introduction. pp. 689–713. *In:* Lahnsteiner, J. (Ed.), *Handbook of Water and Used Water Purification*. Springer International Publishing.

Radoor, S., Karayil, J., Jayakumar, A., Kandel, D. R., Kim, J. T., Siengchin, S. et al. (2024). Recent advances in cellulose- and alginate-based hydrogels for water and wastewater treatment: A review. *Carbohydrate Polymers*, 323: 121339.

Rana, V., Sheokand, M., Jain, K., Dhaka, S., Godara, S. K., Urmaliya, D. N. et al. (2024). Effective and affordable water purification technologies for rural development. pp. 91–106. *In: Water Resources Management for Rural Development*. Elsevier.

Rikta, S. Y. (2019). Application of Nanoparticles for Disinfection and Microbial Control of Water and Wastewater. pp. 159–176. *In: Nanotechnology in Water and Wastewater Treatment*. Elsevier.

Ritu, Verma, K. K., Das, A., and Chandra, P. (2023). Phytochemical-Based Synthesis of Silver Nanoparticle: Mechanism and Potential Applications. *BioNanoScience*, 13(3): 1359–1380.

Rybarczyk, A., Szada, D., Badzińska, W., Degórska, O., Jesionowski, T., and Zdarta, J. (2024). Nanobiocatalysts as efficient tools for wastewater pollutants remediation. In *Advances in Chemical Pollution, Environmental Management and Protection* (Vol. 10, pp. 57–82).

Sachin, Jaishree, Singh, N., Singh, R., Shah, K., and Pramanik, B. K. (2023). Green synthesis of zinc oxide nanoparticles using lychee peel and its application in anti-bacterial properties and CR dye removal from wastewater. *Chemosphere*, 327: 138497.

Saeed, M., Alshammari, Y., Majeed, S. A., and Al-Nasrallah, E. (2020). Chemical Vapour Deposition of Graphene—Synthesis, Characterisation, and Applications: A Review. *Molecules*, 25(17): 3856.

Salwadi, Y., and Kurapati, S. K. (2023). The Role of Nanomaterials in Radioactive Waste Reduction and Decontamination: A Concise Review. pp. 141–157. *In:* Mukherjee, R., Janiak, C., Khan, Z., Chatterjee, S., Roy, B., Ghosh, S. et al. (Eds.), *Recent Trends in Nanotechnology for Sustainable Living and Environment*.

Samad, A., Din, M. I., Ahmed, M., and Ahmad, S. (2021). Synthesis of zinc oxide nanoparticles reinforced clay and their applications for removal of Pb (II) ions from aqueous media. *Chinese Journal of Chemical Engineering*, 32: 454–461.

Sankareswaran, M., Periakaruppan, R., Sasivarnam, M., Danaraj, J., Dhanasekaran, S., and Abomughaid, M. M. (2022). Bio-fabrication of Bio-inspired Silica Nanomaterials from Bryophyllum pinnatum Leaf for Agricultural Applications. *Applied Biochemistry and Biotechnology*, 194(9): 4266–4277.

Sati, A., Ranade, T. N., Mali, S. N., Ahmad Yasin, H. K., and Pratap, A. (2025). Silver Nanoparticles (AgNPs): Comprehensive Insights into Bio/Synthesis, Key Influencing Factors, Multifaceted Applications, and Toxicity—A 2024 Update. *ACS Omega*, 10(8): 7549–7582.

Shadman, S. M., Daneshi, M., Shafiei, F., Azimimehr, M., Khorasgani, M. R., Sadeghian, M. et al. (2019). Aptamer-based electrochemical biosensors. pp. 213–251. *In: Electrochemical Biosensors*.

Sharma, A., Goel, H., Sharma, S., Rathore, H. S., Jamir, I., Kumar, A. et al. (2024). Cutting edge technology for wastewater treatment using smart nanomaterials: Recent trends and futuristic advancements. *Environmental Science and Pollution Research*, 31(48): 58263–58293.

Shayegan, Z., Haghighat, F., and Lee, C.-S. (2020). Carbon-doped TiO2 film to enhance visible and UV light photocatalytic degradation of indoor environment volatile organic compounds. *Journal of Environmental Chemical Engineering*, 8(5): 104162.

Sherugar, P., Naik, N. S., Padaki, M., Nayak, V., Gangadharan, A., Nadig, A. R. et al. (2021). Fabrication of zinc doped aluminium oxide/polysulfone mixed matrix membranes for enhanced antifouling property and heavy metal removal. *Chemosphere*, 275: 130024.

Singh, K. R., Nayak, V., Singh, J., Singh, A. K., and Singh, R. P. (2021). Potentialities of bioinspired metal and metal oxide nanoparticles in biomedical sciences. *RSC Advances*, 11(40): 24722–24746.

Singh, R., Umapathi, A., Patel, G., Patra, C., Malik, U., Bhargava, S. K. et al. (2023). Nanozyme-based pollutant sensing and environmental treatment: Trends, challenges, and perspectives. *Science of The Total Environment*, 854: 158771.

Slavin, Y. N., Asnis, J., Häfeli, U. O., and Bach, H. (2017). Metal nanoparticles: Understanding the mechanisms behind antibacterial activity. *Journal of Nanobiotechnology*, 15(1): 65.

Srivastav, A. L., Patel, N., and Chaudhary, V. K. (2020). Disinfection by-products in drinking water: Occurrence, toxicity and abatement. *Environmental Pollution*, 267: 115474.

Steiner, S. A., Baumann, T. F., Bayer, B. C., Blume, R., Worsley, M. A., MoberlyChan, W. J. et al. (2009). Nanoscale Zirconia as a Nonmetallic Catalyst for Graphitization of Carbon and Growth of Single- and Multiwall Carbon Nanotubes. *Journal of the American Chemical Society*, 131(34): 12144–12154.

Swinehart, D. F. (1962). The Beer-Lambert Law. *Journal of Chemical Education*, 39(7): 333.

Venkatraman, G., Giribabu, N., Mohan, P. S., Muttiah, B., Govindarajan, V. K., Alagiri, M. et al. (2024). Environmental impact and human health effects of polycyclic aromatic hydrocarbons and remedial strategies: A detailed review. *Chemosphere*, 351: 141227.

Xu, J., Cao, Z., Zhang, Y., Yuan, Z., Lou, Z., Xu, X. et al. (2018). A review of functionalized carbon nanotubes and graphene for heavy metal adsorption from water: Preparation, application, and mechanism. *Chemosphere*, 195: 351–364.

Yadav, N., Dubey, A., Shukla, S., Saini, C. P., Gupta, G., Priyadarshini, R. et al. (2017). Graphene Oxide-Coated Surface: Inhibition of Bacterial Biofilm Formation due to Specific Surface–Interface Interactions. *ACS Omega*, *2*(7): 3070–3082.

Yan, X., and Chen, X. (2015). Titanium Dioxide Nanomaterials. pp. 1–38. *In:* Scott, R. A. (Ed.), *Encyclopedia of Inorganic and Bioinorganic Chemistry* (2nd ed.,).

Yang, X., Zhao, R., Zhan, H., Zhao, H., Duan, Y., and Shen, Z. (2024). Modified Titanium dioxide-based photocatalysts for water treatment: Mini review. *Environmental Functional Materials*, 3(1): 1–12.

Yu, C.-H., Betrehem, U. M., Ali, N., Khan, A., Ali, F., Nawaz, S. et al. (2022). Design strategies, surface functionalization, and environmental remediation potentialities of polymer-functionalized nanocomposites. *Chemosphere*, 306: 135656.

Zahmatkesh, S., Bokhari, A., Karimian, M., Zahra, M. M. A., Sillanpää, M., Panchal, H. et al. (2022). A comprehensive review of various approaches for treatment of tertiary wastewater with emerging contaminants: What do we know? *Environmental Monitoring and Assessment*, 194(12): 884.

Zhao, J., Yan, P., Qureshi, A., and Chiang, Y. W. (2023). How do material characteristics and antimicrobial mechanisms affect microbial control and water disinfection performance of metal nanoparticles? *Blue-Green Systems*, 5(2): 75–101.

CHAPTER-10

Role of Nanomaterials in Air Remediation

Savan Tank

1. Introduction

Nowadays, we are facing a major issue of pollution (Saleem et al. 2022), especially air pollution, not only because it contributes to macroclimate change but also because it harms both individual and public health (WHO 2021), increasing rates of disease and mortality. A variety of contaminants greatly influence human illnesses and disorders. PM2.5 and PM10, composed of microscopic particles of different sizes, enter the respiratory system by breathing and can lead to various health issues, including major diseases such as cancer, neurological disorders, and cardiovascular and respiratory ailments. Although O_3 in the atmospheric layer protects humans from ultraviolet rays, high amounts of ground-level O_3 are dangerous and can negatively affect the cardiovascular and respiratory systems. Furthermore, significant health hazards are associated with air pollutants such as SOx, NO_2, VOCs, and polycyclic aromatic hydrocarbons (PAHs). Inhaling high concentrations of carbon monoxide can result in acute poisoning. When absorbed into the body, heavy metals like lead can result in either acute or chronic poisoning, depending on exposure levels. Chronic obstructive pulmonary disease (COPD), asthma, bronchiolitis, lung cancer, cardiovascular disorders, nervous system disorders, and skin diseases are among the health problems brought on by these pollutants. Furthermore, environmental pollution-induced climate change affects the transmission of infectious diseases and intensifies natural disasters. Raising public awareness, enlisting the help of scientific professionals, and coordinating national and international organizations to develop and execute sustainable solutions are all necessary to address this problem (Manisalidis et al. 2020). Nowadays, people spend about 69% of their time at home and 87% of their time indoors. As a result, supporting ideal indoor air quality (IAQ) is essential. Increasing ventilation rates can enhance IAQ, assuming that outdoor air quality is adequate. However, a greater volume of air must be heated or cooled to maintain thermal comfort. The goal of this study was to conduct a cost-benefit analysis of IAQ in residential buildings (Nair et al. 2022). Three sets of materials with high, low, and extremely low pollutant emission levels were selected, along with a case study building. Dynamic thermal simulation was used to evaluate the energy consumption and associated costs for each material set,

Department of Life Science, Bhakta Kavi Narsinh Mehta University, Junagadh, Gujarat.
Email: savantank.env987@gmail.com
ORCiD: https://orcid.org/0000-0002-7561-2875

DOI: 10.1201/9781003558804-10

as well as the ventilation rates needed to reach the same IAQ. The findings show how each energy and financial assumption affects the assessment of affordable capital spending, as well as the range of first capital costs that could be compensated for by reduced operating costs. The discussion also sheds light on the other advantages of preserving adequate IAQ (Babich et al. 2020).

More work and study are being done to develop economical and effective remediation solutions as a result of growing concerns about environmental risks and the negative impacts of hazardous chemical treatments (Malik et al. 2023). Understanding the vital roles that different nanomaterials (NMs) play in reducing air pollution is crucial (Ningthoujam et al. 2022, Jangid et al. 2020, Jiang et al. 2020). For a number of reasons, nanostructured materials are very useful in green chemistry and environmental remediation (Ibrahim et al. 2016). In comparison to bulk materials, they have (1) large surface-to-bulk ratios and high specific surface areas (SSAs); (2) chemical properties that can be tailored for particular reactions, such as oxidation potential, reduction potential, Lewis base, and Lewis acid properties; and (3) flexible textural properties and a higher number of reactive edges, which naturally result in greater surface reactivity. One prominent technique for air purification often employed to reduce different air pollutants is photocatalytic oxidation (Mamaghani et al. 2018a). Nano-spheres, nano-plates/sheets, nanotubes/rods, and nano-aerogels are examples of nanostructured catalyst materials that have special qualities that make them ideal for applications involving the management of air pollution (Ningthoujam et al. 2022). These include the creation of extra active centers for catalytic reactions and a large specific surface area that makes it simple for substrates to diffuse and transport to catalytic sites. These nanomaterials can also be used as catalyst supports, creating highly stable and active catalysts by fostering a synergistic interaction between the catalyst and the support particles. Because they have a greater density of less-coordinated atoms, edges, steps, and kinks than their regular counterparts, metal-based and metal oxide-based nanostructures with higher indices of sides offer more active catalytic sites (Peng and Tsang 2018). This study investigates the application of several nanomaterials, such as nanofilters, nanocatalysts, nanoadsorbents, and nanosensors for air pollution remediation. It works on applying commonly studied nanoparticles and analyzing the factors influencing the effectiveness of these nanomaterials. Additionally, the application of nanosensors for detecting poisonous gases such as NO_2, SO_2, and H_2S is evaluated. The potential negative impacts of nanomaterials, particularly metal-based and carbon-based nanomaterials, are also discussed. Despite significant research on the advancement of nanomaterials for environmental remediation, comprehensive research on recent developments in nanomaterials for air pollution remediation remains limited.

2. Nanomaterials in the Air Environment for Uses in Advanced Technology

2.1 Nano Adsorbents

According to nanotechnology, nanoscale adsorbents, or nano-adsorbents, can be used to efficiently address a number of today's environmental issues, such as poor air quality. These nanoparticles have the ability to absorb a variety of airborne contaminants. To find and adsorb pollutants including bacteria, metal oxides, and volatile organic compounds (VOCs), a wide variety of nano adsorbents have been created and categorized. In order to eliminate and track the release of harmful gases and air pollutants using different nanoadsorbent materials, researchers employ a variety of treatment methods and conditions, which are described in Table 1.

The frequently utilized nanoadsorbents are covered in depth in the sections that follow:

2.1.1 Carbon Nanotubes

The application of nano adsorbents like carbon nanotubes (CNTs) for different metal ions has been the subject of many investigations. Carbon-based nanostructures are perfect for commercial usage as nano adsorbents with improved capacity, affinity, and selectivity because of their remarkable

Table 1: Different treatment techniques/conditions for eliminating and monitoring the discharge of toxic gases and air contaminants by various nano adsorbent materials.

Pollutant	Techniques/Condition/ Modification	Nanomaterial Used	Adsorption Results
CO_2	MWCNTs were changed using a solution of (3-aminopropyl) triethoxysilane in a 2-stage procedure	Carbon nanotubes (CNTs)	The highest adsorption of CO_2 has been achieved by the amine-functionalized CNTs (Gui et al. 2013, Malik et al. 2023)
CO_2	MWCNT functionalized with 3-aminopropyltriethoxysilane (APTS)	Carbon nanotubes	The capacity of adsorption by 100 desorption-adsorption cycles (Su et al. 2011)
Carbon dioxide and CH_3OH gas molecular	Si-doped SWCNT	SWCNTs	The pure adsorption of CH_3OH and CO_2 in Si-doped and B-doped single-walled carbon nanotubes is chemisorption (Azam et al. 2017)
CO_2	3-aminopropyl-triethoxysilane (APTS)-modified CNTs	MWCNTs	96.3 mg/g adsorption ratio by CNT (Lu et al. 2010)
H_2S	Single-step method using ultrasonic-assisted precipitation	ZnO Nanoparticles	The results of ZnO (Zinc Oxide) showed an increase in hydrogen sulfide adsorption ability of 29.50 mg/g^{-1} While no ultrasonic techniques materials synthesized 3.60 mg g^{-1} (Huy et al. 2019)
CO_2	Nanoparticles prepared by wire explosion technique	ZnO Nanoparticle	0.4 mmol/g of CO_2 may be adsorbed on the same pressure of 1 bar and 283 K (Ghost at al 2018)
SO_2	SO_2 separation by iron oxide nanoparticles deposited on clinoptilolite zeolite	Nano Iron Oxide	The competence of 80.3 % (Mahmoodi et al. 2019)
SO_2	ZnO nanoparticles and nanorods prepared	ZnO	Zinc oxide nanomaterials adsorbed by SO_X as per the shape depend on size (Wu et al. 2011)

physical characteristics, which include high pore volume, large pore diameter, and high surface area (Gabal et al. 2020). CNTs are used to regulate the amount of CO_2 and methane released from automobile exhaust and industrial chimneys. Furthermore, these materials can absorb greenhouse gas emissions from power stations and coal mines. Because of their exceptional qualities, like selective adsorption of contaminants, CNTs are useful nanoadsorbents for removing different types of air pollutants. Their large volume, bigger size, and width, along with high surface area, all work together to provide more reactive sites that can attach to particular atmospheric pollutants (Gupta and Saleh 2013). Kyzas and Matis (2015) studied that adding nanoadsorbents as functional groups can increase the number of active sites and improve the adsorptive capability of these nanomaterials.

A (3-aminopropyl) triethoxysilane (APTES) solution was used to modify multi-walled carbon nanotubes (MWCNTs) in a work by Gui et al. (2013). An interaction between the CO_2 molecules and the secondary amine groups in the APTES produced carbamate, which was the product of the chemisorption of CO_2. Figure 1 shows the mechanism of CO_2 adsorption. The APTES increased with longer reflux durations, reaching a maximum loading of 13.75 weight percent, according to thermogravimetric analysis. According to test results, more amino groups were covalently bonded to the MWCNTs' surface as the reflux time rose, resulting in effective CO_2 adsorption sites. With a maximum CO_2 adsorption of 75.4 mg CO_2 per gram of adsorbent, the amine-functionalized CNTs outperformed widely used adsorbents such as SBA-15.

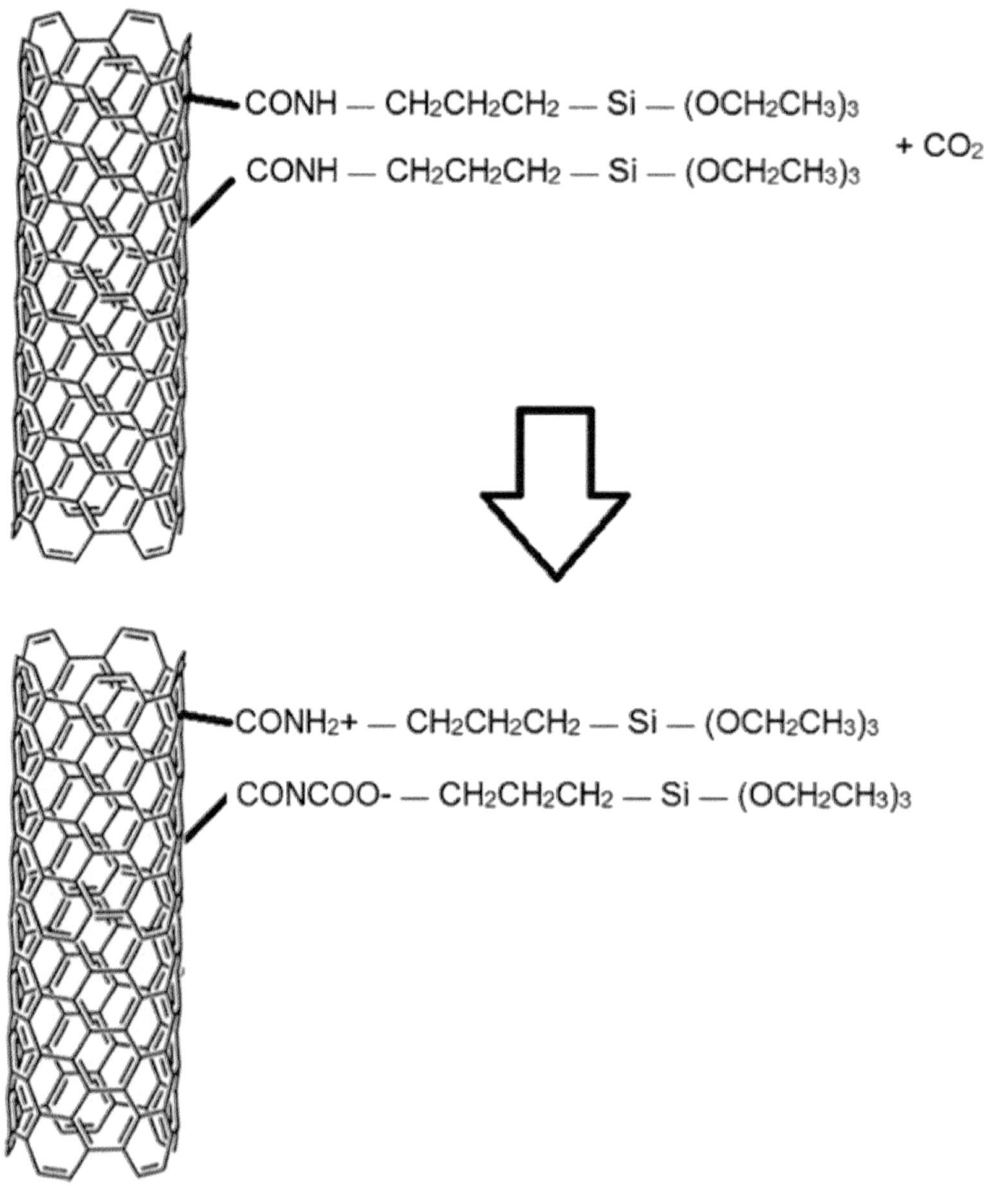

Figure 1: Carbon dioxide adsorption mechanism. Reproduced from Ref. (Gui et al. 2013).

Multi-walled carbon nanotubes (MWCNTs) is gradually increasing the mass load of 3-aminoptopyltriethoxysilen (APTS) in order to examine the MWCNTs performance is cyclic CO_2 adsorption and the related thermodynamic parameters (Su et al. 2011). A successful CO_2 capture method is applying solid APTES-functionalized MWCNTs for CO_2 adsorption. Both pure and APTES-modified carbon nanotubes (CNTs) were evaluated as adsorbents for CO_2 adsorption from air streams in a different study by Lu et al. (2010). The results showed that following modification, CNTs' physicochemical characteristics improved. The modified CNTs' adsorption effectiveness improved with high RH but declined with rising temperature. The CO_2 adsorption isotherms for both pure and APTES-modified CNTs are shown in Figure 2. The modified carbon nanotubes demonstrated improved CO_2 adsorption capabilities compared to APTES-modified silica-based adsorbents reported in the literature, suggesting that these APTES-modified carbon nanotubes are viable adsorbents for CO_2 collection from air streams.

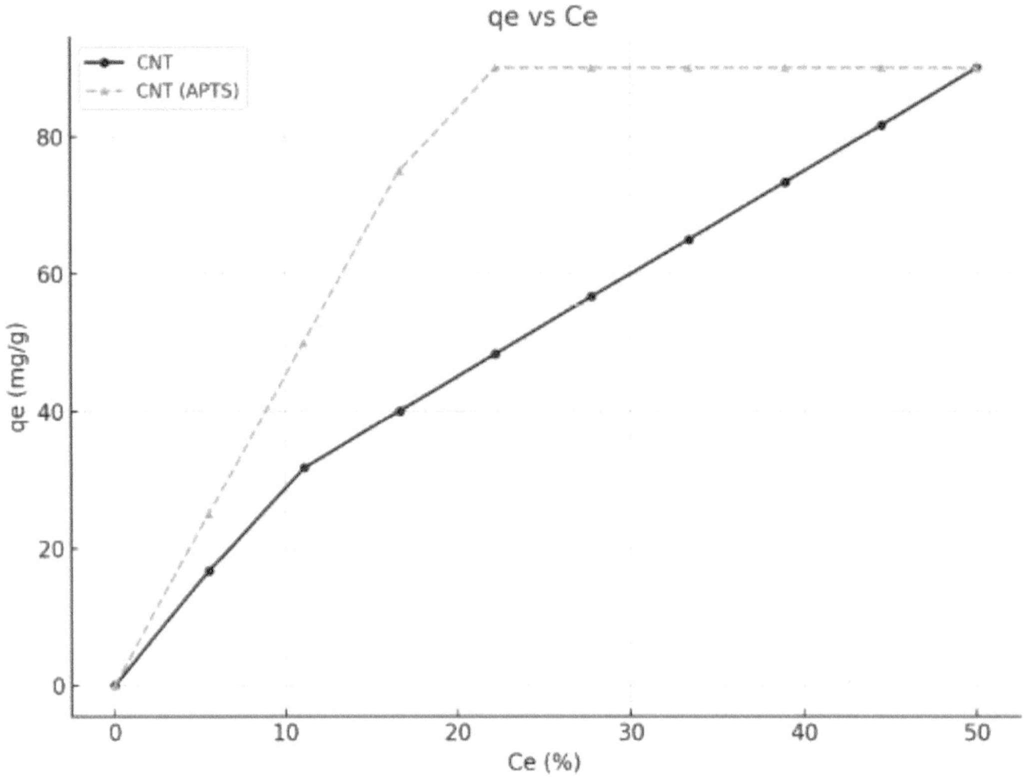

Figure 2: Adsorption isotherms of carbon dioxide with pristine and APTS-modified CNTs. Reproduced from Ref (Lu et al. 2010).

2.1.2 Zinc Oxide

Because nanostructures of ZnO are heat resistant and therefore appropriate for non-catalytic sulfurization reactions, their usage for the removal of hydrogen sulfide (H_2S) from the air, especially at high temperatures via the adsorption mechanism, has been thoroughly investigated (Tran 2016). Huy et al. (2019) developed a one-step method for creating zinc oxide nanoparticles (U–ZnO) that uses ultrasonic assistance instead of surfactants or treatments after heating. The utility of these nanoparticles for adsorption-based H_2S removal from the air was also investigated in the study. The hydrogen sulfide adsorption capacity of U–ZnO was around 29.51 mg/g, which was much higher than the 3.66 mg/g of the nanoparticle that was generated without ultrasonic treatment, according to the findings of the adsorption tests.

In the Figure 3, the breakthrough curves for H_2S adsorption studies utilizing U–ZnO and ZnO materials show how much higher U–ZnO's adsorption capability is. This improved capability demonstrates an environmentally benign, time-efficient, practical, easy, and effective way to create zinc oxide nanostructures that effectively remove hydrogen sulfide and other gaseous pollutants from the atmosphere. In order to manufacture ZnO nanoparticles for carbon dioxide adsorption tests, Ghosh et al. (2018) used a wire explosion technique. The adsorption investigation was carried out at equilibrium pressures up to 1 bar and temperatures between 10 and 25°C in order to examine effects of chamber pressure and charging voltage on zinc oxide morphology. At 283 K and 1 bar equilibrium pressure, These particles displayed an adsorption capacity of roughly 0.4 mmol/g. The study also discovered that when sample temperature dropped and pressure rose, CO_2 adsorption increased. The easy regeneration of the carbon dioxide adsorbed onto the zinc oxide nanoparticles was confirmed by thermodynamic study. Therefore, it is clear that poisonous gases including CO_2, methanol (CH_3OH), and hydrogen sulfide (H_2S) may be effectively removed by carbon nanotubes and

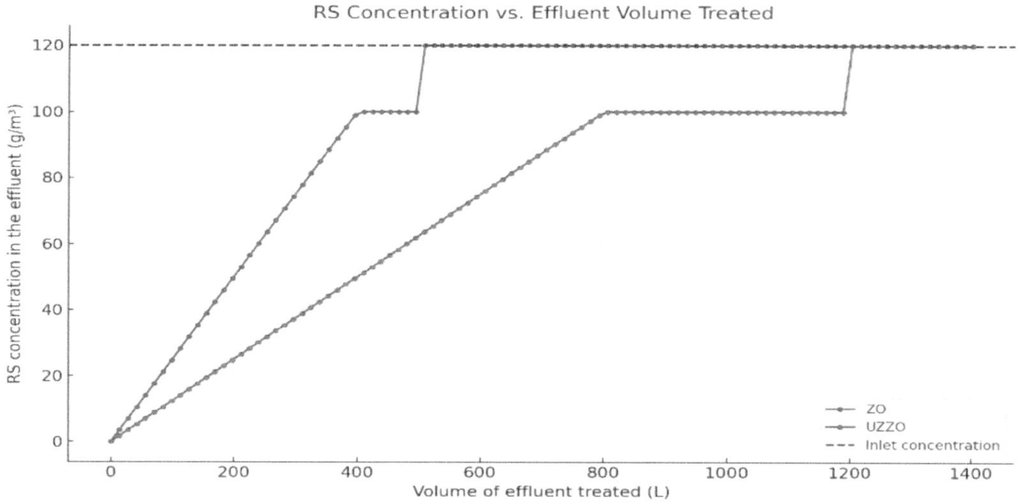

Figure 3: Break-through curves of H_2S adsorption analysis using ZO (material prepared without ultrasonic treatment) and U–ZnO. Reproduced from Ref (Huy et al. 2019).

zinc oxide-based nano adsorbents. Additionally, when compared to their unmodified counterparts, modified CNTs have shown a higher adsorption capability.

2.2 Nanocatalysts

Nanomaterials can be used in a variety of ways to reduce air pollution. One such technique is photocatalytic remediation using semiconducting materials, in which electron-hole pairs are formed when exposed to light with energy equivalent to or greater than the material's band gap (Theerthagiri et al. 2019). The active surface of nanomaterials, where chemical reactions take place, is a crucial aspect of their role as catalysts. The catalyst's active surface area is increased by decreasing its size, which raises the reaction efficiency. In order to create sophisticated nanocatalysts with larger surface areas, nanotechnology can also improve the size, molecular distribution, and structure of nanoparticles (Tai et al. 2020, Salabat and Keshavarz 2019). Because they aid in the reduction of air contaminants, these nanocatalysts hold great promise for air remediation. In comparison to traditional catalysts, the nanocatalyst system enables quick and precise chemical reactions, leading to higher product yields and better catalyst recovery. Table 2 lists several conditions and treatment methods for eliminating air pollutants and hazardous gases with various nanocatalysts.

The frequently utilized nanocatalysts are covered in depth in the sections that follow:

2.2.1 Titanium Dioxide

Titanium dioxide (TiO_2), due to its superior air purification capabilities, stability, non-toxicity, affordability, and availability, has attracted a lot of research interest recently as a semiconductor photocatalyst (Lyu et al. 2014). Effective air decontamination depends on the fabrication method's effect on the surface hydroxyl group density because of the critical function these groups play in pollutant adsorption and the production of hydroxyl radicals (Pelaez et al. 2012). Figure 4 shows how TiO_2 is exposed to ultraviolet light, which causes an electron to move between the valence and conduction bands and leaves a positive hole in its wake. According to Wu et al. (2014), titanium tetrabutoxide was used as the titanium source in the hydrothermal process (190°C for 24 hours) to create surface-fluorinated titanium dioxide nanosheets, with hydrofluoric acid as a capping mixture. The surface-fluorinated TiO_2 catalyst was about twice as effective as P25 (Evonik), a mixture of rutile TiO_2 and anatase (tetragonal crystal), in separating gaseous ammonia (around 60%).

Table 2: Different nanocatalysts used for eliminating toxic gases and air contaminants from the air.

Pollutant	Nanomaterial Used	Techniques/Condition/Modification	Results
Gaseous ammonia separation	Titanium Dioxide	Surface-fluorinated titanium dioxide nanosheets by the hydrothermal technique	60% of the gas was separated (Wu et al. 2014)
NOx Pollutant	Graphitic carbon nitride/titania composite	Graphitic carbon nitride/titanium dioxide combined photocatalysts with various graphitic carbon nitride/ titania were prepared by a facile fabrication route	In NO oxidation, a graphitic carbon nitride/ titanium dioxide composite with a ratio of 1:4 demonstrated enhanced photocatalytic performance. (Ginnakopoulou et al. (2017)
Elemental Mercury	Titanium dioxide nanorods	Titanium dioxide nanorods-supported CrOx, FeOx, MnOx, and MnOx-FeOx-CrOx catalyst	Superior HgO removal efficiencies (Jampaiah et al. (2019)
Sulfur dioxide (SO_2)	Titanium dioxide	With no light irradiation, the major product of sulfur dioxide adsorption has been noted to be SO_2, while the light irradiation caused oxidation along with the development of adsorbed SO_2	It emerged that the firmly adsorbed species had a greater intrinsic adsorption capacity of the tiny nanomaterials (Baltrusaitis et al. 2011)
Rhodamine B dye	Titanium dioxide	TiO_2-infused concrete	Adding tiny amounts of titanium dioxide to cement mixtures enhances the concrete›s ability to selfclean and remove contaminants (Elia et al. 2018)
CO_2	Titanium dioxide	It was discovered that species are highly reliant on the surface structure of titanium dioxide nanoparticles.	CO_2 has a strong interaction with both disks and rods of TiO_2 (Tumuluri et al. 2017)
SO_2	Mn/Copper slag	Under UV light, manganese is supported on copper slag.	The achieved SO_2 elimination rate is over 99% (Rabiee and Mahanpoor 2019)

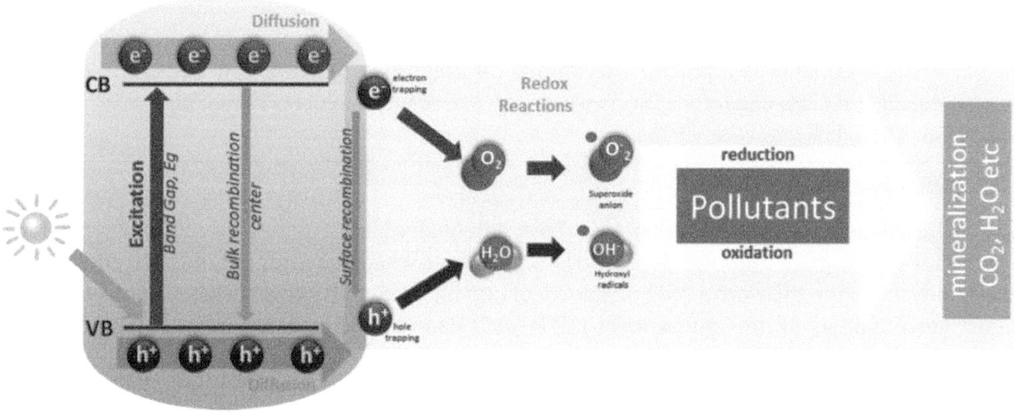

Figure 4: Principal processes in semiconductor photocatalysis. Reproduced from Ref. (Binas et al. 2017).

Although TiO_2 nanoparticles are more photoactive, their reaction rate and separation efficiency are constrained by their smaller surface areas and poor adsorption affinity for air pollutants, such as volatile organic compounds (VOCs). TiO_2 is frequently used with substances that, under real-world circumstances, improve pollutant adsorption in order to enhance performance. Numerous carbon-based materials have been investigated as supports, including graphene oxide (GO), graphitic carbon nitride (g-C_3N_4), carbon fibers, carbon nanotubes (CNTs), and activated carbon. Nowadays, carbon-based nanostructures such as CNTs and graphene nanosheets are frequently employed to increase the photocatalytic performance of TiO_2. Electron-hole recombination is decreased in TiO_2–CNT composites because electrons can move freely through the CNTs (Low et al. 2017). Giannakopoulou et al. (2017) used a simple production process to Blend photocatalysts made of carbon graphite nitride with titania with different g-C_3N_4/TiO_2 ratios. This involved annealing melamine and P25 commercialized titanium dioxide powdered form at 550 degrees Celsius for three hours under argon flow. The 1:4 ratio composite outperformed the parent semiconductors, titanium dioxide and graphitic carbon nitride, in terms of photocatalytic activity for NO oxidation under visible light.

Long-term exposure to formaldehyde (HCHO), a substantial volatile organic compound (VOC), can cause health problems such as skin irritation and nasal tumors. Yu et al. (2018) investigated how titanium dioxide–reduced graphene oxide (TiO_2–rGO), produced by a Graphite particles and titanium tetra butoxide react hydrothermally, degrades formaldehyde when exposed to visible light. After 1 hour and 40 minutes of irradiation, the TiO_2–rGO composite had a 60.0% formaldehyde removal efficiency, reduced graphene oxide had a 20.0% removal efficiency, and titanium dioxide had a 10.0% removal efficiency. By extending the life of charge carriers, reduced graphene oxide improves removal efficiency by promoting photochemical processes and the production of reactive radicals. Silicate–titanium dioxide hybrids were created by Portela et al. (2017) for the photocatalytic oxidation of formaldehyde. Their photocatalytic activity was greatly enhanced by the hybrids' improved formaldehyde adsorption capabilities. Coal burning is linked to the emission of harmful pollutants such as mercury (Hg), which presents serious health and environmental hazards, in addition to volatile organic compounds (VOCs). To investigate the impact of ternary metal oxides on the elimination of elemental mercury (HgO), Jampaiah et al. (2019) created titanium dioxide nanorods supported by several metal oxides (CrOx, FeOx, MnOx, and MnOx-FeOx-CrOx). In the presence of oxygen, the MnOx-FeOx-CrOx/TiO2 catalyst showed a high HgO (Mercury Oxide) separation efficiency (80–83%) at lower temperatures (100–200°C). Nitric oxide (NO) was added, which improved the elimination of HgO even further. Higher catalytic activity was a result of the TiO_2 nanorods' high crystallization and large surface area, which allowed the mixed metal oxide nanoparticles to disperse uniformly. An efficient method for improving Hg0 removal in coal-fired power plants is to combine reducible metal oxides with different supports.

Additionally, "self-cleaning" coatings that can eliminate air pollutants including nitrogen oxides and volatile organic compounds (VOCs) are made using titanium dioxide nanoparticles. The photocatalytic TiO_2 layer is a potential technique for controlling air pollution because it converts these pollutants into less dangerous forms when exposed to ultraviolet (UV) radiation from sunlight (Shen et al. 2016, Haleem et al. 2023).

2.2.2 Other Nanocatalysts

Like TiO_2, zinc oxide (ZnO) nanostructures have attracted a lot of interest for their photocatalytic capabilities because of their various shapes and unique characteristics. Particularly helpful as photocatalysts for the solar-powered breakdown of organic pollutants are ZnO-based nanostructures. Higher surface areas of one-dimensional (1-D) ZnO nanostructures enhance their photocatalytic efficacy in a range of environmental applications, among other benefits over bulk materials (Ong and Law 2018). Rabiee and Mahanpoor (2019) provided a unique catalytic oxidation method utilizing manganese supported on copper slag nanocatalysts under ultraviolet (UV) irradiation for the removal of sulfur dioxide (SO_2) from simulated flue gas. This technique demonstrated promise for industrial use, especially in maximizing the reduction of sulfur dioxide emissions in flue gas desulfurization systems. The photocatalytic properties of various materials, besides ZnO, have also been investigated. The efficiency of gold-based nanocatalysts in reducing air pollution was demonstrated by Moroz et al. (2009), who detailed the production and catalytic characteristics of nanodispersed Au/Al_2O_3 catalysts for low-temperature carbon monoxide (CO) oxidation. In a similar vein, Pu et al. (2020) examined how zeolites work in the well-established method of selective catalytic reduction (SCR) of nitrogen oxides (NOx) using ammonia (NH3). These developments make it clear that TiO_2 is still a highly effective photocatalyst for the separation of formaldehyde and gaseous ammonia. It has been demonstrated that combining TiO_2 with carbon-based nanostructures (such as graphitic carbon nitride and carbon nanotubes) increases photocatalytic efficiency and improves the elimination of dangerous gases. Furthermore, ZnO, copper, and Al_2O_3 nanocatalysts demonstrate the adaptability of nanomaterials in air pollution management and environmental remediation by effectively degrading a variety of organic pollutants, SO_2, and CO, respectively.

2.3 Nanofilters/Nanostructured Membranes

Ultra-small-pore nanostructured membranes are showing promise as a means of reducing air pollution, especially when it comes to removing different gas pollutants from indoor and outdoor air. Optimizing these membranes for effective gas removal has been the subject of numerous investigations (Rubina et al. 2019, Samantara et al. 2019). In air filtration systems, Nanofiber-coated filtration media have already been used, especially for industrial dust particle separation (Muralikrishnan et al. 2014). When it comes to filtering volatile organic carbon (VOC) vapors, such as formaldehyde (HCHO), which is hard to get rid of with traditional photochemical methods because they depend on ultraviolet (UV) light and can cause ozone to form, nanostructured membranes work particularly well (Miyawaki et al. 2012). Alternative methods have been investigated to overcome this difficulty, such as the creation of carbon nanofiber membranes based on electrospun polyacrylonitrile. These membranes have nitrogen-containing functional groups and excellent microporosity, making them efficient adsorption sites for the elimination of formaldehyde. Even at low concentrations, these membranes can absorb a sizable quantity of formaldehyde. Additionally, nanoparticle filters can be used to eliminate bioaerosols, which can lead to a number of health problems. In order to eliminate bioaerosols from the air through air conditioning procedures, silver- and copper-based nanoparticle filters have been widely used (Lee et al. 2010).

2.3.1 Metal Organic Frameworks (MOFs)

Because of their special qualities, which include vast surface areas, rich functions, and variable pore widths, metal-organic frameworks (MOFs) are thought to be excellent options for reducing air pollution (Bian et al. 2018). Combining MOFs with organic polymer matrices has been shown to be a successful pollution control method. MOFs are made up of metal ions or clusters connected to organic ligands and structural changes can readily modify their characteristics. These materials' enormous surface area and tunable functionality make them very effective in capturing gaseous air pollutants like formaldehyde and fine particulate matter (PM2.5).

The presence of unbalanced metal ions and defects in MOFs results in positive charges on their surfaces, which enhance electrostatic interactions and improve the capability to remove particulate matter such as PM2.5, particles with an aerodynamic diameter of less than 2.5 μm (Chen et al. 2017). Bian et al. (2018) introduced a straightforward immersion method to fabricate MOF-infused flexible nanofiber filters, which proved highly efficient in eliminating both formaldehyde and PM2.5. This method facilitated the uniform application of MOFs onto electrospun nanofibers, allowing for the scalable and cost-effective manufacturing of these filters. By incorporating ZIF-67 nanocrystals into the polyacrylonitrile nanofibers, the filtration efficiency for PM2.5 improved from 74.5% to 87.2%, and the rate of formaldehyde removal reached 84%. These MOF-based filtration systems exhibited remarkable long-term stability, maintaining a filtration efficiency greater than 99% for over 30 days.

Table 3: Various nano filters used for dropping the release of toxic gases and air contaminants.

Target Contaminant	**Preparation/ Modification/Treatment**	**Nanomaterial Used**	**Results**
HC, NOx, CO2, and CO	CNTs are contained within a filter.	Carbon Nanotubes (Filter)	Assuming similar CNT loadings were utilized, the proficiencies were demonstrated to range from 5.0 percent to 60.0 percent when collecting HC, NOx, CO2, and CO. (Romero-Guzm´an et al. 2017)
SO2	MOF: CNT-modified filter with UiO-66-NH2	Carbon Nanotubes (Filter)	Excellent SO2 adsorption and ultrafine dust collection (99.99%) in dynamic filtration (Feng et al. 2018)
SO2	The development of nanofibrous filters with augmented MOF loadings (till 60 wt %)	MOF nanofibrous filter	When exposed to a stream of a SO2/nitrogen mixture, this filter could both selectively catch and retain sulfur dioxide (Zhang et al. 2016b)
Gaseous methanol	A simple hydrothermal technique fabricated TiO2-reduced-GO composite	Titanium Dioxide	The study's results are reassuring and indicate that this active filter media is a viable option for developing better solutions to enhance the quality of the environment. (Roso et al. 2015)
Airborne Microbes	Effect of UV light and silver/zinc oxide coating working together	Ag and Zn Nanoparticles	According to the results, an air filter coated with silver or zinc oxide can be produced at low cost and utilized to eliminate germs and other impurities from indoor air (Pokhum et al. 2018)
Formaldehyde and outside particulate matters with an aerodynamic diameter of less than 2.50 μm (PM2.50)	Electrospun polyacrylonitrile nanofiber integration with ZIF-67	MOF nanofiber flexible filter	This filter's performance improved to 87.20 percent (Bian et al. 2018)
Particulate Matter	MOF concentrations are up to 60 wt %.	Metal organic framework filter	89.67% high particulate matter removal efficiency (Zhang et al. 2016a)

In a separate investigation, Zhang et al. (2016) synthesized nanocrystals from four distinct MOF structures and integrated them into nanofibrous filters at high MOF loadings (up to 60 wt%). These filters demonstrated superior performance in filtering particulate matter, underscoring the potential of MOFs as a highly effective material for controlling air pollution. The integration of MOFs with electrospun nanofiber membranes offers a scalable and efficient approach to enhance indoor air quality and mitigate related health risks. The development of nanostructured filters and sensors for air pollution control has seen significant advancements, particularly in nanomaterials in use like MOFs, CNTs, and TiO2 composites. These materials have proved effective in capturing various toxic gases and particulate matter, thus improving indoor and industrial air quality.

2.3.2 Carbon Nanotubes (CNTs)

Carbon nanotubes (CNTs) have emerged as a key solution for lowering contaminant emissions owing to their distinct physical and chemical characteristics, including a high surface area and the capability to adsorb a range of gases. A study by Romero-Guzmán et al. (2017) involved the creation of CNT filters by applying CNTs onto membrane supports with various pore sizes. The findings showed that CNT loading significantly improved filtration efficiency, particularly for pollutants such as nitrogen oxides (NOx). For instance, while capturing NOx, the filtration efficiencies varied from 10% to 95%, influenced by the CNT loading (0.2–1.6 mg cm^2). These CNT filters also proved effective in cutting acid polar gas species (like H_2S and SO_2) and ultrafine dust particles, which are detrimental to human health. Further advancements included the development of a multifunctional filter by Feng et al. (2018), which combined MOFs (UiO-66-NH2) and CNTs for efficient adsorption of acid gases and removal of ultrafine dust. The resulting filter showed an exceptional 99.997% efficiency for ultrafine dust and sulfur dioxide, highlighting its significant potential for indoor air purification.

2.3.3 Another Nanoparticle-Based Filters

Research conducted by Roso et al. (2015) proved that a TiO_2–reduced graphene oxide (GO) composite can effectively eliminate gaseous methanol. The composite's large adsorption capability and enhanced charge transfer make it a promising candidate for air purification. Furthermore, filters coated with silver and zinc oxide nanoparticles have been successfully employed to remove airborne microbial contaminants, such as psychrotrophic bacteria and fungi, which can have a considerable impact on food quality and human health (Pokhum et al. 2018). The synergistic effect of UV light in combination with silver/zinc oxide coatings improves microbial removal efficiency.

2.4 Nanosensors

They represent a novel technology for detecting harmful ecological pollutants. These sensors usually consist of three primary components: the nanomaterial, a detection element for specificity, and a signal transduction method for indicating the presence of pollutants. Nanosensors have been utilized to identify gases such as hydrogen sulfide (H_2S), sulfur oxides (SOx), and nitrogen oxides (NOx), which are vital markers of air quality and pollution levels.

2.4.1 NO_2 Detection Graphene

Reduced graphene oxide (RGO) has been recognized as a promising material for the detection of nitrogen dioxide (NO_2). The high surface area, electrical conductivity, and gas adsorption ability of graphene make it an ideal choice for gas detection applications. Research by Yusuf et al. (2019) has indicated that graphene and reduced graphene oxide can detect NO_2 at low concentrations. Additionally, combining palladium (Pd) and tin dioxide (SnO_2) nanoparticles with reduced graphene oxide has demonstrated enhanced sensitivity and quicker response times for NO_2 detection (Wang et

Table 4: Various nanosensors used for the recognition of toxic gases like NO2, H2S, and SO2.

Detected Gas	Materials/Structure	Concentration (ppm)	Sensitivity/Response	References
NO_2	Graphene-SnO_2 Nanocomposites	1	24.7	Kim et al. (2017)
NO_2	SnO_2/graphene Nanocomposites	5	26, 342	Tammanoon et al. (2015)
NO_2	RGO- Polythiophene thin film	10	26, 36	Bai et al. (2016)
NO_2	RGO/Au Nanocomposite	5	1.33	Zhang et al. (2016b)
NO_2	ZnO Nanorods	100	622	Navale et al. (2017)
SO_2	TiO_2/rGO Nanocomposite	1	10.08	Petryshak et al. (2017a)
H_2S	rGO/hexagonal WO3 Nano-sheets composite	10	45	Shi et al. (2016)
H_2S	Cu nanoparticles decorated SWCNTs Nanotube	5	11%	Asad et al. (2015)
H_2S	SnO_2-rGO Nanofibers	5	34	Choi et al. (2014)

al. 2018). These innovative materials and techniques, such as CNT filters, MOF-based membranes, and nanosensors, are showing considerable promise in air purification and environmental monitoring. As research progresses, it is expected that the scalability and cost-effectiveness of these technologies will improve, paving the way for extensive applications in air quality management in both industrial and residential settings.

2.4.1.1 Metal-oxide based nanostructures for NO_2 detection

To improve sensing capabilities, metal oxides can be designed in a variety of nanostructured forms, such as nanosheets, nanocomposites, nanoflowers, nanotubes, nanoparticles, and nanowires (Vyas et al. 2013). Additionally, the existence of porosity and permeable shell layers makes it easier for electrons to be completely removed and for gases to diffuse, which enables sensing devices to reach higher gas sensitivity (Postica et al. 2017). Qiang et al. (2018) developed a nitrogen dioxide gas sensor that uses porous silicon/tungsten trioxide nanorods functionalized with palladium nanoparticles. Hung et al. (2018) recently developed three sensors using microelectrode chips with zinc oxide and zinc stannate nanowires.

2.4.2 SO_2 Detection

2.4.2.1 Carbon-based nanomaterial in SO_2 detection

Improved surface-to-volume ratios in aligned carbon nanotubes (CNTs) enable effective chemical and physical adsorption of targeted SO_2 gas (Popp et al. 2009, Nehra et al. 2019). Gas sensors based on vertically aligned carbon nanotubes were recently studied by Zouaghi et al. (2018). Chemical vapor deposition was used to produce these patterns on a silicon substrate doped with boron and silicon dioxide. In a previous research by Petryshak et al. (2017b), a cholesteric–nematic mixture intercalated with carbon nanotube walls led to the adsorption of SO_2 molecules onto the carbon nanotubes. The CNTs' conductivity was changed by this adsorption process, which ultimately resulted in a sulfur dioxide detecting signal. A titanium dioxide/graphene film was developed using a layer-by-layer (LbL) self-assembly technique to detect sulfur dioxide at ambient temperatures. When compared to other contaminated gases including NO_2, CO, H_2, C_2H_2, and CH_4, this titanium dioxide/reduced graphene oxide film sensor demonstrated exceptional sensitivity to one ppm of sulfur dioxide gas at environmental temperature.

2.4.2.2 Metal and metal-oxide nanostructure in SO_2 detection

Liu et al. (2018) created ruthenium-catalyzed zinc oxide nanosheets supported by alumina and combined them with a microsensor to detect SO_2. For the purpose of detecting sulfur gas species, Çiftyürek et al. (2016) developed and examined thick films of tungsten and molybdenum in both binary and ternary forms. It was discovered that, out of all the oxide films made, the nanostrontium molybdate made using hydrothermal techniques had the best sensor response. Tyagi et al. (2016) used a sputtering process to create thin layers of tin (IV) oxide. In order to create sulfur dioxide sensors, these films were then functionalized using a variety of metal oxide catalysts, such as vanadium pentoxide, magnesium oxide, nickel oxide, copper oxide, and palladium (II) oxide. To examine the reaction at different operating temperatures, a 500-ppm sulfur dioxide concentration was added to various metal oxides applied to a tin dioxide sensor. Due to the presence of enhanced sites for adsorbed oxygen species on the rough and porous surface of the tin dioxide film, as well as the spill-over effect from nickel (II) oxide nanoclusters to sulfur dioxide molecules, the structure of nickel (II) oxide/tin dioxide showed the highest response at 180°C (Haridas et al. 2011).

2.4.3 H_2S Detection

2.4.3.1 Carbon-based nanomaterial in H_2S detection

Several recent investigations on the detection of hydrogen sulfide have made use of graphene, carbon nanofibers, and reduced graphene oxide. A graphene-based hydrogen sulfide gas sensor functionalized with silver nanoparticles and charged impurities was developed by Ovsianytskyi et al. (2018). The chemical vapor deposition approach was used to create the graphene, and then a wet-chemical process was used to add impurities and silver nanoparticles. Chu et al. (2018) found that using tin oxide-modified r-GO at 125°C produced a sensitivity of roughly 34.31% for hydrogen sulfide at 100 ppm. In different work, zinc oxide–carbon nanofibers were used to create a stable sensor that demonstrated enhanced hydrogen sulfide detection capabilities. It was observed that carbon's protective properties contributed to the enhanced stability of zinc oxide and created oxygen vacancies that facilitated a better sensor response.

2.4.3.2 MetalMetal and metal-oxide nanostructure in H_2S detection

Using ferric oxide nanosheets and a solvothermal technique, Zhang and his associates developed a hydrogen sulfide gas sensor (Zhang et al. 2019). At lower temperatures, the sample's morphology was found to be inconsistent. This was explained by the fact that the nanostructure held both ferric oxide and magnetite, which made it impossible to achieve a homogeneous morphology below 160°C, the reaction temperature. It was also found that the sensor's reaction to hydrogen sulfide decreased for ferric oxide as the operating temperature rose. The researchers created zinc oxide/cupric oxide nanotube arrays in a different study by Li et al. (2018) with the goal of detecting hydrogen sulfide at lower operating temperatures. They found that the nanotube architectures improved gas transport and adsorption because the deposited oxygen and hydrogen sulfide molecules created many active sites. As a result, this study showed increased sensitivity and quick reaction time. Thus, it is clear that sensors based on nanomaterials such as zinc oxide, graphene, and reduced graphene oxide have a high sensitivity for detecting NO_2 gas at levels of parts per billion. Additionally, sensors based on nanomaterials, including graphene, CNT, TiO_2, alumina, and tin (IV) oxide, show remarkable SO_2 gas detection capabilities. Additionally, sensors that use nanomaterials such as cupric oxide, ferric oxide, zinc oxide, reduced graphene oxide, and graphene exhibit good sensing capabilities for H_2S gas detection.

3. The Adverse Effects of NMs on the Atmosphere

There are still a lot of unanswered questions about how nanomaterials (NMs) affect the environment, particularly when they are released into the atmosphere. Engineered nanomaterials may be released into the atmosphere at any point during their life cycle, including production, processing, transportation, handling, application, and disposal (Cucurachi 2019, Garvey et al. 2019, Heidari and Younesi 2020). Vehicle exhaust, explosions, combustion, and the oxidation of atmospheric gases are some of the sources of atmospheric nanomaterials (Silva et al. 2020, Gallego-Hernández et al. 2020). Human health effects may be significantly affected by the nanoparticles present in airborne pollution (Hadrup et al. 2020). It should be noted that particles less than 100 nm are usually categorized as ultrafine particles in atmospheric investigations and combustion emissions research. These materials are biochemically reactive, meaning they can interact with human organs, cells, and subcellular structures due to their varied physical and chemical properties, as well as their enhanced surface area and nanoscale size (Saber et al. 2019). In particular, nanoparticles make up a sizable proportion of airborne particles that are currently unregulated, despite contributing very little to the total mass of particulate matter. The next section looks at how nanomaterials, like metal-based nanoparticles (NMs) and carbon nanotubes (CNTs), affect the air environment negatively.

3.1 Adverse Effects of Metal-Based NMs on the Atmosphere

A substantial part, roughly 10.0 percent of the non-volatile solid nanoparticles, are made of iron and are produced by highly magnetic combustion as well as friction. Automobile exhaust and brake wear release iron and other metal-bearing nanoparticles in large quantities (for example, about $\times 10^8$ magnetic nanoparticles/m^3 of air) along the road, in addition to industrial point sources (Maher et al. 2016). The major particle sizes of the iron-based roadside nanoparticles range from less than 10 to more than 100 nm. According to Calderon et al. (2018), strong magnetic behavior, elevated redox activity, surface charge, and interactions with organic chemicals and transition metals that cause cancer are only a few of the detrimental characteristics of iron-based, traffic-derived, combustion-derived, and friction-derived nanoparticles. These traits may activate critical pathways for organ and cellular damage, including damage to the heart and brain. According to several earlier studies, residents of Metropolitan Mexico City of all ages have magnetic combustion- and friction-derived nanoparticles in their brains (Maher et al. 2016), which have been linked to Alzheimer's disease since the age of 11 months. The creation of nanoparticles for nanomedicine, including clearly inert iron-oxide nanoparticles, provides information about their cellular absorption, metabolism, biodistribution, and cytotoxicity (Bao et al. 2016). Unsettlingly, the intake of nanoparticles is a preferred route for nanoparticle drug administration and is particularly effective from the standpoint of air pollution. Both friction-derived and solid combustion-derived nanoparticles in air pollution have been found by their abundance of highly magnetic, iron-rich particles that oxidize or condense after airborne discharge and often keep a round or oval shape after cooling. According to Maher et al. (2016), the aforementioned magnetic "nanospheres" are made up of a large number of urban airborne particulate matter components. According to the results of the Calderon et al. study (Calderon et al. 2019), early mitochondrial destruction is linked to the abundance of nanoparticles with a diameter of 10–30 nm, endoplasmic reticulum stress in the early stages, and the development of vascular and cardiac dysfunction. As a result, early cardioprotection measures are no longer effective. Failure to compensate for oxidative stress, endoplasmic reticulum stress, and chronic myocardial inflammation—all likely caused by the intrusion of redox-active, magnetic, iron-rich nanoparticles—may hasten the development of pathophysiological cardiovascular states in young adults and children exposed to contaminated environments. The identification of nanoparticles originating from magnetic combustion and friction in the heart tissue of children and young adults added a substantial layer of novel information to our understanding of the pathophysiology of cardiovascular disease and emphasizes how important it is to prioritize the control of particulate air pollution.

3.2 Adverse Effects of Carbon Nanotube on the Atmosphere

One of the main causes of the negative consequences of air pollution has been found to be the substantial particulate matter with a diameter of 2.5 μm, which can be breathed into the human lungs

(Kolosnjaj-Tabi et al. 2015). It has been proved that anthropogenic carbon nanotubes (CNTs) and other carbon-based nanoparticles are the main constituents of the 2.5-μm-diameter particulate matter present in the alveolar macrophages of children in Paris (Kolosnjaj-Tabi et al. 2015). Particulate matter from diesel and soot is also thought to have detrimental effects on health. Kolosnjaj has confirmed that carbon nanotubes (CNTs) are both present and past atmospheric pollutants that can reach human lungs (Kolosnjaj-Tabi et al. 2015). All of the examined samples from asthmatic children in Paris held carbon nanotubes that were breathed, such as those present in automobile emissions and airborne dust (Kolosnjaj-Tabi et al. 2015). The inflammatory reaction that carbon nanotubes cause is depicted.

Moreover, carbon-based particles have the ability to adsorb polycyclic aromatic hydrocarbons, which are known carcinogens (Esteve et al. 2006). Similarly, when combined with ozone, carbon nanotubes can interact with OH radicals and show increased toxicity. CNTs have the potential to increase the generation of reactive oxygen species in cells when they are employed for radical scavenging.

According to a different study, carbon nanotubes were extracted from the lungs of the World Trade Center collapse victims (Wu et al. 2009). However, the test samples were only named using low-magnification transmission electron microscopy (TEM), which could yield mistaken results because pulmonary surfactant filaments could look like filamentous SWCNT bundles. There is a wealth of research on airborne nanoparticles, and the creation of sophisticated nanomaterials keeps expanding. However, nothing is currently known about the existence of anthropogenic carbon nanotubes in indoor and outdoor air. Qualitative studies of breathed particulate matter are also quite uncommon. Only a few techniques, which are not often accessible in clinical settings, can be used to analyze particulate matter, including CNTs, in biological samples. Therefore, methods for detecting airborne carbon nanoparticles that are easy to use and accessible are necessary for clinical research. Anthropogenic carbon nanotubes can probably be detected using some of the same techniques that have already been developed for synthetic carbon nanotubes. Appropriate steps must be taken (such as installing effective air purifiers and filters) to lessen the atmospheric burden of carbon nanotubes if detrimental health consequences are verified (Mohammadian et al. 2019).

4. Prospects

Nanomaterials are essential to air pollution management technology because they lower the amount and emissions of dangerous pollutants into the atmosphere and enhance and preserve air quality. They are being improved and increasingly integrated *in situ* to detect air pollutants. For instance, industrial smokestacks and automobile exhaust systems can use nanofilters to capture and reduce the emissions of harmful pollutants. Carbon nanotube-based filters and metal-organic framework (MOF)-based filters have showed tremendous potential in preventing the release of hazardous gases such as HC, NOx, SO_2, CO_2, and CO. Furthermore, nanosensors have shown remarkable efficacy in detecting even minute quantities of harmful gas leakage. Toxic gases including NO_2, H_2S, and SO_2 can be detected by a variety of sensors that use nanomaterials like graphene, reduced graphene oxide, zinc oxide, and titanium dioxide. Additionally, evidence from a number of studies shows that these catalysts based on nanomaterials have performed better than conventional catalytic systems in terms of effectively separating and breaking down air pollutants, showing that they may be commercially practical. The development or early stages of commercialization of nano-based catalysts for sustainable air filtration are still ongoing. The importance of nanocatalysts as a promising investment opportunity is highlighted by the rising interest in catalysis-based research and the expanding use of these materials for environmental remediation.

Furthermore, in order to produce stable, economical, and environmentally friendly materials, the use of green chemistry principles in the manufacturing of nanomaterials should be further developed. In general, there are several exciting potential uses for nanoparticles in air cleanup.

5. Conclusion

Since air pollution is still a major problem worldwide, more research is being done to create innovative nanomaterials that can efficiently remove different air pollutants, such as nanoadsorbents, nanocatalysts, nanofilters, and nanosensors. This study looks at how common nanoparticles are used and evaluates the variables that affect how well these nanomaterials work. To identify and absorb pollutants such as metal oxides, volatile organic compounds (VOCs), and microorganisms, a variety of nanoadsorbents have been developed. By lowering airborne contaminants, nanocatalysts have shown significant potential in air cleanup. In order to capture different gas pollutants, research has also concentrated on the creation and improvement of nanostructured membranes. Toxic gases such as hydrogen sulfide (H_2S), sulfur dioxide (SO_2), and nitrogen dioxide (NO_2) have been investigated for detection using nanosensors. In order to increase the performance of these sensors, many studies have looked at the fundamental characteristics of gas-sensing materials. By creatively combining nanostructured materials with various features, selectivity, sensitivity, detection limits, and portability have all been improved.

However, there may be environmental hazards associated with the creation and disposal of nanoparticles. This emphasizes the necessity of thorough processes, experimental guidelines, and research plans that consider the special characteristics of nanomaterials as well as their possible toxicity. According to research, nanostructures can have an impact on the environment in a number of ways, including: i) raising pollution levels in the air, water, and soil; ii) building up in ecosystems and having both immediate and long-term consequences; and iii) interfering with species' life cycles. More research into the long-term or accelerated effects of nanomaterials is necessary to better understand their toxicological profile on the environment, as the majority of current studies concentrate on short-term impacts.

Nanomaterials can mix with other pollutants and possibly generate more or less hazardous chemicals because of their highly reactive nature. To create the right criteria, future research should include studies on the disposal of nanomaterials and their effects on the environment. There is still much to learn about the harmful impacts of nanoparticles on both human and animal health. Nevertheless, current toxicity profiles of nanomaterials on human cell lines and lower organisms point to possible health hazards, suggesting that nanomaterials might not be safe for human health. Standardized processes must thus be put in place to guarantee the environmentally responsible use of nanomaterials, supporting sustainable growth from industrial applications to environmental protection.

References

Amrane, A., Assadi, A. A., Nguyen-Tri, P., Nguyen, T. A., and Rtimi, S. (Eds.). (2020). Nanomaterials for Air Remediation. Micro and Nano Technologies.

Asad, Mohsen, Hossein Sheikhi, Mohammad, Pourfath, Mahdi, and Moradi, Mahmood, 2015. High sensitive and selective flexible H2S gas sensors based on Cu nanoparticle decorated SWCNTs. *Sensor. Actuator. B Chem.*, 210: 1–8.

Azam, M. A., Alias, F. M., Tack, L.W., Seman, R. N. A. R., and Taib, M. F. M. (2017). Electronic properties and gas adsorption behaviour of pristine, silicon-, and boron-doped (8, 0) single-walled carbon nanotube: a first principles study. *J. Mol. Graph. Model.*, 75: 85–93.

Babich, F., Demanega, I., Avella, F., and Belleri, A. (2020). Low polluting building materials and ventilation for good air quality in residential buildings: a cost–benefit study. *Atmosphere,* 11 (1): 102.

Bai, Shouli, Guo, Jun, Sun, Jianhua, Tang, Pinggui, Chen, Aifan, Luo, Ruixian, and Li, Dianqing, (2016). Enhancement of NO2-sensing performance at room temperature by graphene-modified polythiophene. *Ind. Eng. Chem. Res.*, 55 (19): 5788–5794.

Baltrusaitis, J., Jayaweera, P. M., and Grassian, V. H. (2011). Sulfur dioxide adsorption on TiO2 nanoparticles: influence of particle size, coadsorbates, sample pretreatment, and light on surface speciation and surface coverage. *J. Phys. Chem. C,* 115: 492–500.

Bao, Yuping, Wen, Tianlong, Samia, Anna Cristina S., Amit, Khandhar, Krishnan, and Kannan, M. (2016). Magnetic nanoparticles: material engineering and emerging applications in lithography and biomedicine. *J. Mater. Sci.*, 51(1): 513–553.

Baukal, Jr., C. E. (2004). Industria l Combustion Pollution and Control. Marcel Dekker, Inc, New York.
Bian, Ye, Wang, Rutao, Wang, Shijie, Yao, Chenyu, Ren, Wei, Chen, Chun, and Zhang, Li, (2018). Metal–organic framework-based nanofiber filters for effective indoor air quality control. *J. Mater. Chem.*, 6 (32): 15807–15814.
Binas, Vassilios, Venieri, Danae, Kotzias, Dimitrios, and George, Kiriakidis, (2017). Modified TiO2 based photocatalysts for improved air and health quality. *Journal of Materiomics,* 3 (1): 3–16.
Chen, Yifa, Zhang, Shenghan, Cao, Sijia, Li, Siqing, Chen, Fan, Yuan, Shuai, and Xu, Cheng, 2017. Roll-to-roll production of metal-organic framework coatings for particulate matter removal. *Adv. Mater.*, 29, 15: 1606221.
Chow, W., and Connor, K. (2020). Managing Hazardous Air Pollutants: State of the Art. CRC Press.
Choi, S., Jang, B., Lee, S., Min, B. K., Rothschild, A., and Kim, I. (2014). Selective detection of acetone and hydrogen sulfide for the diagnosis of diabetes and halitosis using SnO2 nanofibers functionalized with reduced graphene oxide nanosheets. ACS Appl. Mater. *Interfaces,* 6: 2588–2597.
Chu, Jifeng, Wang, Xiaohua, Wang, Dawei, Yang, Aijun, Lv, Pinlei, Wu, Yi et al, 2018. Highly selective detection of sulfur hexafluoride decomposition components H2S and SOF2 employing sensors based on tin oxide changed reduced graphene oxide. *Carbon*, 135: 95–103.
Çiftyürek, Engin, Sabolsky, Katarzyna, Sabolsky, and Edward M. (2016). Molybdenum and tungsten oxide-based gas sensors for high-temperature detection of environmentally hazardous sulfur species. *Sensor. Actuator. B Chem.*, 237: 262–274.
Cucurachi, Stefano, and Carlos Felipe Blanco Rocha, (2019). Life-cycle assessment of engineered nanomaterials. Nanotechnology in Eco-Efficient Construction. Woodhead Publishing, pp. 815–846.
Elia, H. N., Ghosh, A., Akhnoukh, A. K., and Nima, Z. A. (2018). Using nano- and micro-titanium dioxide (TiO2) in concrete to reduce air pollution. *J. Nanomed. Nanotechnol.*, 9: 505.
Esteve, Williams, Budzinski, Helene, and Villenave, Eric. (2006). Relative rate constants for the heterogeneous reactions of NO2 and OH radicals with polycyclic aromatic hydrocarbons adsorbed on carbonaceous particles. Part 2: PAHs adsorbed on diesel particulate exhaust SRM 1650a. *Atmos. Environ.*, 40 (2): 201–211.
Feng, S., Li, X., Zhao, S., Hu, Y., Zhong, Z., Xing, W. et al. (2018). Multifunctional metal organic framework and carbon nanotube-modified filter for combined ultrafine dust capture and SO 2 dynamic adsorption. *Environ. Sci.: Nano.*, 5(12): 3023–3031.
Gabal, E., Chatterjee, S., Ahmed, F. K., and Abd-Elsalam, K. A. (2020). Carbon nanomaterial applications in air pollution remediation. In: Carbon Nanomaterials for Agri-Food and Environmental Applications. *Elsevier*, pp. 133–153.
Gallego-Hern´andez, A. L., Meza-Figueroa, D., Tanori, J., Acosta-Elías, M., Gonz´alez- Grijalva, B., Maldonado-Escalante, J. F. et al. (2020). Identification of inhalable rutile and polycyclic aromatic hydrocarbons (PAHs) nanoparticles in the atmospheric dust. *Environmental Pollution*, p. 114006.
Garvey, T., Moore, E. A., Babbitt, C. W., and Gaustad, G., (2019). Comparing ecotoxicity risks for nanomaterial production and release under uncertainty. *Clean Technol. Environ. Policy*, 21(2): 229–242.
Ghio, A. J., Soukup, J. M., Dailey, L. A., and Madden, M. C. (2020). Air pollutants disrupt iron homeostasis to impact oxidant generation, biological effects, and tissue injury. Free Radic. *Biol. Med.*, 151: 38–55.
Ghosh, S., Ranjan, P., Ramaprabhu, S., and Sarathi, R. (2018). Carbon dioxide adsorption of zinc oxide nanoparticles synthesized by wire explosion technique. *INAE Letters*, 3(4): 197–202.
Giannakopoulou, Tatiana, Papailias, Ilias, Todorova, Nadia, Boukos, Nikos, Liu, Yong, Yu, Jiaguo et al. (2017). Tailoring the energy band gap and edges' potentials of g-C3N4/TiO2 composite photocatalysts for NOx removal. *Chem. Eng. J.,* 310: 571–580.
Gui, M. M., Yap, Y. X., Chai, S. P., and Mohamed, A. R. (2013). Multi-walled carbon nanotubes modified with (3-aminopropyl) triethoxysilane for effective carbon dioxide adsorption. *International Journal of Greenhouse Gas Control,* 14: 65–73.
Gupta, V. K., and Saleh, T. A. (2013). Sorption of pollutants by porous carbon, carbon nanotubes, and fullerene-an overview. *Environ. Sci. Pollut. Res.*, 20 (5): 2828–2843.
Hadrup, N., Saber, A. T., Kyjovska, Z. O., Jacobsen, N. R., Vippola, M., and Sarlin, E. (2020). Pulmonary toxicity of Fe2O3, ZnFe2O4, NiFe2O4, and NiZnFe4O8 nanomaterials: inflammation and DNA strand breaks. *Environ. Toxicol. Pharmacol.*, 74: 103303.
Haleem, A., Javaid, M., Singh, R. P., Rab, S., and Suman, R. (2023). Applications of nanotechnology in medical field: a brief review. *Global Health Journal*, 7(2): 70–77.
Haridas, Divya, Chowdhuri, Arijit, Sreenivas, K., and Gupta, Vinay. (2011). Effect of thickness of platinum catalyst clusters on response of SnO2 thin film sensor for LPG. *Sensor. Actuator. B Chem.*, 153(1): 89–95.
Heidari, A., and Younesi, H. (2020). Synthesis, characterization and life cycle assessment of carbon nanospheres from waste tires pyrolysis over ferrocene catalyst. *Journal of Environmental Chemical Engineering*, 103669.
Hung, Chu Manh, Ha, Viet Phuong, Nguyen Van Duy, Nguyen Duc Hoa, and Nguyen Van Hieu. (2018). Comparative effects of synthesis parameters on the NO2 gas-sensing performance of on-chip grown ZnO and Zn2SnO4 nanowire sensors. *J. Alloys Compd.*, 765: 1237–1242.
Huy, Nguyen Nhat, Vo Thi Thanh Thuy, Hung Thang, Nguyen, Nguyen Thi Thuy et al. (2019). Facile one-step synthesis of zinc oxide nanoparticles by ultrasonic-assisted precipitation method and its application for H2S adsorption in air. *J. Phys. Chem. Solid.*, 132: 99–103.
Ibrahim, Rusul Khaleel, Hayyan, Maan, Mohammed Abdulhakim AlSaadi, Hayyan, Adeeb, and Ibrahim, Shaliza (2016). Environmental application of nanotechnology: air, soil, and water. *Environ. Sci. Pollut. Control Ser.*, 23(14): 13754–13788.

Idrees, Z., and Zheng, L. (2020). Low-cost air pollution monitoring systems: a review of protocols and enabling technologies. *Journal of Industrial Information Integration*, 17: 100123.

Jampaiah, Deshetti, Chalkidis, Anastasios, Sabri, Ylias M., Mayes, Edwin L. H., Reddy, Benjaram M., Bhargava, and Suresh, K. (2019). Low-temperature elemental mercury removal over TiO2 nanorods-supported MnOx-FeOx-CrOx. Catal. Today, 324: 174–182.

Jangid, N. K., Yadav, A., Jadoun, S., Srivastava, A., and Srivastava, M. (2020). Outdoor pollution management by nanotechnology. pp. 258–277. *In:* Impact of Textile Dyes on Public Health and the Environment. IGI Global.

Jiang, Y., Peng, B., Wan, Z., Kim, C., Li, W., and Fortner, J. (2020). Nanotechnology as a key enabler for effective environmental remediation technologies. pp. 197–207. *In:* A New Paradigm for Environmental Chemistry and Toxicology. Springer, Singapore.

Kim, Hyoun Woo, Han, Gil Na, Jung Kwon, Yong, Kang, Sung Yong, Sik Choi et al. (2017). Microwave-assisted synthesis of graphene–SnO2 nanocomposites and their applications in gas sensors. *ACS Appl. Mater. Interfaces,* 9(37): 31667–31682.

Kolosnjaj-Tabi, J., Just, J., Hartman, K. B., Laoudi, Y., Boudjemaa, S., Alloyeau, D. et al. (2015). Anthropogenic carbon nanotubes found in the airways of Parisianchildren. *E-Bio Medicine*, 2(11): 1697–1704.

Kyzas, G. Z., and Matis, K. A. (2015). Nanoadsorbents for pollutants removal: a review. *J. Mol. Liq*., 203: 159–168.

Lee K. J., Shiratori N., Lee G. H., Miyawaki J., Mochida I., Yoon S. H. et al. (2010). (Activated carbon nanofiber produced from electrospun polyacrylonitrile nanofiber as a highly efficient formaldehyde adsorbent. *Carbon*, 48.

Liu, Yingying, Xu, Xinyue, Chen, Ying, Zhang, Yuan, Gao, Xinghua, Xu, Pengcheng, Li et al. (2018). An integrated micro-chip with Ru/Al2O3/ZnO as sensing material for SO2 detection. Sensor. Actuator. *B Chem*., 262: 26–34.

Low, J., Cheng, B. B., and Yu, J. (2017). Surface modification and enhanced photocatalytic CO2 reduction performance of TiO 2 a review. *Appl. Surf. Sci*., 392(15): 658–686.

Lu, C., Wu, B., Chen, W., Lin, Y. K., Bai, H. (2010). Capture of carbon dioxide by modified multiwalled carbon nanotubes. pp. 55–69. *In:* Enviro nano-technology. Elsevier.

Lyu, Jinze, Zhu, Lizhong, Burda, Clemens. (2014). Considerations to improve adsorption and photocatalysis of low concentration air pollutants on TiO2. Catal. *Today*, 225: 24–33.

Maher, Barbara, A., Imad, A. M. Ahmed, Vassil Karloukovski, Donald A. MacLaren, Foulds, Penelope G., David Allsop, Mann et al. (2016). Magnetite pollution nanoparticles in the human brain. *Proc. Natl. Acad. Sci. Unit. States Am*., 113(39): 10797–10801.

Mahmoodi Meimand, M., Javid, N., and Malakootian, M. (2019). Adsorption of sulfur dioxide on clinoptilolite/nano iron oxide and natural clinoptilolite, health scope. *Online ahead of Print*, 8(2): e69158.

Malik, S., Muhammad, K., and Waheed, Y. (2023). Nanotechnology: A revolution in modern industry. *Molecules*, 28(2): 661.

Mamaghani, Alireza Haghighat, Haghighat, Fariborz, Chang-Seo, Lee. (2018a). Photocatalytic degradation of VOCs on various commercial titanium dioxides: impact of operating parameters on removal efficiency and by-products generation. *Build. Environ*., 138: 275–282.

Manisalidis, I., Stavropoulou, E., Stavropoulos, A., and Bezirtzoglou, E. (2020). Environmental and health impacts of air pollution: a review. Frontiers in Public Health 8.

Miyawaki, J., Lee, G. H., Yeh, J., Shiratori, N., Shimohara, T., Mochida, I. et al. (2012). Development of carbon-supported hybrid catalyst for clean removal of formaldehyde indoors. *Catalysis Today*, 185(1): 278–283.

Moroz, B. L., Pyrjaev, P. A., Zaikovskii, V. I., and Bukhtiyarov, V. I. (2009). Nanodispersed Au/ Al2O3 catalysts for low-temperature CO oxidation: results of research activity at the Boreskov Institute of Catalysis. *Catal. Today*, 144(3-4): 292–305.

Muralikrishnan R., Swarnalakshmi M., and Nakkeeran E. (201). Nano particle membrane filtration of vehicular exhaust to reduce air pollution A review. *Int. Res. J. Environ. Sci*., 3: 82–86.

Navale, Y. H., Navale, S. T., Ramgir, N. S., Stadler, F. J., Gupta, S. K., Aswal, D. K. et al. (2017). Zinc oxide hierarchical nanostructures as potential NO2 sensors. *Sensor. Actuator. B Chem*., 251: 551–563.

Nair, A. N., Anand, P., George, A., and Mondal, N. (2022) A review of strategies and their effectiveness in reducing indoor airborne transmission and improving indoor air quality. *Environmental Research*, 213: 113579.

Nehra, Monika, Dilbaghi, Neeraj, Hassan, Ashraf Aly, and Kumar, Sandeep. (2019). Carbon- Based Nanomaterials for the development of sensitive nanosensor platforms. Advances in Nanosensors for Biological and Environmental Analysis. Elsevier, pp. 1–25.

Ningthoujam, R., Singh, Y. D., Babu, P. J., Tirkey, A., and Pradhan, S. (2022) Nanocatalyst in remediating environmental pollutants. *Chemical Physics Impact*, 4: 100064.

Ong, Chin Boon, Law, Yong Ng, Abdul Wahab Mohammad (2018). A review of ZnO nanoparticles as solar photocatalysts: synthesis, mechanisms and applications. *Renew. Sustain. Energy Rev*., 81: 536–551.

Ovsianytskyi, Oleksandr, Nam, Yun-Sik, Tsymbalenko, Oleksandr, Lan, Phan-Thi, Moon, Myoung-Woon, Lee, Kang-Bong, 2018. Highly sensitive chemiresistive H2S gas sensor based on graphene decorated with Ag nanoparticles and charged impurities. *Sensor. Actuator. B Chem*., 257: 278–285.

Pelaez, M., Nolan, N. T., Pillai, S. C., Seery, M.K., Falaras, P., Kontos, A.G. et al. (2012). A review on the visible light active titanium dioxide photocatalysts for environmental applications. *Appl. Catal. B Environ*., 125: 331–349.

Peng, Y. K., and Tsang, S. C. E. (2018). Facet-dependent photocatalysis of nanosize semiconductive metal oxides and progress of their characterization. *Nano Today*, 18: 1534.

Petryshak, Vasyl, Mikityuk, Zinoviy, Vistak, Maria, Gotra, Zenon, Akhmetova, Ardak, W´ojcik, Waldemar, Assembay, Azat, 2017 a b. Highly sensitive active medium of primary converter SO2 sensors based on cholesteric-nematic mixtures, doped by carbon nanotubes. *Przeglad Elektrotechniczny*, 1: 119–122.

Pokhum, Chonlada, Intasanta, Varol, Yaipimai, Wittaya, Subjalearndee, Nakarin, Srisitthiratkul, Chutima, Pongsorrarith, Voraluck et al. (2018). A facile and cost-effective method for removal of indoor airborne psychrotrophic bacterial and fungal flora based on silver and zinc oxide nanoparticles decorated on the fibrous air filter. *Atmospheric Pollution Research*, 9(1): 172–177.

Popp, Alexander, Yilmazoglu, Oktay, Kaldirim, Oktay, and J¨org, J. (2009). Schneider, and Dimitris Pavlidis. "A self-supporting monolith of highly aligned carbon nanotubes as device structure for sensor applications. *Chem. Commun.*, 22: 3205–3207.

Portela, Raquel, Jansson, Ingrid, Su´arez, S., Villarroel, Mirza, S´anchez, Benigno, and Avila, P. (2017). Natural silicate-TiO2 hybrids for photocatalytic oxidation of formaldehyde in gas phase. *Chem. Eng. J.*, 310: 560–570.

Postica, Vasile, Gr¨ottrup, Jorit, Adelung, Rainer, Lupan, Oleg, Mishra, Abhishek Kumar, Nora, H. et al. (2017). Multifunctional materials: a case study of the effects of metal doping on ZnO tetrapods with bismuth and tin oxides. Adv. Funct. Mater. 27(6): 1604676.

Pu, Y., Xie, X., Jiang, W., Yang, L., Jiang, X., and Yao, L. (2020). Low-temperature selective catalytic reduction of NOx with NH3 over zeolite catalysts: a review. *Chin. Chem. Lett.*, 31(10): 2549–2555.

Qiang, Xiaoyong, Hu, Ming, Zhao, Boshuo, Yue, Qin, Yang, Ran, Zhou, Liwei, and Qin, Yuxiang. (2018). Effect of the functionalization of porous silicon/WO3 nanorods with Pd nanoparticles and their enhanced NO2-sensing performance at room temperature. *Materials*, 11(5): 764.

Rabiee, F., and Mahanpoor, K. (2019). Photocatalytic oxidation of SO2 from flue gas in the presence of Mn/copper Slag as a novel nanocatalyst: optimizations by box-behnken design. *Iran. J. Chem. Chem. Eng. (Int. Engl. Ed.)*, 38(3): 69–85.

Romero-Guzm´an, L., Reyes-Guti´errez, L. R., Romero-Guzm´an, E. T., and Savedra-Labastida, E. (2017). Carbon nanotube filters for removal of air pollutants from mobile sources. *J. Miner. Mater. Char. Eng.*, 6(1): 105–118.

Roso, M., Lorenzetti, A., Boaretti, C., Hrelja, D., and Modesti, M. (2015). Graphene/TiO2 based photo-catalysts on nanostructured membranes as a potential active filter media for methanol gas-phase degradation. *Appl. Catal. B Environ.*, 176: 225–232.

Rubina, A., Rubinov´a, O., and Blasinski, P., (2019). Application of nanofilters for ventilation. *International Review of Applied Sciences and Engineering*, 10(1): 51–56.

Saber, A. T., Mortensen, A., Szarek, J., Jacobsen, N. R., Levin, M., Koponen, I. K. et al. (2019). Toxicity of pristine and paint-embedded TiO2 nanomaterials. *Hum. Exp. Toxicol.*, 38(1): 11–24.

Salabat, A., and Keshavarz, A. (2019). Effect of HCl on the structure and catalytic activity of Pt/Al2O3 nanocatalyst prepared in microemulsion system. *Sci. Iran.*, 26 (3): 1925–1930.

Saleem, H., Zaidi, S. J., Ismail, A. F., and Goh, P.S. (2022) Advances of nanomaterials for air pollution remediation and their impacts on the environment. *Chemosphere*, 287: 132083.

Samantara, A. K., Ratha, S., and Raj, S. (2019). Functionalized graphene nanocomposites in air filtration applications. Functionalized Graphene Nanocomposites and Their Derivatives. Elsevier, pp. 65–89.

Shen, W., Zhang, C., Li, Q., Zhang, W., Cao, L. L., and Ye, J. (2014). Preparation of titanium dioxide nanoparticle modified photocatalytic self-cleaning concrete. *J. Clean. Prod.*, 87: 762–765.

Shi, Jinjin, Cheng, Zhixuan, Gao, Liping, Zhang, Yuan, Xu, Jiaqiang, and Zhao, Hongbin, (2016). Facile synthesis of reduced graphene oxide/hexagonal WO3 nanosheets composites with enhanced H2S sensing properties. *Sensor. Actuator. B Chem.*, 230: 736–745.

Silva, L. F., Milanes, C., Pinto, D., Ramirez, O., and Lima, B. D. (2020). Multiple hazardous elements in nanoparticulate matter from a Caribbean industrialized atmosphere. *Chemosphere*, 239: 124776.

Spinazz`e, A., Campagnolo, D., Cattaneo, A., Urso, P., Sakellaris, I. A., and Saraga, D. E. (2020). Indoor gaseous air pollutants determinants in office buildings—the OFFICAIR project. *Indoor Air*, 30 (1): 76–87.

Su, F., Lu, C., and Chen, H. S. (2011). Adsorption, desorption, and thermodynamic studies of CO2 with high-amine-loaded multiwalled carbon nanotubes. *Langmuir*, 27(13): 8090–8098.

Tai, X. H., Lai, C. W., Juan, J. C., and Lee, K. M. (2020). Nanocatalyst-based catalytic oxidation processes. pp. 133–150. *In:* Nanomaterials for Air Remediation. Elsevier.

Tammanoon, Nantikan, Wisitsoraat, Anurat, Sriprachuabwong, Chakrit, Phokharatkul, Ditsayut, Tuantranont, Adisorn, Phanichphant, Sukon et al. (2015). Ultrasensitive NO2 sensor based on ohmic metal–semiconductor interfaces of electrolytically exfoliated graphene/flame-spray-made SnO2 nanoparticles composite operating at low temperatures. *ACS Appl. Mater. Interfaces*, 7(43): 24338–24352

Theerthagiri, J., Duraimurugan, K., Kim, H. S., and Madhavan, J. (2019). Graphitic carbon nitride-based nanostructured materials for photocatalytic applications. *Photocatalytic Functional Materials for Environmental Remediation,* 291–307.

Tran, Dat T. (2016). Synthesis of porous ZnO based materials using an agarose gel template for H 2 S desulfurization. *RSC Adv.*, 6(2): 1339–1345.

Tumuluri, U., Howe, J. D., Mounfield III, W. P., Li, M., Chi, M., Hood, Z. D. et al. (2017). Effect of surface structure of TiO2 nanoparticles on CO2 adsorption and SO2 resistance. *ACS Sustain. Chem. Eng.*, 5(10): 9295–9306.

Tyagi, Punit, Sharma, Anjali, Tomar, Monika, Gupta, Vinay, 2016. Metal oxide catalyst assisted SnO2 thin film based SO2 gas sensor. *Sensor. Actuator. B Chem.*, 224: 282–289.

Vyas, Rishi, Sharma, Sarla, Gupta, Parul, Vijay, Y. K., Prasad, Arun K., Tyagi, A. K. et al. (2013). Enhanced NO2 sensing using ZnO–TiO2 nanocomposite thin films. *J. Alloys Compd.*, 554: 59–63.

Wang, Ziying, Zhang, Tong, Zhao, Chen, Tianyi Han, Teng Fei, Sen Liu et al. (2018). Anchoring ultrafine Pd nanoparticles and SnO2 nanoparticles on reduced graphene oxide for high-performance room temperature NO2 sensing. *J. Colloid Interface Sci.*, 514: 599–608.

WHO (2021) Air Pollution and Health.

Wu, Maoxin, Gordon, Ronald E., Herbert, Robin, Maria padilla, jacqueline moline, david mendelson et al. (2009). Case report: lung disease in world trade center responders exposed to dust and smoke: carbon nanotubes found in the lungs of world trade center patients and dust samples. *Environ. Health Perspect.*, 118 (4): 499–504.

Wu, C. M., Baltrusaitis, J., Gillan, E. G., and Grassian, V. H. (2011). Sulfur dioxide adsorption on ZnO nanoparticles and nanorods. *J. Phys. Chem. C*, 115 (20): 10164–10172.

Wu, H., Ma, J., Li, Y., Zhang, C., and He, H. (2014). Photocatalytic oxidation of gaseous ammonia over fluorinated TiO2 with exposed (0 0 1) facets. *Appl. Catal. B Environ.*, 152: 82–87.

Wu, M. Y., Lo, W. C., Chao, C. T., Wu, M. S., and Chiang, C. K. (2020). Association between air pollutants and development of chronic kidney disease: a systematic review and meta-analysis. *Sci. Total Environ.*, 706: 135522.

Yusuf, M., Kumar, M., Khan, M. A., Sillanpa, M., and Arafat, H. (2019). A review on exfoliation, characterization, environmental and energy applications of graphene and graphenebased composites. *Adv. Colloid Interface Sci.*, 102036.

Zhang, Dongzhi, Liu, Jingjing, Jiang, Chuanxing, and Peng, Li. (2017). High-performance sulfur dioxide sensing properties of layer-by-layer self-assembled titania-modified graphene hybrid nanocomposite. *Sensor. Actuator. B Chem.*, 245: 560–567.

Zhang, H. J., Meng, F. N., Liu, L. Z., and Chen, Y. J. (2019). Convenient route for synthesis of alpha-Fe2O3 and sensors for H2S gas. *J. Alloys Compd.*, 774: 1181–1188.

Zhang, Hao, Li, Qun, Huang, Jinyu, Du, Yu, and Ruan, Shuang Chen. (2016b). Reduced graphene oxide/Au nanocomposite for NO2 sensing at low operating temperature. *Sensors*, 16(7): 1152.

Zhang, Yuanyuan, Yuan, Shuai, Xiao, Feng, Li, Haiwei, Zhou, Junwen, Wang, Bo. (2016a). Preparation of nanofibrous metal–organic framework filters for efficient air pollution control. *J. Am. Chem. Soc.*, 138(18): 5785–578.

Zhao, S., Liu, S., Hou, X., Beazley, R., Sun, Y., and Dong, S. (2020a). Evidence of provincial variability in air pollutants-asthma relations in China. *J. Clean. Prod.*, 242: 118553.

Zhao, Qidong, Li, Xinyong, Zhou, Qiang, Wang, Dan, and Xu, Huixin, (2020b). Nanomaterials developed for removing air pollutants. In: Advanced Nanomaterials for Pollutant Sensing and Environmental Catalysis. *Elsevier*, pp. 203–247.

Zouaghi, Wissem, Hussein, Laith, Mark, D., Thomson, Qamar-ul Islam, Nicoloso, Norbert, Heinlein, Thorsten et al. (2018). Towards gas sensing with vertically aligned carbon nanotubes interrogated by THz radiation pulses. *Lith. J. Phys.*, 58: 1.

CHAPTER-11

Microbial and Smart Nanomaterial Combination for Pollution Abatement

Mital Lodariya,[1] *Debleena Bhattacharya*[1,*] and *Abhilash K.R.*[2,*]

1. Introduction

The world is facing ever-increasing challenges related to mitigating the effects of pollution on the environment. Sources of these wastewater discharges include agricultural practices, effluent from sewage treatment plants, road traffic, oil refineries, and waste disposal sites. Even volcanic eruptions and wildfire incidents of natural origin are considered sources of pollutants (Tyagi et al. 2021). As a result, flora, fauna, and entire ecosystems are negatively impacted by the persistent increase in environmental pollution. To effectively remediate ecological contaminants, a pressing requirement exists to explore cost-effective and environmentally sustainable technologies. Persistent organic pollutants (POPs) like dioxins and sulfur dioxide (SO2), as well as highly reactive and toxic organic compounds (HRTOCs) and polycyclic aromatic hydrocarbons (PAHs), adversely affect human health (Khan et al. 2021). Also, they could increase diseases such as chronic bronchitis, chronic obstructive lung disease, chronic obstructive pulmonary disease (COPD), cancer, skin disorders, and emphysema (Roy et al. 2021). Various traditional physico-chemical methods including condensation, flocculation, froth flotation, sand filtration, valorization, and activated carbon adsorption have been adopted in an attempt to reduce the toxicity of environmental effluents. However, they all contribute to the creation of secondary by-products that are not eco-friendly and demand further efforts to treat (Mondal et al. 2023). Nanotechnology seems to indicate an alternative solution to these problems. Nanocatalysts are made up of nanomaterials which can act as catalysts in the process of remediation. The breakdown of harmful chemicals during the reaction can be successfully done by nanocatalysts (Ningthoujam et al. 2022). The constructive participation of microorganisms alongside advanced nanomaterials represents a paradigmatic shift in treating

[1] Department of Chemical Engineering, Marwadi University, Rajkot, Gujarat, 360003, India. https://orcid.org/0000-0002-4709-6199

[2] National Centre for Sustainable Coastal Management (NCSCM), Ministry of Environment, Forest and Climate Change (MoEFCC), Chennai 600 025, Tamil Nadu, India. https://orcid.org/0000-0002-4542-8516

* Corresponding authors: debleena.b3@gmail.com; abhilaashkr@gmail.com

DOI: 10.1201/9781003558804-11

environmental effluents. Integrated approaches could cover a much wider range of applications at smaller costs and with less environmental damage. The method benefits from a synergistic advantage between biological systems and nanotechnological innovation (Nakum et al. 2022). The green synthesis of nanomaterials using microorganisms and other biological sources has enabled the eco-friendly remediation of pollutants. The unique redox, nontoxic nature, and ferromagnetic properties have made iron nanoparticles highly preferred and effective in the role of a nanocatalyst for the removal of ecological pollutants (Mandeep et al. 2020). Membrane-associated nanomaterials may be useful for improving membrane permeability, mechanical strength, and thermal resistance, and also as innovative approaches for pollutant remediation. Microbial systems comprised of a diverse group of microorganisms have been acknowledged for their capacity to degrade and transform various organic and inorganic contaminants. Nanotechnology has been applied across a wide range of scientific disciplines, including ecological science, engineering, auto technology, skin care products, food production, fabrics, aerospace, security, and health care (Jain et al. 2021). In recognition of their unique characteristics—especially their enormous surface area, versatility, and minuscule ion substance—nanomaterials have been exploited significantly over the past decades for cleaning up harmful substances. Nanoremediation is recognized as a revolutionary ecological clean-up method, in light of its crucial contributions in minimizing emissions, tracking down and recognizing contaminants, and subsequently gaining productive and favorable outcomes in restoration operations (Epelle et al. 2022). Nanoremediation is decisively asserting itself as a superior alternative to traditional, expensive, and labor-intensive methods for soil and groundwater. These traditional methods include heat treatment, chemical deterioration, and washing with hydrophobic cosolvents (Ahmed et al. 2022a). Conventional biological methods like micro- and nano-bioremediation possess various advantages such as producing no by-product pollutants while being economical, effective, and adaptable. The study of the combined application of microorganisms and nanomaterials is listed in Table 1.

2. Microbial Approaches to Pollution Abatement

Bacteria, microalgae, and fungi frequently occur throughout the biosphere and play an important role in bioremediation. They replicate quickly and grow in a variety of environmental conditions, which mainly makes them the primary agents for restoring ecological parity. Either alone or in consortium, these organisms may be employed in synergism in contributing to the restoration of the original ecological equilibrium. They play a central role in the prevention of future pollution and the accelerated biodegradation and biotransformation of toxins owing to their enzymatic capabilities (Maglione et al. 2024). Microorganisms demonstrate incredible capabilities in detoxifying, degrading, and even accumulating both organic and inorganic toxic substances. Due to their high adaptability and metabolic diversity, microorganisms can live in various contaminated natural and artificial environments. Bioremediation is a process that reduces environmental pollutants to harmless levels using biological mechanisms. This process is primarily carried out by specific wild-type or engineered microorganisms, which also produce energy and biomass (Ayilara et al. 2023). Since there is a strong relation between microbial activity and bioremediation, the same factors that impact microbial activity include pH level, temperature, intake of nutrients, aerobic or anaerobic circumstances, and non-biological elements. To reduce environmental toxic effluents, microbial bioremediation capacity will increase due to their multidisciplinary approach, namely genetic engineering and molecular biology. The use of genetically modified organisms (GMOs) offers social and economic benefits by minimizing health hazards, eliminating disposal costs, and a greener environment. To promote circular bio-economy activities with the goal of 'zero-waste,' the research committee is going to promote direct incorporation in biorefinery and remediation concepts to create products containing a higher value (Maglione et al. 2024). Experiments have proven that only fungus, bacteria, and yeast may degrade heavy metals by converting them into less toxic substance. Various microbial techniques, including bioaccumulation, biosorption, oxidation-

Table 1: Different studies of synergistic approaches: Microbial-nanomaterial combinations.

Microbial approach	Nanomaterial's type	Applications	References
B. subtilis strain SL (co-culture with *Pseudomonas aeruginosa* WJ-1)	Native biosurfactant system	Remediation of crude oil pollutants	Zhang et al. 2023
E. asburiae, B. cereus	Immobilized nanoparticles	Immobilization of heavy metals in contaminated sites	Fashola et al. 2020a
Bacteria, fungi, algae	Various nanomaterials	Removal of dyes, plastic, petroleum-based contaminants	Maglione et al. 2024
Ficus benjamina extract-assisted microbial synthesis	Zero-valent silver nanoparticles (AgNPs)	Cadmium (Cd) removal efficency	Mishra et al. 2020
Microbial biofilm (engineered *Bacillus subtilis*)	Nano oxides (Ferric oxide, titanium oxide) immobilized in biochar	Removal of heavy metals such as arsenic (As), cadmium (Cd), and lead (Pb)	Liu et al. 2024
Bioremediation using engineered microbes	Nano zero-valent iron (nZVI)	Removal efficiency of soil and groundwater pollutants	Ahmed et al. 2022b
Bio-augmented microbes	Nano-sized Al_2O_3	Extraction and separation of trace metal ions	Ali et al. 2023a
Bacillus subtilis strain SL (co-cultured with *Pseudomonas aeruginosa* WJ-1)	Native biosurfactant system	Remediation of crude oil pollutants	Wu et al. 2023
Bacterial enzymatic degradation	Biogenic nanoclusters	Reduction of toxic metal ions	Ajith et al. 2021a
Microbial production of nanoparticles	Intracellular and extracellular bio-nanoparticles	Remediation of various industrial pollutants	Behl et al. 2024a
Marine bacteria, fungi, and microalgae	Bio-synthesized metallic nanoparticles	Detoxification of heavy metals and organic pollutants	Upadhayay et al. 2023
Marine microalgae-based bioremediation	Biogenic nanoclusters	Nanoparticle formation and reduction of toxic metal ions	Sharma et al. 2023
Microbes from rhizosphere	Metal chelation nanoparticles	Detoxification and bioremediation of industrial wastewater pollutants	Verma et al. 2024

reduction, and biosurfactant production, have gained attention due to their high removal efficiency of dye, plastic, and petroleum-based contaminants. Some of the studies have applied *B. subtilis* strains NAP1, NAP2, and NAP4 to remediate oil-based paint (Zhang et al. 2023). The strain *Lysinibacillus sphaericus* CBAM5 was isolated from subsurface soil, and it has potential for bioremediation of heavy-metal polluted environments. Various heavy metals like nickel, mercury, copper, chromium, cobalt, and lead have been remediated by *Saccharomyces cerevisiae* and *Cunninghamella elegans* (Peña-Montenegro et al. 2015). Within 4.7 days, *Bacillus subtilis* DM2 may break down petroleum hydrocarbons (C12, C14, C15) by as much as 53.92 percent (Li et al. 2019). Meanwhile, within five days, 65 percent of crude oil may break down using BL-27 strain. For example, *Pseudomonas fluorescens* can bioremediate 18 percent PE efficiency in 270 days (Elahi et al. 2021). Microbes like *E. asburiae* and *B. cereus* can be involved in immobilized processes for heavy metals (Fashola et al. 2020b). Microbes protect themselves during the bioremediation process by forming hydrophobic efflux pumps that protect the outer membrane of the cell. The expansion of genetically engineered microorganisms (GEMs) involves the recognition and subsequent administration of certain gene sequences. GEMs can show increased degradation ability in extensive xenobiotics, and they have high prospective applications in the bioremediation of environmental pollutants (Pant et al. 2021). The development of GEMs for bioremediation has various approaches, for instance: (i) identification of relevant organisms for modification, (ii) metabolic pathway manufacture, augmentation, and regulation, (iii) modification of specific incentives and affinities, and (iv) bio-process evaluation, monitoring, and control (Kulshreshtha 2013). Researchers can engraft genetic modification within living organisms or cross-fertilize with another strain so that it is effective in degrading denatured essences. Conditions in nature, however, may act differently than predicted due to environmental issues like native microbial competition, the inevitability of environmental destruction, and genetic stability. Research in advanced bioremediation technologies must be increased to minimize potential risks for ecosystems and human health, as well as maximize their potency as eco-friendly solutions for the remediation of ecological contamination (Rafeeq et al. 2023). Microorganisms are ubiquitous, and they utilize a broad variety of substrates as carbon sources. They are found in peculiar places where they can absorb a wide range of pollutants. For example, acidophiles live in acidic places, psychrophiles dwell in cold areas, and halophiles can survive in salty regions. Different mechanisms can be employed to remove effluents from the environment using microbes, and these can be divided into two categories, namely mobilization and immobilization. The mobilization process includes enzymatic oxidation, bioleaching, bio-stimulation, bio-augmentation, and enzymatic reduction procedures. However, the immobilization process involves bioaccumulation, complexation, biosorption, and precipitation.

2.1 Enzymatic Oxidative Reaction

In the enzymatic oxidation process, oxidizing effluent converts a higher oxidation rate to a lower oxidation rate. During this process, toxic heavy metals can lose electrons and transfer into less toxic compounds. Moreover, this oxidation process utilizes an enzyme called oxidoreductase that is produced by the bacteria involved and remediates dyes, phenols, and other pollutants that can easily convert into less toxic form or more biodegradable forms by microbes (Unuofin et al. 2019). Microorganisms such as bacteria (*Pseudomonas, Bacillus, Mycobacterium*), fungi (*Trametes versicolor, Phanerochaete chrysosporium*), and actinomycetes produce oxidative enzymes that catalyze these transformations. In the process of microbial oxidation, the key enzyme participants are peroxidases (lignin peroxidase, manganese peroxidase), dioxygenases, tyrosinases, monooxygenases, laccases, and cytochrome P450 enzymes, wherein phenolic compounds, pigments, and PAHs are oxidized by laccases, which are produced by fungi and certain bacteria, via a radical-mediated process that results in polymerization or total mineralization (Sahay et al. 2019, Tian et al. 2020). Though the bioremediation technique offers potential advantages, the additional nutrients released into the ecosystem may augment eutrophication & other adverse environmental effects. Peroxidases are useful for detoxifying industrial wastewater and pesticide-contaminated soil because they employ hydrogen peroxide to break down lignin and several xenobiotics (Sayed et al. 2021). This is especially the case with white rot fungi. Factors affecting enzymatic oxidation used in bioremediation include environmental variables such as pH levels, temperature, oxygen availability, and the presence of cofactors such as hydrogen peroxide and metal ions. Synthetic biology and genetic engineering have provided recent impetus for improvements in microbial bioremediation, leading to higher degradation rates, better substrate selectivity, and optimized enzyme yield. Bioreactor-based systems and methods of enzyme immobilization have also been developed, enhancing enzyme stability and reusability in large-scale applications (Gangola et al. 2018).

2.2 Enzymatic Redox Reaction

In comparison with enzymatic oxidation, this mechanism transforms contaminants into a lower oxidation state that keeps them insoluble. It can be used to change and break down a range of contaminants including dyes, explosives, polyaromatic hydrocarbons, pesticides, and heavy metals. Various studies show successful remediation of toxic compounds like polychlorinated dibenzo-p-dioxins and dibenzofurans (Zacharia 2019). Electron shuttles can promote microbe elimination of various electron acceptors, including nitrate, arsenate, and perchlorate, and may help transform and remove these pollutants from the environment. Within the class of oxidoreductases, dehydrogenases, laccases, peroxidases, and monooxygenases are key to many biochemical functions. In anaerobic bacteria such as *Pseudomonas, Bacillus*, and *Geobacter species*, their enzymatic systems are used to degrade organic contaminants or reduce heavy metals into less toxic forms. Oxidases are enzymes that transfer electrons into molecular oxygen with harmless by-products such as CO_2 and H_2O. Moreover, biological agents may biosynthesize surfactants that enhance the solubility and bioavailability of hydrophobic compounds in order to speed up degradation rates. Environmental or ecological variables such as temperature, aeration supply, pH, and electron donors or acceptors can induce heterogeneity in the machinery of enzymatic reduction. Hexavalent chromium is the precursor to chromium trivalent, reduced using chromate reductase, while azo reductases cleave the azo link to decompose azo compounds (Saxena et al. 2020).

2.3 Bioaugmentation

The bioaugmentation process represents a cheap, rapid, and more effective method for bioremediation through species additions that facilitate the dissolution of toxic contaminants within polluted sites. They may also add external microbes or genetically altered organisms to help support or supplement the resident microbes. The performance of these new strains is dependent upon many parameters, including their capacity to adapt to their new surroundings and compete with the incumbent microorganisms (Babalola et al. 2019). Organisms may naturally not be capable of degrading toxic contaminants; therefore, genetic tactics of resident microbes are used to degrade the toxic pollutants present at a site. When applied to a polluted environment, *Burkholderia sp*. FDS-1 has been shown to degrade nitrophenolic compounds found in pesticide-polluted soil to a less harmful form at a temperature of around 30°C and a slightly acidic pH (Ojuederie et al. 2021). Bioaugmentation depends on microbial survival, nutrient availability, environment, and competition with indigenous microbes. Techniques such as encasing microbes into hydrogel beads, charcoal, or polymer matrices improve microbial activity and stability under challenging conditions (Ayangbenro and Babalola, 2017). Bioaugmentation may also be used in conjunction with biostimulation methods, which consist of providing nutrients or electron donors to promote the growth and activity of microorganisms. Eco-friendly and systematic restoration of polluted environments is possible by bioaugmentation using the natural abilities of certain microbes. This process has been well acknowledged in heavy metal remediation from soil, wastewater treatment, and oil spill management (Ojuederie et al. 2021).

2.4 Biostimulation

Biostimulation includes the bioenvironment modification to stimulate indigenous microorganisms and carry out target bioprocesses productively. This process could be carried out by metabolites, nutrients (like potassium, nitrogen, oxygen, carbon, and phosphorus), electron donors, enzymes, electron acceptors, biosurfactants, etc. The process of biostimulation is adding certain nutrients to debris either by introducing specific microorganisms to effectively break down the waste or by using bacteria that are present naturally in sufficient quantity (Mohamed et al. 2023). The biostimulation process achieves at least two ways to increase the rate of biodegradation of a wider variety of pesticides. First, taxonomic richness could be disclosed as a productive and significant biotransformation rate of trace heavy metals in microbial diversity. In most cases, however, conditionally, microbes produce more catabolic genes. Thus, it is still unclear how the stimulation of catabolic activity with a particular pesticide functions in treatment while saturated with bioremediation and a very high level of metabolic activity of microorganisms in biotreatment wastewaters in different compartments (Helbling et al. 2015). Since native microbes are better competitors than the introduced ones, biostimulation is more effective than bioaugmentation; biostimulation thus goes further to promote the flourishing of the microbial community in the ecosystem. *Bacillus species, Staphylococcus species, Rhodococcus species, Pseudomonas species, Klebsiella species*, and *Citrobacter species* were found to show a high level of bioremediation against heavy metals using biostimulation techniques (Nivetha et al. 2022).

2.5 Bioleaching

Bioleaching helps induce sedimentation of heavy metals into a solid state from the sediment matrix using extremophilic microorganisms. Particularly, sulfur and iron pollutants are remediated by this method. For example, chemolithotroph bacteria namely, *T. thiooxidans*, *L. ferrooxidans*, and *Thiobacillus ferrooxidans* can transfer insoluble metal sulfide to soluble metal sulfide via biochemical oxidation reactions (Bhandari et al. 2023). The application of this process is mainly prospective for detoxification and recovery of industrial debris, heavy metals contaminated soil, and sewage sludge. There is direct and indirect bacterial leaching that can be used for metal release from sulfide minerals. Direct bacterial leaching has shown the physical connection between the surface of mineral sulfide and bacterial cells through enzymatically catalyzed steps (Ai et al. 2019). Indirect bioleaching involves the chemical oxidation of sulfide minerals by a bacterial-produced lixiviant enzyme. Since bacteria catalyse the reoxidation of ferrous iron, direct contact with the mineral surface is redundant; the reaction also occurs in the absence of bacteria (Deng et al. 2017). An array of variables such as temperature, pH, the availability of oxygen and carbon dioxide, and the nutrients in the medium affect the bioleaching process. In autotrophic leaching, oxidizing iron, sulfur reduction, and fixation of carbon dioxide occur using acidophilic microbes, as well as generating end products like ferric iron or H_2SO_4. This process mainly depends on the metabolism and production of primary products used in solubilization, complexation, and oxidation of toxic contamination in soil. In heterotrophic bioleaching, complex lysis, which regulates the metallic ions in the solution by acidolysis, is carried out through the chelation mechanism (Nguyen et al. 2021). Various physiochemical, microbiological, and mineralogical factors contribute to the bioleaching processes, requiring refined control of operational parameters in order to enhance performance.

2.6 Biosorption

Biosorption is considered a systematic and cost-effective physiochemical process that could involve the attachment of heavy metal ions to the biosorbent surface. Some of the methods used for the adsorption of heavy metals include complexation, chelation, ion exchange, physical interaction, and electrostatic force (Xiao et al. 2019). Some microbes are capable of bioremediation through biosorption, including *Streptomyces* sp. K11, *Bacillus anthracis*, and *Rhodococcus erythropolis*. Complexation is primarily accomplished by a variety of ligands, including low-molecular-weight organic acids, siderophores, high-molecular-weight ligands, and toxic metal-binding compounds (Medfu Tarekegn et al. 2020). These ligands primarily interact with environmentally present pollutants like inorganic metals, especially those in solid waste. In the complexation process, environmentally polluted heavy metals are bound with extracellular polymeric substances found on microbial surfaces. Biochar can be employed in the complexation procedure to remediate hexavalent chromium and copper (II) oxide from wastewater (Xie et al. 2020). Siderophores, known as metal chelators containing compounds like hydroxamate, catecholate, and phenolates, are frequently manufactured by pathogens in metal-deficient habitats. They may reduce the disintegration of metallic substances by chelation. Cyanobacteria are reported as effective bioremediation agents due to their production of various types of siderophores such as synechobactin, anachelin, and schizokinen (Arstol and Hohmann-Marriott 2019, Sarmah and Rout 2020). The results showed that lanthanum-modified bone waste had a high efficiency of 91% in removing fluoride, with a biosorption capacity of 8.96 mg/g. Generally, biosorption is meant to test nonliving biomass, as it has greater practicality and zero toxicity with no maintenance, storage, or regeneration conditions (Grabi et al. 2021).

2.7 Bioaccumulation

Bioaccumulation is a natural biological and metabolically active process where microbes can utilize protein to absorb and sequester metal ions in the intracellular space for use in cellular functions (for example, signaling, enzyme catalysis, and stabilizing charges on biomolecules). Inside the intracellular space, heavy metals can be remediated by peptide and protein ligands. This process could be used by an importer to build up translocation pathways across the lipid bilayer (Mir-Tutusaus et al. 2018). Various microbes have been used in the bioaccumulation process in the form of indicators, for instance, bacteria, fungi, algae, mussels, plants, fish, and oysters. Synthetic phytochelatin, or PC, is a metal-binding peptide composed of a metal-binding pattern that follows (Glu-Cys)nGly. Induced by plants, it showed enhanced binding capacity with Cd2+ when expressed in recombinant strains of *Escherichia coli* (Sharma et al. 2022). Bioaccumulation includes absorption into the cell wall or membrane, followed by transport either actively or passively inside the cell (Jaiswar et al. 2015). Transport into cells is facilitated by molecular mimicry and endocytosis, whereas toxicity is influenced by salinity, pH, alkalinity (hardness), and dissolved organic matter. Several organisms are especially involved in the biological accumulation of substantial metals. *Rhizopus arrhizus*, *Pseudomonas putida*, and *Aspergillus niger*, for instance, bioremediate mercury and cadmium (Almeida et al. 2015).

2.8 Precipitation

Insoluble metal cations are translated from soluble to suspended form as a result of biological or chemical interactions; this decreases their mobility and bioavailability. Such reactions may include the precipitate formation in biomineralization when microbial metabolites such as organic acids, sulfides, and phosphates react with metal ions, thus resulting in stable precipitates (Pratush et al. 2018). This process can take place during the biogeochemical cycle to create the deposition of metals (iron and manganese), mineralized manganese and silver, and microfossils, by enzymes and galactosides of secondary metabolites (Maity et al. 2019). For example, when the pH is alkaline, sulfate-reducing bacteria can change organophosphate into orthophosphate. Similarly, SRB can help precipitate toxic metals like lead, cadmium, and arsenic by generating hydrogen sulfide (H_2S), which combines with metal ions to create highly insoluble metal sulfides like PbS and CdS that are easily extracted from the environment (Lin et al. 2023). Several factors, including redox potential, pH, microbial activity, and the availability of ligands, have all been considered in metal binding and affect the solubility of biological immobilization. In other studies, associations between the atmospheric deposition of heavy metals on *Bacillus subtilis* and *Oceanobacillus indicireducens* have been made (Zheng et al. 2024). Biochar, an abundant source of carbon and another matrix capable of influencing the same prevailing understanding, can induce precipitation by providing metal ion binding sites. It could enhance soil quality besides having remedial effects against heavy metal toxicity. Precipitation as an environmentally friendly and sustainable method for remediating contaminated sites is not only bound to precipitation but will also help recover precious metals through biogeochemical pathways (Wang et al. 2023).

3. Smart Nanomaterials for Environmental Pollution Control

Nanomaterials will have tremendous advantages *in situ*. Such nanoparticles, owing to their small size and the properties of the modifying surface layers, can traverse the very small pores found in the soil. The small surface area to volume ratio renders them impotent to form clusters. They can easily adsorb a wide variety of water and gas vapors and are also used in photocatalytic methods for the degradation of pollutants. Metal and metal-oxide nanoparticles are known to significantly adsorb contaminants from contaminant sites (Mondal et al. 2021). Titanium dioxide (TiO_2) nanoparticles exhibit variable catalytic reactivity, chemical behavior, and surface acidity strictly dependent on their surface plane. The semiconductor characteristics of TiO_2 are essential for removing different kinds of natural pollutants, as they use light energy to activate the TiO_2 semiconductors and initiate electron-releasing reactions (Ajith et al. 2021b). Alumina (Al_2O_3) like nanomaterial comes in four different crystalline states and is commonly used as a natural adsorbent due to its high stability. ZnO nanoparticles are used as cheap photocatalysts to remove heavy metal ions. Silver nanoparticles (AgNPs) act as antimicrobial agents to control airborne pathogens, and they are mainly used in wastewater treatment, plant growth, medicine, metal removal, and crop production. Carbon-based nanoparticles, for example, graphene oxide, Carbon Nanotubes (CNTs), activated carbon nanomaterials, and various composites such as engineered nanomaterials, have been used to immobilize pollutants and also adsorb air pollutants like SO_2 and NOx. Zero-valent AgNPs derived from the extract of the leaf of the *Ficus benjamina* tree have been shown to effectively increase Cd removal performance with a high concentration rate (Mishra et al. 2020). Some oxide nanoparticles, for instance ferric oxide and titanium oxide, have extraordinary potential to immobilize and remediate heavy metals like Cd and As from various media samples (Liu et al. 2024). The use of nanoscale zero-valent iron (nZVI) in conjunction with support materials such as zeolites, activated carbon, silica, or polymer membranes, as well as coating surfaces with biopolymers like calcium alginate, chitosan, and gum Arabic, facilitates the separation of nanomaterials from clean soil and prevents mobility. Nano-sized Al_2O_3 has been used for the extraction and concentration of trace metal ions (Ali et al. 2023b). Magnetic nanoparticles, in general, are attracting considerable interest for their high effectiveness, relatively low cost, minimal effect on the environment, improved stability, and maximum reusability, besides acting as catalysts for pollutant remediation. Advanced oxidation processes (AOPs), which employ different oxidants to degrade organic contaminants, use nanomaterials (Wu et al. 2023). Numerous types of nanomaterials are available: metal and metal oxide nanoparticles, carbon-based nanomaterials, polymer-based nanomaterials, and nanocomposites, as mentioned in (Figure 1).

3.1 Metal and Metal Oxide Nanoparticles

Metal oxide nanoparticles are regarded as one of the most promising nanomaterials in many fields due to their special physical and chemical characteristics, such as heat conductivity and heat transfer. Numerous metal oxide nanoparticles have been employed in the oil and gas sector as fillers or additives in drilling and completion fluids, among other applications in nearly every facet of hydrocarbon E&P (Janani et al. 2022). According to the transformation phase nature, the method of metal oxide fabrication is divided into two parts, namely liquid-solid and gas-solid. Nanomagnetite, a certain kind of metal oxide nanoparticle, is characterized by high paramagnetic behavior and is used in the remediation of arsenic-contaminated groundwater using nano iron hydroxide manipulation, α-FeO (OH) (Ramos-Guivar et al. 2021).

3.2 Carbon-Based Nanomaterials

Carbonaceous nanoparticles consist of carbonated compounds and comprise nanotubes, 2D graphene, and carbon quantum dots. Carbon nanoparticles have gained much attention due to their superior ability to carry and deliver a variety of medications to living cells. CNTs have high adsorption sites on the surface area, which may enter the cellular membrane non-invasively thanks to their intrinsic shape (Bakirhan et al. 2017). Among the various nanoparticles, a two-dimensional (one atom thick) carbon nanomaterial, graphene, possesses mechanical strength, a huge surface area, high electron conductivity, and the capacity to receive energy (Joshi et al. 2020). CNTs are one-dimensional carbon nanomaterials mainly used in electrochemical biosensors and fabrication. These possess thermal, mechanical, electrical, and structural properties. They form a broad domain

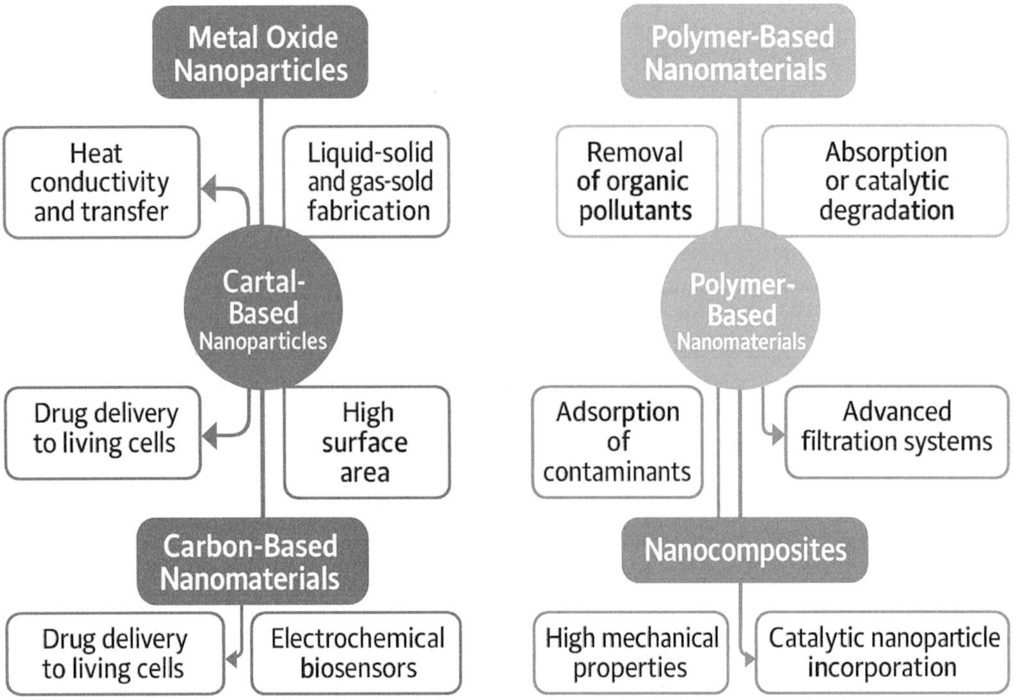

Figure 1: Application of smart nanomaterials in pollutant removal.

of nanotubes and nanofilaments like single-wall and multi-wall carbon nanoparticles that find their applications in several commercial technologies (Aslam et al. 2021).

3.3 Polymer-Based Nanomaterials

Polymeric nanoparticles have the potential to clean up organic pollutants, metals, and dyes from air and water because of the high surface area associated with their flexible nature (Ge et al. 2012). They enable targeted removal from the site of concern for remediation. According to the structure and/or composition, there are four divisions of polymeric nanomaterials: carbon polymer nanocomposites, graphene/polymer nanocomposites, and dendrimer-based nanocomposites (Neha et al. 2021). These nanoparticles can be specifically engineered for the targeted remediation of various environmental contaminants through the processes of absorption, degradation, or neutralization. Such nanomaterials are made of synthetic and natural polymers such as polystyrene, polyethylene glycol (PEG), polylactic acid (PLA), chitosan, and polyvinyl alcohol (PVA). PNPs render remedial applications: removal of metal ions, organic contaminants, and microplastics through adsorption, chelation, and catalytic degradation. Polymer nanofibers and nanogels in advanced filtration systems are capable of adsorbing VOCs and ultrafine particulate matter (PM2.5 and PM10) and greatly aid in improving indoor and outdoor air quality (Agboola et al. 2021).

3.4 Nanocomposites

Nanocomposites are composite materials made by combining nanoparticles with polymeric, ceramic, or metallic matrices that have succeeded as materials for pollution management, mainly thanks to their high mechanical, thermal, and adsorption properties (Moradihamedani et al. 2016). These advanced materials can cope with a spectrum of environmental contaminants, including heavy metals, organic pollutants, and gaseous emissions, owing to their enhanced catalytic effectiveness, large surface area, and tunable functionalization (Mohamad et al. 2020). The incorporation of metal oxides such as TiO_2, FeO_4, and ZnO, and graphite-based nanomaterials such as graphene oxide and CNTs, into biopolymeric matrices that include chitosan and cellulose, can efficiently absorb and degrade contaminants via adsorption, photocatalysis, and redox. Catalytic nanoparticles such as manganese oxides (MnO_2) or platinum can oxidize volatile organic compounds (VOCs) and selectively reduce NOx emissions in nanocomposite membranes and coatings, thus helping reduce air pollution (Abdullah et al. 2021). Biochar- and clay-based hybrids of nanocomposites encapsulate heavy metals and promote microbial decomposition of organic pollutants in soil remediation. Correspondingly, magnetic nanocomposites exhibit excellent sustainability and reusability due to their capacity to enable the fast separation and recovery of contaminants (Soo et al. 2020).

4. Synergistic Approaches: Microbial-Nanomaterial Combinations

An approach to sustainable reduction of environmental toxicity was explored based on the use of microbes in combination with sophisticated nanomaterials, as mentioned in (Figure 2). It is an alternative to chemical and physical forms of nanotechnology: microbial bioprocessing. Bacteria, archaea, fungi, yeast, molds, microalgae, and viruses are now being explored for the possibility of synthesizing bioactive nanostructures with vast applications in several industries (Upadhayay et al. 2023). The microorganisms found in industrial and mining wastewater are capable of extracting metals from released pollutants and transforming them into nanostructures. In a similar way, the process of metal chelation as nanoparticles illustrates the antagonistic relationships between microbes associated with the rhizosphere and pathogenic microbes (Verma et al. 2024). Marine microorganisms can create unique nanostructures through the efficient uptake of minerals from mineral-rich saline water by reduction/precipitation of toxic metal ions into non-toxic metallic nanoclusters to cause metal detoxification. To synthesize metallic nanoparticles by intracellular or extracellular pathways, marine bacteria, fungi, and microalgae have been found to be the best option (Sharma et al. 2023). Microbial production of nanoparticles proceeds via the transformation of positively charged metal ions, which either reside within the cytoplasm or are associated with the cell wall, into fine nuclei. Negatively charged biomolecules can impart the reduction necessary for the nanostructures to form (Behl et al. 2024b).

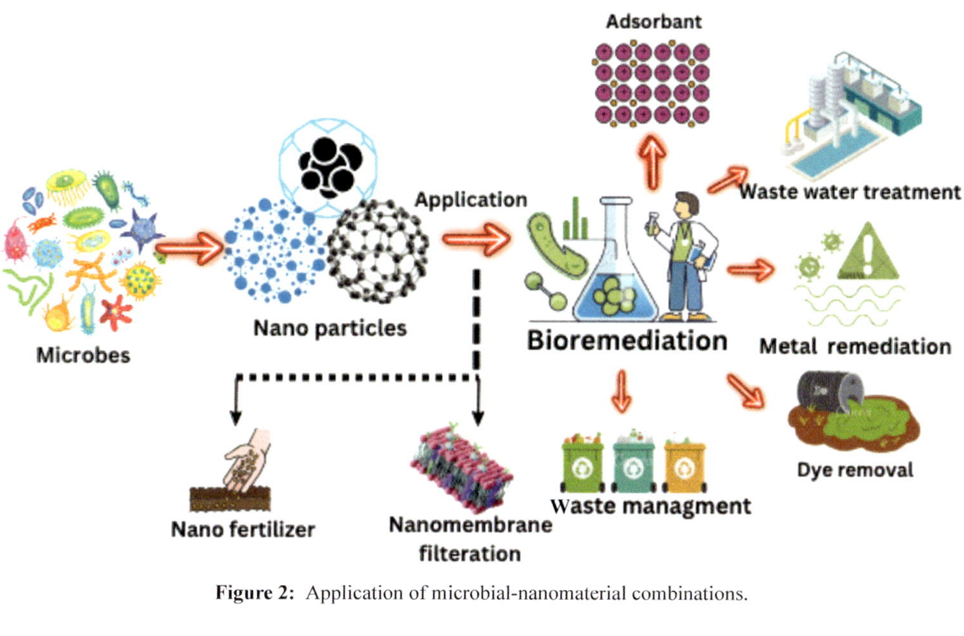

Figure 2: Application of microbial-nanomaterial combinations.

4.1 Biofilm Formation and Nanomaterial Enhancement

The coupling of microbial systems with intelligent nanomaterials represents a breakthrough strategy for enhancing biofilm formation while improving nanomaterial properties for diverse applications. Biofilms, the structured microbial communities embedded within a self-generated extracellular matrix, play an important role in a wide diversity of beneficial and detrimental biological processes (Lahiri et al. 2021). In environmental and industrial biotechnology, biofilms are applied for wastewater treatment, bioremediation, and bioenergy generation, whereas in healthcare and medicine, they are responsible for eliciting antibiotic resistance and associated infections (Liu et al. 2027). This partnership between microbes and nanomaterials can be exploited toward modulating biofilms—enhancing the establishment of biofilm formation in the case of desired applications, or inhibiting it to avoid biofouling and pathogenic colonization. Nanomaterials with potential applications include metal nanoparticles (Ag, Au, ZnO, TiO_2), carbon-based nanostructures (such as graphene oxide and carbon nanotubes), and polymeric nanocomposites, where interaction with microbial cells can alter adhesion, signaling, and EPS production (Mohanta et al. 2023). Smart nanomaterials that incorporate stimuli-responsive properties, such as pH-sensitive nanoparticles, redox-active particulates, and light-responsive nanoparticles, will be able to conceivably apply selective control over the biofilm by stimulating microbial metabolic pathways or disabling quorum-sensing mechanisms to disband biofilms. Moreover, nanomaterials may be functionalized and modified using microbial communities to enhance their stability, biocompatibility, and catalytic activity. Microbial-nanomaterial hybrids may enhance bioremediation and degradation by biogenic enzymes and biofilm-mediated adsorption of organic pollutants and heavy metals (Al-Wrafy et al. 2022). Microbial-nanomaterial composites enhance organic pollutant and heavy metal removal in bioremediation by enzymatic hydrolysis and biofilm-mediated adsorption. In the biomedical field, antimicrobial nanomaterial-engineered biofilms are employed as future-generation wound dressings, which prevent bacterial infection but induce tissue regeneration. Microbial consortia also help in the functionalization and modification of nanomaterials, which in turn enhances their biocompatibility, catalytic activity, stability, and overall performance (Malaekeh-Nikouei et al. 2020). By means of biofilm-mediated adsorption and enzymatic hydrolysis, microbial-nanomaterial hybrids help degrade contaminants in bioremediation, thereby enabling heavy metal detoxification and organic pollution degradation. Antimicrobial nanoparticle-designed biofilms have been developed for applications in biomedicine as smart wound dressings, preventing bacterial infection while providing tissue regeneration (Ouyang et al. 2017). Moreover, the biosynthesis of nanomaterials by bacteria—such as gold or silver nanoparticles—is a green and sustainable alternative for nanotechnology, whereby the utilization of harmful chemical synthesis is diminished. Nanomaterial enhancement—a synergistic interaction—is also included, whereby microbial interactions contribute to nanoparticle dispersion, stability, and reactivity (Li et al. 2020). For example, biofilms made with nanoparticles have enhanced mechanical properties, electrical conductivity, and biocompatibility for use in biosensors, drug delivery systems, and tissue engineering scaffolds. Microbial-nanomaterial interactions via biofilm-mediated nutrient cycling and pathogenic suppression enhance soil remediation, plant growth, and disease control. Based on enzymes, proteins, and extracellular vesicles, functionalization of nanomaterials with microbial components further increases their applications in targeted drug delivery, environmental sensing, and biocatalysis (Wang et al. 2019). Notwithstanding such exciting uses, problems in maximizing the stability, scalability, and environmental safety of microbial-nanomaterial hybrids remain unresolved. Nanomaterials' long-term ecological effects and possible harm to beneficial bacteria call for thorough risk analysis and legislative structures (Hou et al. 2020). Future studies should focus on the creation of bio-inspired nanomaterials that replicate natural microbial interactions, which can be precisely regulated over biofilm dynamics and nanomaterial functions (Zhang et al. 2017, Qain et al. 2017). Analytical techniques, including omics, microfluidics, and real-time imaging, will provide deeper insight into nanomaterial-microbe interactions that could delineate the rational design of next-generation biohybrids (Li et al. 2019a).

4.2 Electron Transfer Processes in Microbial-Nanomaterial Systems

Application of microbes with intelligent nanomaterials is very effective for enhancing the electron transfer processes in bioelectrochemical applications, especially in MFCs and environmental remediation. Microbial-nanomaterial systems utilize conductive nanomaterials such as graphene oxide, carbon nanotubes, and metal nanoparticles to improve extracellular electron transfer (EET) and, therefore, microbial metabolism and energy conversion efficiency (Dhanker et al. 2021). These nanomaterials get incorporated into biofilms to facilitate direct electron transfer via conductive networks, exploiting the connection between bacterial cells and electrodes; this further reduces electron transfer resistance and improves the output power of bioelectrochemical devices (Tauseef et al. 2023). Microbial-nanomaterial hybrids harness the bioremediation role by escalating pollutant degradation through improving redox reactions, with nanomaterials serving as electron shuttles between microbial cells and contaminants (Pereira et al. 2015). The synergistic interaction evidenced in bioelectronics, biocatalysis, and sustainable energy solutions opens the gate for electron transfer applications in microbial-smart nanomaterial combinations.

4.3 Enzymatic Activity Enhancement by Nanomaterials

The synergistic combination of high-performance nanomaterials with microbial systems can significantly enhance enzymatic activities and broaden their range of applications in bioremediation, biosensing technology, and industrial catalytic processes. Gold, silver, and carbon nanoparticles possess the capability of adsorbing onto microbial enzymes, thus improving their stability, reactivity, and selectivity for particular substrates. They can provide a suitable environment for enzyme immobilization, prevent enzymes from environmental stress, and facilitate electron transfer processes, thus elevating enzymatic efficiency (Razzaghi et al. 2022). The integration of microbes with nanomaterials enhances not only enzymatic activity but also contributes to the regeneration of catalytic systems in biotechnological applications—for instance, metal-organic frameworks and CNTs that promote activity by providing a controlled microenvironment, controlled interaction of substrate with an enzyme, and so on (Tan et al. 2023). This consortia establishment thus brings forth a promising approach toward the augmentation of enzyme processes in a diversity of industrial and environmental settings (Zhang et al. 2021).

4.4 Modulation of Microbial Stress Responses

This synergistic approach combines microbial systems with smart nanomaterials that impart augmented genetic adaptation and evolution under the influence of nanomaterials. Such approaches utilize the important properties of nanomaterials, including high surface area, reactivity, and interactive effects in microbial cells, to facilitate microbial evolution. Many involve nanomaterials as selective stressors that drive mutations and genetic adaptations for resistance or survival—evolutionary processes (Duran et al. 2019). These interactions can be the basis for developing novel biotechnological applications, including bioremediation, biosensing, and drug delivery, by harnessing the adaptive capabilities of microbes along with the functional properties of nanomaterials (Lee et al. 2022, Omran et al. 2020). Co-adaptation of microbes and nanomaterials may provide solutions to agricultural, industrial, and health issues without extensive intervention.

4.5 Genetic Adaptation and Evolution in the Presence of Nanomaterials

The synergistic approaches combine microbial systems with smart nanomaterials to endow the capacity for enhanced genetic adaptation and evolution under the influence of nanomaterials. Such approaches use the important features of nanomaterials, including high surface area, reactivity, and microbial cell interaction, to mediate microbial evolution. Many involve nanomaterials as selective stressors that drive mutations and genetic adaptations for resistance or survival—evolutionary processes (Duran et al. 2019). Such interactions may provide the basis for the engineering of novel biotechnological applications in the future, including bioremediation, biosensing, and drug delivery, through the harnessing of the adaptive properties of microbes in conjunction with the functional properties of nanomaterials (Lee et al. 2022, Omran et al. 2020). The concurrent adaptation of nanomaterials and microbes may solve problems in agriculture, industry, and medicine without the necessity of massive intervention.

5. Challenges and Future Opportunities

Microbial bioremediation using intelligent nanomaterials is an extremely promising field for the decontamination of pollution but is equally rich in future challenges and opportunities. Chief among these is the effect certain nanomaterials have on microbial communities, seeking to cross interdiction toward widespread use in bioremediation. Several unanswered questions have also arisen through considerations of nanomaterial stability and persistence when subjected to a range of environmental scenarios, regarding their long-term impacts on the livelihoods of organisms. Mechanistic elements of interactions between microbes and engineered nanomaterials are poorly understood and necessitate in-depth studies to optimize the synergistic paradigm. Another hurdle is integrating nanomaterials into microbial systems on a larger scale and ensuring their commercial viability; this is mainly because large-scale application would require a sustainable production platform, economically. Furthermore, the currently unresponsive regulatory provision for the use of nanotechnology in environmental remediation applications further stymies their large-scale acceptance. Further perspectives should develop around designing nanoparticles that are eco-friendly and enhance microbial activity without having side effects from applications. Other advanced techniques include the use of omics, synthetic biology, and machine learning to engineer microbes with better compatibility with nanomaterials for improved bioremediation efficiency. Similarly, field trials and pilot studies would be highly valuable in counter-verifying laboratory conclusions and judging direct applicability. Public awareness and policy development must also be increased to assure safe and responsible technologies for implementing nanomaterials and microbes in hybrid approaches. With such interdisciplinary collaborations coupled with technological advancements, microbial bioremediation along with smart nanomaterials will advance further in addressing the challenges of pollution in a sustainable manner.

5.1 Compatibility and Stability

The successful introduction of nanomaterials and microorganisms within certain realms of biotechnology, environmental cleanup, and medical interventions depends on the compatibility and stability of such materials. Compatibility is determined by specific physicochemical properties such as size, shape, surface charge, and chemical composition of nanomaterials, along with the metabolic adaptability of the microorganisms. Some nanomaterials support the growth and functionality of microorganisms, while others are harmful and therefore inhibit growth and functionality (Yamini et al. 2023). Stability in such interactions relies on conditions that support the maintenance of functional relationships between nanomaterials and microbes over a long time. The surface functionalization of nanomaterials, encapsulation of microbes, and optimization of the environmental conditions for continuous bio-nano interaction can prevent instability (Simonin et al. 2017). Stability and compatibility remain salient because, for such systems to find practical applications in wastewater treatment, bioremediation, and targeted drug delivery, such systems must remain stable under laboratory and natural conditions.

5.2 Environmental Toxicological Issues

The interaction between nanomaterials and microorganisms has multiple ecotoxicological and environmental implications. Although these nanomaterials may enhance microbial functions and adaptability, the interaction has a broad range of unforeseen ecological effects. Engineered nanomaterials can influence microbial community behaviors and metabolism, disrupt natural biogeochemical cycles, and be cytotoxic, causing oxidative stress and consequently leading to DNA damage in non-target organisms (Sahu et al. 2022). Also, their persistence in soil and aquatic ecosystems poses long-term risks, including the risky transfer of engineered traits to pathogenic microbes, bringing antimicrobial resistance (Komire et al. 2024). Nano-microbe hybrids that are discharged into the environment must undergo careful risk assessment to mitigate such negative impacts that threaten biodiversity, ecosystem stability, and even human health (Seena et al. 2023). Sustainable design and regulations of nano-bio interactions are urgently needed to realize the benefits of this approach while mitigating the threats that are likely to ensue.

5.3 Scalability and Cost-effectiveness

The process of joining nanomaterials and microorganisms can be seen as giving greater and more proportionate possibilities for scale-up and cheaper implementation in bioremediation, agriculture, and medicine. This is mainly due to the self-replicating nature of microorganisms, whose use also requires less continuous input of biological agents into the system, thus improving overall process efficiency, while nanomaterials improve microbial activity, stability, and performance. Hence, the development of nanotechnology opened doors to producing cost-effective and eco-friendly nanomaterials from agricultural and industrial waste, thus saving costs (Manganayi et al. 2024). Yet mass application entails several challenges, including mass production of biocomposites, maintaining biocompatibility of host organisms, as well as regulatory compliance issues. The advent of cheaper biosynthesized nanomaterials and stable microbial strains holds vast promise for green solutions for both environmental and industrial applications (Pete et al. 2021).

5.4 Regulatory and Policy Limitations

Utilization of nanomaterials and microorganisms in various applications, such as bioremediation, agriculture, and drug discovery, is confronted with several regulatory and policy limitations. One of the challenges is the dearth of well-defined guidelines on best practices for preparing, utilizing, or characterizing the risks posed to the environment by nanomaterials in microbial systems. While regulatory agencies like the U.S. Environmental Protection Agency (EPA) and the European Chemicals Agency (ECHA) advocate for risk assessment frameworks, there are still no coordinated approaches for identifying and measuring the interactions between microbes and engineered nanomaterials (Zhong et al. 2024). Further, the biosafety concern is raised by accepting the potential issues such as genetic modifications and horizontal gene transfer among the microbial communities exposed to nanomaterials in the present era (Xin et al. 2025). Additional policy gaps related to assessing long-term effects on the environment and health hinder their large-scale industrial applications. There is a need for more rigorous international legislation and policy harmonization for responsible and sustainable exploitation of microbial-nanomaterial technologies (Xing et al. 2023).

6. Conclusion

The integration of smart nanomaterials with bioremediation technology has created a new innovative area of activity for pollution control. This nanohybrid, by combining the bioremediation potential and active characteristics of nanomaterials, enables improved degradation/transformation and pollution immobilization. In this way, many pollutants are detoxified as nanomaterials enhance the efficiency of those processes through increased pollutant bioavailability, protection of microbial cells, and greater adsorption and degradation capability. While the increased toxicity, scalability, and expense of nanomaterials will continue to pose challenges to this design, there is room for viable solutions to emerge from the continuous development of green nanomaterials and optimization of the interactions between microbes and nanomaterials, which this synergy of their technologies holds.

The development of this combined technology opens the door to new pathways for sustainable approaches to deal with complex environmental issues like water purification, soil remediation, and industrial waste treatment. Such applications will set the trend for many innovations in the field and future use in pollution control aimed at minimizing pollution and the subsequent risks it poses to human and ecosystem health.

Abbreviations

POPs	Persistent Organic Pollutants
SO_2	Sulfur Dioxide
HRTOCs	Highly Reactive and Toxic Organic Compounds
PAHs	Polycyclic Aromatic Hydrocarbons
COPD	Chronic Obstructive Pulmonary Disease
GMOs	Genetically Modified Organisms
GEMs	Genetically Engineered Microorganisms
EPS	Extracellular Polymeric Substances
EET	Extracellular Electron Transfer
MFCs	Microbial Fuel Cells
AOPs	Advanced Oxidation Processes
nZVI	Nanoscale Zero-Valent Iron
AgNPs	Silver Nanoparticles
CNTs	Carbon Nanotubes
TiO_2	Titanium Dioxide
ZnO	Zinc Oxide
Al_2O_3	Aluminum Oxide
PCs	Phytochelatins
PEG	Polyethylene Glycol
PLA	Polylactic Acid
PVA	Polyvinyl Alcohol
VOCs	Volatile Organic Compounds
PM2.5/PM10	Particulate Matter (size $\leq$ 2.5/10 microns)
UiO-66-NH_2	University of Oslo Metal–Organic Framework (Amino Modified)

References

Abdullah, N., Yusof, N., Jye, L. W., Jaafar, J., Misdan, N., and Ismail, A. F. (2021). Removal of lead (II) by nanofiltration-ranged thin film nanocomposite membrane incorporated UiO-66-NH2: Comparative removal performance between hydraulic-driven and osmotic-driven membrane process. *Journal of the Taiwan Institute of Chemical Engineers*, 128: 354–369.

Adeleke, B. S., Akinola, S. A., Adedayo, A. A., Glick, B. R., and Babalola, O. O. (2022). Synergistic relationship of endophyte-nanomaterials to alleviate abiotic stress in plants. *Frontiers in Environmental Science*, 10: 1015897.

Agboola, O., Fayomi, O. S. I., Ayodeji, A., Ayeni, A. O., Alagbe, E. E., Sanni, S. E. et al. (2021). A review on polymer nanocomposites and their effective applications in membranes and adsorbents for water treatment and gas separation. *Membranes*, 11(2): 139.

Ahmed, S. N., Mishra, S. S., Sahu, J. K., Shroff, S., Naik, P. P., Baitharu, I. et al. (2022a). Application of nanomaterials in environmental pollution abatement and their impact on ecological sustainability: Recent status and future perspective. *Biotechnological Advances for Microbiology, Molecular Biology, and Nanotechnology*, 629–654.

Ahmed, T., Scholz, M., Al-Falahi, A., and Rajab, T. K. (2022b). Nanoremediation for soil and groundwater decontamination: Opportunities, challenges, and future prospects. *Environmental Technology & Innovation*, 26: 102342. https://doi. org/10.1016/j.eti.2022.102342.

Ai, C., Yan, Z., Chai, H., Gu, T., Wang, J., Chai, L. et al. (2019). Increased chalcopyrite bioleaching capabilities of extremely thermoacidophilic Metallosphaera sedula inocula by mixotrophic propagation. *Journal of Industrial Microbiology and Biotechnology*, 46(8): 1113–1127. https://doi.org/10.1007/s10295-019-02193-3.

Ajith, A., Kumar, A., and Sharma, P. (2021a). Photocatalytic degradation of organic pollutants using titanium dioxide nanoparticles: A review. *Journal of Environmental Chemical Engineering*, 9(5): 106182. https://doi.org/10.1016/j.jece.2021.106182.

Ajith, M. P., Aswathi, M., Priyadarshini, E., and Rajamani, P. (2021b). Recent innovations of nanotechnology in water treatment: A comprehensive review. *Bioresource Technology*, 342: 126000.

Ali, A., Khan, Z., and Khan, S. (2023a). Application of nano-sized aluminum oxide for the separation and concentration of trace metal ions. Environmental Science and Pollution Research, 30(8): 9921–9935. https://doi.org/10.1007/s11356-023-24563-8.

Ali, Q., Zia, M. A., Kamran, M., Shabaan, M., Zulfiqar, U., Ahmad, M. et al. (2023b). Nanoremediation for heavy metal contamination: A review. *Hybrid Advances*, 4: 100091.

Al-Wrafy, F. A., Al-Gheethi, A. A., Ponnusamy, S. K., Noman, E. A., and Fattah, S. A. (2022). Nanoparticles approach to eradicate bacterial biofilm-related infections: A critical review. *Chemosphere*, 288: 132603.

Årstøl, E., and Hohmann-Marriott, M. F. (2019). Cyanobacterial siderophores—physiology, structure, biosynthesis, and applications. *Marine Drugs*, 17(5): 281.

Aslam, M. M. A., Kuo, H. W., Den, W., Usman, M., Sultan, M., and Ashraf, H. (2021). Functionalized carbon nanotubes (CNTs) for water and wastewater treatment: preparation to application. *Sustainability*, 13(10): 5717.

Ayangbenro, A. S., and Babalola, O. O. (2017). A new strategy for heavy metal polluted environments: a review of microbial biosorbents. *International Journal of Environmental Research and Public Health*, 14(1): 94.

Ayilara, M. S., and Babalola, O. O. (2023). Bioremediation of environmental wastes: the role of microorganisms. *Frontiers in Agronomy*, 5: 1183691.

Babalola O. O., Aremu B. R., and Ayangbenro A. S. (2019). Draft genome sequence of heavy metal-resistant bacillus cereus NWUAB01. *Microbiol. Resour. Announc*., 8 (7): e01706–e01718. doi: 10.1128/MRA.01706-18.

Bakirhan, N. K., Uslu, B., and Ozkan, S. A. (2017). Sensitive and selective assay of antimicrobials on nanostructured materials by electrochemical techniques. pp. 55–83. *In: Nanostructures for Antimicrobial Therapy*. Elsevier.

Behl, K., Jaiswal, P., and Pabbi, S. (2024a). Recent advances in microbial and nano-formulations for effective delivery and agriculture sustainability. *Biocatalysis and Agricultural Biotechnology*, 103180.

Behl, N., Gupta, R., and Sharma, A. (2024b). Microbial synthesis of nanostructures and their environmental applications. *Microbial Biotechnology*, 11(1): 12–25. https://doi.org/10.1016/j.mib.2024.01.008.

Bhandari, G., Gupta, S., Chaudhary, P., Chaudhary, S., and Gangola, S. (2023). Bioleaching: a sustainable resource recovery strategy for urban mining of e-waste. pp. 157–175. *In: Microbial Technology For Sustainable E-Waste Management*. Cham: Springer International Publishing.

Deng, Y., Liu, X. D., Liu, H. W., Jiang, H. D., Xu, L. F., Xiao, Y. H. et al. (2017). Bioleaching of cadmium from contaminated paddy fields by consortium of autotrophic and indigenous cadmium-tolerant bacteria. *Solid State Phenomena*, 262: 617–621.

Deshpande, B. D., Agrawal, P. S., Yenkie, M. K. N., and Dhoble, S. J. (2020). Prospective of nanotechnology in degradation of waste water: A new challenges. *Nano-Structures & Nano-Objects*, 22: 100442.

Dhanker, R., Hussain, T., Tyagi, P., Singh, K. J., and Kamble, S. S. (2021). The emerging trend of bio-engineering approaches for microbial nanomaterial synthesis and its applications. *Frontiers in Microbiology*, 12: 638003.

Dhawan, A., Singh, S., Kumar, A., and Shanker, R. (Eds.). (2018). Nanobiotechnology: *Human Health and the Environment*. CRC Press.

Elahi, A., Bukhari, D. A., Shamim, S., and Rehman, A. (2021). Plastics degradation by microbes: A sustainable approach. *Journal of King Saud University-Science*, 33(6): 101538. https://doi.org/10.1016/j.jksus.2021.101538.

Epelle, E. I., Okoye, P. U., Roddy, S., Gunes, B., and Okolie, J. A. (2022). Advances in the applications of nanomaterials for wastewater treatment. *Environments*, 9(11): 141.

Fashola, M. O., Ngole-Jeme, V. M., and Babalola, O. O. (2020a). Heavy metal immobilization potential of indigenous bacteria isolated from gold mine tailings. *International Journal of Environmental Research*, 14: 71–86.

Fashola, M. O., Ngole-Jeme, V. M., and Babalola, O. O. (2020b). Heavy metal pollution from gold mines: Environmental effects and bacterial strategies for resistance. *International Journal of Environmental Research and Public Health*, 17(3): 561. https://doi.org/10.3390/ijerph17030561.

Fernández-Triana, I., Rubilar, O., Parada, J., Fincheira, P., Benavides-Mendoza, A., Durán, P. et al. (2024). Metal nanoparticles and pesticides under global climate change: assessing the combined effects of multiple abiotic stressors on soil microbial ecosystems. *Science of The Total Environment*, 942: 173494.

Gangola, S., Sharma, A., Bhatt, P., Khati, P., and Chaudhary, P. (2018). Presence of esterase and laccase in Bacillus subtilis facilitates biodegradation and detoxification of cypermethrin. *Scientific Reports*, 8(1): 12755. DOI https://doi.org/10.1038/s41598-018-31082-5.

Ge, F., Ye, H., Li, M. M., and Zhao, B. X. (2012). Efficient removal of cationic dyes from aqueous solution by polymer-modified magnetic nanoparticles. *Chemical Engineering Journal*, 198: 11–17.

Grabi, H., Derridj, F., Lemlikchi, W., and Guénin, E. (2021). Studies of the potential of a native natural biosorbent for the elimination of an anionic textile dye Cibacron Blue in aqueous solution. *Scientific Reports*, 11(1): 9705.

Helbling, D. E. (2015). Bioremediation of pesticide-contaminated water resources: the challenge of low concentrations. *Current Opinion in Biotechnology*, 33: 142–148.

Hou, J., Li, T., Miao, L., You, G., Xu, Y., and Liu, S. (2020). Dynamic responses of community structure and microbial functions of periphytic biofilms during chronic exposure to TiO 2 NPs. *Environmental Science: Nano*, 7(2): 665–675.

Jain, K., Patel, A. S., Pardhi, V. P., and Flora, S. J. S. 2021. Nanotechnology in wastewater management: a new paradigm towards wastewater treatment. *Molecules*, 26(6): 1797.

Jaiswar, S., Kazi, M. A., and Mehta, S. (2015). Bioaccumulation of heavy metals by freshwater algal species of Bhavnagar, Gujarat, India. *Journal of Environmental Biology*, 36(6): 1361.

Janani, R., Gurunathan, B., Sivakumar, K., Varjani, S., Ngo, H. H., and Gnansounou, E. (2022). Advancements in heavy metals removal from effluents employing nano-adsorbents: way towards cleaner production. *Environmental Research*, 203: 111815.

Joshi, A., and Kim, K. H. (2020). Recent advances in nanomaterial-based electrochemical detection of antibiotics: Challenges and future perspectives. *Biosensors and Bioelectronics*, 153: 112046.

Khan, S., Naushad, M., Al-Gheethi, A., and Iqbal, J. (2021). Engineered nanoparticles for removal of pollutants from wastewater: Current status and future prospects of nanotechnology for remediation strategies. *Journal of Environmental Chemical Engineering*, 9(5): 106160.

Komire, K., Tiwari, M., Gupta, P. C., Bahorun, T., and Sharma, N. (2024). Plant-Microbe Synergies. NANOBIOSTIMULANTS: *Emerging Strategies for Agricultural Sustainability*, 197.

Kulshreshtha, S. 2013. Genetically engineered microorganisms: a problem solving approach for bioremediation. *J. Bioremed. Biodegr.*, 4(4): 1–2.

Lahiri, D., Nag, M., Sheikh, H. I., Sarkar, T., Edinur, H. A., Pati, S. et al. (2021). Microbiologically-synthesized nanoparticles and their role in silencing the biofilm signaling cascade. *Frontiers in Microbiology*, 12: 636588.

Lee, Y. C., Moon, J. Y., Lee, Y. C., and Moon, J. Y. (2020). Introduction to nanotechnology and bionanotechnology. *Introduction to Bionanotechnology*, 1–14.

Li, K., Qian, J., Wang, P., Wang, C., Fan, X., Lu, B. et al. (2019a). Toxicity of three crystalline TiO2 nanoparticles in activated sludge: bacterial cell death modes differentially weaken sludge dewaterability. *Environmental Science & Technology*, 53(8): 4542–4555.

Li, K., Qian, J., Wang, P., Wang, C., Lu, B., Jin, W. et al. (2020). Responses of freshwater biofilm formation processes (from colonization to maturity) to anatase and rutile TiO2 nanoparticles: effects of nanoparticles aging and transformation. *Water Research*, 182: 115953.

Li, S. W., Liu, M. Y., and Yang, R. Q. (2019b). Comparative genome characterization of a petroleum-degrading *Bacillus subtilis* Strain DM2. *International Journal of Genomics*, 2019(1): 7410823.

Lin, H., Zhou, M., Li, B., and Dong, Y. (2023). Mechanisms, application advances and future perspectives of microbial-induced heavy metal precipitation: a review. *International Biodeterioration & Biodegradation*, 178: 105544.

Liu, C., Guo, J., Yan, X., Tang, Y., Mazumder, A., Wu, S. et al. (2017). Antimicrobial nanomaterials against biofilms: an alternative strategy. *Environmental Reviews*, 25(2): 225–244.

Liu, Y., Wang, X., Zhang, H., Li, J., and Chen, Y. (2024). Engineered Bacillus subtilis Biofilm@Biochar living materials for in-situ sensing and bioremediation of heavy metal ions pollution. *Journal of Hazardous Materials*, 465: 133119. https://doi.org/10.1016/j.jhazmat.2023.133119.

Lozano, R., Ceulemans, K., Alonso-Almeida, M., Huisingh, D., Lozano, F. J., Waas, T. et al. (2015). A review of commitment and implementation of sustainable development in higher education: results from a worldwide survey. *Journal of Cleaner Production*, 108: 1–18.

Maglione, G., Zinno, P., Tropea, A., Mussagy, C. U., Dufossé, L., Giuffrida, D. et al. (2024). Microbes' role in environmental pollution and remediation: a bioeconomy focus approach. *AIMS Microbiology*, 10(3): 723.

Maity, J. P., Chen, G. S., Huang, Y. H., Sun, A. C., and Chen, C. Y. (2019). Ecofriendly heavy metal stabilization: microbial induced mineral precipitation (MIMP) and biomineralization for heavy metals within the contaminated soil by indigenous bacteria. *Geomicrobiology Journal*, 36(7): 612–623.

Malaekeh-Nikouei, B., Bazzaz, B. S. F., Mirhadi, E., Tajani, A. S., and Khameneh, B. (2020). The role of nanotechnology in combating biofilm-based antibiotic resistance. *Journal of Drug Delivery Science and Technology*, 60: 101880.

Mandeep, and Shukla, P. (2020). Microbial nanotechnology for bioremediation of industrial wastewater. *Frontiers in Microbiology*, 11: 590631.

Manganyi, M. C., Dikobe, T. B., and Maseme, M. R. (2024). Exploring the Potential of Endophytic Microorganisms and Nanoparticles for Enhanced Water Remediation. *Molecules*, 29(12): 2858.

Medfu Tarekegn M., Zewdu Salilih F., Ishetu A. I. (2020). Microbes used as a tool for bioremediation of heavy metal from the environment. *Cogent Food Agric.*, 6 (1): 1783174. doi: 10.1080/23311932.2020.1783174.

Mir-Tutusaus, J. A., Baccar, R., Caminal, G., and Sarrà, M. (2018). Can white-rot fungi be a real wastewater treatment alternative for organic micropollutants removal? A review. *Water Research*, 138: 137–151.

Mishra, D., and Srivastava, M. (2020). Nano-Materials as Photocatalysts for Degradation of Environmental Pollutants.

Mishra, S., Kumar, R., and Singh, S. (2020). Green synthesis of zero-valent silver nanoparticles using Ficus benjamina leaf extract and their application for cadmium removal. *Journal of Environmental Management*, 260: 110132. https://doi.org/10.1016/j.jenvman.2020.110132.

Mohamad, A. N., Aziz, F., Yusof, N., Jaafar, J., Wan Salleh, W. N., Jye, L. W. et al. (2023). Recent advances in heavy metal removal by thin film nanocomposite membrane. *Water Supply*, 23(9): 3635–3659.

Mohamed, B. A., and Samer, M. (2023). Biobutanol production from agricultural wastes. pp. 181–200. *In: Valorization of Wastes for Sustainable Development*. Elsevier. https://doi.org/10.1016/B978-0-323-95417-4.00008-1.

Mohamad, M., Abd Razak, J., Mohamad, N., Ahmad, S. H., Junid, R., and Puspitasari, P. (2020, October). A short review on polyurethane-based nanocomposites for various applications. In *IOP Conference Series: Materials Science and Engineering* (Vol. 957, No. 1, p. 012029). IOP Publishing.

Mohanta, Y. K., Chakrabartty, I., Mishra, A. K., Chopra, H., Mahanta, S., Avula, S. K. et al. (2023). Nanotechnology in combating biofilm: A smart and promising therapeutic strategy. *Frontiers in Microbiology*, 13: 1028086.

Mondal, P., Nandan, A., Ajithkumar, S., Siddiqui, N. A., Raja, S., Kola, A. K. et al. (2023). Sustainable application of nanoparticles in wastewater treatment: Fate, current trend & paradigm shift. *Environmental Research*, 232: 116071.

Moradihamedani, P., Kalantari, K., Abdullah, A. H., and Morad, N. A. (2016). High efficient removal of lead (II) and nickel (II) from aqueous solution by novel polysulfone/Fe3O4–talc nanocomposite mixed matrix membrane. *Desalination and Water Treatment*, 57(59): 28900–28909.

Nakum, J., and Bhattacharya, D. (2022). Various green nanomaterials used for wastewater and soil treatment: a mini-review. *Frontiers in Environmental Science*, 9: 724814.

Neha, R., Adithya, S., Jayaraman, R. S., Gopinath, K. P., Pandimadevi, M., Praburaman, L. et al. (2021). Nano-adsorbents an effective candidate for removal of toxic pharmaceutical compounds from aqueous environment: A critical review on emerging trends. *Chemosphere*, 272: 129852.

Nguyen, T. H., Won, S., Ha, M. G., Nguyen, D. D., and Kang, H. Y. (2021). Bioleaching for environmental remediation of toxic metals and metalloids: A review on soils, sediments, and mine tailings. *Chemosphere*, 282: 131108. doi: 10.1016/j.chemosphere.2021.131108.

Ningthoujam, R., Singh, Y. D., Babu, P. J., Tirkey, A., Pradhan, S., and Sarma, M. (2022). Nanocatalyst in remediating environmental pollutants. *Chemical Physics Impact*, 4: 100064.

Nivetha, N., Srivarshine, B., Sowmya, B., Rajendiran, M., Saravanan, P., Rajeshkannan, R. et al. (2023). A comprehensive review on bio-stimulation and bio-enhancement towards remediation of heavy metals degeneration. *Chemosphere*, 312: 137099.

Ojuederie, O. B., Chukwuneme, C. F., Samuel, O., Olanrewaju, M. A., Adegboyega, T. T., and Babalola, O. O. (2021). Contribution of microbial inoculants in sustainable maintenance of human health, including test methods and evaluation of safety of microbial pesticide microorganisms. *Biopesticides: Botanicals and Microorganisms for Improving Agriculture and Human Health*, 207–240.

Omran, B. A., and Omran, B. A. (2020). Fundamentals of nanotechnology and nanobiotechnology. *Nanobiotechnology: A Multidisciplinary Field of Science*, 1–36.

Ouyang, K., Yu, X. Y., Zhu, Y., Gao, C., Huang, Q., and Cai, P. (2017). Effects of humic acid on the interactions between zinc oxide nanoparticles and bacterial biofilms. *Environmental Pollution*, 231: 1104–1111.

Pant, G., Garlapati, D., Agrawal, U., Prasuna, R. G., Mathimani, T., and Pugazhendhi, A. (2021). Biological approaches practised using genetically engineered microbes for a sustainable environment: a review. *Journal of Hazardous Materials*, 405: 124631.

Peña-Montenegro, T. D., Lozano, L., and Dussán, J. (2015). Genome sequence and description of the mosquitocidal and heavy metal tolerant strain Lysinibacillus sphaericus CBAM5. *Standards in Genomic Sciences*, 10: 1–10.

Pereira, L., Mehboob, F., Stams, A. J., Mota, M. M., Rijnaarts, H. H., and Alves, M. M. (2015). Metallic nanoparticles: microbial synthesis and unique properties for biotechnological applications, bioavailability and biotransformation. *Critical Reviews in Biotechnology*, 35(1): 114–128.

Pete, A. J., Bharti, B., and Benton, M. G. (2021). Nano-enhanced bioremediation for oil spills: a review. *ACS ES&T Engineering*, 1(6): 928–946.

Pratush, A., Kumar, A., and Hu, Z. (2018). Adverse effect of heavy metals (As, Pb, Hg, and Cr) on health and their bioremediation strategies: a review. *International Microbiology*, 21: 97–106.

Qian, J., Li, K., Wang, P., Wang, C., Shen, M., Liu, J. et al. (2017). Toxic effects of three crystalline phases of TiO2 nanoparticles on extracellular polymeric substances in freshwater biofilms. *Bioresource Technology*, 241: 276–283.

Rafeeq, H., Afsheen, N., Rafique, S., Arshad, A., Intisar, M., Hussain, A. et al. (2023). Genetically engineered microorganisms for environmental remediation. *Chemosphere*, 310: 136751.

Ramos-Guivar, J. A., Flores-Cano, D. A., and Caetano Passamani, E. (2021). Differentiating nanomaghemite and nanomagnetite and discussing their importance in arsenic and lead removal from contaminated effluents: A critical review. *Nanomaterials*, 11(9): 2310.

Razzaghi, M., Homaei, A., Vianello, F., Azad, T., Sharma, T., Nadda, A. K. et al. (2022). Industrial applications of immobilized nano-biocatalysts. *Bioprocess and Biosystems Engineering*, 1–20.

Roy, A., Sharma, A., Yadav, S., Jule, L. T., and Krishnaraj, R. (2021). Nanomaterials for remediation of environmental pollutants. *Bioinorganic Chemistry and Applications*, 2021(1): 1764647.

Sahay, R. (2021). Synthetic applications of laccase and its catalytic potentials. *International Journal of Advanced Engineering Research and Science*, 8(6). https://dx.doi.org/10.22161/ijaers.86.12

Sahu, J., Vaishnav, A., and Singh, H. B. (Eds.). (2022). Plant-microbe Interactions: Harnessing Next-generation Molecular Technologies for Sustainable Agriculture. *CRC Press.*

Sarmah, P., and Rout, J. (2020). Role of algae and cyanobacteria in bioremediation: prospects in polyethylene biodegradation. pp. 333–349. *In:* Advances in cyanobacterial biology. *Academic Press.*

Saxena, G., Kishor, R., and Bharagava, R. N. (2020). Application of microbial enzymes in degradation and detoxification of organic and inorganic pollutants. *Bioremediation of Industrial Waste for Environmental Safety: Volume I: Industrial Waste and Its Management*, 41–51. https://doi.org/10.1007/978-981-13-1891-7_3.

Sayed K., Baloo L., and Sharma N. K. (2021). Bioremediation of total petroleum hydrocarbons (TPH) by bioaugmentation and biostimulation in water with floating oil spill containment booms as bioreactor basin. *Int. J. Environ. Res. Public Health,* 18(5): 2226. doi: 10.3390/ijerph18052226.

Seena, S., Kumar, S., and Rai, A. (Eds.). (2023). Nanoparticles and plant-microbe interactions: An environmental perspective. *Elsevier Science & Technology.*

Sharma, B., Tiwari, S., Kumawat, K. C., and Cardinale, M. (2023). Nano-biofertilizers as bio-emerging strategies for sustainable agriculture development: Potentiality and their limitations. *Science of The Total Environment*, 860: 160476.

Sharma, P., Dutta, D., Udayan, A., Nadda, A. K., Lam, S. S., and Kumar, S. (2022). Role of microbes in bioaccumulation of heavy metals in municipal solid waste: Impacts on plant and human being. *Environmental Pollution*, 305: 119248.

Simonin, M., and Richaume, A. (2015). Impact of engineered nanoparticles on the activity, abundance, and diversity of soil microbial communities: a review. *Environmental Science and Pollution Research*, 22: 13710–13723.

Soo, K. W., Wong, K. C., Goh, P. S., Ismail, A. F., and Othman, N. (2020). Efficient heavy metal removal by thin film nanocomposite forward osmosis membrane modified with geometrically different bimetallic oxide. *Journal of Water Process Engineering*, 38: 101591.

Tan, W. Y., Gopinath, S. C., Anbu, P., Yaakub, A. R. W., Subramaniam, S., Chen, Y. et al. (2023). Bio-enzyme hybrid with nanomaterials: A potential cargo as sustainable biocatalyst. *Sustainability*, 15(9): 7511.

Tauseef, A., Hisam, F., Hussain, T., Caruso, A., Hussain, K., Châtel, A. et al. (2023). Nanomicrobiology: emerging trends in microbial synthesis of nanomaterials and their applications. *Journal of Cluster Science*, 34(2): 639–664.

Tian, Q., Dou, X., Huang, L., Wang, L., Meng, D., Zhai, L. et al. (2020). Characterization of a robust cold-adapted and thermostable laccase from Pycnoporus sp. SYBC-L10 with a strong ability for the degradation of tetracycline and oxytetracycline by laccase-mediated oxidation. *Journal of Hazardous Materials*, 382: 121084. https://doi.org/10.1016/j.jhazmat.2019.121084.

Tyagi, B., and Kumar, N. (2021). Bioremediation: Principles and applications in environmental management. In *Bioremediation for environmental sustainability* (pp. 3–28). Elsevier.

Unuofin, J. O., Okoh, A. I., and Nwodo, U. U. (2019). Aptitude of oxidative enzymes for treatment of wastewater pollutants: a laccase perspective. *Molecules*, 24(11): 2064.

Upadhayay, V. K., Chitara, M. K., Mishra, D., Jha, M. N., Jaiswal, A., Kumari, G. et al. (2023). Synergistic impact of nanomaterials and plant probiotics in agriculture: A tale of two-way strategy for long-term Verma, R., Kaushik, A., and Pandey, S. (2024). Microbial-metal chelation interactions for the detoxification of industrial pollutants. *Environmental Science and Technology*, 58(2): 330–348. https://doi.org/10.1021/acs.est.3c025.

Wang, P., Li, K., Qian, J., Wang, C., Lu, B., Tian, X. et al. (2019). Differential toxicity of anatase and rutile TiO 2 nanoparticles to the antioxidant enzyme system and metabolic activities of freshwater biofilms based on microelectrodes and fluorescence in situ hybridization. *Environmental Science: Nano*, 6(8): 2626–2640.

Wang, S., Fang, L., Dapaah, M. F., Niu, Q., and Cheng, L. (2023). Bio-remediation of heavy metal-contaminated soil by microbial-induced carbonate precipitation (MICP)—a critical review. *Sustainability*, 15(9): 7622.

Wu, B., Xiu, J., Yu, L., Huang, L., Yi, L., and Ma, Y. (2023). Degradation of crude oil in a co-culture system of Bacillus subtilis and Pseudomonas aeruginosa. *Frontiers in Microbiology*, 14: 1132831. https://doi.org/10.3389/fmicb.2023.1132831.

Xiao, F., Cheng, J., Cao, W., Yang, C., Chen, J., and Luo, Z. (2019). Removal of heavy metals from aqueous solution using chitosan-combined magnetic biochars. *Journal of Colloid and Interface Science*, 540: 579–584.

Xie, Q., Liu, N., Lin, D., Qu, R., Zhou, Q., and Ge, F. (2020). The complexation with proteins in extracellular polymeric substances alleviates the toxicity of Cd (II) to Chlorella vulgaris. *Environmental Pollution*, 263: 114102.

Xin, Y., Huang, C., Zeng, J., Zhang, W., Zhou, Y., Xu, Y. et al. (2025). Biochemical mechanism underlying the synthesis of PbS nanoparticle and its in-situ photo effect on Shinella zoogloeoides PQ7. *Journal of Hazardous Materials*, 482: 136652.

Xing, S. F., Tian, H. F., Yan, Z., Song, C., and Wang, S. G. (2023). Stability and biomineralization of cadmium sulfide nanoparticles biosynthesized by the bacterium Rhodopseudomonas palustris under light. *Journal of Hazardous Materials*, 458: 131937.

Yamini, V., Shanmugam, V., Rameshpathy, M., Venkatraman, G., Ramanathan, G., Garalleh, H. A. et al. (2023). Environmental effects and interaction of nanoparticles on beneficial soil and aquatic microorganisms. *Environmental Research*, 236: 116776.

Zacharia J. T. (2019). Degradation pathways of persistent organic pollutants (POPs) in the environment. *Persistent Organic Pollutants*, 17–30. doi: 10.5772/intechopen.79645

Zhang, C., Lohwacharin, J., and Takizawa, S. (2017). Properties of residual titanium dioxide nanoparticles after extended periods of mixing and settling in synthetic and natural waters. *Scientific Reports*, 7(1): 9943.

Zhang, L., Zhang, M., Yang, Q., Wang, L., and Yang, S. (2023). Biodegradation of crude oil by a newly isolated *Pseudomonas aeruginosa* WJ-1 and *Bacillus subtilis* SL co-culture system. Frontiers in Microbiology, 14: 1132831. https://doi.org/10.3389/fmicb.2023.1132831.

Zhang, X., Li, G., Chen, G., Wu, D., Wu, Y., and James, T. D. (2021). Enzyme mimics for engineered biomimetic cascade nanoreactors: mechanism, applications, and prospects. *Advanced Functional Materials*, 31(50): 2106139.

Zheng, X., Lin, H., Du, D., Li, G., Alam, O., Cheng, Z. et al. (2024). Remediation of heavy metals polluted soil environment: A critical review on biological approaches. *Ecotoxicology and Environmental Safety*, 284: 116883.

Zhong, C., Ren, Y., Guo, Y. Y., Lu, A., and Liu, J. (2024). Photoelectron-Promoted Sulfate Reduction for Heavy Metal Removal without Organic Carbon Addition. *Environmental Science & Technology*, 58(49): 21680–21691.

CHAPTER-12

Smart Nanomaterials for Environmental Monitoring and Remediation

Pareshkumar G. Moradeeya,[1] *Ianca Oliveira Borges,*[2] *Gustavo Henrique Denzin Tonoli*[2] and *Geoffrey S. Simate*[1,*]

1. Introduction

Consistently worsening environmental quality is a direct outcome of the ever-increasing global population, rapid urbanisation, and industrialisation (Anil Kumar et al. 2024). The Industrial Revolution was an era of technological progress that resulted in the exploitation of natural resources and significantly affected the environment (Usman and Balsalobre-Lorente 2022). The continuously growing human population, agricultural advancement, and urbanisation have maximised the utilisation of natural resources, resulting in their overexploitation and consequent environmental deterioration (Moradeeya et al. 2022a). Currently, pollution is a primary environmental hazard that is closely linked to the rise in healthcare expenditures, the incidence of early mortality, and the decline in global economic productivity (Escolà-Gascón et al. 2022, Remoundou and Koundouri 2009). Numerous studies indicate that developing countries are afflicted by various pollutants in the air, including carbon monoxide (CO), chlorofluorocarbons, volatile organic compounds (VOCs), hydrocarbons (HxCy), and nitrogen oxides (NOx) (Zhang et al. 2014, Ren et al. 2017, Bhaskaran et al. 2023, Peláez et al. 2020). Water and soil are compromised due to sewage discharge, industrial effluents, indiscriminate pesticide and fertiliser application, heavy metals, chlorinated compounds, and oil spills, which are significant contributors to environmental degradation (Adil et al. 2023, Dheri et al. 2007, Meena et al. 2024, Islam et al. 2018). Figure 1 illustrates anthropogenic activities and their impact on key components of the environment.

Pesticide pollution denotes the environmental contamination by pesticide chemicals, which can adversely affect non-target organisms, while also polluting soil, air, and water bodies via runoff from treated fields. This has profound consequences for ecosystems and could endanger human wellness as a result of contaminated food (Kaur et al. 2024, Moradeeya et al. 2022a). Pharmaceutical

[1] School of Chemical and Metallurgical Engineering, University of the Witwatersrand, Johannesburg, South Africa. https://orcid.org/0000-0003-2780-2532

[2] Department of Forest Sciences, Federal University of Lavras, P.O. Box 3037, Lavras, 37200-900, MG, Brazil. https://orcid.org/0000-0003-4370-4705, https://orcid.org/0000-0002-6502-8974

* Corresponding author: Geoffrey.Simate@wits.ac.za, https://orcid.org/0000-0001-9447-2663

DOI: 10.1201/9781003558804-12

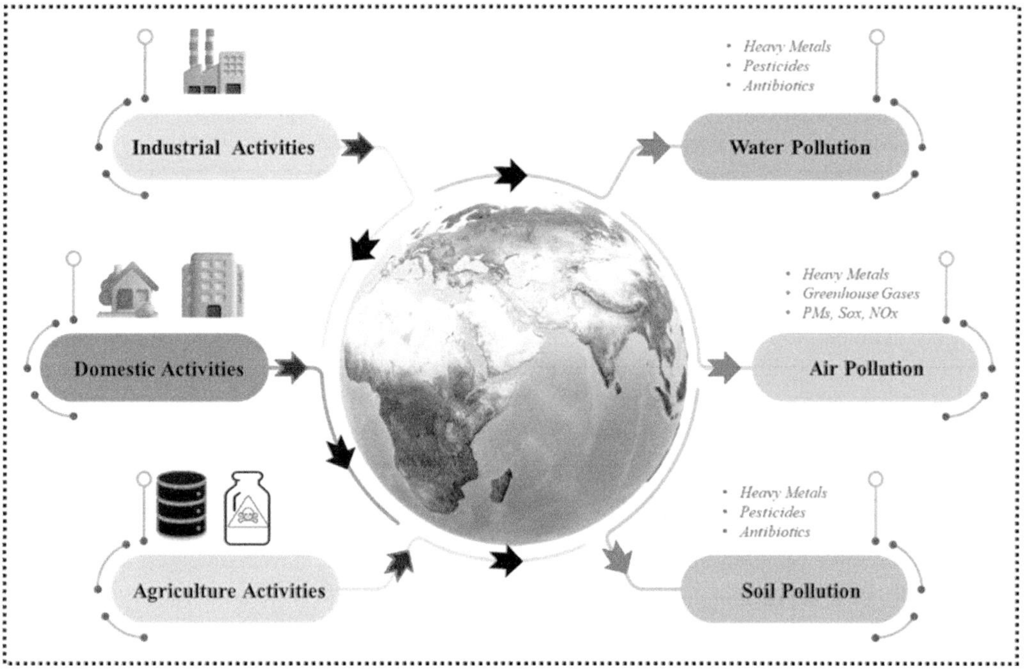

Figure 1: Anthropogenic activities and their impacts on key components of the environment.

pollution refers to the release of drugs into the environment, primarily through wastewater runoff. This phenomenon can profoundly affect ecosystems by disrupting wildlife reproduction, fostering antibiotic resistance, and modifying aquatic organism behaviours, even at minimal concentrations. Such pollution may arise from improper medication disposal, excretion by humans and animals, and leaks from manufacturing facilities (Kayode-Afolayan et al. 2022, Ngqwala and Muchesa 2020, Kostich et al. 2008). Heavy metal pollution denotes the environmental contamination by toxic metals predominantly resulting from anthropogenic activities including industrial processes, mining, agriculture, and inadequate waste disposal. This pollution can severely impact ecosystems by accumulating in soil, water, and air, adversely affecting flora and fauna through bioaccumulation, thereby causing detrimental health effects and disrupting the ecological equilibrium (Dagdag et al. 2023, Dheri et al. 2007, Adil et al. 2023).

Given that contaminants predominantly exist as mixes, there is a necessity for systems adept in monitoring, identifying, and remediating minimal concentrations of contaminants in essential components of the ecosystem. Nanoscience facilitates the creation and manipulation of materials at the molecular scale, enabling fabrication of unique functions capable of identifying particular pollutants within a mixture (Khan et al. 2019b, Baig et al. 2021, Hairom et al. 2021). Monitoring devices for diverse environmental parameters can mitigate various risks to human safety (Hairom et al. 2021, Thakur and Kumar 2022, Saleem et al. 2022). The new characteristics of nanoparticles will facilitate the development of extremely miniaturised and precise monitoring devices for detecting pollutants (Zhang and Fang 2010). Nevertheless, a knowledge gap persists concerning the application of nanomaterials for green development. This chapter delineates the potential of synthesised smart nanomaterials for the removal and monitoring of noxious compounds in the environment.

2. Environmental Monitoring Using Smart Nanomaterials

Recently, the utilisation of nanotechnology for sustainable management has been pivotal in mitigating water and soil contamination via various detection techniques (Hairom et al. 2021, Liang et al. 2016, Walekar et al. 2017). The Raman spectroscopy technique proves to be greatly applicable in assessing the mobility of toxic substances and the potential bioavailability of hazardous compounds that contribute to pollution (Vaisakh et al. 2023). Raman spectroscopy serves as a valuable tool for obtaining structural and molecular insights into the compounds responsible for pollution, effectively illustrating the potential risks associated with these pollutants (Ong et al. 2020, Aarthi et al. 2021). Currently, the uses of nanomaterials in the fabrication of wearable devices are drawing significant attention, particularly in the context of environmental applications (Yang et al. 2023).

Prior to the emergence of analytical nanotechnology, numerous methods were established for detecting harmful compounds. These include chromatography techniques such as gas chromatography, high performance liquid chromatography, gas chromatography–mass spectrometry, liquid chromatography–mass spectrometry, inductively coupled plasma optical emission spectroscopy, inductively coupled plasma mass spectrometry, enzymatic–spectrophotometric methods, capillary electrophoresis, among others. However, these instruments and methods present several drawbacks, including high maintenance costs, the necessity for highly purified samples, time-intensive analyses, elevated analysis costs, complex operations, required separation and purification steps, and extensive sample pre-treatment, as noted by Du et al. (2021). Given the limitations of existing techniques, many efforts have been directed towards developing innovative and alternative detection methods for harmful substances that are subtle, accurate, and manageable.

Recent reviews have highlighted advancements in flexible and wearable sensors utilising nanomaterials, emphasising their various uses in healthcare, gas monitoring, food and beverages, and construction safety (Pawar et al. 2024, Mamun et al. 2020, Fu et al. 2024b, Qiao et al. 2022, Howard et al. 2022). Figure 2 displays smart nanomaterials-based devices designed for the monitoring of toxic compounds. This section examines recent studies on sensors enhanced by nanomaterials, utilising chemical transduction methods for functionality.

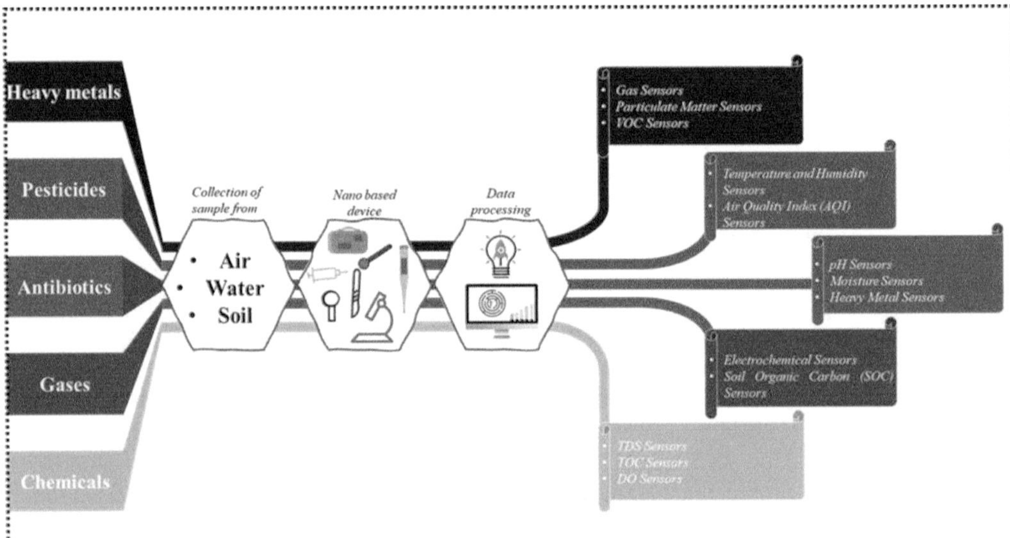

Figure 2: Devices incorporating smart nanomaterials for the detection and monitoring of toxic compounds.

2.1 Sensors for Water Quality Monitoring

2.1.1 Pesticides

Pollution harming the environment, agriculture, and aquatic ecosystems has resulted from the uncontrolled and overabundant use of pesticides, which has contaminated food. In developing countries in particular, there is a serious public health problem with the immediate and long-term health effects of agrochemicals and food consumption. These chemical pesticides pose serious risks to human health, with potential carcinogenic, cytotoxic, and mutagenic properties (Ahmad et al.

2024). The research by El-Sheikh et al. (2022) investigated the remaining concentrations of 40 pesticides in fruits and vegetables obtained from farmers' markets. El-Nahhal and El-Nahhal (2021) observed that around 113 pesticide residues were detected in drinking water samples collected from 31 countries globally. Degrendele et al. (2022) examined the occurrence of 30 currently used pesticides in two distinct locations in South Africa. Electrochemical sensing for herbicide (aclonifen) using gadolinium niobate nanoparticles (NPs), with actual river water and soil sample analysis, was conducted by Gopi et al. (2021). Abraham and Vasantha (2020) illustrated the development of a chemical sensor for the hazardous compound fenuron through the utilisation of a hollow polypyrrole composite film. The researchers Singha et al. (2017) found that Azinphos-methyl pesticide could be detected at ppb levels in the aqueous phase using a newly synthesised metal–organic framework (MOF). The detection of organophosphate insecticide was accomplished using an immunosensor that was based on Au (gold) NPs and polyaniline/carboxylated multiwall carbon nanotubes–chitosan nanocomposite film (Sun et al. 2013). An immunosensor utilising Prussian blue–multiwalled carbon nanotubes–chitosan nanocomposite film and Au NPs was able to detect carbofuran (Sun et al. 2012).

2.1.2 Pharmaceuticals and API

Hospital wastewater comprises a complex mixture of chemicals, in addition to pharmaceutical residues, significant contributors of silver and mercury. The complex composition endangers the aquatic ecology in numerous ways, including physically, chemically, and biologically (Akhter et al. 2024). One of the main concerns about antibiotic traces in nature is the possibility that they could affect the human microbiome and lead to the development and spread of bacteria that are tolerant to antibiotics (Akhter et al. 2024, Eapen et al. 2024). Ohlsen et al. (2003) identified the presence of antibiotics, specifically ciprofloxacin at concentrations of up to 0.051 mg/L and erythromycin at levels reaching 0.027 mg/L, in sewage samples. According to Li et al. (2023b), ciprofloxacin and clarithromycin were identified in drainage. Water and lettuce samples obtained in Ghana showed elevated levels of antibiotics, according to the research of Azanu et al. (2018).

A fluorescence probe made with boron-doped carbon quantum dots was used to sensitively detect the antibiotic amoxicillin, as demonstrated by Zhang et al. (2020). A ratiometric fluorescent sensor array utilising lanthanide-functionalised MOF was developed by Xie et al. (2021) for the identification and quantification of antibiotics in water. An aptamer-labelled fluorescent nanoprobe integrated with shell-isolated nanoparticle-enhanced fluorescence spectroscopy has been suggested for the purpose of selectively and sensitively detecting tetracycline (TC), as performed by Wang et al. (2020). Li et al. (2020a) developed zeolite imidazolate framework–8 (ZIF-8) for the concurrent capture and fluorescence detection of TC. A 3D hierarchical dual-metal–organic framework was developed for the detection of TC residues by Li et al. (2020b). Kumar et al. (2020) created a nanocomposite to detect and degrade ciprofloxacin. A carbon dot (CD) was synthesised using a straightforward hydrothermal method from Curcuma amada for the detection of fluoroquinolones and TCs, as reported by Korah et al. (2022). Chen et al. (2020) developed an aptaprobe for detecting sulfadimethoxine residues in water and fish. A fluorescent probe was created by Chen et al. (2021) that uses multiple co-doped CDs to selectively and highly sensitively determine sulfathiazole.

2.1.3 Heavy Metals

Water pollution caused by heavy metals stems from various human activities, including the combustion of fossil fuels, vehicle emissions, mining operations, agricultural practices, and the incineration of solid and liquid waste (Sall et al. 2020). The negative effects of toxic metals on human health are significantly associated with their ability to disrupt antioxidant defence mechanisms, mainly by interacting with intracellular glutathione or sulfhydryl groups of antioxidant enzymes and various other enzyme systems (Jomova et al. 2024). The research by Obasi et al. (2019) on the prevalence and assessment of toxic metals in the water resources of the mining regions of Abakaliki, Southeastern Nigeria, disclosed alarming levels of multiple metals, all surpassing the WHO recommended thresholds for potable water. 3D manganese dioxide (MnO2) nanowire/Ni foam was applied as the electrode for the copper (II) ion (Cu(II)) sensor developed by Hao et al. (2016). Cobalt oxide nanosheets modified indium tin oxide electrodes were employed for the electrochemical assessment of various trace Pb(II) (Yu et al. 2017). Deshmukh et al. (2017) used magnetic nanoparticles for the individual detection of heavy metals (mercury (II) ion (Hg(II)), lead (II) ion (Pb(II)), and cadmium (II) ion (Cd(II))). Chimezie et al. (2017) developed reduced graphene oxide (rGO)/iron (II, III) oxide (Fe3O4) for the trace detection of arsenic (III) ion (As(III)) in water samples. Riman et al. (2015) used bismuth oxide (Bi2O3)-modified graphite screen-printed electrodes (SPE) for the detection of heavy metals (Cd(II) and Pb(II)).

2.2 Air Quality Monitoring

2.2.1 Gases Pollutants

Human activities, including the use of solvents in pesticides, fossil fuel evaporation, emissions from coal-fired power plants, vehicle exhaust, industrial processes, volatile organic compounds from forests, and the burning of biomass and biofuels, are significant contributors to the release of toxic VOC into the atmosphere (Chaturvedi et al. 2023, Bhaskaran et al. 2023, Mangotra and Singh 2024). Particulate Matter (PM) pollution encompasses particles that are 10 μm in diameter or smaller, referred to as PM_{10}, as well as very fine particles that typically measure 2.5 μm or smaller (Manisalidis et al. 2020). Direct contact with fine particulate from inefficient fuel combustion is expected to result in 2.3 (1.6–3.1) million premature deaths globally each year (Chowdhury et al. 2023). Moreover, nitrogen oxides (Nox), sulphur dioxide (SO_2), VOCs, dioxins, and polycyclic aromatic hydrocarbons (PAHs) are identified as contaminants in the air that threaten human health (Manisalidis et al. 2020). The study conducted by Piñeiro et al. (2021) examined VOCs and PAHs from a waterproof coal tar membrane, assessing their impact on indoor air quality. As stated by Fan et al. (2012), the amount of fuel used for domestic flights in China in 2010 amounted to 12.12 million metric tonnes. The emissions recorded were 4600 tonnes of HxCy, 39,700 tonnes of CO, 154,100 tonnes of NOx, 38.21 million tonnes of carbon dioxide (CO_2), and 9700 tonnes of SO_2. Tungsten trioxide nanorods were used for nitrogen dioxide (NO_2) gas sensing (Liu et al. 2009). A nickel oxide/ zinc oxide (NiO/ZnO) biomorphic nanocomposite was used for the detection of n-butanol, as demonstrated by Zeng et al. (2020). Lin et al. (2010) fabricated indium oxide (In_2O_3)/ZnO composite films for investigated NOx sensing features. Gas sensors have been developed utilising single-crystalline stannic oxide (SnO2) nanobelts, demonstrating sensitivity to environmental pollutants such as CO and NO2, in addition to ethanol for applications in breath analysis and food control, as reported by Comini et al. (2002). ZnO sensors used for gas-sensing characteristics for ethyl alcohol and hydrogen sulphide (H2S) were examined by Wang et al. (2006). Kim et al. (2005) prepared WO2.72 nanorod films for the detection of ammonia vapour. Liang et al. (2008) improved the vanadium pentoxide-doped titanium dioxide (TiO2) solid-state potentiometric SO2 sensor. Gong et al. (2008) developed a micromachined nanocrystalline SnO2 reinforced with single-walled carbon nanotubes (SWCNTs), for the purpose of hydrogen gas detection. SnO2-based sensors for the detection of SO2 were developed by Berger et al. (1997).

2.3 Soil Contamination Detection

2.3.1 Heavy Metals

The extraction of mineral resources through mining and industrial processing, as well as their use in various sectors such as industry, agriculture, and economic development, has led to increased movement of these components within ecological systems and disruption to the biogeochemical processes. Agricultural soils are often polluted by extreme amounts of heavy metals, which have a detrimental effect on the health of plants (Angon et al. 2024). Kumar et al. (2019) reviewed heavy metals in soils across India. An ion exchange membrane modified with Au NPs and utilising nanoparticles-modified ion exchange membrane–laser-induced breakdown spectroscopy (LIBS) was employed to efficiently extract Pb and Cd ions from soil solution, followed by detection post-separation (Fu et al. 2024a). A straightforward method for enhancing self-synthetic Ag NPs using LIBS was introduced by Zhou et al. (2022) to enable sensitive analysis of selective metals (Fe, Mn, Cu, and Zn) in soil samples. Ma et al. (2023) introduced a technique that utilises LIBS in conjunction with filter paper enhanced by bimetallic nanoparticles for the highly sensitive detection of Hg(II), trivalent chromium, and Pb(II). Huang et al. (2019) used Au NPs and double ligand exchange for detection of Hg(II). Wang et al. (2017) presented a cost-effective and user-friendly SPE for the detection of Cd(II). The detection of soil Cd(II), Pb(II), Cu(II), and Hg(II) using square wave anodic stripping voltammetry was demonstrated by Lv et al. (2023).

2.3.2 Pesticides

Pesticides serve mainly to protect plants, but an important consideration is their conversion into toxic metabolites. Consequently, improper or excessive use of these substances can increase residual uncertainty, obscure risks associated with transformed metabolites, and pose potential threats to non-target organisms. Pesticides can directly harm aquatic flora and fauna and may also affect human health through numerous exposure pathways (Mitra et al. 2024). Degrendele et al. (2022) examined the occurrence of nine currently used pesticides in soil across two different agricultural locations in South Africa, where the highest concentrations of chlorpyrifos, carbaryl, and tebuconazole were found.

Chen et al. (2018) synthesised polydopamine gold nanowaxberry, which was utilised for the practical detection of pesticides and other chemicals in water, fruits, and soil. Li et al. (2021) conducted a study that demonstrates how Au nanorods can be used as a surface-enhanced Raman scattering (SERS) platform to detect pesticides in soil. Gong et al. (2020) developed 3D SERS substrates via magnetic force-assisted self-assembly of nanoparticles in arrayed holes. These substrates demonstrated the capability to detect 4-aminothiophenol and were utilised to identify hexachlorobenzene residue in soil. Rubira et al. (2022) utilised Ag NPs as SERS substrates to detect herbicides. He et al. (2019a, b) utilised Au NPs as SERS substrates to detect deltamethrin, chlorpyrifos, and carbofuran residues in soil.

2.4 Food Contamination Detection

A reliable and efficient method for assessing food quality is critically required, as various components in food, including heavy metals, mycotoxins, pesticides, and food-borne pathogens, create serious threats to people's well-being (Tarannum et al. 2014). The absorption of noxious substances by plants from the soil can lead to the contamination of not just fruits and vegetables, but also seafood (Anjum and Rana 2023, Angon et al. 2024). Heavy metals originating from industrial zones have the potential to infiltrate the soil, subsequently making their way into the food chain (Zhao et al. 2024). When pesticides are used to protect plants, they end up in the food chain, which means that humans are exposed to these toxins. This exposure is linked to a wide variety of health problems, such as immunological suppression, impaired cognitive function, hormone disruption, cancer, and reproductive abnormalities (Onyeaka et al. 2024, Rather et al. 2017).

MnO_2 was synthesised using the co-precipitation method, aimed at investigating the interaction with aflatoxin-B1 and ochratoxin-A through UV-visible spectroscopy (Singh et al. 2021). The study conducted by Verma et al. (2020) focusses on a naphthalimide-gold-based nanocomposite designed for the radiometric detection of okadaic acid in shellfish. A colorimetric sensor array utilising Au NPs has been developed for the swift identification of various pesticides in a range of fruits, vegetables, and traditional Chinese herbs, as prepared and tested by Zhao et al. (2023). Xu et al. (2023) designed, synthesised, and employed a hydrophobic cobalt anion-based benzyl-functionalised MIL for monitoring trace low-level antibiotics in food. Li et al. (2023a) conducted a study utilising an Au@Ag-labelled SERS lateral flow assay for detection of casein and α-lactalbumin in milk.

2.5 Integrated Environmental Sensing Systems

The Triboelectric Nanogenerator (TENG) technology has been used to facilitate the harvesting of environmental vibration energy as well as the sensing of environmental parameters, resulting in self-powered environmental monitoring systems (Liu et al. 2024). The Internet of Things (IoT) has made a significant impact on our daily lives due to the widespread use of sensors. However, it has also created challenges in the development of a Wireless Sensor Network (WSN) that assures a reliable power supply and dependable wireless communication (Ahmad and Zhang 2021, Khan et al. 2019a). A Piezoelectric Nanogenerator (PENG) utilising a core–shell heterostructure of barium titanate and polyvinylidene fluoride composite fibres, coated with barium titanate@Ag, has been developed for energy harvesting and wireless sensing applications (Xia et al. 2024). Zhang et al. (2022) presented a fully self-powered multi-parameter wireless sensing system utilising a TENG. A wind-driven self-powered wireless environmental sensing system was introduced by Liu et al. (2020), featuring nine TENG units integrated into a 3D-printed shell, serving as a power source for two environmental monitoring systems. Mamalis and Gerakidis (2024) introduced an innovative combined strategy that utilises both WSN and Vehicular Ad-Hoc Networks (VANET) clustered structures to effectively collect and process environmental parameters along long eNodeB/RSU-enabled roads and highways, traversing extensive rural regions. Radio Frequency Identification technology, which can wirelessly process substantial volumes of information, is attracting interest alongside the progress of IoT technology (Roh et al. 2024).

Karthick and Chandrasekaran (2021) presented a novel Commercial Building Energy Management System that uses an IoT-based Smart Compact Energy Meter to successfully track and control energy usage and address power quality issues. The Smart Home Energy Management System introduced a novel approach to enhancing energy efficiency in homes by leveraging the integration of IoT technology and Machine Learning algorithms, as demonstrated by Karuna et al. (2024). A proposed system utilises IoT technology for smart monitoring of water quality in various water bodies, incorporating cloud computing and deep learning methodologies (Ajith et al. 2020). Hamid et al. (2020) studied the design of the proposed Smart Water Quality Monitoring System and evaluated the factors affecting pH and temperature in swimming pools through the application of DOE and ANOVA statistical tools.

3. Nanomaterials for Environmental Remediation

Nanomaterials in environmental applications demonstrate varied properties and specificity that aid in the removal of harmful pollutants (Moradeeya et al. 2022a). Nanomaterials, with their elevated surface area and reactivity, demonstrate greater efficacy as adsorbents compared to traditional methods, rendering them valuable for wastewater treatment (Moradeeya et al. 2017, Chormare et al. 2023). The application of engineered nanomaterials for environmental remediation, referred to as nanoremediation, has expanded as a technology that provides mutually beneficial solutions for the rapid and effective identification and elimination of pollutants from contaminated environmental compartments. Figure 3 demonstrates the application of smart nanomaterials in effectively reducing and mitigating pollution.

3.1 Water Remediation Technologies

Adsorption is a physicochemical approach that helps materials stick to the surfaces of adsorbents. This is done through chemical bonding and physical attraction between the solid adsorbent and toxicants, which results in the formation of a layer of adsorbate (Idris et al. 2023, Gupta and Saleh 2013). The efficacy of any adsorption process relies on an adsorbent possessing a substantial surface area, suitable functionalities, and diverse porous materials. Activated carbon, biochar, biomass, bionanomaterials, clays, zeolites, metal oxides, polymers, and metal–organic frameworks have been

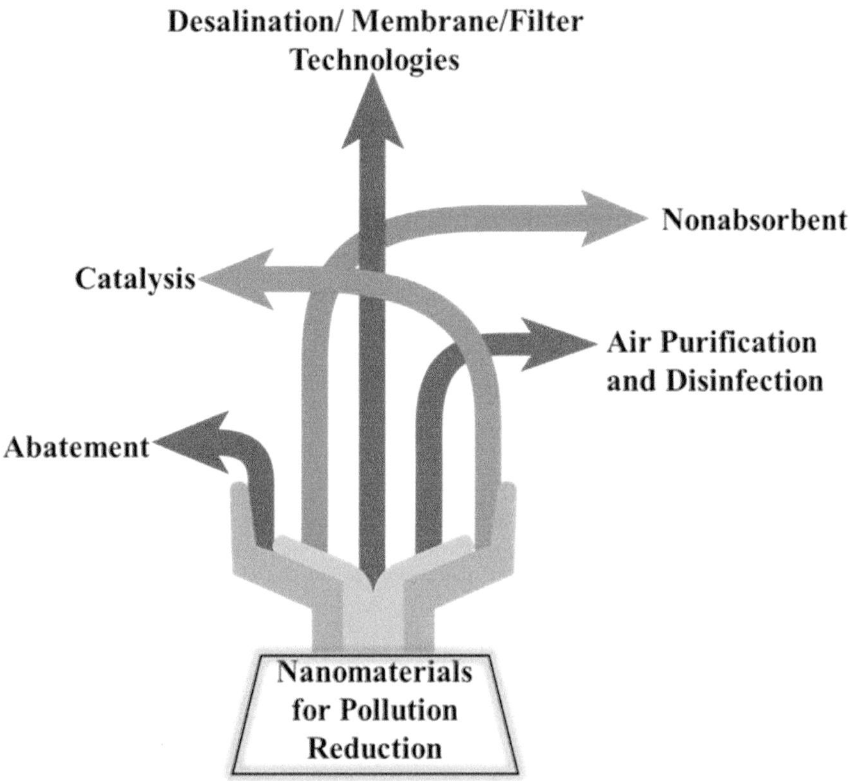

Figure 3: Application of smart nanomaterials in reducing and mitigating environmental pollution.

engineered, each exhibiting varying degrees of effectiveness in eliminating toxic pollutants (Nama et al. 2024, Sahoo et al. 2024, Najafi et al. 2021, Wang et al. 2013, Moradeeya et al. 2017, Diagboya and Dikio 2018, Adegoke et al. 2020, Chormare et al. 2023, Gupta and Saleh 2013).

Persistent organic pollutants are acknowledged as chemicals of considerable concern due to their poisonousness, persistence, ability for long-range transport, and propensity to bioaccumulate in organisms. These chemicals can traverse significant distances and persist within living creatures (Akhtar et al. 2021, Krithiga et al. 2022). The degradation of new refractory pollutants, especially those with elevated concentrations and intricate structures, presents considerable problems for conventional wastewater treatment technology. Photocatalysis technology has been extensively studied in advanced oxidation processes for the degradation of organic pollutants in water treatment (Moradeeya et al. 2022a). Photocatalysis, highlighted for its environmentally friendly and energy-efficient properties, has attracted considerable attention owing to its various benefits, including environmental sustainability, availability, stability, strong photoactivity, and reusability, rendering it a highly versatile approach for the effective treatment of wastewater (Moradeeya et al. 2022a, b, Rathod et al. 2018).

The integration of nanoparticles into membrane frameworks aims to improve filtration efficiency by facilitating smaller pore sizes and selectively eliminating contaminants such as salts, microbes, and dissolved organic matter (Chadha et al. 2022, Ying et al. 2017, Rashid and Ralph 2017, Ma et al. 2017, Kalash et al. 2020). Ag, Au, and Zn NPs are employed to eliminate bacteria in water filtration due to their antimicrobial properties (Mondal et al. 2024). Alongside the enhancement in separation efficiency, various properties are incorporated into the composite membranes associated with nanoparticles, including catalytic, antibacterial, and antifouling characteristics (Saleem and Zaidi 2020).

3.2 Soil Remediation Approaches

A range of technologies, including adsorption, immobilisation, Fenton-like oxidation, and the integration of nanotechnology with bioremediation, have been employed for the remediation of soil contaminants (Dhanapal et al. 2024). Nanoremediation employs nanoparticles or nanomaterials to address soil pollution, contributing to areas such as sustainable crop production, enhancing crop yield, minimising nutrient loss, and managing diseases. Focussing on the immobilisation of toxic metals (Pb(II), Cu(II), Ni(II), and Zn(II)) in polluted soils, Matos et al. (2017) examined the effectiveness of using Carbon Nanotubes (CNTs). Su et al. (2016) developed a biochar-supported Nano Zerovalent Iron (nZVI) material for the *in-situ* remediation of Cr(VI). Singh and Lee (2016) investigated the effects of TiO2 in the removal of Cd from contaminated soil. The study conducted by Cao et al. (2018) investigates the potential of nZVI for the removal of Cd(II), Pb(II), and Zn(II) from contaminated soils. Soil immobilisation of Cr(VI) was achieved by Xu and Zhao (2007) using stabilised nZVI. The impact of nZVI on the degradation of DDT and its ecotoxicological implications for collembola (Folsomia candida) and ostracods (Heterocypris incongruens) were examined by El-Temsah and Joner (2013).

Danila et al. (2020) found that micro-ZVI- and nZVI-treated soils retained As and Cu well but desorbed Pb. Liu et al. (2021) created a porous bone char to support nZVI and utilised it for the remediation of Cr(VI) contamination in water and paddy soil under anaerobic conditions. The synergistic effects of various materials on the immobilisation of toxic metals were studied by Munir et al. (2020).

3.3 Air Pollution Control

Activated carbon, carbon-based nanoparticles, and composites have shown significant potential as effective adsorbents because of their functionality for harmful gases (Saleem et al. 2022). Gui et al. (2013) conducted a study on modified Multi-Walled Carbon Nanotubes (MWCNTs) to enhance CO2 adsorption. The study conducted by Azam et al. (2017) investigates the electronic properties and adsorption behaviour of CO_2 and methanol on pristine, silicon-doped, and boron-doped SWCNTs. Huy et al. (2019) described a straightforward one-step approach employing ultrasonic-assisted precipitation, devoid of surfactants and subsequent heat treatments, to synthesise ZnO nanoparticles and their use in the adsorption of H2S from air. The study conducted by Meimand et al. (2019) focusses on the adsorption of SO2 on clinoptilolite/nano iron oxide and natural clinoptilolite.

In the presence of visible light, the graphitic carbon nitride (g-C_3N_4)/TiO_2 composite demonstrated photocatalytic activity for NO oxidation as reported by Giannakopoulou et al. (2017). Improved visible-light photocatalytic performance was reported for nitrogen-doped, silver-doped, and N–Ag co-doped TiO_2 synthesized via an *in situ* solvothermal method, demonstrating efficient gas-phase ammonia removal (Sirivallop et al. 2020). The degradation of benzene in the gaseous phase was effectively achieved through the application of $ZnO+Zn_2TiO_4$ multicomponent oxide thin films as photocatalysts (Hernandez-Garcia et al. 2016). The growth of TiO_2 and Cadmium Oxide (CdO) + Cadmium Titanate ($CdTiO_3$) multicomponent oxide thin films was conducted using various Ti/Cd ratios through the sol–gel method, aimed at degrading benzene in the gas phase under different relative humidity conditions in a batch reactor at room temperature (Hernández-García et al. 2015). The composite films of TiO_2/Aluminium Oxide (Al_2O_3)/Silicon Dioxide (SiO_2) exhibit notable photocatalytic activity in the degradation of gaseous acetaldehyde, as reported by Yuan et al. (2008). Electrospun nanofibrous membranes have become a topic of interest as a possible platform because of their complex porosity structure, reduced fibre sizes, effective internal connectivity, and the ease with which active compounds can be incorporated for antimicrobial purposes (Wang et al. 2024). Sepahvand et al. (2021) developed a nanofilter utilising cellulose nanofibres combined with chitosan, subsequently assessing its effectiveness for the separation of PM and CO_2. Wu et al. (2022) integrated silver nanowires into bacterial cellulose filters using an *in-situ* cultivation method to create multifunctional air filters capable of effectively removing particulate matter (PM2.5 and

PM10). The efficient and straightforward production of MOF/cellulose composite paper for the elimination of air pollutants is achieved through the *in-situ* synthesis of MOF/chitosan hydrogel, as demonstrated by Nie et al. (2022). Su et al. (2018) developed a multifunctional cellulose/ZIF-8 air filter aimed at the removal of $PM_{0.3}$.

4. Challenges and Limitations of Smart Nanomaterials

4.1 Toxicity and Environmental Impact

The ongoing release of nanomaterials into the environment, driven by advancements in nanotechnology, poses significant threats to ecological systems, jeopardising biodiversity, natural resources, and human health (Pervez and Yadav 2023). The engineered nanomaterials that are released from direct use or from products containing nanomaterials hold greater significance for ecotoxicological studies and risk assessment compared to pristine engineered nanomaterials (Nowack et al. 2012). Nanotoxicology possesses the capacity to address the health and environmental issues linked to nano-sized compounds, which may exhibit high reactivity and pose possible risks upon interaction with living organisms and the natural world. The physicochemical factors, routes of exposure, biodistribution, molecular determinants, genotoxicity, and regulatory considerations are incorporated into the research of nanotoxicology, as described by Qamar et al. (2024).

4.2 Scalability and Cost-Effectiveness

The difficulties of scalability in nanoparticle production underscore the need for mechanised and accessible synthesis methods to facilitate commercial uptake. Additionally, the difficulties related to the incorporation of nanomaterials into existing systems are analysed, focussing on interoperability and handling techniques in traditional production processes. The capacity for conversion and the significant obstacles associated with the scalability of nanomaterials require sustained commitment, including material selection, large-scale manufacturing, and multifunctional integration. Recent advancements in functional nanofibres and nanotechnology for large-scale manufacturing indicate that wearable electronics and smart textiles could greatly enhance human-machine interfaces and medical devices. The efficacy of nanomanufacturing is significantly dependent on strong collaboration between academic institutions and industry. This collaboration is crucial for being apprised of current demands and future obstacles, as well as for developing goods that may be effortlessly incorporated into the industrial sector. The choice of a suitable process that integrates nanomaterial production with optimal characteristics, few impurities, and scalability is essential.

5. Conclusion

This chapter presents an overview of recent advancements in the development of sensors employing nanomaterials for environmental monitoring, focussing on their transmission processes. These sensors possess considerable promise to monitor individualised micro-environments, facilitating the early identification of possible environmental hazards. Wearable chemical detection devices employing nanomaterials have been successfully produced, demonstrating its utility in detecting environmental factors such as gases, soil, and water contaminated with diverse chemical agents and dangers. Moreover, nanocatalysts and nanomaterials can effectively and cost-efficiently eliminate inorganic, organic, pharmaceutical, and heavy metal pollutants from water sources. Considering the significance of reducing production costs for broader application in wastewater treatment, it is crucial that forthcoming research prioritises improving the economic viability of these nanomaterials and evaluating their interaction mechanisms within water treatment systems. Moreover, it is imperative to conduct a comprehensive investigation of their possible toxicity to human health and the environment; meticulous evaluations of their harmfulness are vital to ensure safer usage. Further investigations are required to evaluate the relative efficiency of these nanomaterials in terms of energy consumption and resource utilisation, and to identify potentially earth-abundant elements warranting further research.

References

Aarthi, A., Bindhu, M. R., Umadevi, M., Parimaladevi, R., Sathe, G. V., Al-Mohaimeed, A. M. et al. (2021). Evaluating the detection efficacy of advanced bimetallic plasmonic nanoparticles for heavy metals, hazardous materials and pesticides of leachate in contaminated groundwater. *Environmental Research*, 201: 111590.

Abraham, D. A., and Vasantha, V. S. (2020). Hollow polypyrrole composite synthesis for detection of trace-level toxic herbicide. *ACS omega*, 5(34): 21458–21467.

Adegoke, K. A., Agboola, O. S., Ogunmodede, J., Araoye, A. O., and Bello, O. S. (2020). Metal-organic frameworks as adsorbents for sequestering organic pollutants from wastewater. *Materials Chemistry and Physics*, 253: 123246.

Adil, N., Ashraf, K., Munir, M., Mohiuddin, M., Abbasi, A., Riaz, U. et al. (2023). Pesticides, heavy metals and plasticizers: contamination and risk assessment of drinking-water quality. *Sustainability*, 15(17): 13263.

Ahmad, M. F., Ahmad, F. A., Alsayegh, A. A., Zeyaullah, M., AlShahrani, A. M., Muzammil, K. et al. (2024). Pesticides impacts on human health and the environment with their mechanisms of action and possible countermeasures. *Heliyon*.

Ahmad, T., and Zhang, D. (2021). Using the internet of things in smart energy systems and networks. *Sustainable Cities and Society*, 68: 102783.

Ajith, J. B., Manimegalai, R., and Ilayaraja, V. (2020, February). An IoT based smart water quality monitoring system using cloud. In *2020 International conference on emerging trends in information technology and engineering (ic-ETITE)* (pp. 1-7). IEEE.

Akhtar, A. B. T., Naseem, S., Yasar, A., and Naseem, Z. (2021). Persistent organic pollutants (POPs): sources, types, impacts, and their remediation. *Environmental Pollution and Remediation*, 213–246.

Akhter, S., Bhat, M. A., Ahmed, S., and Siddiqui, W. A. (2024). Antibiotic residue contamination in the aquatic environment, sources and associated potential health risks. *Environmental Geochemistry and Health*, 46(10): 387.

Angon, P. B., Islam, M. S., Das, A., Anjum, N., Poudel, A., and Suchi, S. A. (2024). Sources, effects and present perspectives of heavy metals contamination: Soil, plants and human food chain. *Heliyon*, 10(7).

Anil Kumar, M., Moradeeya, P. G., Manikanda Bharath, K., Jakulin Divya Mary, P., and Giridharan, K. S. (2024). Pyrolytic Conversion of Heterogenic Natural Waste Biomass from Rural Communities with Concomitant Valorization. pp. 259–287. *In: Circular Bioeconomy Perspectives in Sustainable Bioenergy Production*. Singapore: Springer Nature Singapore.

Anjum, S., and Rana, S. (2023). Impact of environmental pollutants on agriculture and food system. pp. 133–151. *In: Advanced Microbial Techniques in Agriculture, Environment, and Health Management*. Academic Press.

Anthony A, A., Adekunle C, F., and Thor A, S. (2018). Residual antibiotics, antibiotic resistant superbugs and antibiotic resistance genes in surface water catchments: Public health impact. *Physics and Chemistry of the Earth*, 105: 177–183.

Azam, M. A., Alias, F. M., Tack, L. W., Seman, R. N. A. R., and Taib, M. F. M. (2017). Electronic properties and gas adsorption behaviour of pristine, silicon-, and boron-doped (8, 0) single-walled carbon nanotube: A first principles study. *Journal of Molecular Graphics and Modelling*, 75: 85–93.

Azanu, D., Styrishave, B., Darko, G., Weisser, J. J., and Abaidoo, R. C. (2018). Occurrence and risk assessment of antibiotics in water and lettuce in Ghana. *Science of the Total Environment*, 622: 293–305.

Baig, N., Kammakakam, I., and Falath, W. (2021). Nanomaterials: A review of synthesis methods, properties, recent progress, and challenges. *Materials advances*, 2(6): 1821–1871.

Berger, F., Fromm, M., Chambaudet, A., and Planade, R. (1997). Tin dioxide-based gas sensors for SO2 detection: a chemical interpretation of the increase in sensitivity obtained after a primary detection. *Sensors and Actuators B: Chemical*, 45(3): 175–181.

Bhaskaran, A., Sharma, D., Roy, S., and Singh, S. A. (2023). Technological solutions for NOx, SOx, and VOC abatement: Recent breakthroughs and future directions. *Environmental Science and Pollution Research*, 30(40): 91501–91533.

Cao, Y., Zhang, S., Zhong, Q., Wang, G., Xu, X., Li, T. et al. (2018). Feasibility of nanoscale zero-valent iron to enhance the removal efficiencies of heavy metals from polluted soils by organic acids. *Ecotoxicology and Environmental Safety*, 162: 464–473.

Chadha, U., Selvaraj, S. K., Thanu, S. V., Cholapadath, V., Abraham, A. M., Manoharan, M. et al. (2022). A review of the function of using carbon nanomaterials in membrane filtration for contaminant removal from wastewater. *Materials Research Express*, 9(1): 012003.

Chaturvedi, S., Kumar, A., Singh, V., Chakraborty, B., Kumar, R., and Min, L. (2023). Recent advancement in organic aerosol understanding: a review of their sources, formation, and health impacts. *Water, Air, & Soil Pollution*, 234(12): 750.

Chen, D., Zhu, X., Huang, J., Wang, G., Zhao, Y., Chen, F. et al. (2018). Polydopamine@ gold nanowaxberry enabling improved SERS sensing of pesticides, pollutants, and explosives in complex samples. *Analytical chemistry*, 90(15): 9048–9054.

Chen, L., Liu, Y., Cheng, G., Fan, Z., Yuan, J., He, S. et al. (2021). A novel fluorescent probe based on N, B, F co-doped carbon dots for highly selective and sensitive determination of sulfathiazole. *Science of the Total Environment*, 759: 143432.

Chen, X. X., Lin, Z. Z., Yao, Q. H., and Huang, Z. Y. (2020). A practical aptaprobe for sulfadimethoxine residue detection in water and fish based on the fluorescence quenching of CdTe QDs by poly (diallyldimethylammonium chloride). *Journal of Food Composition and Analysis*, 91: 103526.

Chimezie, A. B., Hajian, R., Yusof, N. A., Woi, P. M., and Shams, N. (2017). Fabrication of reduced graphene oxide-magnetic nanocomposite (rGO-Fe3O4) as an electrochemical sensor for trace determination of As (III) in water resources. *Journal of Electroanalytical Chemistry*, 796: 33–42.

Chormare, R., Moradeeya, P. G., Sahoo, T. P., Seenuvasan, M., Baskar, G., Saravaia, H. T. et al. (2023). Conversion of solid wastes and natural biomass for deciphering the valorization of biochar in pollution abatement: A review on the thermo-chemical processes. *Chemosphere*, 139760.

Chowdhury, S., Pillarisetti, A., Oberholzer, A., Jetter, J., Mitchell, J., Cappuccilli, E. et al. (2023). A global review of the state of the evidence of household air pollution's contribution to ambient fine particulate matter and their related health impacts. *Environment international*, 173: 107835.

Comini, E., Faglia, G., Sberveglieri, G., Pan, Z., and Wang, Z. L. (2002). Stable and highly sensitive gas sensors based on semiconducting oxide nanobelts. *Applied Physics Letters*, 81(10): 1869–1871.

Dagdag, O., Quadri, T. W., Haldhar, R., Kim, S. C., Daoudi, W., Berdimurodov, E. et al. (2023). An overview of heavy metal pollution and control. *Heavy Metals in the Environment: Management Strategies for Global Pollution*, 3–24.

Danila, V., Kumpiene, J., Kasiuliene, A., and Vasarevičius, S. (2020). Immobilisation of metal (loid) s in two contaminated soils using micro and nano zerovalent iron particles: Evaluating the long-term stability. *Chemosphere*, 248: 126054.

Degrendele, C., Klánová, J., Prokeš, R., Příbylová, P., Šenk, P., Šudoma, M. et al. (2022). Current use pesticides in soil and air from two agricultural sites in South Africa: Implications for environmental fate and human exposure. *Science of the Total Environment*, 807: 150455.

Deshmukh, S., Kandasamy, G., Upadhyay, R. K., Bhattacharya, G., Banerjee, D., Maity, D. et al. (2017). Terephthalic acid capped iron oxide nanoparticles for sensitive electrochemical detection of heavy metal ions in water. *Journal of Electroanalytical Chemistry*, 788: 91–98.

Dhanapal, A. R., Thiruvengadam, M., Vairavanathan, J., Venkidasamy, B., Easwaran, M., and Ghorbanpour, M. (2024). Nanotechnology approaches for the remediation of agricultural polluted soils. *ACS Omega*, 9(12): 13522–13533.

Diagboya, P. N., and Dikio, E. D. (2018). Silica-based mesoporous materials; emerging designer adsorbents for aqueous pollutants removal and water treatment. *Microporous and Mesoporous Materials*, 266: 252–267.

Du, H., Xie, Y., and Wang, J. (2021). Nanomaterial-sensors for herbicides detection using electrochemical techniques and prospect applications. *TrAC Trends in Analytical Chemistry*, 135: 116178.

Eapen, J. V., Thomas, S., Antony, S., George, P., and Antony, J. (2024). A review of the effects of pharmaceutical pollutants on humans and aquatic ecosystem. *Exploration of Drug Science*, 2(5): 484–507.

El-Nahhal, I., and El-Nahhal, Y. (2021). Pesticide residues in drinking water, their potential risk to human health and removal options. *Journal of Environmental Management*, 299: 113611.

El-Sheikh, E. S. A., Ramadan, M. M., El-Sobki, A. E., Shalaby, A. A., McCoy, M. R., Hamed, I. A. et al. (2022). Pesticide residues in vegetables and fruits from farmer markets and associated dietary risks. *Molecules*, 27(22): 8072.

El-Temsah, Y. S., and Joner, E. J. (2013). Effects of nano-sized zero-valent iron (nZVI) on DDT degradation in soil and its toxicity to collembola and ostracods. *Chemosphere*, 92(1): 131–137.

Escolà-Gascón, Á., Micó-Sanz, J. L., and Casero-Ripollés, A. (2022). Global evidence of environmental and lifestyle effects on medical expenditures across 154 countries. *Preventive medicine reports*, 30: 102036.

Fan, W., Sun, Y., Zhu, T., and Wen, Y. (2012). Emissions of HC, CO, NOx, CO2, and SO2 from civil aviation in China in 2010. *Atmospheric Environment*, 56: 52–57.

Fu, X., Gou, Y., Li, G., Zhao, S., Ma, S., and Zhao, C. (2024a). Detection of available heavy metals in soil using gold nanoparticles-modified ion exchange membrane with laser-induced breakdown spectroscopy. *Microchemical Journal*, 110922.

Fu, Y., Liu, T., Wang, H., Wang, Z., Hou, L., Jiang, J. et al. (2024b). Applications of nanomaterial technology in biosensing. *Journal of Science: Advanced Materials and Devices*, 100694.

Giannakopoulou, T., Papailias, I., Todorova, N., Boukos, N., Liu, Y., Yu, J. et al. (2017). Tailoring the energy band gap and edges' potentials of g-C3N4/TiO2 composite photocatalysts for NOx removal. *Chemical Engineering Journal*, 310: 571–580.

Gong, J., Sun, J., and Chen, Q. (2008). Micromachined sol–gel carbon nanotube/SnO2 nanocomposite hydrogen sensor. *Sensors and Actuators B: Chemical*, 130(2): 829–835.

Gong, T., Huang, Y., Wei, Z., Huang, W., Wei, X., and Zhang, X. (2020). Magnetic assembled 3D SERS substrate for sensitive detection of pesticide residue in soil. *Nanotechnology*, 31(20): 205501.

Gopi, P. K., Mutharani, B., Chen, S. M., Chen, T. W., Eldesoky, G. E., Ali, M. A. et al. (2021). Electrochemical sensing base for hazardous herbicide aclonifen using gadolinium niobate (GdNbO4) nanoparticles-actual river water and soil sample analysis. *Ecotoxicology and Environmental Safety*, 207: 111285.

Gui, M. M., Yap, Y. X., Chai, S. P., and Mohamed, A. R. (2013). Multi-walled carbon nanotubes modified with (3-aminopropyl) triethoxysilane for effective carbon dioxide adsorption. *International Journal of Greenhouse Gas Control*, 14: 65–73.

Gupta, V. K., and Saleh, T. A. (2013). Sorption of pollutants by porous carbon, carbon nanotubes and fullerene-an overview. *Environmental Science and Pollution Research*, 20: 2828–2843.

Hairom, N. H. H., Soon, C. F., Mohamed, R. M. S. R., Morsin, M., Zainal, N., Nayan, N. et al. (2021). A review of nanotechnological applications to detect and control surface water pollution. *Environmental Technology & Innovation*, 24: 102032.

Hamid, S. A., Rahim, A. M. A., Fadhlullah, S. Y., Abdullah, S., Muhammad, Z., and Leh, N. A. M. (2020, August). IoT based water quality monitoring system and evaluation. pp. 102–106. *In: 2020 10th IEEE International Conference on Control System, Computing and Engineering (ICCSCE)*. IEEE.

Hao, S., Li, J., Li, Y., Zhang, Y., and Hu, G. (2016). Facile synthesis of a 3D MnO 2 nanowire/Ni foam electrode for the electrochemical detection of Cu (ii). *Analytical Methods*, 8(24): 4919–4925.

He, Y., Xiao, S., Dong, T., and Nie, P. (2019a). Gold nanoparticles for qualitative detection of deltamethrin and carbofuran residues in soil by surface enhanced Raman scattering (SERS). *International Journal of Molecular Sciences*, 20(7): 1731.

He, Y., Xiao, S., Dong, T., and Nie, P. (2019b). Gold nanoparticles with different particle sizes for the quantitative determination of chlorpyrifos residues in soil by SERS. *International Journal of Molecular Sciences*, 20(11): 2817.

Hernandez-Garcia, F. A., Torres-Delgado, G., Castanedo-Perez, R., and Zelaya-Angel, O. (2016). Gaseous benzene degradation by photocatalysis using ZnO+ Zn 2 TiO 4 thin films obtained by sol-gel process. *Environmental Science and Pollution Research*, 23: 13191–13199.

Hernández-García, F. A., Torres-Delgado, G., Castanedo-Pérez, R., and Zelaya-Ángel, O. (2015). Photodegradation of gaseous C6H6 using CdO+ CdTiO3 and TiO2 thin films obtained by sol–gel technique. *Journal of Photochemistry and Photobiology A: Chemistry*, 310: 52–59.

Howard, J., Murashov, V., Cauda, E., and Snawder, J. (2022). Advanced sensor technologies and the future of work. *American Journal of Industrial Medicine*, 65(1): 3–11.

Huang, D., Liu, X., Lai, C., Qin, L., Zhang, C., Yi, H. et al. (2019). Colorimetric determination of mercury (II) using gold nanoparticles and double ligand exchange. *Microchimica Acta*, 186: 1–8.

Huy, N. N., Thuy, V. T. T., Thang, N. H., Thuy, N. T., Khoi, T. T., and Van Thanh, D. (2019). Facile one-step synthesis of zinc oxide nanoparticles by ultrasonic-assisted precipitation method and its application for H2S adsorption in air. *Journal of Physics and Chemistry of Solids*, 132: 99–103.

Idris, M. O., Yaqoob, A. A., Ibrahim, M. N. M., Ahmad, A., and Alshammari, M. B. (2023). Introduction of adsorption techniques for heavy metals remediation. pp. 1–18. *In: Emerging Techniques for Treatment of Toxic Metals from Wastewater*. Elsevier.

Islam, M. N., Taki, G., Jung, Y. J., Jung, S. K., and Park, J. H. (2018). Remediation of gulf war oil spill contaminated soil by a subcritical water extraction process: oil removal, recovery, and degradation. *Soil and Sediment Contamination: An International Journal*, 27(2): 120–130.

Jomova, K., Alomar, S. Y., Nepovimova, E., Kuca, K., and Valko, M. (2024). Heavy metals: Toxicity and human health effects. *Archives of Toxicology*, 1–57.

Kalash, K., Kadhom, M., and Al-Furaiji, M. (2020). Thin film nanocomposite membranes filled with MCM-41 and SBA-15 nanoparticles for brackish water desalination via reverse osmosis. *Environmental Technology & Innovation*, 20: 101101.

Karthick, T., and Chandrasekaran, K. (2021). Design of IoT based smart compact energy meter for monitoring and controlling the usage of energy and power quality issues with demand side management for a commercial building. *Sustainable Energy, Grids and Networks*, 26: 100454.

Karuna, G., Ediga, P., Akshatha, S., Anupama, P., Sanjana, T., Mittal, A. et al. (2024). Smart energy management: real-time prediction and optimization for IoT-enabled smart homes. *Cogent Engineering*, 11(1).

Kaur, R., Choudhary, D., Bali, S., Bandral, S. S., Singh, V., Ahmad, M. A. et al. (2024). Pesticides: An alarming detrimental to health and environment. *Science of The Total Environment*, 170113.

Kayode-Afolayan, S. D., Ahuekwe, E. F., and Nwinyi, O. C. (2022). Impacts of pharmaceutical effluents on aquatic ecosystems. *Scientific African*, 17: e01288.

Khan, A. A., Mahmud, A., Zhang, S., Islam, S., Voss, P., and Ban, D. (2019a). A self-powered multi-broadcasting wireless sensing system realized with an all-in-one triboelectric nanogenerator. *Nano Energy*, 62: 691–699.

Khan, I., Saeed, K., and Khan, I. (2019b). Nanoparticles: Properties, applications and toxicities. *Arabian Journal of Chemistry*, 12(7): 908–931.

Kim, Y. S., Ha, S. C., Kim, K., Yang, H., Choi, S. Y., Kim, Y. T. et al. (2005). Room-temperature semiconductor gas sensor based on nonstoichiometric tungsten oxide nanorod film. *Applied Physics Letters*, 86(21).

Korah, B. K., Chacko, A. R., Mathew, S., John, B. K., Abraham, T., and Mathew, B. (2022). Biomass-derived carbon dots as a sensitive and selective dual detection platform for fluoroquinolones and tetracyclines. *Analytical and Bioanalytical Chemistry*, 414(17): 4935–4951.

Kostich, M. S., and Lazorchak, J. M. (2008). Risks to aquatic organisms posed by human pharmaceutical use. *Science of the Total Environment*, 389(2–3): 329–339.

Krithiga, T., Sathish, S., Renita, A. A., Prabu, D., Lokesh, S., Geetha, R. et al (2022). Persistent organic pollutants in water resources: Fate, occurrence, characterization and risk analysis. *Science of The Total Environment*, 831: 154808.

Kumar, S., Karfa, P., Majhi, K. C., and Madhuri, R. (2020). Photocatalytic, fluorescent BiPO4@ Graphene oxide based magnetic molecularly imprinted polymer for detection, removal and degradation of ciprofloxacin. *Materials Science and Engineering: C*, 111: 110777.

Kumar, V., Sharma, A., Kaur, P., Sidhu, G. P. S., Bali, A. S., Bhardwaj, R. et al. (2019). Pollution assessment of heavy metals in soils of India and ecological risk assessment: A state-of-the-art. *Chemosphere*, 216: 449–462.

Li, C., Yang, W., Zhang, X., Han, Y., Tang, W., Yue, T. et al. (2020b). A 3D hierarchical dual-metal–organic framework heterostructure up-regulating the pre-concentration effect for ultrasensitive fluorescence detection of tetracycline antibiotics. *Journal of Materials Chemistry C*, 8(6): 2054–2064.

Li, C., Zhang, X., Wen, S., Xiang, R., Han, Y., Tang, W. et al. (2020a). Interface engineering of zeolite imidazolate framework-8 on two-dimensional Al– metal– organic framework nanoplates enhancing performance for simultaneous capture and sensing tetracyclines. *Journal of Hazardous Materials*, 395: 122615.

Li, J., Xu, J., Pan, Y., Zhu, Y., Wang, Y., Chen, S. et al. (2023a). Au@Ag-labeled SERS lateral flow assay for highly sensitive detection of allergens in milk. *Food Science and Human Wellness*, 12(3): 912–919.

Li, M., Ji, D., Pamudji, M. S., Lui, K. H., Zhao, Y., Zhao, G. et al. (2023b). Occurrence, risk, and treatment of ciprofloxacin and clarithromycin in drainage. *Chemical Engineering Journal*, 466: 142968.

Li, X., Lin, X., Lin, S., Zhou, S., Fang, G., Zhao, H. et al. (2021). From dilute to multiple layers: bottom-up self-assembly of rough gold nanorods as SERS platform for quantitative detection of thiram in soil. *Advanced Materials Interfaces*, 8(12), 2100412.

Liang, W., Lin, H., Chen, J., and Chen, C. (2016). Utilization of nanoparticles in microfluidic systems for optical detection. *Microsystem Technologies*, 22: 2363–2370.

Liang, X., Zhong, T., Quan, B., Wang, B., and Guan, H. (2008). Solid-state potentiometric SO2 sensor combining NASICON with V2O5-doped TiO2 electrode. *Sensors and Actuators B: Chemical*, 134(1): 25–30.

Lin, C. Y., Fang, Y. Y., Lin, C. W., Tunney, J. J., and Ho, K. C. (2010). Fabrication of NOx gas sensors using In2O3–ZnO composite films. *Sensors and Actuators B: Chemical*, 146(1): 28–34.

Liu, D., Chen, B., An, J., Li, C., Liu, G., Shao, J. et al. (2020). Wind-driven self-powered wireless environmental sensors for Internet of Things at long distance. *Nano Energy*, 73: 104819.

Liu, K., Li, F., Tian, Q., Nie, C., Ma, Y., Zhu, Z. et al. (2021). Aáhighlyáporousáanimalábone-derivedáchar with a superiority of promoting nZVI for Cr (VI) sequestration in agricultural soils. *Journal of Environmental Sciences*, 104: 27–39.

Liu, L., Zhao, X., Hu, T., Liang, F., Guo, B., and Tao, K. (2024). Deep-learning-assisted self-powered wireless environmental monitoring system based on triboelectric nanogenerators with multiple sensing capabilities. *Nano Energy*, 132: 110301.

Liu, Z., Miyauchi, M., Yamazaki, T., and Shen, Y. (2009). Facile synthesis and NO2 gas sensing of tungsten oxide nanorods assembled microspheres. *Sensors and Actuators B: Chemical*, 140(2): 514–519.

Lv, H., Zhang, G., Yang, W., Dai, X., Huang, Y., Ni, J. et al. (2023). Portable anti-fouling electrochemical sensor for soil heavy metal ions detection based on the screen-printed carbon electrode modified with silica isoporous membrane. *Journal of Electroanalytical Chemistry*, 930: 117141.

Ma, D., Peh, S. B., Han, G., and Chen, S. B. (2017). Thin-film nanocomposite (TFN) membranes incorporated with super-hydrophilic metal–organic framework (MOF) UiO-66: toward enhancement of water flux and salt rejection. *ACS Applied Materials & Interfaces*, 9(8): 7523–7534.

Ma, S., Cao, F., Wen, X., Xu, F., Tian, H., Fu, X. et al. (2023). Detection of heavy metal ions using laser-induced breakdown spectroscopy combined with filter paper modified with PtAg bimetallic nanoparticles. *Journal of Hazardous Materials*, 443: 130188.

Mamalis, B., and Gerakidis, S. (2024). A Combined Environmental Monitoring Framework based on WSN Clustering and VANET Edge Computation Offloading. *arXiv preprint arXiv:2411.10560*.

Mangotra, A., and Singh, S. K. (2024). Volatile organic compounds: A threat to the environment and health hazards to Living Organisms–A Review. *Journal of Biotechnology*.

Manisalidis, I., Stavropoulou, E., Stavropoulos, A., and Bezirtzoglou, E. (2020). Environmental and health impacts of air pollution: a review. *Frontiers in Public Health*, 8: 14.

Matos, M. P., Correia, A. A. S., and Rasteiro, M. G. (2017). Application of carbon nanotubes to immobilize heavy metals in contaminated soils. *Journal of Nanoparticle Research*, 19: 1–11.

Meimand, M. M., Javid, N., and Malakootian, M. (2019). Adsorption of sulfur dioxide on clinoptilolite/nano iron oxide and natural clinoptilolite. *Health Scope*, 8(2).

Mitra, S., Saran, R. K., Srivastava, S., and Rensing, C. (2024). Pesticides in the environment: Degradation routes, pesticide transformation products and ecotoxicological considerations. *Science of The Total Environment*, 173026.

Mondal, S. K., Chakraborty, S., Manna, S., and Mandal, S. M. (2024). Antimicrobial nanoparticles: current landscape and future challenges. *RSC Pharmaceutics*.

Moradeeya, P. G., Kumar, M. A., Sharma, A., and Basha, S. (2022b). Conductive polymer layered semiconductor for degradation of triclopyr acid and 2, 4-dichlorophenoxyacetic acid from aqueous stream using coalesce adsorption-photocatalysis technique. *Chemosphere*, 298: 134360.

Moradeeya, P. G., Kumar, M. A., Thorat, R. B., Rathod, M., Khambhaty, Y., and Basha, S. (2017). Nanocellulose for biosorption of chlorpyrifos from water: chemometric optimization, kinetics and equilibrium. *Cellulose*, 24: 1319–1332.

Moradeeya, P. G., Sharma, A., Kumar, M. A., and Basha, S. (2022a). Titanium dioxide based nanocomposites–current trends and emerging strategies for the photocatalytic degradation of ruinous environmental pollutants. *Environmental Research*, 204: 112384.

Munir, M. A. M., Liu, G., Yousaf, B., Ali, M. U., Cheema, A. I., Rashid, M. S. et al. (2020). Bamboo-biochar and hydrothermally treated-coal mediated geochemical speciation, transformation and uptake of Cd, Cr, and Pb in a polymetal (iod) s-contaminated mine soil. *Environmental Pollution*, 265: 114816.

Najafi, H., Farajfaed, S., Zolgharnian, S., Mirak, S. H. M., Asasian-Kolur, N., and Sharifian, S. (2021). A comprehensive study on modified-pillared clays as an adsorbent in wastewater treatment processes. *Process Safety and Environmental Protection*, 147: 8–36.

Nama, M., Satasiya, G., Sahoo, T. P., Moradeeya, P. G., Sadukha, S., Singhal, K. et al. (2024). Thermo-chemical behaviour of Dunaliella salina biomass and valorising their biochar for naphthalene removal from aqueous rural environment. *Chemosphere*, 353: 141639.

Ngqwala, N. P., and Muchesa, P. (2020). Occurrence of pharmaceuticals in aquatic environments: A review and potential impacts in South Africa. *South African Journal of Science*, 116(7–8): 1–7.

Nie, J., Xie, H., Zhang, M., Liang, J., Nie, S., and Han, W. (2020). Effective and facile fabrication of MOFs/cellulose composite paper for air hazards removal by virtue of in situ synthesis of MOFs/chitosan hydrogel. *Carbohydrate Polymers*, 250: 116955.

Nowack, B., Ranville, J. F., Diamond, S., Gallego-Urrea, J. A., Metcalfe, C., Rose, J. et al. (2012). Potential scenarios for nanomaterial release and subsequent alteration in the environment. *Environmental Toxicology and Chemistry*, 31(1): 50–59.

Obasi, P. N., and Akudinobi, B. E. B. (2019). Heavy metals occurrence, assessment and distribution in water resources of the lead–zinc mining areas of Abakaliki, Southeastern Nigeria. *International Journal of Environmental Science and Technology*, 16: 8617–8638.

Ohlsen, K., Ternes, T., Werner, G., Wallner, U., Löffler, D., Ziebuhr, W. et al. (2003). Impact of antibiotics on conjugational resistance gene transfer in Staphylococcus aureus in sewage. *Environmental Microbiology*, 5(8): 711–716.

Ong, T. T., Blanch, E. W., and Jones, O. A. (2020). Surface Enhanced Raman Spectroscopy in environmental analysis, monitoring and assessment. *Science of The Total Environment*, 720: 137601.

Onyeaka, H., Ghosh, S., Obileke, K., Miri, T., Odeyemi, O. A., Nwaiwu, O. et al. (2024). Preventing chemical contaminants in food: Challenges and prospects for safe and sustainable food production. *Food Control*, 155: 110040.

Pawar, K. K., Kumar, A., Mirzaei, A., Kumar, M., Kim, H. W., and Kim, S. S. (2024). 2D nanomaterials for realization of flexible and wearable gas sensors: A review. *Chemosphere*, 141234.

Peláez, L. M. G., Santos, J. M., de Almeida Albuquerque, T. T., Reis Jr, N. C., Andreão, W. L., and de Fátima Andrade, M. (2020). Air quality status and trends over large cities in South America. *Environmental Science & Policy*, 114: 422–435.

Pervez, A., and Yadav, M. (2023). Nanomaterials in the environment: impacts and challenges. pp. 389–414. *In: Nanoparticles and Plant-Microbe Interactions*. Academic Press.

Piñeiro, R., Jimenez-Relinque, E., Nevshupa, R., and Castellote, M. (2021). Primary and secondary emissions of VOCs and PAHs in indoor air from a waterproof coal-tar membrane: diagnosis and remediation. *International Journal of Environmental Research and Public Health*, 18(23): 12855.

Qamar, W., Gulia, S., Athar, M., Ahmad, R., Imam, M. T., Chandra, P. et al. (2024). An insight into impact of nanomaterials toxicity on human health. *PeerJ*, 12: e17807.

Qiao, X., He, J., Yang, R., Li, Y., Chen, G., Xiao, S. et al. (2022). Recent advances in nanomaterial-based sensing for food safety analysis. *Processes*, 10(12): 2576.

Rashid, M. H. O., and Ralph, S. F. (2017). Carbon nanotube membranes: synthesis, properties, and future filtration applications. *Nanomaterials*, 7(5): 99.

Rather, I. A., Koh, W. Y., Paek, W. K., and Lim, J. (2017). The sources of chemical contaminants in food and their health implications. *Frontiers in pharmacology*, 8: 830.

Rathod, M., Moradeeya, P. G., Haldar, S., and Basha, S. (2018). Nanocellulose/TiO 2 composites: preparation, characterization and application in the photocatalytic degradation of a potential endocrine disruptor, mefenamic acid, in aqueous media. *Photochemical & Photobiological Sciences*, 17: 1301–1309.

Remoundou, K., and Koundouri, P. (2009). Environmental effects on public health: An economic perspective. *International journal of environmental research and public health*, 6(8): 2160–2178.

Ren, L., Yang, W., and Bai, Z. (2017). Characteristics of major air pollutants in China. *Ambient air pollution and health impact in China*, 7–26.

Riman, D., Jirovsky, D., Hrbac, J., and Prodromidis, M. I. (2015). Green and facile electrode modification by spark discharge: Bismuth oxide-screen printed electrodes for the screening of ultra-trace Cd (II) and Pb (II). *Electrochemistry Communications*, 50: 20–23.

Roh, S., Nguyen, T. D., and Lee, J. S. (2024). Applications of Nanomaterials in RFID Wireless Sensor Components. *Applied Sciences*, 14(3): 1216.

Rubira, R. J., Camacho, S. A., Constantino, C. J., and Sanchez-Cortes, S. (2022). Increasing the sensitivity of surface-enhanced Raman scattering detection for s-triazine pesticides by taking advantage of interactions with soil humic substances. *Journal of Raman Spectroscopy*, 53(1): 40–48.

Sahoo, T. P., Satasiya, G., Moradeeya, P. G., Saravaia, H. T., and Kumar, M. A. (2024). Removal of fluoroquinolone antibiotic and sulfonated dye by functionalized Persea americana seed powder: Appraisal on phase transfer kinetics, equilibrium, economics, and applications in rural settings. *Environmental Research*, 261: 119727.

Saleem, H., and Zaidi, S. J. (2020). Developments in the application of nanomaterials for water treatment and their impact on the environment. *Nanomaterials*, 10(9): 1764.

Saleem, H., Zaidi, S. J., Ismail, A. F., and Goh, P. S. (2022). Advances of nanomaterials for air pollution remediation and their impacts on the environment. *Chemosphere*, 287: 132083.

Sall, M. L., Diaw, A. K. D., Gningue-Sall, D., Efremova Aaron, S., and Aaron, J. J. (2020). Toxic heavy metals: impact on the environment and human health, and treatment with conducting organic polymers, a review. *Environmental Science and Pollution Research*, 27: 29927–29942.

Sepahvand, S., Bahmani, M., Ashori, A., Pirayesh, H., Yu, Q., and Dafchahi, M. N. (2021). Preparation and characterization of air nanofilters based on cellulose nanofibers. *International Journal of Biological Macromolecules*, 182: 1392–1398.

Šimatović, A., and Udiković-Kolić, N. (2020). Antibiotic resistance in pharmaceutical industry effluents and effluent-impacted environments. *Antibiotic Resistance in the Environment: A Worldwide Overview*, 101–122.

Singh, A. K., Lakshmi, G. B. V. S., Dhiman, T. K., Kaushik, A., and Solanki, P. R. (2021). Bio-active free direct optical sensing of aflatoxin B1 and ochratoxin A using a manganese oxide nano-system. *Frontiers in Nanotechnology*, 2: 621681.

Singh, J., and Lee, B. K. (2016). Influence of nano-TiO2 particles on the bioaccumulation of Cd in soybean plants (Glycine max): A possible mechanism for the removal of Cd from the contaminated soil. *Journal of Environmental Management*, 170: 88–96.

Singha, D. K., Majee, P., Mondal, S. K., and Mahata, P. (2017). Highly Selective Aqueous Phase Detection of Azinphos-Methyl Pesticide in ppb Level Using a Cage-Connected 3D MOF. *ChemistrySelect*, 2(20): 5760–5768.

Sirivallop, A., Areerob, T., and Chiarakorn, S. (2020). Enhanced visible light photocatalytic activity of N and Ag doped and co-doped TiO2 synthesized by using an *in-situ* solvothermal method for gas phase ammonia removal. *Catalysts*, 10(2): 251.

Su, H., Fang, Z., Tsang, P. E., Fang, J., and Zhao, D. (2016). Stabilisation of nanoscale zero-valent iron with biochar for enhanced transport and *in-situ* remediation of hexavalent chromium in soil. *Environmental Pollution*, 214: 94–100.

Su, Z., Zhang, M., Lu, Z., Song, S., Zhao, Y., and Hao, Y. (2018). Functionalization of cellulose fiber by in situ growth of zeolitic imidazolate framework-8 (ZIF-8) nanocrystals for preparing a cellulose-based air filter with gas adsorption ability. *Cellulose*, 25: 1997–2008.

Sun, X., Du, S., and Wang, X. (2012). Amperometric immunosensor for carbofuran detection based on gold nanoparticles and PB-MWCNTs-CTS composite film. *European Food Research and Technology*, 235: 469–477.

Sun, X., Qiao, L., and Wang, X. (2013). A novel immunosensor based on Au nanoparticles and polyaniline/multiwall carbon nanotubes/chitosan nanocomposite film functionalized interface. *Nano-Micro Letters*, 5: 191–201.

Tarannum, N., Gautam, A., Chauhan, T., and Kumar, D. (2024). Nanomaterial based sensors for detection of food contaminants: a prospect. *Sensing Technology*, 2(1): 2373196.

Tejashwini, D. M., Deepa, H. M., Deshmukh, V. V., and Naik, Y. V. (2024). Challenges and Opportunities in Nanomaterials for Energy Storage. *Advanced Nanomaterials for Energy Storage Devices*, 165–181.

Thakur, A., and Kumar, A. (2022). Recent advances on rapid detection and remediation of environmental pollutants utilizing nanomaterials-based (bio) sensors. *Science of The Total Environment*, 834: 155219.

Usman, M., and Balsalobre-Lorente, D. (2022). Environmental concern in the era of industrialization: can financial development, renewable energy and natural resources alleviate some load?. *Energy Policy*, 162: 112780.

Vaisakh, P. S., Adarsh, U. K., Amrutha, K., Warrier, A. K., Kartha, V. B., and Unnikrishnan, V. K. (2023). Integrated LIBS-Raman spectroscopy: A comprehensive approach to monitor microplastics and heavy metal contamination in water resources. *Environmental Research*, 231: 116198.

Verma, M., Chaudhary, M., Singh, A., Kaur, N., and Singh, N. (2020). Naphthalimide-gold-based nanocomposite for the ratiometric detection of okadaic acid in shellfish. *Journal of Materials Chemistry B*, 8(36): 8405–8413.

Wang, C., Chu, X., and Wu, M. (2006). Detection of H2S down to ppb levels at room temperature using sensors based on ZnO nanorods. *Sensors and Actuators B: Chemical*, 113(1): 320–323.

Wang, G., Xu, Z., Qi, Y., Fang, Y., Ning, G., and Ye, J. (2024). Electrospun nanofibrous membranes with antimicrobial activity for air filtration. *Chinese Chemical Letters*, 109503.

Wang, H., Zhao, G., Yin, Y., Wang, Z., and Liu, G. (2017). Screen-printed electrode modified by bismuth/fe3o4 nanoparticle/ionic liquid composite using internal standard normalization for accurate determination of Cd (II) in Soil. *Sensors*, 18(1): 6.

Wang, S., Sun, H., Ang, H. M., and Tadé, M. O. (2013). Adsorptive remediation of environmental pollutants using novel graphene-based nanomaterials. *Chemical Engineering Journal*, 226: 336–347.

Wang, X., Zhang, L., Hao, A., Shi, Z., Dai, C., Yang, Y. et al. (2020). Silica-coated silver nanoparticles decorated with fluorescent CdTe quantum dots and DNA aptamers for detection of tetracycline. *ACS Applied Nano Materials*, 3(10): 9796–9803.

Wu, A., Hu, X., Ao, H., Chen, Z., Chu, Z., Jiang, T. et al. (2022). Rational design of bacterial cellulose-based air filter with antibacterial activity for highly efficient particulate matters removal. *Nano Select*, 3(1): 201–211.

Xia, J., Lu, H., Chen, G., Lin, D., Yang, W., Liu, C. et al. (2024). High Performance Piezoelectric Nanogenerator by Fiber Microstructure Engineering toward Self-Powered Wireless Sensing System. *Nano Energy*, 109901.

Xie, R., Yang, P., Liu, J., Zou, X., Tan, Y., Wang, X. et al. (2021). Lanthanide-functionalized metal-organic frameworks based ratiometric fluorescent sensor array for identification and determination of antibiotics. *Talanta*, 231: 122366.

Xu, X., Gao, J., Zhang, Y., and Zhang, L. (2023). Tailored novel multifunctional benzyl-functionalized magnetic ionic liquid for rapid and efficient monitoring of trace fluoroquinolones in food samples. *Food Chemistry*, 404: 134654.

Xu, Y., and Zhao, D. (2007). Reductive immobilization of chromate in water and soil using stabilized iron nanoparticles. *Water Research*, 41(10): 2101–2108.

Yang, M., Ye, Z., Ren, Y., Farhat, M., and Chen, P. Y. (2023). Recent advances in nanomaterials used for wearable electronics. *Micromachines*, 14(3): 603.

Ying, Y., Ying, W., Li, Q., Meng, D., Ren, G., Yan, R. et al. (2017). Recent advances of nanomaterial-based membrane for water purification. *Applied Materials Today*, 7: 144–158.

Yu, L., Zhang, P., Dai, H., Chen, L., Ma, H., Lin, M. et al. (2017). An electrochemical sensor based on Co 3 O 4 nanosheets for lead ions determination. *RSC Advances*, 7(63): 39611–39616.

Yuan, J., Hu, H., Chen, M., Shi, J., and Shangguan, W. (2008). Promotion effect of Al2O3–SiO2 interlayer and Pt loading on TiO2/nickel-foam photocatalyst for degrading gaseous acetaldehyde. *Catalysis Today*, 139(1–2): 140–145.

Zeng, Q. R., Feng, J. T., Lin, X. C., Zhao, Y., Liu, H., Wang, S. Z. et al. (2020). One-step facile synthesis of a NiO/ZnO biomorphic nanocomposite using a poplar tree leaf template to generate an enhanced gas sensing platform to detect n-butanol. *Journal of Alloys and Compounds*, 815: 150550.

Zhang, C., Chen, J., Kuang, H., Wu, J., Li, J., Zhang, K. et al. (2022). Self-powered autonomous wireless sensor system with multivariable sensing capability. *Nano Energy*, 104: 107939.

Zhang, X., Ren, Y., Ji, Z., and Fan, J. (2020). Sensitive detection of amoxicillin in aqueous solution with novel fluorescent probes containing boron-doped carbon quantum dots. *Journal of Molecular Liquids*, 311: 113278.

Zhang, Y., Wang, X., Simpson, I. J., Barletta, B., Blake, D. R., Meinardi, S. et al. (2014). Ambient CFCs and HCFC-22 observed concurrently at 84 sites in the Pearl River Delta region during the 2008–2009 grid studies. *Journal of Geophysical Research: Atmospheres*, 119(12): 7699–7717.

Zhao, D., Wang, P., and Zhao, F. J. (2024). Toxic Metals and Metalloids in Food: Current Status, Health Risks, and Mitigation Strategies. *Current Environmental Health Reports*, 1–16.

Zhao, T., Liang, X., Guo, X., Yang, X., Guo, J., Zhou, X. et al. (2023). Smartphone-based colorimetric sensor array using gold nanoparticles for rapid distinguishment of multiple pesticides in real samples. *Food Chemistry*, 404: 134768.

Zhou, R., Tang, Z., Liu, K., Zhang, W., Liu, K., Li, X. et al. (2022). High sensitivity analysis of soil by laser-induced breakdown spectroscopy with Ag nanoparticles. *Optics & Laser Technology*, 155: 108386.

CHAPTER-13

Next-Generation Smart Nanomaterials for Sustainable Environment

Kusum Sharma,[1,*] *Lebanan Keisham,*[1] *Megha Barot,*[2,3] *Arpan Modi*[4] and *Azamal Husen*[5]

1. Introduction

One of the most revolutionary scientific discoveries of the twenty-first century is nanotechnology, which provides previously unheard-of control over matter at the atomic and molecular level. Because of this accuracy, it is possible to create nanostructures with distinct physicochemical properties that set them apart from their bulk counterparts. Nanomaterials have become extremely adaptable in a variety of fields, including environmental remediation, agriculture, food technology, biotechnology, and medicine, because of their unique attributes (Thakur et al. 2018). Their uses include antibacterial agents (Islam et al. 2022), functional food additives (Chen et al. 2023), wastewater treatment (Zahra et al. 2020), and environmental monitoring (Rassaei et al. 2011).

The demand for eco-friendly and sustainable technologies has increased due to increasing concerns about resource depletion, ecological deterioration, and climate change. Among these, smart nanomaterials (SNMs), a novel class of synthetic nanomaterials with adaptive and stimuli-responsive properties, provide flexible ways to address environmental issues (Aflori 2021). These materials offer precision-based interventions in resource optimization, sustainable farming, and pollution management by effectively interacting with biological systems and abiotic environmental elements. They offer promising solutions in different sectors (Figure 1). Both bottom-up (biological) and top-down (physical/chemical) methods can be used to synthesize nanomaterials (Tripathy et al. 2023). Moreover, the synthesis of SNMs doesn't use harmful chemicals and uses biological

[1] Department of Life Science, Rabindranath Tagore University, Raisen, Madhya Pradesh-464993, India. ORCID ID: 0000-0002-7135-3861, ORCID ID: 0000-0002-6535-071X

[2] Research and Development Cell, Parul University; Department of Environmental Science, Parul Institute of Applied Sciences, Parul University, Vadodara, Gujarat, India. ORCID ID: 0000-0002-9680-0760

[3] Department of Environmental Science, Parul Institute of Applied Sciences, Parul University, Vadodara, Gujarat, India.

[4] Axenic Agritech Private Limited, Ahmedabad, Gujarat, India. ORCID ID: 0000-0001-8949-6660

[5] Wolaita Sodo University, PO Box 138 Wolaita, Ethiopia. ORCID ID: 0000-0002-9120-5540

* Corresponding author: kusum.sharma@aisectuniversity.ac.in

DOI: 10.1201/9781003558804-13

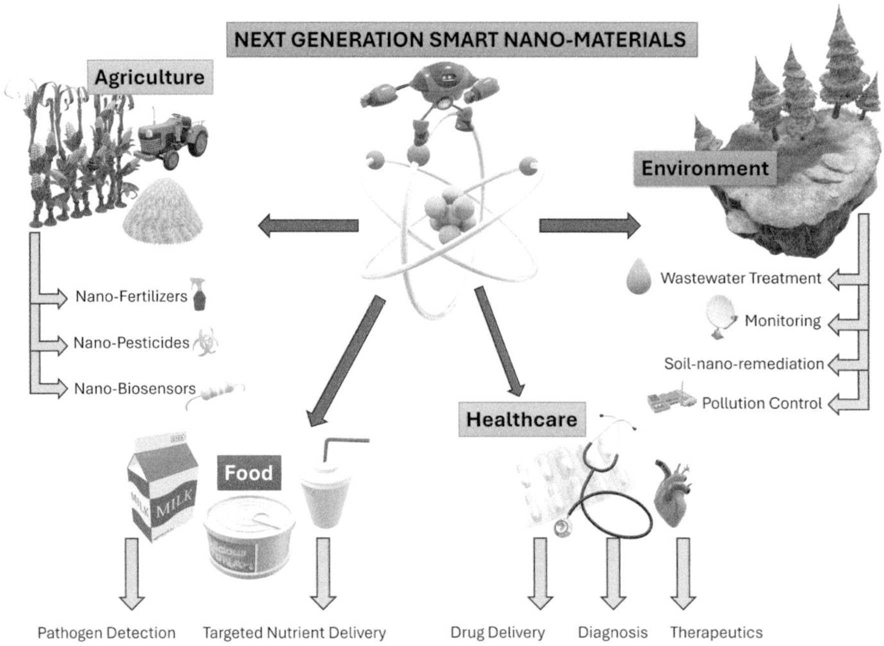

Figure 1: Application of smart nanomaterials in different sectors viz. Agriculture, Food, Healthcare and Environment.

processes like plants, bacteria, fungi, or algae—the biogenic or green synthesis—which has drawn a lot of interest. The manufacture of silver nanoparticles using leaf extracts from *Brassica juncea* and *Azadirachta indica*, for instance, was effectively demonstrated by Akinola et al. (2020). These extracts showed strong antibacterial activity and soil-enhancing properties.

In agriculture, smart pesticide delivery platforms are incorporating SNMs increasingly into crop management plans. Leaching, volatilization, and microbial degradation are common causes of significant losses from traditional fertilizer and pesticide application techniques, contaminating the environment (Boudh and Singh 2018). Using technological advances, it is possible to encapsulate agrochemicals into nanocarriers, which offer regulated and prolonged release, addressing these inefficiencies. Pest control against *Helicoverpa armigera* was considerably increased by biopolymeric nano-pesticides based on chitosan and alginate (Kumar et al. 2022). Additionally, SNMs are essential parts of technology used in precision agriculture. Highly sensitive detection of traces of infections, heavy metals, and nutritional imbalances is possible with advanced nanosensors functionalized with biological recognition elements. Zhao et al. (2025) developed a graphene oxide-based nanosensor capable of detecting lead ions in irrigation water at concentrations as low as 10 parts per billion (ppb). Research is also being done on using nanocarriers to transfer genetic material into plants. In order to promote gene expression in *Nicotiana tabacum* without integrating into the host genome, Zhou et al. (2024) employed gold-capped mesoporous silica nanoparticles. This technique has the potential to completely transform trait-specific plant breeding.

All of these technological advancements highlight the enormous potential of next-generation smart nanomaterials to promote sustainable development in the fields of agriculture and the environment. With an emphasis on their contributions to environmental sustainability, this chapter examines the synthesis techniques, mechanistic applications, and system-level integration of SNMs. In order to establish SNMs as essential instruments for creating resilient, resource-efficient, and ecologically harmonious ecosystems, emphasis is given on green synthesis techniques, and smart pesticide delivery.

Table1: Classification and Functional Applications of Smart Nanomaterials (SNMs).

Type of SNMs	Examples	Key Properties	Applications	References
Metal and metal oxide nanoparticles	TiO_2, ZnO, Fe_3O_4	Photocatalytic, antimicrobial, magnetic	Water purification, heavy metal removal, pollutant degradation	Singh et al. 2023
Carbon-based nanomaterials	Graphene, CNTs, Fullerenes	High electrical conductivity, adsorption capacity, mechanical strength	Environmental sensing, contaminant adsorption, soil enhancement	Kulkarni et al. 2024
Polymeric nanomaterials	Smart polymers, stimuli-responsive hydrogels	Reversible changes with pH, temperature, enzymes	Controlled release of agrochemicals, pollutant capture	Kumar et al. 2022
Composite nanomaterials	Metal-polymer hybrids, metal-carbon composites	Multifunctional (adsorption and catalysis)	Wastewater treatment, smart delivery systems, multifunctional environmental remediation	Zahra et al. 2020, Zhao et al. 2024

2. Classification and Synthesis of Smart Nanomaterials

The composition and functional characteristics of smart nanomaterials (SNMs) allow for a broad classification, with each category providing unique benefits for agricultural and environmental applications. Due to their photocatalytic, antimicrobial, and magnetic qualities, metal and metal oxide nanoparticles such as titanium dioxide (TiO_2), zinc oxide (ZnO), and magnetite (Fe_3O_4) have attracted a lot of interest (Table 1). These properties make them ideal for heavy metal removal, water purification, and pollutant degradation (Singh et al. 2023).

2.1 Classification

Carbon-based nanomaterials, such as graphene, carbon nanotubes (CNTs), and fullerenes, are useful for environmental sensing because of their remarkable electrical conductivity, mechanical strength, and adsorption capabilities (Kulkarni et al. 2024). Environmental factors such as pH, temperature, or other changes can cause polymeric nanomaterials—especially those made of smart or stimuli-responsive polymers—to experience reversible physical or chemical changes (Table 1). This characteristic makes it possible to selectively absorb or release agrochemicals and contaminants, increasing the efficiency of input utilization and lowering ecological risks (Kumar et al. 2022). The beneficial qualities of each element are integrated into composite nanomaterials, which are created by merging two or more material types. According to Zahra et al. (2020) and Zhao et al. (2024), these hybrids are being investigated for advanced applications in wastewater treatment, smart delivery systems, and environmental monitoring. They demonstrate multifunctionality, including simultaneous adsorption and catalysis. Given the variety of SNM types, a flexible toolkit for creating focused, effective, and long-lasting solutions that address contemporary environmental and agricultural issues is made available.

2.2 Synthesis Approaches

Smart nanomaterials' (SNMs') physicochemical properties and future uses are largely determined by their manufacture. For the manufacturing of nanoparticles, conventional physical and chemical synthesis techniques like chemical reduction, laser ablation, and ball milling have been extensively employed. But these methods frequently need a lot of energy, dangerous chemicals, and toxic byproducts, which raise questions about safety and the environment (Kumar et al. 2022). On the other hand, because of their affordability, scalability, and lack of negative effects on the environment, green synthesis techniques, which make use of biological systems including plant extracts, fungi, bacteria, and algae, have become increasingly popular (Singh et al. 2023). These environmentally friendly methods adhere to the principles of green chemistry by doing away with the requirement for hazardous reducing chemicals and high temperatures that are usually needed in conventional synthesis. It has been reported that *Ocimum sanctum* (Tulsi) extract was used to successfully biosynthesize silver nanoparticles, which showed notable antibacterial activity and size homogeneity (Shahzadi et al. 2025). The adaptability of green synthesis, which creates metal nanoparticles with particular morphologies utilizing a range of natural precursors like citrus fruits, tea leaves, and microbial biomass, was also highlighted by Bahrulolum et al. (2021).

According to a study by Singh et al. (2024), *Sargassum muticum* algae have the ability to biosynthesize gold nanoparticles, and their high biomass yield and quick development make them useful for large-scale manufacturing. Furthermore, fungi like *Fusarium oxysporum* have been used to produce metal oxide nanoparticles (Gaber et al. 2024) extracellularly because of their high metal ion tolerance and ability to secrete stabilizing enzymes (Sharma et al. 2022). These advances not only highlight the sustainability of green synthesis techniques but also the mounting evidence in favor of their use in agriculture, biomedicine, and environmental remediation.

3. Environmental Applications of Smart Nanomaterials

3.1 Water and Wastewater Treatment

The unique physicochemical characteristics of smart nanomaterials (SNMs), such as high surface area, reactivity, and catalytic potential, make them effective for enhancing water and wastewater treatment (Ahmed et al. 2022). Among these, zero-valent iron nanoparticles (nZVI) have been extensively studied for their capacity to immobilize hazardous heavy metals and break down chlorinated organic compounds (Yan et al. 2024). Further, as per the study of Zhang et al. 2023, nZVI was used at a vinyl chloride-contaminated site and reported over 90% degrading efficiency in the upper and middle soil layers, proving groundwater remediation. Similar to this, Ahmed et al. (2021) developed ZVI–chitosan hybrid beads that successfully eliminated arsenic from tainted water, bringing concentrations below WHO-permissible limits. Additionally, Antony et al. (2022) investigated the removal of excess iron from roof-harvested rainwater using encapsulated and immobilized nZVI. Their findings confirmed the application of nZVI in low-cost decentralized water filtration by showing a notable improvement in water quality. Furthermore, Liu et al. (2022) found that nZVI and *Paracoccus* sp. bacteria worked in concert to remove nitrate, increasing denitrification efficiency from 66.9% to 85.2%, indicating the use of an integrated nano-bioremediation approach. Apart from nZVI, titanium dioxide (TiO_2) nanoparticles are widely used for the photocatalytic breakdown of organic contaminants in the presence of visible and ultraviolet light (Kholief et al. 2024). Modified TiO_2-based photocatalysts were evaluated by Xu et al. (2025), who reported that metal and non-metal doping greatly increased photocatalytic effectiveness. Another noteworthy study is the fabrication of Fe–Ag co-doped TiO_2 nanoparticles by Rao et al. (2021), which were applied to the elimination of pesticide residues from flumioxazin. The potential of these co-doped photocatalysts to clean pesticide-contaminated agricultural runoff was demonstrated by their total breakdown under sun irradiation in under 60 minutes. All of these studies support the vital role SNMs play in facilitating scalable, sustainable, and effective water purification systems. The next generation of environmental remediation technologies cannot function without them due to their high reactivity, reusability, and compatibility with both biological and physical systems.

3.2 Air Pollution Control

Airborne pollutants that contribute to respiratory ailments, smog formation, and ecosystem damage include nitrogen oxides (NO_x), volatile organic compounds (VOCs), and particulate matter (PM) is defined as microscopic solid and liquid particles suspended in the atmosphere, including fine ($PM_2.^2$) and coarse ($PM_1{}^1$) particles that are extremely dangerous to the environment and human health. These pollutants are among the most urgent environmental hazards. The enormous surface area, adjustable catalytic activity, and multifunctional characteristics of smart nanomaterials (SNMs) make them innovative and effective tools for reducing air pollution. In the selective catalytic reduction (SCR) of NO_x at low temperatures, MnO_x–CeO_x hybrid nanoparticles based on nitrogen-doped reduced graphene oxide (N-rGO) have shown exceptional catalytic performance. Zhang et al. (2020) demonstrated that these catalysts were suitable for industrial flue gas treatment since they demonstrated over 90% NO conversion at 150°C and excellent resistance to SO_2 and H_2O poisoning.

The potential of Mn–TiO_2@GO nanocomposites in passive air purification systems for urban settings was demonstrated by Li et al. (2023), with excellent NO removal efficiency when exposed to visible light (Amin et al. 2025). The enormous surface area, hydrophobic nature, and porous 3D architectures of SNMs, like graphene aerogels, have drawn great interest for their potential to mitigate volatile organic compounds. Huang et al. (2023) showed that chemically reduced graphene aerogels could adsorb significant amounts of acetic acid and formaldehyde vapors, surpassing conventionally activated carbon by about 40% (Nassar et al. 2021). This suggests that these aerogels could be used in indoor air filtration systems. Furthermore, iron oxide–graphene nanohybrids have been created to improve amphiphilic VOC adsorption. These nanohybrids were found to have a 15-fold higher adsorption capacity for benzene and toluene than commercial carbon filters (Yadav et al. 2022).

Graphene-enhanced ceramic air filters have demonstrated the ability to capture over 95% of airborne particles larger than 1 μm while offering self-cleaning functionality through integrated nanonetworks (Reimers et al. 2023). The operating lifespan of these filters in industrial and medical settings can be greatly increased by thermally degrading bioaerosols and VOC residues (Yang et al. 2024). These results demonstrate the versatility of SNMs in addressing a range of air pollutants—by means of adsorption, filtration, and catalytic decomposition—and emphasize their critical role in the creation of environmentally friendly and effective next-generation air purification technologies.

3.3 Soil Remediation

Smart nanomaterials (SNMs), which provide strategies including pollutant immobilization, adsorption, and catalytic destruction, are becoming more and more important in the rehabilitation of contaminated soils. Since they have a strong affinity for heavy metals, iron oxide nanoparticles are well known for lowering the bioavailability of metals in soil systems. Applying phytogenic iron oxide nanoparticles (PIONs) to soils contaminated with cadmium (Cd) decreased the exchangeable Cd fraction by up to 91% in oxic conditions, as shown by Lin et al. (2019). The production of insoluble hydroxides and increased soil pH, which fixed the metal in less mobile forms, were the causes of the immobilization, according to Lin et al. (2019).

SNMs can also be added to biochar matrices to create hybrid nanocomposites that enhance soil health while also remediating pollutants. This is another promising strategy. One study found that applying biochar-supported nano-zero-valent iron (nZVI@BC) to soils contaminated with several metals significantly increased the activity of soil catalase (Zhang et al. 2023). The residual percentages of Cu, Zn, As, Cd, and Pb increased by 7.44% to 12.75%, indicating a considerable decrease in phytoavailability and effective immobilization of heavy metals (Zhang et al. 2023).

Mazarji et al. (2023) have reviewed the synergy between biochar and nanomaterials in detail. They highlighted how the immobilization of heavy metals like Pb, Cd, Cu, Zn, Cr, and As was greatly improved by biochar composites combined with FeO_4, FeS nanoparticles, graphene oxide, and chitosan-based nanostructures. Additionally, these biochar-based nanomaterials promoted enzymatic and microbial activity in soils, supporting ecological balance and long-term fertility (Mazarji et al. 2023). Furthermore, Thabet et al. (2024) investigated the use of magnesium oxide nanoparticles embedded in wheat husk biochar for the rehabilitation of soils contaminated with lead. Soil restoration and agricultural productivity were highlighted by decreased Pb bioavailability by more than 80% and markedly increased seed germination and biomass in test crops (Naz et al. 2025).

3.4 Agricultural Enhancements

In agriculture, smart nanomaterials (SNMs) are being used more and more to improve crop performance, particularly in the face of abiotic stress. SNMs improve nutrient use efficiency and lower agrochemical losses by enabling controlled release, targeted delivery, and stress tolerance through their roles as nanopesticides, growth enhancers, and nanofertilizers. Zinc oxide nanoparticles (ZnO NPs) have shown significant promise in enhancing plant vigor and seed germination under stress. In a recent study, ZnO NPs (10 mg/L) seed priming improved germination, chlorophyll content, and α-amylase activity in wheat (Triticum aestivum) under salinity stress (Rani and Jha 2021). Similarly, in drought conditions, rice (Oryza sativa) seeds primed with 25 ppm ZnO NPs showed enhanced plant height, biomass accumulation, and chlorophyll levels (Kumar et al. 2023). Furthermore, ZnO NPs have demonstrated encouraging results in legumes. Applying ZnO NPs at 50 ppm to *Pisum sativum* (pea) in saline circumstances greatly increased root length, dry weight, and pod number, according to Mustafa et al. (2024). Additionally, the treatment decreased oxidative damage, suggesting enhanced pathways for stress tolerance (Liu et al. 2024).

Further study evaluated ZnO NPs in *Phaseolus vulgaris* (common bean) under salt stress using soil application, foliar spray, and nano-priming (Gupta et al. 2024). It has been reported that nano-priming was the most effective because it improved nutrient balance and decreased the buildup of reactive oxygen species (ROS) (Gupta and Bharati 2024). Furthermore, the use of a chitosan-based nanopesticide on *Brassica juncea* increased yield by 20% over the untreated control while also controlling *Alternaria* infection (Giri et al. 2023). SNMs are next-generation agricultural inputs that reduce reliance on synthetic chemicals because of their dual function in promoting growth and suppressing disease (Singh et al. 2023).

4. Challenges and Future Perspectives

Developing and implementing next-generation smart nanomaterials (SNMs) for environmentally friendly applications is fraught with difficulties. A major issue is the absence of established procedures for the large-scale synthesis, characterization, and use of SNMs, which frequently results in inconsistent environmental behavior and material performance.

Furthermore, there are valid concerns about the toxicological uncertainties surrounding their long-term interactions with human and ecological systems. Although green synthesis techniques show promise, the complexity of bioreduction mechanisms and the diversity of biological sources continue to limit their scale and reproducibility. The cost- and resource-intensive nature of fabricating composite and functionalized SNMs is another important barrier that limits their availability, particularly in rural and low-income areas. The fate, transformation, and accumulation of SNMs in soil, water, and biotic systems are currently poorly understood from an environmental standpoint, requiring further study in environmental nanotoxicology and life cycle evaluations. Regulatory gaps further restrict the appropriate commercialization and field-level use of nano-enabled products.

Future studies must also concentrate on eco-design concepts, using materials that are organically sourced or biodegradable to cause the least amount of ecological disturbance. The development of SNMs that are not only effective but also safe, economical, and scalable will require interdisciplinary cooperation in the fields of materials science, microbiology, agronomy, and policymaking. In the upcoming decades, SNMs have the potential to completely rethink environmental sustainability with innovative green nanotechnology, inclusive policy frameworks, and prudent risk management.

5. Conclusion

Next-generation smart nanomaterials are an important development in the search for environmentally friendly solutions. Soil remediation, climate-resilient agriculture, and water and air purification have all benefited from their unique physicochemical characteristics, stimuli-responsive behaviors, and multifunctional capacities. Through the integration of precise functionality and green synthesis methodologies, SNMs provide environmentally friendly substitutes for traditional technologies, reducing their impact on the environment and increasing system efficiency. SNMs' adaptability and capacity to address the intricate problems brought on by pollution, resource depletion, and climate change are demonstrated by their integration into nanopesticides, sensors, bioremediation platforms, and nanofertilizers. However, current obstacles pertaining to scalability, safety, governmental control, and socioeconomic accessibility must be removed in order to fully realize their advantages. Moving SNMs from lab innovation to practical impact will require advancing collaborative research, supporting sustainable design, and fostering open risk assessment. Smart nanomaterials will surely be essential instruments in creating resilient ecosystems and attaining long-term environmental sustainability as we transition to a more circular and green economy.

References

Aflori, M. (2021). Smart nanomaterials for biomedical applications—a review. *Nanomaterials*, 11(2): 396.

Ahmed, S. F., Mofijur, M., Ahmed, B., Mehnaz, T., Mehejabin, F., Maliat, D. et al. (2022). Nanomaterials as a sustainable choice for treating wastewater. *Environmental Research*, 214: 113807.

Ahmed, F., Islam, M. A., Kumar, M., Hossain, M., Bhattacharya, P., Islam, M. T. et al. (2021). First detection of SARS-CoV-2 genetic material in the vicinity of COVID-19 isolation Centre in Bangladesh: variation along the sewer network. *Science of the Total Environment*, 776: 145724.

Akinola, S. A., Adeyemo, R. O., and Adebayo, I. A. (2022). Overview and emergence of nanobiotechnology in plants. pp. 173–197. *In: Diverse applications of nanotechnology in the biological sciences*. Apple Academic Press.

Amin, H., Tayyab, A., Umair, M., Naveed, M. R., Yaseen, H. R., Qasim, M. et al. (2025). Recent Progress in Wastewater Treatment: Exploring the Roles of Zero-Valent Iron and Titanium Dioxide Nanoparticles. *Indus Journal of Bioscience Research*, 3(4): 97–107.

Antony, J., V, M., Raphael, V. P., and P, V. (2022). Application of encapsulated and immobilized nano zero-valent iron for iron removal from roof-harvested rainwater. *Water Supply*, 22(12): 8957–8971.

Bahrulolum, H., Nooraei, S., Javanshir, N., Tarrahimofrad, H., Mirbagheri, V. S., Easton, A. J. et al. (2021). Green synthesis of metal nanoparticles using microorganisms and their application in the agrifood sector. *Journal of Nanobiotechnology*, 19: 1–26.

Boudh, S., and Singh, J. S. (2018). Pesticide contamination: environmental problems and remediation strategies. pp. 245–269. *In: Emerging and Eco-Friendly Approaches For Waste Management*. Singapore: Springer Singapore.

Chen, J., Guo, Y., Zhang, X., Liu, J., Gong, P., Su, Z. et al. (2023). Emerging nanoparticles in food: Sources, application, and safety. *Journal of Agricultural and Food Chemistry*, 71(8): 3564–3582.

Gaber, S. E., Hashem, A. H., El-Sayyad, G. S., and Attia, M. S. (2024). Antifungal activity of myco-synthesized bimetallic ZnO-CuO nanoparticles against fungal plant pathogen Fusarium oxysporum. *Biomass Conversion and Biorefinery*, 14(20): 25395–25409.

Gupta, A., Bharati, R., Kubes, J., Popelkova, D., Praus, L., Yang, X. et al. (2024). Zinc oxide nanoparticles application alleviates salinity stress by modulating plant growth, biochemical attributes and nutrient homeostasis in Phaseolus vulgaris L. *Frontiers in Plant Science*, 15: 1432258.

Islam, N., and Saikia, B. K. (2022). An overview on atmospheric carbonaceous particulate matter into carbon nanomaterials: A new approach for air pollution mitigation. *Chemosphere*, 303: 135027.

Kholief, M. G., Hesham, A. E. L., Hashem, F. S., and Mohamed, F. M. (2024). Synthesis and utilization of titanium dioxide nano particle (TiO2NPs) for photocatalytic degradation of organics. *Scientific Reports*, 14(1): 11327.

Kulkarni, R., Lingamdinne, L. P., Koduru, J. R., Karri, R. R., Kailasa, S. K., Mubarak, N. M. et al. (2024). Exploring the recent cutting-edge applications of CNTs in energy and environmental remediation: Mechanistic insights and remarkable performance advancements. *Journal of Environmental Chemical Engineering*, 113251.

Kumar, R., Kumar, N., Rajput, V. D., Mandzhieva, S., Minkina, T., Saharan, B. S. et al. (2022). Advances in biopolymeric nanopesticides: A new eco-friendly/eco-protective perspective in precision agriculture. *Nanomaterials*, 12(22): 3964.

Kumar, M., Kaushik, D., Kumar, A., Gupta, P., Proestos, C., Oz, E. et al. (2023). Green synthesis of copper nanoparticles from Nigella sativa seed extract and evaluation of their antibacterial and antiobesity activity. *International Journal of Food Science and Technology*, 58(6): 2883–2892.

Liu, R. U. I., Bian, Y. A. N., Liu, L. I. N., Liu, L., Liu, X., and Ma, S. (2022). Molecular pathways associated with oxidative stress and their potential applications in radiotherapy. *International Journal of Molecular Medicine*, 49(5): 65.

Liu, X., Wei, J., Wu, Y., Zhang, J., Xing, L., Zhang, Y. et al. (2022). Performances and mechanisms of microbial nitrate removal coupling sediment-based biochar and nanoscale zero-valent iron. *Bioresource Technology*, 345: 126523.

Mustafa, G., Chaudhari, S. K., Manzoor, M., Batool, S., Hatami, M., and Hasan, M. (2024). Zinc oxide nanoparticles mediated salinity stress mitigation in Pisum sativum: a physio-biochemical perspective. *BMC Plant Biology*, 24(1): 835.

Nageswara Rao, T., Prashanthi, Y., Ahmed, F., Kumar, S., Arshi, N., Rajasekhar Reddy, G. et al. (2021). Photocatalytic applications of Fe–Ag Co-doped TiO2 nanoparticles in removal of flumioxazin pesticide residues in water. *Frontiers in Nanotechnology*, 3: 652364.

Nassar, G., Daou, E., Najjar, R., Bassil, M., and Habchi, R. (2021). A review on the current research on graphene-based aerogels and their applications. *Carbon Trends*, 4: 100065.

Naz, T., Iqbal, M. M., Raza, B., Mubeen, M. A., Nadeem, M. A., Al-Ghamdi, A. A. et al. (2025). Green remediation of lead (pb) from Pb-toxic soil by combined use of silicon nanomaterials and leguminous *Lens culinaris* L. plants. *Scientific Reports*, 15(1): 4366.

Rassaei, L., Marken, F., Sillanpää, M., Amiri, M., Cirtiu, C. M., and Sillanpää, M. (2011). Nanoparticles in electrochemical sensors for environmental monitoring. *TrAC Trends in Analytical Chemistry*, 30(11): 1704–1715.

Reimers, A., Bouhanguel, A., Greve, E., Möller, M., Saure, L. M., Kaps, S. et al. (2023). Multifunctional, self-cleaning air filters based on graphene-enhanced ceramic networks. *Device*, 1(4).

Shahzadi, S., Fatima, S., Shafiq, Z., and Janjua, M. R. S. A. (2025). A review on green synthesis of silver nanoparticles (SNPs) using plant extracts: a multifaceted approach in photocatalysis, environmental remediation, and biomedicine. *RSC Advances*, 15(5): 3858–3903.

Sharma, S., Kumari, P., Thakur, P., Brar, G. S., Bouqellah, N. A., and Hesham, A. E. L. (2022). Synthesis and characterization of Ni0. 5Al0. 5Fe2O4 nanoparticles for potent antifungal activity against dry rot of ginger (Fusarium oxysporum). *Scientific Reports*, 12(1): 20092.

Singh, H. (2023). Enhancement of plant secondary metabolites by genetic manipulation. pp. 59–90. *In: Genetic Manipulation of Secondary Metabolites In Medicinal Plant*. Singapore: Springer Nature Singapore.

Singh, S., Paswan, S. K., Kumar, P., Singh, R. K., and Kumar, L. (2023). Heavy metal water pollution: an overview about remediation, removal and recovery of metals from contaminated water. *Metals in Water*, 263–284.

Singh, S. K., Sharma, A., Sharma, L., and Sundaram, S. (2024). Green Synthesis of Algal Nanoparticles: Harnessing Nature's Biofactories for Sustainable Nanomaterials. pp. 257–284. *In: Biogenic Wastes-Enabled Nanomaterial Synthesis: Applications in Environmental Sustainability*. Cham: Springer Nature Switzerland.

Thabet, M., Abd El-Monaem, E. M., Alharbi, W. R., Mohamoud, M., Abdel-Aty, A. H., Ibrahim, I. et al. (2024). Adsorption and photocatalytic degradation activities of a hybrid magnetic mesoporous composite of α-Fe2O3 nanoparticles embedded with sheets-like MgO. *Journal of Water Process Engineering*, 60: 105192.

Thakur, S., Thakur, S., and Kumar, R. (2018). Bio-nanotechnology and its role in agriculture and food industry. *J. Mol. Genet. Med.*, 12(324): 1747–0862.

Tripathy, S., Rodrigues, J., and Shimpi, N. G. (2023). Top-down and Bottom-up Approaches for Synthesis of Nanoparticles. *Nanobiomaterials Perspect. Med. Appl. Diagn. Treat. Dis.*, 145: 92–130.

Xu, Y., Chiam, S. L., Leo, C. P., and Hu, Z. (2025). Recent Advances in Photocatalytic TiO2-based Membranes for Eliminating Water Pollutants. *Separation & Purification Reviews*, 1–18.

Yadav, N., Gaikwad, R. P., Mishra, V., and Gawande, M. B. (2022). Synthesis and photocatalytic applications of functionalized carbon quantum dots. *Bulletin of the chemical society of Japan*, 95(11): 1638–1679.

Yan, Z., Ouyang, J., Wu, B., Liu, C., Wang, H., Wang, A. et al. (2024). Nonmetallic modified zero-valent iron for remediating halogenated organic compounds and heavy metals: A comprehensive review. *Environmental Science and Ecotechnology*, 100417.

Zahra, Z., Habib, Z., Chung, S., and Badshah, M. A. (2020). Exposure route of TiO2 NPs from industrial applications to wastewater treatment and their impacts on the agro-environment. *Nanomaterials*, 10(8): 1469.

Zhang, L., Jing, D., Jiang, N., Rojalin, T., Baehr, C. M., Zhang, D. et al. (2020). Transformable peptide nanoparticles arrest HER2 signalling and cause cancer cell death *in vivo*. *Nature Nanotechnology*, 15(2): 145-153.

Zhang, P., Wang, C., Guo, Z., Hong, J., and Wang, F. (2023). Effect of polyvinyl alcohol fibers on mechanical properties of nano-SiO2-reinforced geopolymer composites under a complex environment. *Nanotechnology Reviews*, 12(1): 20230142.

Zhao, H., Chen, H., Yang, F., Liu, N., Deng, H., Cai, H. et al. (2024). Explainability for large language models: A survey. *ACM Transactions on Intelligent Systems and Technology*, 15(2): 1-38.

Zhao, J., Wang, Y., and Liu, B. (2025). Doping Detection Based on the Nanoscale: Biosensing Mechanisms and Applications of Two-Dimensional Materials. *Biosensors*, 15(4): 227.

Zhou, D., Yan, X., Xiao, L., Wang, J., and Wei, J. (2024). Gold capped mesoporous bioactive glass guides bone regeneration via BMSCs recruitment and drug adaptive release. *Chemical Engineering Journal*, 487: 150546.

CHAPTER-14

Challenges of Using Emerging Nanomaterials for Environmental Applications

Tanvi Nayak,[1] *Amanullakhan Pathan*[2,*] and *Riddhi P. Dave*[3]

1. Introduction

In recent years, the application of nanotechnology has become a cornerstone in advancing various environmental management solutions. Nanomaterials, due to their unique physicochemical properties, such as high surface area, reactivity, and quantum effects, offer potential for improving environmental quality by addressing challenges related to pollution control, resource recovery, and environmental remediation (Ali et al. 2025, Niazi and Gu 2009, Zhang et al. 2016). From the removal of hazardous contaminants from water and air to the detoxification of soils and restoration of ecosystems, nanomaterials have shown considerable promise in enhancing the effectiveness and efficiency of traditional environmental technologies (Zhao et al. 2022). For example, nanoparticles such as zero-valent iron (ZVI), titanium dioxide (TiO_2), and carbon-based nanomaterials have been widely studied for their role in contaminant removal, pollutant sensing, and wastewater treatment (Kataria et al. 2025, Zhang et al. 2016). In particular, their high reactivity, small particle size, and ability to easily interact with a wide range of contaminants make them ideal candidates for addressing issues of soil and water pollution. Similarly, carbon nanotubes and graphene oxide have demonstrated potential for adsorbing heavy metals, organic pollutants, and other harmful substances, facilitating resource recovery from waste streams (Zhou et al. 2021, Nayak et al. 2023). Furthermore, the use of nanomaterials in energy-efficient applications, such as in photovoltaics and hydrogen production, has also gained significant attention due to their ability to contribute to sustainability goals (Osama et al. 2024). Despite the great promise that emerging nanomaterials hold for environmental applications, several scientific, technical, and regulatory challenges must be addressed before their full-scale deployment in real-world scenarios (Sohail et al. 2019). These challenges span across various stages of nanomaterial development, from synthesis and characterization to their interaction with the environment and biological systems. The introduction of nanomaterials into environmental

[1] M. B. Patel Science College, Anand, Gujarat 388001, India. 10009-0008-3249-243X
[2] Department of Chemistry, Shri Sarvajanik Science College (PG), Mehsana, Gujarat 384001, India. 20000-0002-4941-4525
[3] School of Chemical Science, Central University of Gujarat, Gandhinagar 382030, India. 30009-0006-2418-7625
* Corresponding author: amankhan255@gmail.com

DOI: 10.1201/9781003558804-14

systems requires careful consideration of their properties and behavior in complex, heterogeneous environments, which often differ significantly from controlled laboratory conditions (Desai 2012).

1.1 Synthesis and Characterization Challenges

The synthesis and characterization of nanomaterials are central to their successful application in environmental technologies. However, the very properties that make nanomaterials advantageous—such as high surface area, reactivity, and the ability to interact with diverse pollutants—also present significant challenges in terms of their production and the ability to accurately assess their structure and performance in environmental applications. These challenges are multifaceted and encompass variability in material properties, reproducibility of synthesis methods, scalability, and the selection of characterization techniques that can provide a comprehensive understanding of the material's behavior in both laboratory and field settings. This section will explore these issues in greater detail (Ali et al. 2016, Laux et al. 2018).

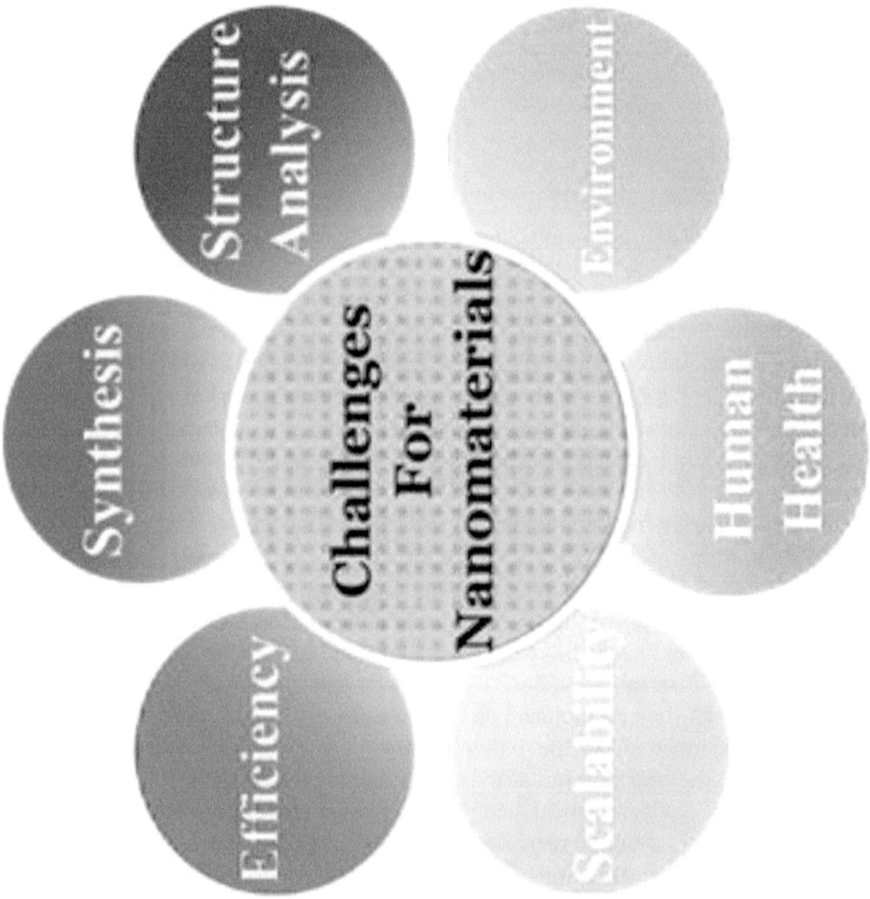

Figure 1: Challenges of using emerging nanomaterials for environmental applications.

1.1.1 Variability in Nanomaterial Properties

One of the fundamental challenges in nanomaterial synthesis is the variability in their properties based on factors such as size, shape, surface chemistry, and method of preparation. Unlike bulk materials, nanomaterials exhibit unique and often unpredictable behaviors due to the increased surface-to-volume ratio, which leads to enhanced reactivity and changes in optical, magnetic, and chemical properties (Åberg et al. 2021, Bell et al. 2014). Nanoparticles can exist in different morphologies such as spheres, rods, disks, and tubes, each of which can significantly affect their interaction with environmental contaminants. For example, spherical nanoparticles typically exhibit different reactivity and adsorption capacities compared to their rod-shaped counterparts, even if they are composed of the same material (Xie et al. 2022, 2009). This variability complicates the prediction of nanomaterials' performance in environmental applications, where consistency in material properties is crucial. Surface functionalization further exacerbates this variability. The surface of a nanoparticle can be modified to enhance its interaction with specific pollutants, thereby increasing its efficiency in environmental remediation (Wang et al. 2019). However, the functionalization process is highly sensitive to the synthesis conditions, leading to inconsistencies in the functional groups and surface charge, which in turn can alter the material's reactivity and stability (Ahmad et al. 2022). Additionally, surface coatings and stabilizers, which are often used to prevent nanoparticle aggregation in suspension, can also affect the material's ability to interact with contaminants, adding another layer of complexity to their design and application (Gambinossi et al. 2015).

1.1.2 Reproducibility and Batch-to-Batch Variability

A critical challenge in the synthesis of nanomaterials is ensuring the reproducibility of results across different batches. Nanomaterials are often synthesized using various methods such as chemical vapor deposition (CVD), sol-gel processes, hydrothermal synthesis, or electrochemical deposition. Each of these methods has its own set of parameters such as temperature, pressure, precursor concentration, and reaction time that can influence the size, morphology, and surface chemistry of the nanoparticles (Mülhopt et al. 2018, Saravanan et al. 2022, Pathan et al. 2018). Even small changes in these synthesis conditions can lead to significant differences in the final product, leading to inconsistencies in the properties of the material (Onamade et al. 2025).

In environmental applications, such variability can be problematic, as it can impact the effectiveness of the nanomaterial in real-world conditions. For instance, inconsistencies in the size and surface charge of nanoparticles could affect their adsorption efficiency for pollutants such as heavy metals or organic contaminants, leading to variable remediation outcomes (Richard et al. 2022). Achieving high reproducibility is crucial for both laboratory research and scaling up nanomaterials for practical deployment in large-scale environmental remediation projects.

1.1.3 Challenges in High-Throughput Production

While laboratory-scale synthesis of nanomaterials has made significant advances, scaling up production to meet the demands of environmental applications remains a significant challenge. High-throughput production of nanomaterials is often hindered by issues related to cost, scalability, and maintaining consistent material properties. Traditional methods such as chemical vapor deposition or hydrothermal synthesis, although effective for producing high-quality nanomaterials, are often expensive and not easily scalable (Zavanelli 2021, Yang et al. 2011). For example, CVD requires high temperatures and precise control over reaction parameters, which makes the process costly and energy-intensive (Ali et al. 2016). On the other hand, wet chemical methods, such as sol-gel synthesis and precipitation, offer more cost-effective approaches, but they often struggle with achieving the high purity and uniformity required for consistent performance in environmental applications (Kumar et al. 2014, Pathan et al. 2022). Furthermore, these methods can result in the production of nanoparticles with varying sizes or morphologies, which may not be ideal for a specific environmental remediation process. A promising avenue for addressing scalability is the use of green synthesis methods, which involve the use of natural, renewable materials such as plant extracts or bacteria to facilitate the synthesis of nanomaterials (Shuaixuan et al. 2022, Vaghela et al. 2018). These methods are often more sustainable and can be performed under milder conditions. However, challenges remain in achieving consistent yields and material properties, and these processes are typically less efficient than conventional methods (Hosseingholian et al. 2023).

1.1.4 Characterization of Nanomaterials

Characterizing the physical and chemical properties of nanomaterials is essential for understanding their behavior and performance in environmental applications. However, the very small size and heterogeneity of nanoparticles make them challenging to characterize accurately. Several techniques have been developed to characterize nanomaterials, including scanning electron microscopy (SEM), transmission electron microscopy (TEM), atomic force microscopy (AFM), dynamic light scattering (DLS), and X-ray photoelectron spectroscopy (XPS). Each of these methods provides specific information about the nanomaterial's morphology, size distribution, surface properties, and chemical composition (Khan et al. 2019, Baer et al. 2008, Lin et al. 2014, Praveen et al. 2024). The size and morphology of nanoparticles play a critical role in their effectiveness for environmental remediation. Nanoparticles with smaller sizes have a larger surface area for interactions with pollutants, but they may also exhibit more pronounced agglomeration tendencies, which could reduce their reactivity and overall performance (Baral et al. 2024, Sharma et al. 2024). Techniques like TEM and SEM are widely used to observe the morphology and distribution of nanoparticles. However, these methods can suffer from resolution limitations, especially for nanoparticles smaller than 5 nm, where atomic-scale features become difficult to resolve. The surface properties of nanomaterials, including charge and functionalization, significantly influence their interaction with pollutants and their stability in suspension. Zeta potential measurements can be used to assess the surface charge, which impacts particle aggregation and interaction with contaminants (Zhou et al. 2010). However, surface functionalization, such as the attachment of functional groups to enhance pollutant uptake, can be complex to characterize accurately, especially when functional groups are present in low quantities or heterogeneously distributed across particles. Techniques like XPS and Fourier-transform infrared spectroscopy (FTIR) are often used to probe surface chemistry, but interpreting the results can be challenging, particularly when dealing with highly diverse or irregularly functionalized nanoparticles (Karolina et al. 2021). Achieving stable dispersion of nanoparticles in environmental media, such as water or soil, is another critical aspect of nanomaterial characterization. Nanoparticles tend to aggregate due to van der Waals forces, which can significantly impact their reactivity and bioavailability (Niazi et al. 2009). Techniques such as DLS and nanoparticle tracking analysis (NTA) can provide insight into the size distribution and aggregation state of nanoparticles in suspension. However, the interpretation of aggregation behavior in complex media, such as natural waters with dissolved organic matter, remains a challenge (Yang et al. 2011).

1.1.5 Standardization of Characterization Protocols

The lack of standardized methods for nanomaterial characterization remains a major obstacle to their widespread use in environmental applications. Although various characterization techniques are available, there is no universally accepted set of protocols for evaluating the properties of nanomaterials, which complicates comparisons across studies and hinders the reproducibility of results (Sharifi et al. 2022). This lack of standardization is particularly problematic when considering the regulatory landscape, as there is no clear consensus on the specific metrics that should be used to assess the safety and efficacy of nanomaterials in environmental contexts (Pereira et al. 2018). Efforts are underway to develop standardized guidelines for nanomaterial characterization, and organizations such as the International Organization for Standardization (ISO) and the American Society for Testing and Materials (ASTM) have made strides in addressing this issue (ISO 2023). However, much work remains to be done to ensure that these standards are widely adopted and implemented across research and industrial sectors.

1.1.6 Cost and Scalability of Synthesis Techniques

The high cost associated with synthesizing high-quality nanomaterials remains one of the primary obstacles to their widespread adoption in environmental applications. Many of the methods currently used to produce nanomaterials, particularly those that yield high-purity products, are energy-intensive and require expensive precursors and equipment (Paliwal et al. 2014). Moreover, some of these methods, such as CVD, are not easily scalable for commercial or industrial applications, further limiting their practicality for large-scale deployment in environmental remediation processes. To overcome these challenges, researchers have been investigating more cost-effective and scalable synthesis methods. For example, bottom-up approaches such as sol-gel, hydrothermal synthesis, and biosynthesis using natural precursors have been explored as more sustainable alternatives. These methods are typically more environmentally friendly and can potentially reduce production costs, but they still face challenges in terms of achieving the required purity and uniformity needed for reliable environmental applications (Ahmed 2022).

The synthesis and characterization of nanomaterials present significant challenges in their application for environmental remediation. Variability in material properties, reproducibility of synthesis methods, scalability of production, and the need for standardized characterization techniques are all critical barriers to the effective use of nanomaterials in real-world environmental solutions. Overcoming these challenges will require continued advances in nanomaterial synthesis techniques, as well as the development of more reliable and consistent characterization methods. Ultimately, the successful integration of nanomaterials into environmental applications will depend on the ability to produce high-quality, reproducible materials with well-characterized properties that are optimized for specific remediation goals.

2. Environmental and Health Risks Associated with Nanomaterials

2.1 Toxicity of Nanomaterials: Mechanisms and Potential Risks

The environmental and health risks of nanomaterials are primarily driven by their unique physicochemical properties, such as size, surface area, shape, and surface chemistry, which can significantly influence their behavior in biological systems and the environment (Ray et al. 2009). Due to their small size, nanomaterials exhibit properties that differ from bulk materials, making them highly reactive and capable of interacting with biological cells, tissues, and ecosystems in complex ways. These properties can present both benefits in environmental applications and challenges regarding their toxicity (Shikha et al. 2023). One of the most important factors contributing to the toxicity of nanomaterials is their size. At the nanoscale, the materials' large surface-to-volume ratio increases their reactivity, potentially enhancing interactions with biological molecules such as proteins, lipids, and DNA (Schrand et al. 2010). Smaller nanoparticles are more likely to penetrate biological membranes, enter cells, and distribute to various organs. For instance, nanoparticles smaller than 100 nm have been shown to enter the bloodstream via the lungs and gastrointestinal tract (Klaine et al. 2012). These particles can also cross the blood-brain barrier, which raises concerns about their potential to cause neurological damage (Liu et al. 2013).

Surface charge and functionalization are crucial factors that influence the toxicity of nanomaterials. Nanomaterials that are positively charged tend to aggregate more readily, whereas those with negative charges are less likely to aggregate but may interact more strongly with cellular membranes (Rai et al. 2020). Surface functionalization can also impact toxicity. For example, coating nanoparticles with organic or inorganic molecules can enhance or reduce their toxic potential by altering their interactions with biological systems (Sharifi et al. 2012). A common issue is that functionalized nanoparticles can release harmful substances, such as metals or other reactive components, as they degrade over time. The shape of nanomaterials also plays a key role in their toxicity. Nanomaterials can have various shapes, including spherical, rod-shaped, and tubular structures, each with different interactions in biological systems. For example, elongated nanomaterials like carbon nanotubes (CNTs) have been shown to exhibit asbestos-like toxicity, causing respiratory and cardiovascular issues due to their ability to penetrate deep into the lungs and induce chronic inflammation. Rod-shaped nanoparticles have also been reported to cause greater cellular damage than spherical nanoparticles due to their higher surface area per unit volume and their ability to disrupt cell membranes (Huang et al. 2019).

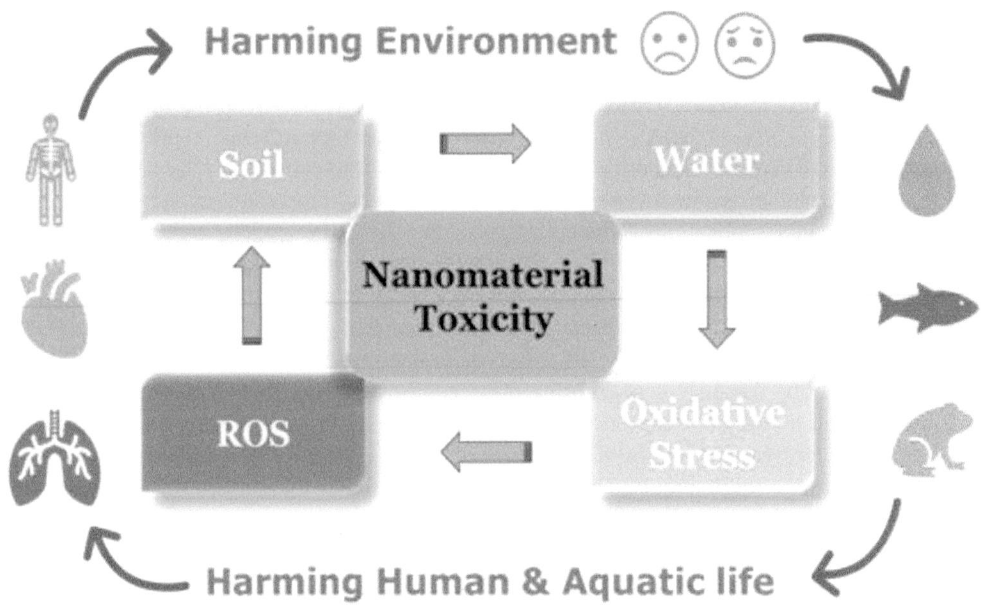

Figure 2: Harmful effects of nanomaterials.

Nanomaterials can induce oxidative stress by generating reactive oxygen species (ROS), which are highly reactive molecules capable of damaging cellular structures, including lipids, proteins, and DNA. ROS are produced when nanomaterials interact with cellular components such as lipids in cell membranes, leading to lipid peroxidation and disrupting the integrity of the cellular membrane (Wang et al. 2017). This oxidative damage can result in inflammation, cell death, and even mutations, which contribute to long-term health problems, such as cancer. Nanomaterials, particularly those made from metals or metal oxides, can catalyze the generation of reactive oxygen species (ROS) upon exposure to biological systems. ROS are highly reactive molecules that can cause cellular damage, leading to oxidative stress, DNA damage, and inflammation. These oxidative reactions can lead to adverse effects on living organisms, including mutagenesis, carcinogenesis, and neurodegenerative diseases. For example, titanium dioxide (TiO_2) and zinc oxide (ZnO) nanoparticles have been shown to induce ROS production, leading to cytotoxicity in human lung and skin cells (Rim et al. 2013). Nanomaterials can trigger an immune response by interacting with immune cells such as macrophages and dendritic cells. These interactions can activate pro-inflammatory pathways and lead to chronic inflammation, which can damage tissues over time (Jiang et al. 2009). In particular, nanomaterials with high surface reactivity or charge tend to accumulate in tissues, causing persistent inflammation. This inflammation can disrupt normal cell functions and contribute to diseases like cardiovascular disorders, pulmonary diseases, and neurodegenerative conditions (Liu et al. 2013). Due to their small size, nanomaterials can enter cells via different mechanisms, including endocytosis, phagocytosis, and pinocytosis. Once inside the cell, nanomaterials can disrupt cellular functions by interacting with organelles such as the mitochondria, nucleus, and endoplasmic reticulum (Zhao et al. 2012). For instance, the accumulation of nanomaterials in mitochondria can impair energy production, leading to cellular dysfunction and, in extreme cases, cell death (Yang et al. 2011). Furthermore, some nanomaterials can directly interfere with DNA by inducing strand breaks or forming adducts with nucleic acids, potentially leading to genetic mutations and cancer (Liu et al. 2010).

2.2 Bioaccumulation and Environmental Persistence of Nanomaterials

2.2.1 Bioaccumulation in Aquatic and Terrestrial Ecosystems

Bioaccumulation is a critical issue for nanomaterials, as their small size enables them to be taken up by a wide range of organisms, including microorganisms, plants, and animals. The risk of bioaccumulation is particularly high in aquatic ecosystems, where nanoparticles can be introduced via industrial wastewater, runoff, or accidental spills. Studies have shown that nanoparticles can be absorbed by aquatic organisms such as fish, algae, and invertebrates (Klaine et al. 2012). Once taken up by organisms, nanomaterials can accumulate in tissues and potentially enter the food chain, leading to risks for higher organisms, including humans.

2.2.2 Aquatic Bioaccumulation and Transport

In aquatic environments, the transport and fate of nanomaterials are influenced by several factors, including particle size, surface charge, and the presence of natural organic matter (NOM). The interaction of nanoparticles with dissolved organic matter and other environmental factors can affect their dispersion, aggregation, and bioavailability (Schrand et al. 2010). For example, carbon-based nanoparticles, such as fullerenes, have been shown to accumulate in the gills and organs of fish, leading to sub-lethal effects on growth and reproduction (Zhao et al. 2022). Similarly, metal nanoparticles like silver (AgNPs) and gold (AuNPs) have been found to bioaccumulate in aquatic organisms, potentially leading to long-term environmental toxicity.

2.2.3 Persistence and Degradation of Nanomaterials

Nanomaterials are often more stable in environmental conditions than bulk materials, which can contribute to their persistence in ecosystems. For instance, titanium dioxide and silver nanoparticles have been shown to remain intact in soil and aquatic environments for extended periods, potentially leading to chronic exposure to organisms (Lee et al. 2013). In addition, the degradation of certain nanomaterials, particularly those made of metals, can lead to the release of toxic metal ions that further contribute to environmental contamination (Yang et al. 2011, Pathan et al. 2017). This persistence raises concerns about the long-term ecological consequences of releasing nanomaterials into the environment, particularly when they accumulate over time.

2.2.4 Soil Contamination and Terrestrial Bioaccumulation

Nanomaterials can also accumulate in soil and be taken up by terrestrial plants. Soil particles have been found to adsorb and aggregate with nanoparticles, which may affect the mobility and bioavailability of these particles in soil systems (Li et al. 2010). Nanoparticles such as silver, zinc oxide, and copper oxide have been shown to be absorbed by plants, leading to changes in growth, photosynthesis, and nutrient uptake (Zhang et al. 2022). The impact of nanomaterials on terrestrial ecosystems is particularly concerning, as these materials can potentially enter the food chain via plant consumption by herbivores.

2.3 Potential Health Risks of Nanomaterials in Humans

One of the primary routes of human exposure to nanomaterials is through inhalation. Nanoparticles in the air, either from industrial emissions or environmental contamination, can be inhaled into the lungs, where they may cause respiratory problems. Studies have shown that inhalation of nanomaterials can cause inflammation, oxidative stress, and damage to lung tissues. For example, carbon nanotubes (CNTs), which are often used in environmental applications, have been shown to cause respiratory toxicity and fibrosis in animal models, akin to the effects of asbestos exposure (Rim et al. 2013). Nanomaterials can also enter the human body through dermal exposure (e.g., through skin contact with nanomaterial-containing products) or oral ingestion (e.g., through contaminated food or water). Dermal exposure to nanoparticles has been shown to result in skin irritation and may facilitate the entry of nanoparticles into the bloodstream, leading to systemic toxicity (Jiang et al. 2020). Oral exposure to nanoparticles is another significant route, particularly through contaminated water or food sources. The gastrointestinal tract may absorb certain nanoparticles, leading to toxicity in the liver, kidneys, and other organs (Kose et al. 2023). In particular, silver nanoparticles, commonly used for antimicrobial purposes, have been shown to accumulate in the liver and kidneys, resulting in adverse health effects.

Nanomaterials can also induce cardiovascular toxicity. Inhaled nanoparticles have been shown to translocate from the lungs into the bloodstream and accumulate in the heart and blood vessels, causing oxidative damage, inflammation, and endothelial dysfunction. Additionally, nanomaterials can accumulate in organs such as the liver, kidneys, and spleen, potentially causing nephrotoxicity, hepatotoxicity, and other systemic health effects. Recent studies suggest that long-term exposure to nanoparticles, particularly metal-based ones, may lead to chronic health issues such as cardiovascular disease, liver dysfunction, and kidney damage (Yanping et al. 2021). Another critical concern is the potential neurotoxicity of nanomaterials. Several studies have demonstrated that certain nanoparticles can cross the blood-brain barrier, leading to potential damage to brain cells. For instance, carbon-based nanoparticles, including fullerenes and multi-walled carbon nanotubes (MWCNTs), have been shown to induce neuroinflammation and oxidative stress in animal models, with potential long-term effects on cognitive function and behavior (Zhao Y. et al. 2022). The ability of nanoparticles to accumulate in the brain presents a unique risk, as the long-term effects on neural development and function remain largely unknown.

2.4 Regulatory and Risk Assessment Challenges

The lack of comprehensive regulatory frameworks for the safe use of nanomaterials in environmental and health contexts is a major issue. Current regulations often fail to address the unique risks posed by nanomaterials due to their size-dependent properties. The U.S. Environmental Protection Agency (EPA), European Chemicals Agency (ECHA), and other regulatory bodies primarily regulate nanomaterials under existing chemical regulations, which were designed for bulk materials and may not be sufficient to assess the risks associated with nanoscale substances (Fadel et al. 2015). Nanomaterials' toxicity and bioaccumulation potential necessitate the development of new, nanomaterial-specific regulatory frameworks. A critical challenge in risk assessment is the lack of standardized methods for testing the toxicity and environmental impact of nanomaterials. The properties of nanomaterials are highly variable and depend on factors such as size, shape, surface charge, and chemical composition (De Jong et al. 2022). As a result, traditional testing methods for bulk chemicals, such as acute toxicity tests, may not be applicable to nanomaterials. There is an urgent need for standardized protocols that can evaluate the toxicity, bioaccumulation, and long-term environmental impacts of nanomaterials.

2.5 Strategies to Mitigate Risks

To mitigate the risks associated with nanomaterials, it is essential to develop safer nanomaterials with controlled properties that minimize toxicity. This includes the use of "green" or "safe-by-design" nanomaterials, which are engineered to have minimal environmental and health impacts. Additionally, improved risk assessment frameworks and regulations tailored specifically for nanomaterials are necessary to ensure their safe use in environmental applications (Schrand et al. 2010). Collaboration among researchers, regulatory agencies, and industry stakeholders will be crucial in developing and implementing these solutions.

3. Scalability and Practical Deployment of Nanomaterials

Nanomaterials have garnered significant interest in their potential to address pressing environmental challenges, such as pollution remediation, resource recovery, and water treatment. Despite their promise, the transition from laboratory-scale applications to large-scale deployment faces numerous challenges. These challenges arise not only from the inherent properties of nanomaterials but also from the need for efficient, cost-effective, and environmentally safe production and deployment strategies. This section delves into the critical aspects of scaling up the production and deployment of nanomaterials for environmental applications, with a particular focus on material stability, distribution, scalability of production methods, and economic viability.

3.1 Stability and Durability of Nanomaterials in Environmental Conditions

One of the fundamental challenges in deploying nanomaterials for environmental remediation is ensuring their stability and durability under real-world conditions. Environmental systems, whether aquatic, terrestrial, or atmospheric, are inherently complex and dynamic, subject to fluctuations in temperature, pH, salinity, pressure, and the presence of other pollutants or biological entities. These environmental stressors can alter the physical and chemical properties of nanomaterials, potentially compromise their effectiveness, and lead to unforeseen ecological or health risks. Nanomaterials, particularly those composed of metals or metal oxides, are prone to degradation in environmental conditions. For example, silver nanoparticles (AgNPs), commonly used in environmental applications for their antimicrobial properties, can be oxidized to release toxic silver ions (Ag^+), which may reduce their effectiveness over time (Gao et al. 2014, Shao et al. 2011). Similarly, nanoparticles of titanium dioxide (TiO_2) and zinc oxide (ZnO) can undergo photo-induced degradation when exposed to sunlight, altering their surface characteristics and reactivity (Kim et al. 2021, Mampilly et al. 2024). Such degradation processes not only affect the performance of nanomaterials but also raise concerns about the release of potentially harmful degradation by-products into the environment. One promising strategy to enhance the stability of nanomaterials is surface modification or functionalization. By altering the surface chemistry of nanoparticles, it is possible to increase their resistance to environmental degradation and improve their interaction with target pollutants. For example, the functionalization of carbon nanotubes (CNTs) with various organic ligands can improve their dispersibility in aqueous environments and reduce their tendency to aggregate (Rai et al. 2020). Similarly, coating nanoparticles with biocompatible or inert materials can protect them from degradation and enhance their stability in harsh environmental conditions (Wang et al. 2017). However, surface modifications must be carefully designed to avoid unintended consequences, such as reduced effectiveness in pollutant capture or increased toxicity to non-target organisms. For instance, while functional coatings can improve nanoparticle stability, they can also alter the surface charge, leading to changes in the particles' interactions with biological and environmental systems (Schrand et al. 2010).

3.2 Challenges in Achieving Uniform Distribution in Heterogeneous Environments

A critical challenge in the practical deployment of nanomaterials for environmental remediation is achieving uniform distribution of nanoparticles in heterogeneous environments. Environmental systems, such as soil and water bodies, are complex and non-uniform, with varying compositions, particle sizes, and biological activity. Nanomaterials must be effectively dispersed and transported within these systems to achieve optimal performance.

The transport and mobility of nanomaterials are influenced by their size, surface charge, and interactions with surrounding media. In water, nanoparticles must overcome aggregation tendencies and interact with dissolved organic matter, which can significantly affect their mobility (Klaine et al. 2012). For example, nanoparticles with a high surface charge tend to aggregate, forming larger clusters that are less mobile and less effective in pollutant removal (Rai et al. 2020). To address this challenge, researchers are exploring the use of surfactants or stabilizing agents that can improve nanoparticle dispersion in aqueous systems, thereby enhancing their ability to reach target pollutants. In soil environments, the challenge of achieving uniform nanoparticle distribution is compounded by the complexity of soil structure, which consists of pores, organic matter, and mineral particles. Nanomaterials may become trapped in soil aggregates or adsorbed onto soil particles, reducing their effectiveness in remediation efforts (Zhang et al. 2016). Various strategies, such as the use of nanoparticle suspensions or the development of nanoparticle-based delivery systems, are being explored to improve the mobility and distribution of nanomaterials in soil (Rai et al. 2020). While ensuring the effective distribution of nanomaterials is critical for their performance in environmental applications, it also raises concerns regarding the bioavailability and potential bioaccumulation of nanoparticles in ecosystems. The same properties that make nanoparticles highly reactive and efficient in pollutant removal also increase the risk of unintended interactions with biota, leading to potential toxicity and accumulation in food chains (Klainc et al. 2012). Therefore, a delicate balance must be struck between ensuring effective nanoparticle dispersion and mitigating the risks of bioaccumulation.

3.3 Scalability of Nanomaterial Production Methods

The scalability of nanomaterial production methods is another significant barrier to the widespread deployment of nanomaterials in environmental applications. While laboratory-scale synthesis methods can produce high-quality nanomaterials with well-defined properties, these methods are often expensive, energy-intensive, and difficult to scale up for industrial use (Shao et al. 2011).

3.3.1 Bottom-Up vs. Top-Down Approaches

Nanomaterials can be synthesized using either top-down or bottom-up approaches. The top-down approach involves breaking down bulk materials into nanoscale particles through mechanical, chemical, or laser-based methods, while the bottom-up approach involves assembling nanoparticles from smaller building blocks such as atoms or molecules (Schrand et al. 2010). Bottom-up methods, such as chemical vapor deposition (CVD) or sol-gel processes, can produce high-purity, well-defined nanomaterials, but these methods are often difficult to scale up due to their complexity and cost. On the other hand, top-down approaches, such as grinding or milling, are more scalable but may lead to materials with less uniform size distributions and potentially reduced functionality. Hybrid approaches that combine both top-down and bottom-up techniques are also being explored to improve scalability while maintaining material quality (Shao et al. 2011).

3.3.2 High-Throughput and Cost-Effective Production

For large-scale applications, the cost of nanomaterial production must be reduced to make environmental remediation technologies viable. High-throughput production methods, which allow for the rapid synthesis of large quantities of nanomaterials, are being investigated to increase scalability and reduce costs (Rai et al. 2020). Techniques such as continuous flow reactors, electrochemical deposition, and green chemistry methods are being explored for their potential to produce nanomaterials in an environmentally sustainable and cost-effective manner. The cost of raw materials, energy consumption, and the need for specialized equipment remain significant barriers to large-scale production. Strategies for reducing the cost of nanomaterials include optimizing synthesis processes to minimize waste, using low-cost precursors, and exploring waste materials (such as agricultural by-products or industrial residues) as feedstocks for nanomaterial production (Zhang et al. 2014).

3.4 Economic Viability and Commercialization of Nanomaterial-Based Environmental Technologies

The economic viability of nanomaterial-based environmental technologies is another critical factor that influences their large-scale deployment. While nanomaterials hold significant promise for addressing environmental pollution, their high production costs, coupled with the need for long-term efficacy and stability, pose challenges to their commercialization. For nanomaterial-based environmental technologies to become commercially viable, a comprehensive cost-benefit analysis must be conducted. This analysis should consider not only the direct costs of nanomaterial production and deployment but also the potential savings associated with environmental remediation, resource recovery, and the long-term benefits of sustainable solutions (Wang et al. 2019, Bhasin et al. 2024). For example, the use of nanomaterials in wastewater treatment may offer significant advantages in terms of efficiency and performance compared to conventional methods, potentially leading to cost savings in terms of energy consumption, chemicals, and labor. The successful commercialization of nanomaterials for environmental applications will also depend on market penetration and public acceptance. Public perception of nanotechnology, influenced by concerns about safety, toxicity, and environmental impact, may hinder the widespread adoption of nanomaterial-based solutions (Ankita and Mahesh 2021). Transparent communication regarding the safety, benefits, and risks associated with nanomaterials, along with robust regulatory frameworks, will be essential to gaining public trust and encouraging investment in nanomaterial-based environmental technologies. Regulatory frameworks are crucial in determining the feasibility and success of nanomaterial-based applications. Governments must create and enforce clear regulations regarding the use, disposal, and environmental monitoring of nanomaterials, while ensuring that these regulations do not stifle innovation (Mansi et al. 2019). International cooperation in developing global standards and regulations for nanomaterials will be essential to ensure that nanotechnologies can be deployed in a safe and coordinated manner.

3.5 Strategies for Overcoming Scalability Challenges

To address the challenges associated with the scalability and practical deployment of nanomaterials, a range of strategies are being explored:

3.5.1 Development of Safer-by-Design Nanomaterials

Designing nanomaterials with specific, controlled properties that ensure long-term stability and minimal environmental impact is a key strategy for scaling up their use. This includes developing materials with controlled surface chemistry and improved resistance to degradation under real-world conditions (Schrand et al. 2010).

3.5.2 Optimization of Synthesis Methods

Scalable and cost-effective synthesis methods are essential for ensuring that nanomaterials can be produced in large quantities at a reasonable cost. Researchers are exploring low-cost, high-throughput production methods that minimize energy consumption and material waste, while ensuring consistent quality (Rai et al. 2020).

3.5.3 Collaboration Between Academia, Industry, and Governments

To overcome the barriers to large-scale deployment, collaboration between academic researchers, industry players, and government agencies is crucial. Governments can support the development of nanomaterial-based environmental technologies through funding, regulatory support, and the development of public-private partnerships.

3.5.4 Comprehensive Life Cycle Assessments

Conducting life cycle assessments (LCAs) for nanomaterial-based environmental technologies will provide a holistic view of their environmental, economic, and social impacts, helping to identify opportunities for improvement and optimize deployment strategies (Kim et al. 2012).

4. The Need for Interdisciplinary Collaboration and Standardization

As nanomaterials continue to show significant promise in addressing environmental challenges such as pollution control, water treatment, and resource recovery, the successful deployment of these materials requires more than just scientific and technical advancements. The integration of nanomaterials into real-world applications involves addressing a range of complex issues that require expertise from various fields. These challenges include ensuring safety, scalability, environmental impact assessments, regulatory compliance, and effective management of risks. Consequently, the development and implementation of nanomaterials for environmental applications necessitate robust interdisciplinary collaboration and the establishment of standardized protocols that span multiple disciplines, such as materials science, environmental engineering, toxicology, regulatory science, and policymaking. This section discusses the importance of interdisciplinary collaboration and standardization in overcoming the challenges associated with the deployment of nanomaterials for environmental applications. It highlights the roles of diverse stakeholders ranging from scientists to regulators and emphasizes the need for collective efforts to ensure the safe, effective, and sustainable use of nanotechnology.

4.1 Interdisciplinary Collaboration: A Necessity for Success

The integration of nanomaterials into environmental technologies cannot be accomplished by isolated research efforts or the work of a single discipline. Nanotechnology is inherently interdisciplinary, drawing from materials science, chemistry, physics, biology, and environmental engineering. To ensure the successful deployment of nanomaterials in environmental applications, it is imperative to foster collaboration across various fields to address the multifaceted challenges associated with their development, safety, and scaling up. Materials scientists and environmental engineers are pivotal in designing nanomaterials that are tailored for specific environmental applications, such as pollutant removal or resource recovery. Collaborative efforts between these disciplines are essential to ensure that nanomaterials possess the necessary properties for functionality in real-world environmental conditions, such as stability, reactivity, and compatibility with various environmental media (Kumar et al. 2014). Environmental engineers also play a key role in adapting laboratory-based nanomaterial applications to large-scale systems, ensuring that technologies are scalable and cost-effective (Rai et al. 2020). To fully understand the potential risks associated with nanomaterials, toxicologists and environmental health experts must work closely with materials scientists and engineers. The toxicity of nanomaterials can vary significantly depending on their size, surface charge, and composition (Sharma et al. 2024). Understanding the biological interactions of nanomaterials is critical for assessing their potential health risks and environmental impacts. Collaboration between toxicologists and materials scientists can help develop "safe-by-design" nanomaterials that minimize adverse effects on human health and ecosystems (Niazi et al. 2009).

The successful implementation of nanomaterial-based environmental technologies requires the establishment of clear regulatory frameworks that ensure public safety, environmental protection, and compliance with legal standards. Regulatory science experts and policymakers must work alongside researchers and industry leaders to develop guidelines for the safe manufacture, use, and disposal of nanomaterials. Such collaboration can help ensure that nanomaterials are deployed responsibly while fostering innovation. Furthermore, regulatory bodies must stay updated on the latest scientific findings to ensure that nanomaterials are assessed using the most current and accurate methodologies (Schrand et al. 2010). Nanotechnology's widespread adoption hinges not only on scientific and technical advancements but also on public perception and societal acceptance. Social scientists play a critical role in understanding public concerns regarding the safety, ethics, and environmental impact of nanomaterials. Engaging with the public and addressing concerns through transparent communication and public education is essential to ensure the acceptance and sustainable development of nanomaterial-based technologies. Furthermore, collaborations between social scientists, ethicists, and industry stakeholders can help identify potential risks and societal challenges associated with the deployment of nanomaterials in environmental applications. Several successful examples of interdisciplinary collaboration in nanotechnology development highlight the importance of working across different fields. For instance, the collaboration between environmental engineers and toxicologists in the development of carbon nanotubes (CNTs) for water purification has led to a deeper understanding of the material's environmental behavior and toxicity. By jointly assessing the materials' efficacy, stability, and potential risks, researchers have improved CNTs' functionality and safety profiles (Shao et al. 2011). Similarly, the integration of regulatory scientists into research teams developing nanomaterials for environmental applications has led to the creation of safety protocols and regulatory guidelines that ensure the responsible development and use of these materials.

4.2 Standardization of Nanomaterial Testing and Risk Assessment

While the interdisciplinary collaboration of various scientific disciplines is essential for the development of nanomaterials, standardization is equally critical to ensure the effectiveness, safety, and scalability of these materials in environmental applications. Standardized protocols are necessary for the synthesis, characterization, testing, and evaluation of nanomaterials to ensure consistency and reproducibility across research efforts and real-world applications. The lack of standardized methods for testing and assessing nanomaterials has hindered the broad adoption of these technologies in environmental remediation. One of the key challenges in the development of nanomaterials is the variability in their properties due to differences in synthesis methods, raw materials, and environmental conditions. Standardized synthesis protocols are needed to ensure that nanomaterials produced across different research groups or industrial settings exhibit consistent properties, such as size, surface charge, morphology, and composition. Standardized methods for characterizing nanomaterials such as dynamic light scattering (DLS), transmission electron microscopy (TEM), and X-ray diffraction (XRD) are essential for assessing and comparing the properties of nanomaterials and ensuring their reproducibility (Schrand et al. 2010). Another area where standardization is critical is toxicity testing and environmental risk assessment. Currently, there is no globally accepted standardized protocol for assessing the environmental and health risks of nanomaterials (Wang et al. 2020). Different methodologies and endpoints for toxicity testing can yield inconsistent results, making it difficult to compare the safety profiles of different nanomaterials or to determine their suitability for environmental applications. International bodies, such as the International Organization for Standardization (ISO) and the Organization for Economic Co-operation and Development (OECD), are working to establish standardized protocols for nanomaterial toxicity testing, which would include methods for assessing bioaccumulation, toxicity in aquatic organisms, and interactions with soil microorganisms.

In addition to toxicity testing, standardized environmental monitoring techniques and life cycle assessment (LCA) protocols are necessary to assess the potential environmental impacts of nanomaterials throughout their life cycle, from production to disposal. LCA is an important tool for evaluating the sustainability of nanomaterial-based environmental technologies, as it can provide insights into their energy consumption, material use, waste generation, and long-term environmental impact (Kim et al. 2012). The development of standardized LCA frameworks for nanomaterials will enable more accurate assessments and comparisons of different nanomaterial-based remediation technologies. The absence of standardized risk assessment frameworks for nanomaterials is another key challenge in the deployment of nanomaterials for environmental applications. The risk assessment of nanomaterials requires a comprehensive approach that considers their behavior in different environmental media, their interactions with organisms, and their potential for bioaccumulation and long-term ecological effects. The development of internationally recognized risk assessment protocols will help ensure that the potential risks of nanomaterials are adequately evaluated before they are used in large-scale environmental applications. Such frameworks would involve testing for toxicity, environmental behavior, and transport in environmental systems, as well as assessing long-term ecological consequences (Rai et al. 2020).

4.3 Global Cooperation and Harmonization of Standards

In addition to national and regional efforts, global cooperation is necessary to harmonize standards for nanomaterials. The global nature of the nanotechnology market and the cross-border deployment of environmental technologies require international cooperation to develop unified standards and regulations. Institutions such as the OECD, ISO, and the World Health Organization (WHO) play crucial roles in fostering international dialogue and promoting the harmonization of testing and safety protocols for nanomaterials. Furthermore, international collaborations on research, data sharing, and best practices can facilitate the development of more effective nanomaterial-based environmental solutions. One notable example of international collaboration in nanomaterials research is the OECD's Working Party on Nanomaterials (WPN), which brings together experts from different countries to develop safety guidelines and testing protocols for nanomaterials. Similarly, the ISO Technical Committee 229 on Nanotechnologies is working to establish international standards for the characterization and testing of nanomaterials, including standardized methods for assessing their toxicity and environmental impact. These international efforts help ensure that nanomaterials developed in one region can be safely and effectively used in other parts of the world (Falkner and Jaspers 2012).

4.4 Towards Sustainable and Safe Deployment

To realize the full potential of nanomaterials in addressing global environmental challenges, it is essential to foster interdisciplinary collaboration and establish standardized protocols for their development, testing, and deployment. Collaboration between materials scientists, toxicologists, engineers, regulatory bodies, and social scientists is crucial for developing safe, effective, and sustainable nanomaterials. At the same time, the standardization of testing, risk assessment, and life cycle analysis methods will provide the consistency and reproducibility needed to scale up nanomaterial-based environmental technologies and ensure their safety. With global cooperation and a commitment to developing clear regulatory frameworks, nanomaterials can become a powerful tool in the fight against environmental pollution and resource depletion.

5. Conclusion

Nanomaterials represent a transformative class of materials that hold immense potential for addressing a wide array of environmental challenges, including pollution control, water purification, resource recovery, and sustainable remediation. Their unique properties, such as high surface area, reactivity, and tunability, make them ideal candidates for a variety of environmental applications, potentially providing innovative solutions to some of the world's most pressing environmental issues. However, despite their promising applications, the successful deployment of nanomaterials in real-world environmental contexts is beset by numerous challenges that must be overcome to ensure their effectiveness, safety, and long-term sustainability. As discussed throughout this chapter, several key obstacles stand in the way of realizing the full potential of nanomaterials in environmental applications. These include issues related to material synthesis, variability in properties, stability, scalability, environmental and health risks, and the complexity of integrating nanomaterials into large-scale systems. While progress has been made in addressing some of these challenges, others remain unresolved, necessitating a concerted effort from researchers, industry leaders, policymakers, and regulatory bodies. One of the most critical aspects of overcoming these challenges is the need for interdisciplinary collaboration. The complexity of nanomaterials and the diverse set of challenges they present require expertise from various scientific and engineering disciplines. Materials scientists, environmental engineers, toxicologists, and regulatory scientists must work together to design nanomaterials that are both effective in environmental remediation and safe for human health and ecosystems. Furthermore, social scientists and ethicists must be engaged in discussions surrounding public perception, ethical considerations, and the societal implications of nanotechnology, ensuring that technological advancements align with public interests and values. Additionally, standardization plays a pivotal role in the successful application of nanomaterials. The development of universally accepted protocols for synthesis, characterization, toxicity testing, and life cycle assessment of nanomaterials is essential for ensuring their reproducibility, safety, and consistency. Standardization will not only help in comparing different nanomaterial technologies but also in addressing regulatory uncertainty, providing clear guidelines for their safe manufacture, use, and disposal. Standardized risk assessment frameworks will allow stakeholders to evaluate the potential impacts of nanomaterials on human health and the environment with greater confidence and clarity. The importance of global cooperation cannot be overstated. Nanotechnology is inherently global, and its environmental applications are no exception. International collaboration between governments, scientific communities, and industry is essential to establish harmonized standards, share research findings, and create regulatory frameworks that facilitate the safe and widespread adoption of nanomaterials. Efforts by international organizations such as the OECD, ISO, and WHO are crucial in guiding the responsible development and use of nanomaterials across borders.

In short, while the potential of nanomaterials in solving global environmental problems is enormous, realizing this potential will require more than just technological innovation. It will require a comprehensive approach, one that combines interdisciplinary research, standardized practices, and global cooperation. By fostering collaboration across scientific disciplines and sectors, ensuring that rigorous standards and safety protocols are in place, and promoting clear regulatory frameworks, nanomaterials can be integrated into environmental technologies that are not only effective but also safe, sustainable, and widely adopted. As the research and development of nanomaterials continue to evolve, it is essential that these efforts are carried out in a coordinated, responsible, and forward-thinking manner to ensure that the benefits of nanotechnology are realized in a way that promotes environmental protection and human well-being. Ultimately, with careful attention to the scientific, technical, and societal challenges discussed in this chapter, nanomaterials have the potential to significantly contribute to global efforts to tackle some of the most urgent environmental issues of our time.

References

Åberg, C., Piattelli, V., Montizaan, D., and Salvati, A. (2021). Sources of variability in nanoparticle uptake by cells. *Nanoscale*, 13(41): 17530–17546.

Ahmad, F., Salem-Bekhit, M.M., Khan, F., Alshehri, S., Khan, A., Ghoneim, M. M. et al. (2022). Unique properties of surface-functionalized nanoparticles for bio-application: functionalization mechanisms and importance in application. *Nanomaterials*, 12(8): 1333.

Ahmed, H. (2022). The Challenges Behind Scaling Up Nanomaterials.

Ali Talha Khalil, Arshad Islam. (2025). Expanding Nanobiotechnology - Applications and Commercialization, CRC Press.

Ali, A., Zafar, H., Zia, M., ul Haq, I., Phull, A. R., Ali, J. S., and Hussain, A. (2016). Synthesis, characterization, applications, and challenges of iron oxide nanoparticles. *Nanotechnology, Science and Applications*, pp. 49–67.

Baer, D. R., Amonette, J. E., Engelhard, M. H., Gaspar, D. J., Karakoti, A. S., Kuchibhatla, S. et al. (2008). Characterization challenges for nanomaterials. *Surface and Interface Analysis: An International Journal Devoted to the Development And Application of Techniques for the Analysis of Surfaces, Interfaces and Thin Films*, 40(3-4): 529–537.

Baral, B., Altaee, A., Simeonidis, K., and Samal, A. K. (2024). Shape and size dependent nanostructures for environmental applications. *Frontiers in Chemistry*, 12: 1362033.

Bell, I. R., Ives, J. A., and Wayne, B. J. (2014). Nonlinear effects of nanoparticles: biological variability from hormetic doses, small particle sizes, and dynamic adaptive interactions. *Dose-Response*, 12(2): dose-response.

Bhasin, C. P., Pathan, A., and Patel, R. V. (2024). An Evaluation of Carbon Nanotube-based and Activated Carbon-based Nanocomposites for Fluoride and Other Pollutant Removal from Water: A Review. *Current Nanomaterials*, 9(1): 16–40.

Cheng, Y., Chen, Z., Yang, S., Liu, T., Yin, L., Pu, Y. et al. (2021). Nanomaterials-induced toxicity on cardiac myocytes and tissues, and emerging toxicity assessment techniques. *Science of The Total Environment*, 800: 149584.

Cross, R. K., Bossa, N., Stolpe, B., Loosli, F., Sahlgren, N. M., Clausen, P. A. et al. (2022). Reproducibility of methods required to identify and characterize nanoforms of substances. *NanoImpact*, 27: 100410.

De Jong, W. H., Geertsma, R. E., and Borchard, G. (2022). Regulatory safety evaluation of nanomedical products: key issues to refine. *Drug Delivery and Translational Research*, 12(9): 2042–2047.

Desai, N. (2012). Challenges in development of nanoparticle-based therapeutics. *The AAPS Journal*, 14(2): 282–295.

Fadel, T. R., Steevens, J. A., Thomas, T. A., and Linkov, I. (2015). The challenges of nanotechnology risk management. *Nano Today*, 10(1): 6–10.

Falkner, R., and Jaspers, N., (2012). Regulating nanotechnologies: risk, uncertainty and the global governance gap. *Global Environmental Politics*, 12(1): 30–55.

Gambinossi, F., Mylon, S. E., and Ferri, J. K. (2015). Aggregation kinetics and colloidal stability of functionalized nanoparticles. *Advances in Colloid and Interface Science*, 222: 332–349.

Gao, C., Hu, Y., Wang, M., Chi, M., and Yin, Y. (2014). Fully alloyed Ag/Au nanospheres: combining the plasmonic property of Ag with the stability of Au. *Journal of the American Chemical Society*, 136(20): 7474–7479.

Gohar, O., Khan, M. Z., Bibi, I., Bashir, N., Tariq, U., Bakhtiar, M. et al. (2024). Nanomaterials for advanced energy applications: Recent advancements and future trends. *Materials & Design*, 241: 112930.

Hosseingholian, A., Gohari, S. D., Feirahi, F., Moammeri, F., Mesbahian, G., Moghaddam, Z. S. et al. (2023). Recent advances in green synthesized nanoparticles: From production to application. *Materials Today Sustainability*, 24: 100500.

Huang, J., Cao, C., Liu, J., Yan, C., and Xiao, J. (2019). The response of nitrogen removal and related bacteria within constructed wetlands after long-term treating wastewater containing environmental concentrations of silver nanoparticles. *Science of the Total Environment*, 667: 522–531.

Jagadeesh, P., Rangappa, S. M., and Siengchin, S. (2024). Advanced characterization techniques for nanostructured materials in biomedical applications. *Advanced Industrial and Engineering Polymer Research*, 7(1): 122–143.

Jiang, B., Kauffman, A. E., Li, L., McFee, W., Cai, B., Weinstein, J. et al. (2020). Health impacts of environmental contamination of micro-and nanoplastics: a review. *Environmental Health and Preventive Medicine*, 25: 1–15.

Jiang, J., Oberdörster, G., and Biswas, P. (2009). Characterization of size, surface charge, and agglomeration state of nanoparticle dispersions for toxicological studies. *Journal of Nanoparticle Research*, 11: 77–89.

Kataria, N., Garg, V. K., Han, C., and Rene, E. R. eds. (2025). *Microplastic Pollution: Occurrence, Health Risk and Challenges*. CRC Press.

Khan, I., Saeed, K., and Khan, I. (2019). Nanoparticles: Properties, applications and toxicities. *Arabian Journal of Chemistry*, 12(7): 908–931.

Kim, B., Murayama, M., Colman, B. P., and Hochella, M.F. (2012). Characterization and environmental implications of nano-and larger TiO 2 particles in sewage sludge, and soils amended with sewage sludge. *Journal of Environmental Monitoring*, 14(4): 1128–1136.

Klaine, S. J., Koelmans, A. A., Horne, N., Carley, S., Handy, R. D., Kapustka, L. et al. (2012). Paradigms to assess the environmental impact of manufactured nanomaterials. *Environmental Toxicology & Chemistry*, 31(1).

Kose, O., Béal, D., Motellier, S., Pelissier, N., Collin-Faure, V., Blosi, M. et al. (2023). Physicochemical Transformations of Silver Nanoparticles in the Oro-Gastrointestinal Tract Mildly Affect Their Toxicity to Intestinal Cells *In Vitro*: An AOP-Oriented Testing Approach. *Toxics*, 11(3): 199.

Kumar, S., Ahlawat, W., Bhanjana, G., Heydarifard, S., Nazhad, M. M., and Dilbaghi, N. (2014). Nanotechnology-based water treatment strategies. *Journal of Nanoscience and Nanotechnology*, 14(2): 1838–1858.

Laux, P., Riebeling, C., Booth, A. M., Brain, J. D., Brunner, J., Cerrillo, C. (2018). Challenges in characterizing the environmental fate and effects of carbon nanotubes and inorganic nanomaterials in aquatic systems. *Environmental Science: Nano*, 5(1): 48–63.

Lee, K. J., Browning, L. M., Nallathamby, P. D. and Xu, X. H. N. (2013). Study of charge-dependent transport and toxicity of peptide-functionalized silver nanoparticles using zebrafish embryos and single nanoparticle plasmonic spectroscopy. *Chemical Research in Toxicology*, 26(6): 904–917.

Li, Z., Greden, K., Alvarez, P. J., Gregory, K. B. and Lowry, G. V. (2010). Adsorbed polymer and NOM limits adhesion and toxicity of nano scale zerovalent iron to E. coli. *Environmental science & technology*, 44(9): 3462–3467.

Lin, P. C., Lin, S., Wang, P. C., and Sridhar, R. (2014). Techniques for physicochemical characterization of nanomaterials. *Biotechnology Advances*, 32(4): 711–726.

Liu, J., Sonshine, D. A., Shervani, S., and Hurt, R. H. (2010). Controlled release of biologically active silver from nanosilver surfaces. *ACS Nano*, 4(11): 6903–6913.

Liu, Y., Zhao, Y., Sun, B., and Chen, C. (2013). Understanding the toxicity of carbon nanotubes. *Accounts of Chemical Research*, 46(3): 702–713.

Mampilly, R. B., Bhatt, S. H., Modi, N. J., and Pathan, A. (2024). Adsorption of eriochrome Black-T dye by batch investigations using waste tea@ Fe NPs as low-cost adsorbent. *Int. J. Thin Film Sci. Technol.*, 13: 17–25.

Mishra, M., Dashora, K., Srivastava, A., Fasake, V. D., and Nag, R. H. (2019). Prospects, challenges and need for regulation of nanotechnology with special reference to India. *Ecotoxicology and Environmental Safety*, 171: 677–682.

Mülhopt, S., Diabaté, S., Dilger, M., Adelhelm, C., Anderlohr, C., Bergfeldt, T. et al. (2018). Characterization of nanoparticle batch-to-batch variability. *Nanomaterials*, 8(5): 311.

Nayak, T., and Pathan, A. (2023). Environmental remediation and application of carbon-based nanomaterials in the treatment of heavy metal-contaminated water: A review. *Materials Today: Proceedings*.

Niazi, J. H., and Gu, M. B. (2009). Toxicity of metallic nanoparticles in microorganisms-a review. *Atmospheric and Biological Environmental Monitoring*, pp. 193–206.

Onamade, A.O., Asaju, O. A., and Adewumi, B. J. (2025). Challenges and future prospects of smart nanomaterials for environmental remediation. *Smart Nanomaterials for Environmental Applications*, pp. 757–780.

Paliwal, R., Babu, R. J., and Palakurthi, S. (2014). Nanomedicine scale-up technologies: feasibilities and challenges. *Aaps Pharmscitech*, 15(6): 1527–1534.

Pathan, A., Bhatt, S. H., Vajapara, S., and Bhasin, C. P. (2022). Solar light induced photo catalytic properties of α-Fe2O3 nanoparticles for degradation of methylene blue dye. *Int. J. Thin Film Sci. Technol.*, 11: 213–224.

Pathan, A. A., Desai, K. R., and Bhasin, C. P. (2017). Improved photocatalytic properties of NiS nanocomposites prepared by displacement method for removal of rose bengal dye. *Current Nanomaterials*, 2(3): 169–176.

Pathan, A. A., Desai, K. R., Vajapara, S., and Bhasin, C. P. (2018). Conditional optimization of solution combustion synthesis for pioneered La2O3 nanostructures to application as future CMOS and NVMS generations. *Advances in Nanoparticles*, 7(01): 28–35.

Pereira, S. P., Jesus, F., Aguiar, S., de Oliveira, R., Fernandes, M., Ranville, J. et al. (2018). Phytotoxicity of silver nanoparticles to Lemna minor: Surface coating and exposure period-related effects. *Science of the Total Environment*, 618: 1389–1399.

Rai, P. K., Lee, J., Brown, R. J., and Kim, K. H. (2021). Micro-and nano-plastic pollution: Behavior, microbial ecology, and remediation technologies. *Journal of cleaner production*, 291: 125240.

Rathore, A., and Mahesh, G. (2021). Public perception of nanotechnology: A contrast between developed and developing countries. *Technology in Society*, 67: 101751.

Ray, P. C., Yu, H., and Fu, P. P. (2009). Toxicity and environmental risks of nanomaterials: challenges and future needs. *Journal of Environmental Science and Health Part C*, 27(1): 1–35.

Rim, K. T., Song, S. W., and Kim, H. Y. (2013). Oxidative DNA damage from nanoparticle exposure and its application to workers' health: a literature review. *Safety and Health at Work*, 4(4): 177–186.

Saravanan, A., Kumar, P. S., Hemavathy, R. V., Jeevanantham, S., Jawahar, M. J., Neshaanthini, J. P. et al. (2022). A review on synthesis methods and recent applications of nanomaterial in wastewater treatment: Challenges and future perspectives. *Chemosphere*, 307: 135713.

Schrand, A. M., Rahman, M. F., Hussain, S. M., Schlager, J. J., Smith, D. A., and Syed, A. F. (2010). Metal-based nanoparticles and their toxicity assessment. *Wiley Interdisciplinary Reviews: Nanomedicine and Nanobiotechnology*, 2(5): 544–568.

Shao, Q., Wu, P., Gu, P., Xu, X., Zhang, H., and Cai, C. (2011). Electrochemical and spectroscopic studies on the conformational structure of hemoglobin assembled on gold nanoparticles. *The Journal of Physical Chemistry B*, 115(26): 8627–8637.

Sharifi, S., Behzadi, S., Laurent, S., Forrest, M. L., Stroeve, P., and Mahmoudi, M. (2012). Toxicity of nanomaterials. *Chemical Society Reviews*, 41(6): 2323–2343.

Sharifi, S., Mahmoud, N. N., Voke, E., Landry, M. P., and Mahmoudi, M. (2022). Importance of standardizing analytical characterization methodology for improved reliability of the nanomedicine literature. *Nano-Micro Letters*, 14(1): 172.

Sharma, R., Sharma, N., Prashar, A., Hansa, A., Lajayer, B. A., and Price, G. W. (2024). Unraveling the plethora of toxicological implications of nanoparticles on living organisms and recent insights into different remediation strategies: A comprehensive review. *Science of the Total Environment*, 906: 167697.

Singh, S., Prasad, S. M., and Bashri, G. (2023). Fate and toxicity of nanoparticles in aquatic systems. *Acta Geochimica*, 42(1): 63–76.

Sohail, M. I., Waris, A. A., Ayub, M. A., Usman, M., ur Rehman, M. Z., Sabir, M. et al. (2019). Environmental application of nanomaterials: a promise to sustainable future. In *Comprehensive Analytical Chemistry* (Vol. 87: 1–54). Elsevier.

Vaghela, H., Shah, R., and Pathan, A. (2018). Palladium nanoparticles mediated through bauhinia variegata: Potent in vitro anticancer activity against mcf-7 cell lines and antimicrobial assay. *Current Nanomaterials*, 3(3): 168–177.

Wang, E., Huang, Y., Du, Q., and Sun, Y. (2017). Silver nanoparticle induced toxicity to human sperm by increasing ROS (reactive oxygen species) production and DNA damage. *Environmental Toxicology and Pharmacology*, 52: 193–199.

Wang, H., Gu, X. K., Zheng, X., Pan, H., Zhu, J., Chen, S. et al. (2019). Disentangling the size-dependent geometric and electronic effects of palladium nanocatalysts beyond selectivity. *Science Advances*, 5(1): eaat6413.

Wieszczycka, K., Staszak, K., Woźniak-Budych, M. J., Litowczenko, J., Maciejewska, B. M. and Jurga, S. (2021). Surface functionalization–The way for advanced applications of smart materials. *Coordination Chemistry Reviews*, 436: 213846.

Xie, X., and Shen, W. (2009). Morphology control of cobalt oxide nanocrystals for promoting their catalytic performance. *Nanoscale*, 1(1): 50–60.

Xie, Y., Ou, P., Wang, X., Xu, Z., Li, Y. C., Wang, Z. et al. (2022). High carbon utilization in CO2 reduction to multi-carbon products in acidic media. *Nature Catalysis*, 5(6): 564–570.

Yang, X., Gondikas, A. P., Marinakos, S. M., Auffan, M., Liu, J., Hsu-Kim, H. et al. (2012). Mechanism of silver nanoparticle toxicity is dependent on dissolved silver and surface coating in *Caenorhabditis elegans*. *Environmental Science & Technology*, 46(2): 1119–1127.

Ying, S., Guan, Z., Ofoegbu, P. C., Clubb, P., Rico, C., He, F. et al. (2022). Green synthesis of nanoparticles: Current developments and limitations. *Environmental Technology & Innovation*, 26: 102336.

Zavanelli, N., Kim, J., and Yeo, W.H. (2021). Recent advances in high-throughput nanomaterial manufacturing for hybrid flexible bioelectronics. *Materials*, 14(11): 2973.

Zhang, Y., Wu, B., Xu, H., Liu, H., Wang, M., He, Y. et al. (2016). Nanomaterials-enabled water and wastewater treatment. *NanoImpact*, 3: 22–39.

Zhao, C. M., and Wang, W. X. (2012). Size-dependent uptake of silver nanoparticles in Daphnia magna. *Environmental Science & Technology*, 46(20): 11345–11351.

Zhao, R., Wang, H. H., Gao, J., Zhang, Y. J., Li, X., Zhou, J. J. et al. (2022). Plant volatile compound methyl benzoate is highly effective against *Spodoptera frugiperda* and safe to non-target organisms as an eco-friendly botanical-insecticide. *Ecotoxicology and Environmental Safety*, 245: 114101.

Zhao, Y., Yang, Q., Liu, D., Liu, T., and Xing, L. (2022). Neurotoxicity of nanoparticles: Insight from studies in zebrafish. *Ecotoxicology and Environmental Safety*, 242: 113896.

Zhou, Q., Pan, J., Deng, S., Xia, F., and Kim, T. (2021). Triboelectric nanogenerator-based sensor systems for chemical or biological detection. *Advanced Materials*, 33(35): 2008276.

Zhou, X., Xu, W., Liu, G., Panda, D., and Chen, P. (2010). Size-dependent catalytic activity and dynamics of gold nanoparticles at the single-molecule level. *Journal of the American Chemical Society*, 132(1): 138–146.

CHAPTER-15

Past, Present and Future of Nanomaterials in Pollution Management

Birhanu Abera,[1,2,*] *Bawoke Mekuye,*[3] *Gedefaw Mebratie*[2,3] and *Abayeneh Girma*[4]

1. Introduction

With pollutants impacting air, water, and soil ecosystems at a never-before-seen scale, environmental pollution has become one of the 21st century's most urgent global issues (Chandra 2025). When handling complicated contamination scenarios, traditional remediation strategies frequently encounter difficulties in terms of efficiency, cost-effectiveness, and scalability (Kuppan et al. 2024). By providing creative solutions through engineered nanomaterials—materials with at least one dimension between 1 and 100 nanometers that have distinct physicochemical properties—nanotechnology has transformed environmental management in this context (Soni 2022). Nanomaterials are highly useful for pollution control applications due to their unique properties, which include a high surface area-to-volume ratio, quantum effects, and customizable surface chemistry (Prasad et al. 2025). Compared to traditional approaches, these characteristics allow nanomaterials to interact with contaminants at molecular levels, enabling more effective detection, degradation, and removal (Thakur and Kumar 2022). For example, nanoparticles can have surface areas that are many orders of magnitude larger than those of their bulk counterparts, which means that there are more active sites for adsorption and chemical reactions (Mane et al. 2024). Nanomaterials for environmental applications have historically developed along an interesting path. Environmental applications emerged as a secondary issue in early research conducted in the 1970s and 1980s, which was primarily focused on fundamental material qualities (Hughes et al. 2025). But a paradigm shift occurred in the 1990s when researchers realized how effective carbon-based nanomaterials, such as fullerenes and carbon nanotubes, could be in cleaning up the environment (Ayanda et al. 2024).

[1] College of Natural and Computational Science, Department of Physics, Injibara University, Injibara, Ethiopia. https://orcid.org/0000-0001-8856-0720
[2] College of Science, Department of Physics, Bahir Dar University, Bahir Dar, Ethiopia. https://orcid.org/0000-0002-8621-4653
[3] College of Natural and Computational Science, Department of Physics, Mekdela Amba University, Gimba, Ethiopia. https://orcid.org/0000-0002-7214-3735
[4] College of Natural and Computational Science, Department of Biology, Mekdela Amba University, Gimba, Ethiopia. https://orcid.org/0000-0001-6155-315X
* Corresponding author: birhanuabera643@gmail.com

DOI: 10.1201/9781003558804-15

Research on synthetic nanomaterials intended primarily for pollution control, especially in water treatment and soil remediation, increased at the turn of the century (Susaimanickam et al. 2024).

Nanomaterials have a variety of uses in several remediation sectors in modern environmental research (Asghar et al. 2024). They enable sophisticated water treatment procedures, such as the use of graphene oxide for the adsorption of heavy metals and titanium dioxide nanoparticles for the photocatalytic destruction of organic pollutants (Badoni et al. 2025). Nanofiber filters for the removal of particulate matter and nanocatalysts for automobile emissions are beneficial for air quality management (Zubair et al. 2025). Nanoscale zerovalent iron (nZVI) for *in-situ* groundwater treatment and nanocomposites for immobilization of contaminants have revolutionized soil remediation (Yang et al. 2024). Despite these impressive developments, there are still several obstacles in the way of the broad adoption of nanomaterial-based solutions. Research is still being conducted on issues such as possible ecotoxicity, long-term environmental fate, and large-scale production costs (Kirubakaran et al. 2025). To address the unique features of nanotechnology applications in environmental cleanup, regulatory frameworks are still being developed (Alizadeh et al. 2025). Furthermore, to guarantee truly sustainable solutions, the energy needs and life cycle effects of nanomaterial production and deployment must be carefully taken into account (Jamwal et al. 2025). Next-generation smart nanomaterials with improved functionality and environmental compatibility are the direction the field is heading toward (Reddy et al. 2024). Technologies such as Artificial Intelligence (AI)-designed nanocomposites, biogenic nanoparticles produced using green chemistry techniques, and self-healing nanomaterials promise to overcome present constraints while creating new opportunities (Tafti et al. 2025). Another fascinating area is the incorporation of nanomaterials with cutting-edge technologies such as the Internet of Things (IoT) for real-time environmental monitoring (Islam et al. 2024).

Through three temporal lenses—past advancements that founded the field, present applications showing practical impact, and future approaches that could revolutionize environmental remediation—this chapter offers a thorough analysis of nanomaterials in pollution control. We aim to give readers a comprehensive grasp of how nanotechnology is changing the way we address environmental issues by examining historical turning points, current scientific advancements, and developing trends.

2. The Past of Nanomaterials in Pollution Management

From accidental discoveries to specifically designed solutions, the use of nanomaterials in environmental remediation has experienced remarkable progression. Although nanotechnology's environmental applications have only recently acquired substantial traction, its foundation in pollution management dates back to the middle of the twentieth century (Zhang et al. 2024). The catalytic capabilities of nanoscale metal oxides, especially titanium dioxide (TiO_2), for decomposing organic contaminants under ultraviolet light were the main focus of early research in the 1970s and 1980s (Greco et al. 2024). These groundbreaking investigations established the foundation for sophisticated photocatalytic degradation mechanisms that are still in use today (Zhang et al. 2023).

The discovery of fullerenes (C_{60}) in 1985 marked a significant advancement, and the identification of carbon nanotubes (CNT) in 1991 brought forth a new class of carbon-based nanomaterials with remarkable adsorption properties (Islam et al. 2024). Due to their large surface area and functionalizable structures, researchers soon realized that they might be used to purify water, especially to remove organic pollutants and heavy metals (Dey et al. 2024). Additionally, silicon quantum dots emerged in the 1990s; originally created for optoelectronics, they later found use in pollution degradation and sensing (Mondal et al. 2024). The creation of synthetic nanomaterials especially intended for environmental cleanup in the early 2000s signaled a sea change. With its unmatched effectiveness in breaking down chlorinated solvents like trichloroethylene (TCE) and immobilizing heavy metals like chromium (Cr (VI)), nanoscale zerovalent iron (nZVI) has become a groundwater treatment breakthrough (Chen and Qian 2024). The practicality of nZVI was confirmed by field-scale applications in the mid-1900s, which resulted in its widespread use in the cleaning of contaminated sites (Sánchez Hidalgo et al. 2025). At the same time, developments in metal-organic frameworks (MOFs) produced extremely porous nanomaterials that could selectively adsorb liquid-phase contaminants and gases (Xie et al. 2024). The variety of nanomaterials and their range of applications increased exponentially throughout the 2010s. Due to its mechanical robustness, antibacterial qualities, and adjustable surface chemistry, graphene oxide (GO) and reduced graphene oxide (rGO) were thoroughly investigated for use in water treatment membranes (Tiwary et al. 2024). Using their surface plasmon resonance to break down pollutants in the presence of visible light, plasmonic nanoparticles such as gold and silver were investigated concurrently for improved photocatalytic systems. Hybrid nanocomposites, which combine multiple nanomaterials (such as ZnO-graphene or TiO_2-CNTs) to improve pollution removal efficiency, became popular in the latter part of the decade (Zeng et al. 2025).

Advances in scalable synthesis techniques and a better understanding of the interplay between nanomaterials and their surroundings propelled industrial adoption to accelerate in the 2020s (Ikumapayi and Laseinde 2024). In recent years, there has also been a focus on designing sustainable nanomaterials, such as green synthesis methods that use microbial or plant extracts to reduce their environmental impact (Mandal et al. 2025). As regulatory frameworks began to catch up, organizations such as the EU Commission and the US EPA published guidelines for the use of nanomaterials in environmental applications (Chávez-Hernández et al. 2024). With continuous research maximizing their efficacy, safety, and cost-effectiveness for worldwide deployment, nanomaterials are becoming essential to sophisticated remediation techniques (Mandal et al. 2025).

3. The Present Role of Nanomaterials in Pollution Management

Because they are more effective, selective, and versatile than traditional techniques, nanomaterials have become essential in contemporary environmental rehabilitation. They use special qualities, including high reactivity, a large surface area, and customizable surface chemistry, to reduce contamination in soil, water, and air. With the help of performance data and recent case studies, this part offers a comprehensive overview of the applications of nanomaterials today (Liu et al. 2024).

3.1 Water Remediation Technologies

Water treatment has been transformed by nanomaterials thanks to advanced filtration techniques, catalytic degradation, and adsorption (Figure 1 and Table 1). They are especially useful for tackling complicated water contamination issues due to their capacity to target certain contaminants at low concentrations (Tripathy et al. 2024, Alam 2025).

3.1.1 Adsorption-based Removal

Carbon-based nanomaterials, such as graphene oxide (GO), carbon nanotubes (CNTs), and activated carbon nanocomposites, are commonly used to remove organic contaminants (such as dyes and medications) and heavy metals (such as Pb^{2+}, Hg^{2+}, and Cd^{2+}) from water (Sneha et al. 2025). Functionalization with the -OH, -COOH, or $-NH_2$ groups increases their efficacy by enhancing the binding ability. Furthermore, because of their ultrahigh porosity and adaptable pore architectures, metal-organic frameworks (MOFs) have drawn interest for selective adsorption and are very effective in removing particular pollutants (Yan et al. 2024).

3.1.2 Catalytic and Photocatalytic Degradation

By photocatalyzing under ultraviolet or visible light, nanoparticles such as TiO_2 and ZnO degrade organic contaminants, such as insecticides and medications (Ahmad et al. 2025). Particularly efficient at breaking down chlorinated compounds and converting dangerous Cr(VI) into less hazardous Cr(III) are iron-based nanocatalysts such as nZVI and Fe_3O_4 (AG et al. 2025). By using

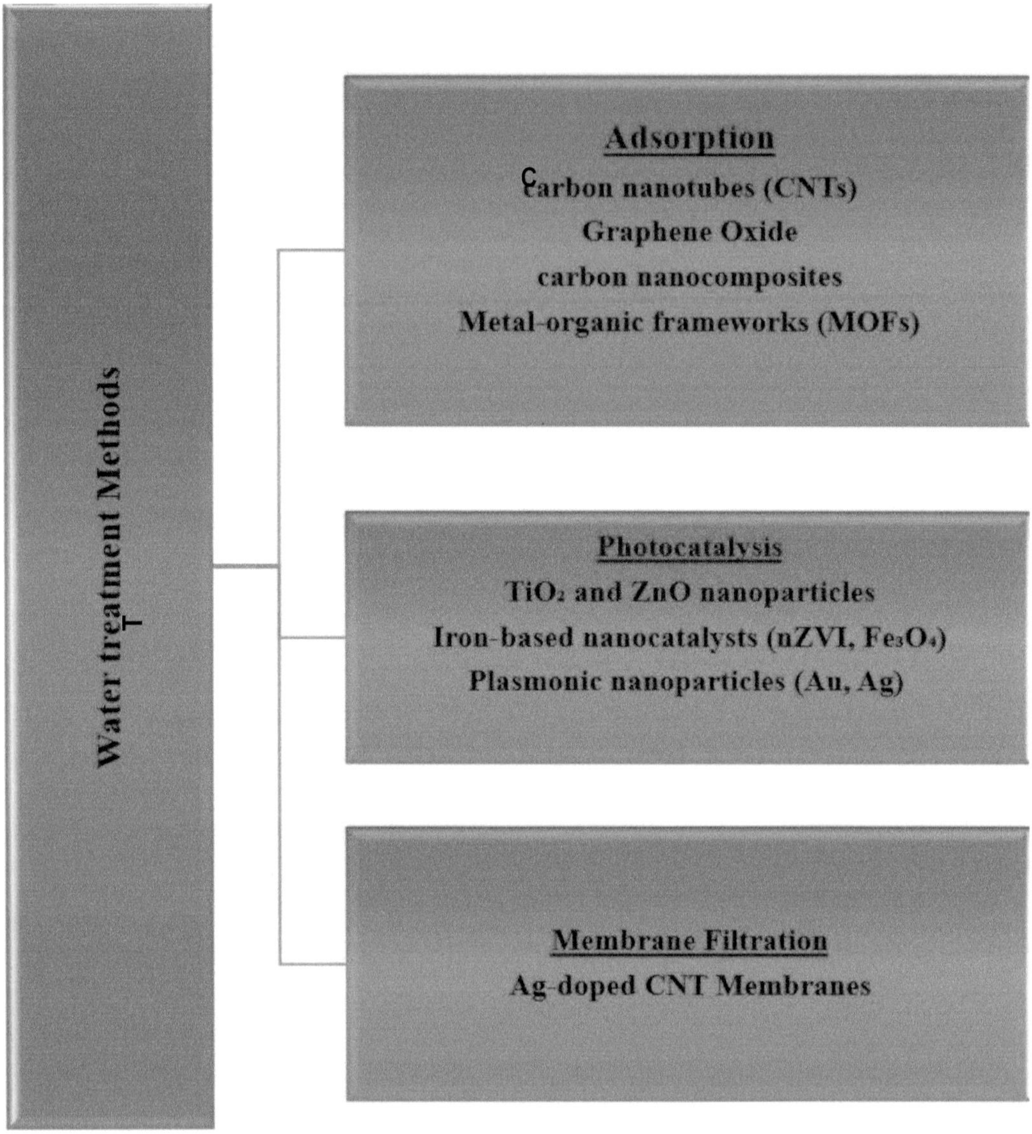

Figure 1: Water treatment methods.

Table 1: Nanomaterials, target contaminant, mechanism and efficiency in water remediation.

Nanomaterial	Target contaminant	Mechanism	Efficiency	Reference
Graphene Oxide (GO)	Heavy metals (Pb^{2+}, Hg^{2+})	Adsorption, ion exchange	> 95% removal	Tiwari et al. 2023
TiO_2 Nanoparticles	Organic dyes, drugs	Photocatalytic oxidation	80–99% degradation	Inamdar et al. 2024
nZVI	TCE, Cr(VI)	Reductive degradation	> 90% reduction	Gao et al. 2023
Ag-doped CNT Membranes	Bacteria, viruses	Size exclusion, antimicrobial	99.9% rejection	Leal et al. 2023

localized surface plasmon resonance to increase photocatalytic activity, plasmonic nanoparticles, such as gold (Au) and silver (Ag), further improve degradation efficiency (Hussain et al. 2025).

3.1.3 Membrane and Filtration Technologies

Water treatment procedures have been greatly improved using nanocomposite membranes that contain graphene oxide, carbon nanotubes, or silver nanoparticles. These membranes increase pathogen disinfection, efficiently remove microplastics, and improve desalination efficiency (Ben Dassi and Chamam 2025). These technologies offer a reliable way to deal with a variety of water contamination issues by incorporating nanoparticles into filter systems.

3.2 Air Pollution Control

In order to reduce airborne contaminants such as particulate matter (PM), volatile organic compounds (VOC), and greenhouse gases (GHGs), nanotechnology is essential (Odubo and Kosoe 2024).

Effective pollutant conversion and degradation are made possible by nanocatalysts, which are essential for emission control (Figure 2 and Table 2). To convert toxic exhaust gases, including nitrogen oxides (NO_x), carbon monoxide (CO), and hydrocarbons, into less toxic nitrogen (N_2), carbon dioxide (CO_2), and water (H_2O), platinum (Pt), palladium (Pd), and rhodium (Rh) nanoparticles are frequently used in catalytic converters in automotive applications (Chadha et al. 2022). Additionally, when exposed to light, titanium dioxide (TiO_2)-based photocatalysts have shown considerable potential in improving indoor air quality by breaking down volatile organic compounds (VOCs), including formaldehyde (Alam 2025).

Nanofiber filters have become a very successful way to remove particulate matter. To improve the performance of filtration, metal oxide nanoparticles such as copper oxide (CuO) and zinc oxide (ZnO) are frequently placed in electrospun polymer nanofibers, such as polyacrylonitrile (PAN) and polyvinylidene fluoride (Zhou et al. 2022). These cutting-edge filters are useful both for industrial and urban air cleaning, since they can capture tiny particulate matter ($PM_{2.5}$ and PM_{10}) with over 99% effectiveness (Choi 2025). When combined, these nanotechnology-based strategies offer creative and incredibly effective ways to reduce air pollution.

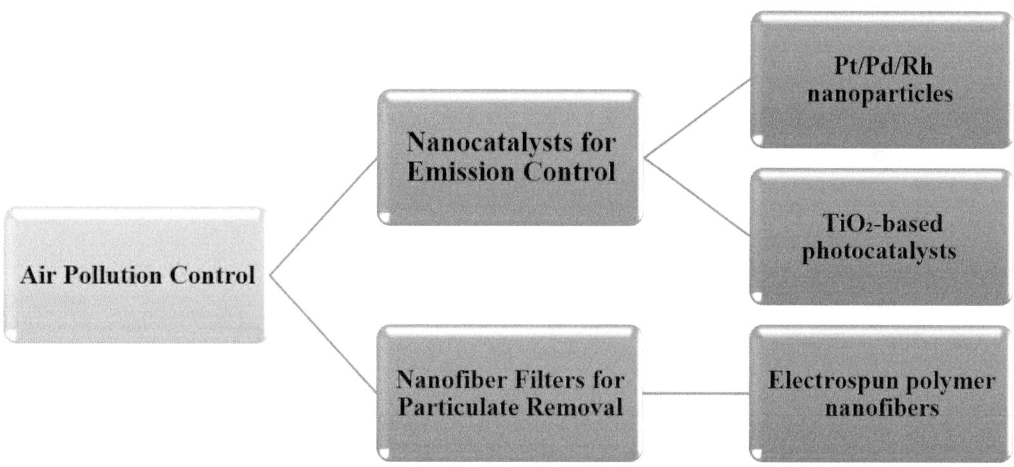

Figure 2: Air pollution control methods.

Table 2: Nanomaterials in air pollution control.

Nanomaterial	Application	Target pollutant	Performance	References
Pt-Pd/Rh Nanoparticles	Automotive catalysts	NO_x, CO, VOCs	>90% conversion	Alex et al. 2022
TiO_2-Coated Filters	Indoor air purification	Formaldehyde, VOCs	85-95% degradation	Yan et al. 2023
ZnO-PAN Nanofibers	PM filtration	$PM_{2.5}$, PM_{10}	99% capture	Aamer et al. 2021

3.3 Soil and Groundwater Remediation

In contrast to conventional excavation and pump-and-treat techniques, nanomaterials allow for *in situ* and *ex situ* remediation of contaminated soils and groundwater (Figure 3 and Table 3) (Liu et al. 2024).

The remarkable ability of nanoscale zerovalent iron (nZVI) to degrade contaminants has made it a potent instrument for environmental cleanup. Through reductive dechlorination, it efficiently breaks down chlorinated solvents like trichloroethylene (TCE) and perchloroethylene (PCE), turning them into non-toxic hydrocarbons (Rastogi et al. 2025). Furthermore, ZVI immobilizes harmful heavy metals such as arsenic (As (III)) and hexavalent chromium (Cr (VI)) by reducing them to less dangerous and soluble forms by reductive precipitation (Verma et al. 2024).

Using nanoclays and biochar composites to stabilize the environment is another interesting strategy. Strong adsorption capabilities for pesticides and heavy metals are demonstrated by nanoclays such as montmorillonite and kaolinite, which efficiently remove these contaminants from soil and water systems (Yang et al. 2024). Furthermore, biochar-nanoparticle hybrids are an efficient dual-purpose solution for sustainable agriculture and pollution management, since they improve nutrient retention, increase soil fertility, and immobilize contaminants (Mandal et al. 2025).

These nanotechnology-based approaches for degradation and nanoclays/biochar composites for stabilization combine to provide incredibly effective and long-lasting ways to reduce environmental contamination. Their use in soil restoration, groundwater remediation, and industrial waste treatment shows how adaptable and successful nanomaterials are in solving difficult pollution problems (Visentin et al. 2024).

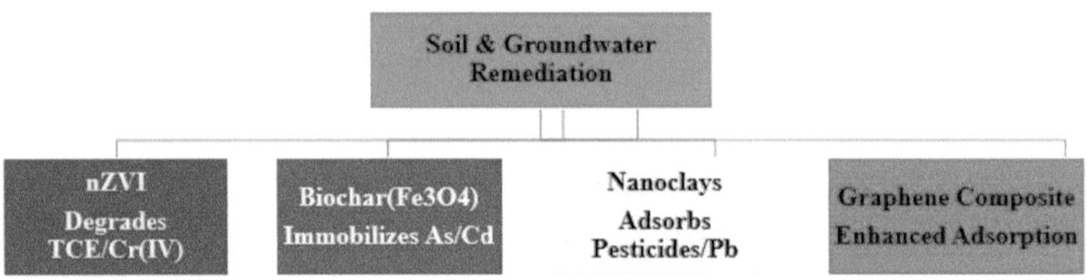

Figure 3: Soil and groundwater remediation.

Table 3: Nanomaterials in soil remediation.

Nanomaterial	Contaminant treated	Mechanism	Efficiency	References
nZVI	TCE, Cr(VI)	Reductive dechlorination	70-95% degradation	Barka et al. 2023
Biochar-Fe_3O_4 Hybrid	As(III), Cd(II)	Adsorption, stabilization	>80% immobilization	Murtaza et al. 2022
Graphene-Montmorillonite	Pesticides, Pb(II)	Enhanced adsorption	90% removal	Yang et al. 2024

4. Challenges and Limitations

Although nanoparticles have great potential for environmental remediation, a number of significant obstacles prevent their widespread use (Table 4). For successful implementation, multidisciplinary solutions are needed to address these constraints, which cross the technical, financial, environmental, and regulatory sectors (Kirubakaran et al. 2023).

Table 4: Challenge, impact, and potential solutions of nanomaterials.

Challenge	Impact	Potential solutions	References
High production costs	Limits large-scale deployment	Green synthesis methods	Mandal et al. 2025
Aggregation in real systems	Reduces effective surface area (O'Carroll et al. 2023)	Surface modification with stabilizers	Yusuf et al. 2025
Difficulty in recovery/reuse	Increases operational costs (Gupta et al. 2024)	Magnetic nanocomposite designs	Date and Jaspal 2024

4.1 Environmental and Health Risks

Although nanomaterials exhibit extraordinary efficacy in reducing pollution attributable to their elevated reactivity and extensive surface area, these identical characteristics engender considerable anxiety about their prospective ecotoxicological implications (Peng et al. 2024). Empirical investigations suggest that specific nanoparticles, such as silver (Ag), zinc oxide (ZnO), and carbon nanotubes (CNTs), may present environmental and health hazards. For example, they have the potential to disrupt vital microbial communities that underpin ecosystem functionality (Jonathan et al. 2024), bioaccumulate within aquatic organisms, potentially infiltrating and amplifying through trophic levels (Tiwari et al. 2024), and elicit oxidative stress in biological cells through the generation of reactive oxygen species (Zhang et al. 2024).

4.2 Technical and Economic Barriers

Despite promising applications in environmental remediation, nanomaterials face significant technical and economic barriers that hinder widespread adoption, including high production costs, aggregation issues, and challenges in recovery and reuse. Addressing these limitations through innovative solutions such as green synthesis, surface modifications, and magnetic nanocomposite designs is crucial for sustainable and cost-effective implementation (Bhaskar et al. 2025).

4.3 Regulatory and Standardization Gaps

Advances in nanomaterial technology have led to many regulatory and standardization gaps because traditional systems cannot follow the pace of innovation. One of the main problems is that different protocols are used to assess nanomaterial risk, and this makes it difficult to evaluate risks consistently (Amutha et al. 2024).

5. Emerging Nanomaterials and their Future Applications

The advancement of environmental nanotechnology is driven by the emergence of smart materials, self-healing technology, and AI in its design (Agarwal et al. 2025). The new types of nanomaterials being developed are set to overcome present challenges and make it possible to do more in the monitoring, treatment, and prevention of pollution (Table 5). An important development is the creation of smart nanosensors that link nanomaterials with wireless devices to detect and sense pollutants at very low levels, monitor several agents at one time, and send data wirelessly through IoT networks (Zhang et al. 2024). Technological examples such as sensors that use quantum dots

Table 5: Emerging nanomaterials and their future applications.

Nanomaterial	Potential application	Key advantage	Current Status	References
MXenes	Electrocatalytic water treatment	• Ultrahigh conductivity (10,000 S/cm) • Tunable surface chemistry	Pilot-scale testing for industrial wastewater	Purbayanto et al. 2025
Quantum dots	Heavy metal sensing	• Single-molecule detection capability • Multiplexed analysis	Field prototypes under evaluation	Didarian and Azar 2025
Biogenic nanoparticles	Green soil remediation	• Plant/microbe-mediated synthesis • Enhanced biocompatibility	Commercialization phase for agricultural use	Mandal et al. 2025
MOF-based hybrids	Carbon capture and storage	• Record porosity (7,000 m2/g) • Selective CO_2 adsorption	Demonstration plants operational	Sher et al. 2024
Plasmonic meta-materials	Solar-enhanced photocatalysis	• 90% solar energy utilization • Broad-spectrum pollutant degradation	Lab-scale validation complete	Zhang and Ma 2025

indicate great sensitivity, rapid responses, and the ability to calibrate by learning from data (El-Azazy et al. 2024). Another interesting prospect is the development of materials that can heal themselves, helping to maintain functionality—for example, photocatalysts repairing surface gaps and the activity of enzyme-nanomaterial hybrids due to their biological traits (Ji et al. 2024). Thanks to these inventions, materials often last longer, as in the case of graphene-TiO2 compounds that maintain 95% efficiency over many cycles (Devarajan 2025). Now, AI is changing the way nanomaterials are developed, using machine learning for prediction and robotic tools to speed up the synthesis process (Jiang et al. 2022). These technologies, when combined, will likely reshape environmental cleanup by providing safer, more efficient, and custom-made solutions to the world's pollution dilemmas.

5.1 Converging Technologies for Holistic Solutions

The next step in environmental remediation is to combine novel nanomaterials with other modern technologies, resulting in stronger and more effective solutions than each part alone. For example, nanobiohybrid systems have been created that merge materials made by nature with synthetic nanomaterials. In this way, pollutants are treated more cleanly and effectively (Malik et al. 2023). With digital twins, it is possible to simulate and test new strategies for nanoremediation before using them in real life, saving money on unexpected errors (Wang et al. 2025). At the same time, ways to recycle, reuse, and regenerate nanomaterials are being developed according to the circular economy to keep waste and resource usage as low as possible (Brar et al. 2022).

According to industry predictions, there will be a 50% decrease in costs for cleaning up persistent organic pollutants (Li et al. 2025), real-time citywide pollution networks will appear (Guo et al. 2023), and zero-waste systems in treatment will become more common (Yazdani and Lakzian 2023). Because of these advances, nanomaterials are now essential for reaching sustainable goals related to clean water, smart consumption, and action on climate (Mukherjee and Pradeep 2023). With the arrival of this new era in environmental technology, we can see that innovative nanomaterials will help us deal with the most serious ecological issues.

6. Conclusion

From basics to expert solutions, the science behind nanomaterials has changed the way we handle issues in pollution control. Early research on photocatalytic metal oxides and carbon-based nanoparticles led to the development of the field, which now relies on engineered nanoparticles for water treatment, air purification, and remediation of contaminated soil. Using nanotechnology today, contaminants in water can be removed with over 95% efficiency, air purified with 90% success, and long-lasting soil pollutants can be greatly reduced. Still, problems such as environmental risks, costly production, and the lack of proper safety rules must be addressed before use can be considered safe.

The next big developments in nanomaterials for environmental remediation are smart nanosensors, materials that can repair themselves, and designs based on artificial intelligence (AI) for maximum benefits. In addition to green synthesis and circular economy efforts, these developments are expected to strengthen both the usefulness and sustainability of products using nanomaterials. Realizing such potential calls for partnerships between scientists, policymakers, and key players in the industry. By integrating the latest technology with responsible development, nanomaterials will continue to improve pollution management and support the global effort to preserve the environment. The experience of what has been done, what is being done, and what is emerging in nanotechnology points to its strong ability to help us create a healthier and greener world.

References

Aamer, H., Kim, S. B., Oh, J. M., Park, H., and Jo, Y. M. (2021). ZnO impregnated polyacrylonitrile nanofiber filters against various phases of air pollutants. *Nanomaterials*, 11(9): 2313. https://doi.org/10.3390/nano11092313.

AG, M., Kaleeswari, R. K., and S, M. (2025). Innovative nanoremediation techniques for soil contamination: exploring metal based nanomaterial approaches-a review. *Soil and Sediment Contamination: An International Journal*, 1–69. https://doi.org/10.1080/15320383.2025.2477818.

Agarwal, M. B., Mehrotra, M., Panday, M. K., Mittal, P., Jadon, V. K., and Agarwal, S. (2025). Nano AI synergy: advancements in smart materials, devices and intelligent systems. pp. 69–93. *In: NanoMind: Exploring Synergies in Nanotechnology and Machine Learning*. Cham: Springer Nature Switzerland. https://doi.org/10.1007/978-3-031-77296-2_4.

Ahmad, M., Sadiq, M. T., Zafar, K., Abbas, G., Asif, S., Kousar, N. et al. (2025). Photocatalytic Degradation of Deltamethrin in Drinking Water under Visible Light by Using Zno and Tio2.

Alam, K. (2025). Development and Production of Bimetallic Nanomaterials by Means of a Spinning Disc Reactor for Environmental and Industrial Applications.

Alex, K. V., Kamakshi, K., Silva, J. P. B., Sathish, S., and Sekhar, K. C. (2022). Automobile exhaust nanocatalysts. pp. 529–560. *In: Nanotechnology in the Automotive Industry*. Elsevier. https://doi.org/10.1016/B978-0-323-90524-4.00031-1.

Alizadeh, M., Musazade, E., Qaderi, S., Qarachal, J. F., Siahpoush, S., Abbod, M. et al. (2025). The environmental and anthropogenic impacts of nanoparticles on forest trees. *Journal of Nanoparticle Research*, 27(6): 1–27. https://doi.org/10.1007/s11051-025-06339-6.

Amutha, C., Gopan, A., Pushbalatatha, I., Ragavi, M., and Reneese, J. A. (2024). Nanotechnology and Governance: Regulatory Framework for Responsible Innovation. pp. 481–503. *In: Nanotechnology in Societal Development. Singapore: Springer Nature Singapore*. https://doi.org/10.1007/978-981-97-6184-5_14.

Asghar, N., Hussain, A., Nguyen, D. A., Ali, S., Hussain, I., Junejo, A. et al. (2024). Advancement in nanomaterials for environmental pollutants remediation: a systematic review on bibliometrics analysis, material types, synthesis pathways, and related mechanisms. *Journal of Nanobiotechnology*, 22(1): 26. https://doi.org/10.1186/s12951-023-02151-3.

Ayanda, O. S., Mmuoegbulam, A. O., Okezie, O., Durumin Iya, N. I., Mohammed, S. A. E., James, P. H. et al. (2024). Recent progress in carbon-based nanomaterials: critical review. *Journal of Nanoparticle Research*, 26(5): 106. https://doi.org/10.1007/s11051-024-06006-2.

Badoni, A., Thakur, S., Vijayan, N., Swart, H. C., Bechelany, M., Chen, Z. et al. (2025). Recent progress in understanding the role of graphene oxide, TiO2 and graphene oxide-TiO2 nanocomposites as multidisciplinary photocatalysts in energy and environmental applications. *Catalysis Science & Technology*. https://doi.org/10.1039/D4CY01334D.

Barka, E., Noutsopoulos, C., Galani, A., Panagou, I., Kalli, M., Koumaki, E. et al. (2023). Removal of contaminants of emerging concern from wastewater using an integrated column system containing zero valent iron nanoparticles. *Water*, 15(3): 598. https://doi.org/10.3390/w15030598.

Ben Dassi, R., and Chamam, B. (2025). Nanomaterial-enhanced membranes for advanced water and wastewater treatment: a comprehensive review. *Reviews in Environmental Science and Bio/Technology*, 1–35. https://doi.org/10.1007/s11157-025-09723-9.

Bhaskar, Y., Goel, I., Ranga, V., and Singh, C. K. (2025). Challenges and Fate of Bio-nanomaterials in Industrial Applications. *Bio-Nanomaterials in Environmental Remediation: Industrial Applications*, 273–294. https://doi.org/10.1002/9783527848546.ch11.

Brar, K. K., Magdouli, S., Othmani, A., Ghanei, J., Narisetty, V., Sindhu, R. et al. (2022). Green route for recycling of low-cost waste resources for the biosynthesis of nanoparticles (NPs) and nanomaterials (NMs)-A review. *Environmental Research*, 207: 112202. https://doi.org/10.1016/j.envres.2021.112202.

Chadha, U., Selvaraj, S. K., Ashokan, H., Hariharan, S. P., Mathew Paul, V., Venkatarangan, V. et al. (2022). Complex nanomaterials in catalysis for chemically significant applications: from synthesis and hydrocarbon processing to renewable energy applications. *Advances in Materials Science and Engineering*, 2022(1): 1552334. https://doi.org/10.1155/2022/1552334.

Chandra, S. (2025). Integrating Artificial Intelligence with Sustainable Practices for Effective Climate Change Mitigation and Environmental Protection. pp. 151–170. *In:* Environmental Monitoring Technologies for Improving Global Human Health. *IGI Global Scientific Publishing*. https://doi.org/10.4018/979-8-3693-8532-6.ch007.

Chávez-Hernández, J. A., Velarde-Salcedo, A. J., Navarro-Tovar, G., and Gonzalez, C. (2024). Safe nanomaterials: from their use, application, and disposal to regulations. *Nanoscale Advances*, 6(6): 1583–1610. https://doi.org/10.1039/D3NA01097J

Chen, H., and Qian, L. (2024). Performance of field demonstration nanoscale zero-valent iron in groundwater remediation: A review. *Science of the Total Environment*, 912: 169268. https://doi.org/10.1016/j.scitotenv.2023.169268.

Choi, K. (2025). Characterizing the Relationship Between Volatile Organic Compounds and Air Pollutants via Exposure Pathways Originating from Emission Sources in Large Industrial Cities. https://doi.org/10.20944/preprints202505.0511.v1.

Date, M., and Jaspal, D. (2024). Dyes and heavy metals: removal, recovery and wastewater reuse-a review. *Sustainable Water Resources Management*, 10(2): 90. https://doi.org/10.1007/s40899-024-01073-8.

Devarajan, Y. (2025). Nanomaterials-Based Wastewater Treatment: Addressing Challenges and Advancing Sustainable Solutions. *BioNanoScience*, 15(1): 1–14. https://doi.org/10.1007/s12668-024-01780-8.

Dey, S., Kumari, P., Samal, P. P., Rao, V. S., and Dey, B. (2025). Engineered mesoporous silica: a robust solution for inorganic and organic pollutant removal from water. *Environmental Monitoring and Assessment*, 197(4): 1–27. https://doi.org/10.1007/s10661-025-13854-5.

Didarian, R., and Azar, M. T. (2025). Microfluidic biosensors: revolutionizing detection in DNA analysis, cellular analysis, and pathogen detection. *Biomedical Microdevices*, 27(1): 10. https://doi.org/10.1007/s10544-025-00741-6.

El-Azazy, M., Osman, A. I., Nasr, M., Ibrahim, Y., Al-Hashimi, N., Al-Saad, K. et al. (2024). The interface of machine learning and carbon quantum dots: From coordinated innovative synthesis to practical application in water control and electrochemistry. *Coordination Chemistry Reviews*, 517: 215976. https://doi.org/10.1016/j.ccr.2024.215976.

Gao, Y., Yang, X., Lu, X., Li, M., Wang, L., and Wang, Y. (2022). Kinetics and mechanisms of Cr (VI) removal by nZVI: influencing parameters and modification. Catalysts, 12(9): 999. https://doi.org/10.3390/catal12090999.

Greco, E., De Spirt, A., Miani, A., Piscitelli, P., Trombin, R., Barbieri, P. et al. (2025). Nanomaterials in Photocatalysis: An In-Depth Analysis of Their Role in Enhancing Indoor Air Quality. *Applied Sciences*, 15(3): 1629. https://doi.org/10.3390/app15031629.

Guo, R., Zhang, Q., Yu, X., Qi, Y., and Zhao, B. (2023). A deep spatio-temporal learning network for continuous citywide air quality forecast based on dense monitoring data. *Journal of Cleaner Production*, 414: 137568. https://doi.org/10.1016/j.jclepro.2023.137568.

Hughes, K. J., Ganesan, M., Tenchov, R., Iyer, K. A., Ralhan, K., Diaz, L. L. et al. (2025). Nanoscience in Action: Unveiling Emerging Trends in Materials and Applications. *ACS omega*, 10(8): 7530–7548. https://doi.org/10.1021/acsomega.4c10929.

Hussain, S., Sattar, A., Abbas, Z., and Ullah, Z. (2025). Investigating the role of quantum confinement and surface plasmon resonance in enhancing water purification efficiency of nanomaterial-based membranes. *Kashf Journal of Multidisciplinary Research*, 2(02): 233–250. https://doi.org/10.71146/kjmr289.

Ikumapayi, O. M., and Laseinde, O. T. (2024). Nanomanufacturing in the 21st Century: A Review of Advancements, Applications and Future Prospects. *Journal Européen des Systèmes Automatisés*, 57(4): https://doi.org/10.18280/jesa.570430.

Inamdar, A. K., Inamdar, S. A., Birajdar, C. T., Bhale, J., Rajmane, S. V., Shinde, B. H. et al. (2024). Applications of titanium dioxide (TiO2) nanoparticles in photocatalysis. *Science of Advanced Materials*, 16(7): 757–771. https://doi.org/10.1166/sam.2024.4694.

Islam, M. M., Hossain, I., and Martin, M. H. H. (2024). The Role of IoT and Artificial Intelligence in Advancing Nanotechnology: A Brief Review. *Control Systems and Optimization Letters*, 2(2): 204–210. https://doi.org/10.59247/csol.v2i2.124.

Jamwal, A., Hooda, K., Chatterjee, M., Singh, G., and Barnwal, R. P. (2025). Role of Nanoscience and Nanotechnology in Sustainable Development. pp. 119–140. *In:* Role of Science and Technology for Sustainable Future: Volume 2-Applied Sciences and Technologies. Singapore: Springer Nature Singapore. https://doi.org/10.1007/978-981-97-5177-8_7.

Ji, W., Liu, J., Sha, C., Yong, Y. C., Jiang, Y., and Fang, Z. (2024). Nanomaterial-biological hybrid systems: Advancements in solar-driven CO2-to-chemical conversion. *Green Carbon*. https://doi.org/10.1016/j.greenca.2024.05.005.

Jiang, Y., Salley, D., Sharma, A., Keenan, G., Mullin, M., and Cronin, L. (2022). An artificial intelligence enabled chemical synthesis robot for exploration and optimization of nanomaterials. *Science Advances*, 8(40): eabo2626. https://doi.org/10.1126/sciadv.abo2626.

Jonathan, E. M., and Agbini, O. A. (2024). Exploring carbon nanotubes as antimicrobial agents: efficacy, toxicity, challenges, and future prospects. *Journal of Applied Sciences and Environmental Management*, 28(9): 2601–2613. https://doi.org/10.4314/jasem.v28i9.3.

Kirubakaran, D., Wahid, J. B. A., Karmegam, N., Jeevika, R., Sellapillai, L., Rajkumar, M., and SenthilKumar, K. J. (2025). A comprehensive review on the green synthesis of nanoparticles: advancements in biomedical and environmental applications. *Biomedical Materials & Devices*, 1–26. https://doi.org/10.1007/s44174-025-00295-4.

Kirubakaran, D., Wahid, J. B. A., Karmegam, N., Jeevika, R., Sellapillai, L., Rajkumar, M. et al. (2025). A comprehensive review on the green synthesis of nanoparticles: advancements in biomedical and environmental applications. *Biomedical Materials & Devices*, 1–26. https://doi.org/10.1007/s44174-025-00295-4.

Kuppan, N., Padman, M., Mahadeva, M., Srinivasan, S., and Devarajan, R. (2024). A comprehensive review of sustainable bioremediation techniques: Eco friendly solutions for waste and pollution management. *Waste Management Bulletin*. https://doi.org/10.1016/j.wmb.2024.07.005.

Leal, L. L. M. (2024). Application of Ag-doped carbon nanotubes for wastewater treatment: antimicrobial and catalytic activity (Master's thesis, Instituto Politecnico de Braganca (Portugal)).

Li, L., Chen, C., Li, D., Breivik, K., Abbasi, G., and Li, Y. F. (2023). What do we know about the production and release of persistent organic pollutants in the global environment? *Environmental Science: Advances*, 2(1): 55–68. https://doi.org/10.1039/D2VA00145D.

Liu, X., Sathishkumar, K., Zhang, H., Saxena, K. K., Zhang, F., Naraginiti, S. et al. (2024). Frontiers in environmental cleanup: Recent advances in remediation of emerging pollutants from soil and water. *Journal of Hazardous Materials Advances*, 100461. https://doi.org/10.1016/j.hazadv.2024.100461.

Malik, S., Muhammad, K., and Waheed, Y. (2023). Nanotechnology: a revolution in modern industry. *Molecules*, 28(2): 661. https://doi.org/10.3390/molecules28020661.

Mane, P. V., Rego, R. M., Yap, P. L., Losic, D., and Kurkuri, M. D. (2024). Unveiling cutting-edge advances in high surface area porous materials for the efficient removal of toxic metal ions from water. *Progress in Materials Science*, 101314. https://doi.org/10.1016/j.pmatsci.2024.101314.

Mondal, J., Lamba, R., Yukta, Y., Yadav, R., Kumar, R., Pani, B. et al. (2024). Advancements in semiconductor quantum dots: expanding frontiers in optoelectronics, analytical sensing, biomedicine, and catalysis. *Journal of Materials Chemistry C*. https://doi.org/10.1039/D4TC01396D.

Mukherjee, S., and Pradeep, T. (2023). Nanomaterials-enabled technologies for clean water and their sustainability aspects. pp. 16–31. *In: Industrial Applications of Nanoparticles*. CRC Press. https://doi.org/10.1201/9781003183525-2.

Murtaza, G., Ahmed, Z., Dai, D. Q., Iqbal, R., Bawazeer, S., Usman, M. et al. (2022). A review of mechanism and adsorption capacities of biochar-based engineered composites for removing aquatic pollutants from contaminated water. *Frontiers in Environmental Science*, 10: 1035865. https://doi.org/10.3389/fenvs.2022.1035865.

Nguyen, T. A. (2025). Revolutionizing Lead (II) Ion Removal from Water: Eco-Friendly Composite Film with Graphene Oxide and Bacterial Cellulose. *Trends in Sciences*, 22(5): 9589-9589. https://doi.org/10.48048/tis.2025.9589.

Odubo, T. C., and Kosoe, E. A. (2024). Sources of air pollutants: impacts and solutions. pp. 75–121. *In: Air Pollutants in the Context of One Health: Fundamentals, Sources, and Impacts*. Cham: Springer Nature Switzerland. https://doi.org/10.1007/698_2024_1127.

Prasad, R. D., Prasad, N. R., Prasad, N., Prasad, R. S., Prasad, S. R., Shrivastav, O. P. et al. (2025). A Review on Aspects of Nanotechnology in Environmental Science and Engineering. *ES General*, 7: 1397. https://doi.org/10.30919/esg1397.

Purbayanto, M. A., Presser, V., Skarżyński, K., Słoma, M., Naguib, M., and Jastrzębska, A. M. (2025). MXenes: multifunctional materials for the smart cities of tomorrow. *Advanced Functional Materials*, 35(10): 2409953. https://doi.org/10.1002/adfm.202409953.

Rame, R., Purwanto, P., and Sudarno, S. (2024). Industry 5.0 and sustainability: An overview of emerging trends and challenges for a green future. *Innovation and Green Development*, 3(4): 100173. https://doi.org/10.1016/j.igd.2024.100173.

Rastogi, S., Sethi, P., Panwar, N., Yadav, P., Verma, R., and Madan, A. (2025). Harnessing Nanotechnology to Transform Waste Reclamation for Enhanced Efficiency and Environmental Impact. pp. 349-392. *In: AI Technologies for Enhancing Recycling Processes. IGI Global Scientific Publishing*. https://doi.org/10.4018/979-8-3693-7282-1.ch016.

Reddy, B. N., Saravanan, S., Manjunath, V., and Reddy, P. R. S. (2024). Review on Next-Gen Healthcare: The Role of MEMS and Nanomaterials in Enhancing Diagnostic and Therapeutic Outcomes. *Biomaterials Connect*, 1(1): 1–10. https://doi.org/10.69709/BIOMATC.2024.131006.

Sánchez Hidalgo, G. C., Ábrego-Bonilla, J., Deago, E., and Ortega Del Rosario, M. D. L. A. (2025). Global Perspectives on Groundwater Decontamination: Advances and Challenges of the Role of Permeable Reactive Barriers. *Hydrology* (2306-5338), 12(4). https://doi.org/10.3390/hydrology12040098.

Sher, F., Hayward, A., El Guerraf, A., Ziani, I., Hrnjić, H., Boškailo, E. et al. (2024). Advanced metal-organic frameworks for superior carbon capture, high-performance energy storage and environmental photocatalysis-A critical review. *Journal of Materials Chemistry A*. https://doi.org/10.1039/D4TA03877K.

Sneha M, P., Biju, M., Thomas, A., and Balachandran, P. (2025). Carbon-based Nanomaterials: A Promising Tool for Sensing Toxic Metal Ions from Degraded Soil. *Nanotechnology-based Sustainable Agriculture*, 305–332. https://doi.org/10.1002/9783527849949.ch11.

Soni, R. A., Rizwan, M. A., and Singh, S. (2022). Opportunities and potential of green chemistry in nanotechnology. *Nanotechnology for Environmental Engineering*, 7(3): 661–673. https://doi.org/10.1007/s41204-022-00233-5.

Susaimanickam, A., Selvaraj, D., and Manickam, P. (2024). Nanoparticle-based bioremediation of organic and inorganic substances from water bodies. pp. 135-162. *In:* Bioremediation of Emerging Contaminants in Water. Volume 2. *American Chemical Society*. https://doi.org/10.1021/bk-2024-1476.ch007.

Tafti, M. F., Fayyaz, Z., Aghamollaei, H., Jadidi, K., and Faghihi, S. (2025). Drug delivery strategies to improve the treatment of corneal disorders. Heliyon. https://doi.org/10.1016/j.heliyon.2025.e41881.

Tewari, C., Tatrari, G., Kumar, S., Pathak, M., Rawat, K. S., Kim, Y. N. et al. (2023). Can graphene-based composites and membranes solve current water purification challenges-a comprehensive review. *Desalination*, 567: 116952. https://doi.org/10.1016/j.desal.2023.116952.

Thakur, A., and Kumar, A. (2022). Recent advances on rapid detection and remediation of environmental pollutants utilizing nanomaterials-based (bio) sensors. *Science of the Total Environment*, 834: 155219. https://doi.org/10.1016/j.scitotenv.2022.155219.

Tiwary, S. K., Singh, M., Chavan, S. V., and Karim, A. (2024). Graphene oxide-based membranes for water desalination and purification. *npj 2D Materials and Applications*, 8(1): 27. https://doi.org/10.1038/s41699-024-00462-z.

Tripathy, J., Mishra, A., Pandey, M., Thakur, R. R., Chand, S., Rout, P. R. et al. (2024). Advances in nanoparticles and nanocomposites for water and wastewater treatment: a review. *Water*, 16(11): 1481. https://doi.org/10.3390/w16111481.

Verma, Y., Verma, A., Bhaskaralingam, A., Dhiman, P., Wang, T., Kumar, A. et al. (2025). Application of Zero-Valent Iron and Its Derivatives in the Removal of Toxic Metal Ions from Groundwater. *Water*, 17(10): 1524. https://doi.org/10.3390/w17101524.

Visentin, C., Braun, A. B., Reginatto, C., Cecchin, I., Vanzetto, G. V., and Thomé, A. (2024). Are contaminated soil and groundwater remediation with nanoscale zero-valent iron sustainable? An analysis of case studies. *Environmental Pollution*, 352: 124167. https://doi.org/10.1016/j.envpol.2024.124167.

Wang, A. J., Li, H., He, Z., Tao, Y., Wang, H., Yang, M. et al. (2024). Digital twins for wastewater treatment: A technical review. *Engineering*, 36: 21–35. https://doi.org/10.1016/j.eng.2024.04.012.

Xie, Y., Zhang, T., Wang, B., and Wang, W. (2024). The Application of Metal-Organic Frameworks in Water Treatment and Their Large-Scale Preparation: A Review. *Materials*, 17(9): 1972. https://doi.org/10.3390/ma17091972.

Yan, P., Feng, D., and Liu, S. (2024). Experimental study on the biodegradability of petroleum wastewater and improving the performance of a moving-bed biofilm reactor. *International Journal of Environmental Science and Technology*, 1–22. https://doi.org/10.1007/s13762-024-06149-8.

Yan, X., Pang, Y., Niu, K., Hu, B., Zhu, Z., Tan, Z. et al. (2025). Wearable Sensors for Plants: Status and Prospects. *Biosensors*, 15(1): 53. https://doi.org/10.3390/bios15010053.

Yang, M., Zhang, X., and Sun, Y. (2024). Remediation of Cr (VI) Polluted Groundwater Using Zero-Valent Iron Composites: Preparation, Modification, Mechanisms, and Environmental Implications. *Molecules*, 29(23): 5697. https://doi.org/10.3390/molecules29235697.

Yang, X., Zhou, Y., Hu, J., Zheng, Q., Zhao, Y., Lv, G. et al. (2024). Clay minerals and clay-based materials for heavy metals pollution control. *Science of The Total Environment*, 176193. https://doi.org/10.1016/j.scitotenv.2024.176193.

Yazdani, S., and Lakzian, E. (2023). Conservation; waste reduction/zero waste. pp. 131–152. *In:* Pragmatic engineering and lifestyle. *Emerald Publishing Limited*. https://doi.org/10.1108/978-1-80262-997-220231007.

Yusuf, E. O., Amber, I., Officer, S., and Oluyemi, G. F. (2025). Controlled application of nanoparticles for remediation in oil and gas application: strategies, challenges, and innovations. *Energies*, 18(4): 991. https://doi.org/10.3390/en18040991.

Zeng, G., Li, G., Yuan, W., Liu, J., Wu, Y., Li, M. et al. (2025). Ag nanoparticle-mediated LSPR effect and electron transfer for enhanced oxidative degradation process of g-C3N5 nanoflowers. *Separation and Purification Technology*, 353: 128484. https://doi.org/10.1016/j.seppur.2024.128484.

Zhang, Q., Zuo, Z., and Ma, D. (2023). Plasmonic nanomaterials for solar-driven photocatalysis. *Chemical Communications*, 59(50): 7704–7716. https://doi.org/10.1039/D3CC01512B.

Zhang, X., Xiong, S., Sathiyaseelan, A., Zhang, L., Lu, Y., Chen, Y. . et al. (2024). Recent advances in photocatalytic nanomaterials for environmental remediation: strategies, mechanisms, and future directions. *Chemosphere*, 143142. https://doi.org/10.1016/j.chemosphere.2024.143142.

Zhou, Y., Liu, Y., Zhang, M., Feng, Z., Yu, D. G., and Wang, K. (2022). Electrospun nanofiber membranes for air filtration: A review. *Nanomaterials*, 12(7): 1077. https://doi.org/10.3390/nano12071077.

Zubair, M. O., Zobair, G., Sharma, J., and Dhale, D. A. (2025). Improvement of Air Quality Using Nanomaterials and Nanotechnology. *Applications of Nanomaterials in Civil and Environmental Engineering: Towards Sustainable Development and Environment Remediation*, 291. https://doi.org/10.1007/978-981-96-3953-3_17.

CHAPTER-16

Sustainable Algal Nanomaterials for Microplastics and Lanthanide Remediation

Ruchika Agarwal,[1] *Sana Parveen,*[1] *Rajnikant Patidar*[2] and *Shailesh Kumar Patidar*[1,*]

1. Introduction

Population explosion and globalization lead to the discharge of a vast number of toxic pollutants into the environment. Various pollutants, including heavy metals, herbicides, pesticides, microplastics (MPs), and particulate matter, pose significant threats to the balance of the ecosystem. Among these, plastic and metal pollution have emerged as some of the most dangerous environmental challenges affecting biotic and abiotic components of the ecosystem. Biotic degradation, which includes biodegradation and biodisintegration, along with abiotic degradation factors such as UV radiation, heat, and mechanical stress, leads to the formation of MPs from large plastic polymers (Zhang et al. 2021). Plastic particles less than 5 mm in length are called MPs. There are diverse sources of MP contamination, like municipal waste, industrial refuse, construction waste, and many others. MPs harm soil, water, lakes, rivers, reservoirs, oceans, and the organisms residing in these ecosystems (Rahman et al. 2021). A huge increase in plastic production has been observed from 225 million tons in 2004 to 367 million tons in 2020, with a 4% annual increase since 2012. This has led scientists to call this time the "era of plastic" (García Rellán et al. 2023). Petrochemical plastics such as polyethylene terephthalate (PET), polyethylene (PE), polypropylene (PP), polystyrene (PS), and polyvinyl chloride (PVC) comprise over 80% of plastic waste (Tiwari et al. 2020). Toxicological and experimental studies show that exposure to MPs leads to various toxic effects like ROS imbalance, metabolic disorder, inflammatory response, neural damage, as well as reproductive and developmental toxicity (Li et al. 2023). On the other hand, heavy metals like Cd, Pb, Cr, and others, when introduced into the environment, cause harm to the ecosystem as well as human health. Intake of these metals causes toxicity to human health, which includes neurotoxicity, anxiety, hypertension,

[1] Department of Environmental Science, School of Earth Sciences, Central University of Rajasthan, NH-8 Bandarsindri, Kishangarh, Ajmer, 305817, Rajasthan, India.
[2] Department of Geology, Faculty of Earth Science, Mohanlal Sukhadia University, Udaipur, 313001, Rajasthan, India.
* Corresponding author: shailesh.patidar@curaj.ac.in, shaileshpatidar84@gmail.com

DOI: 10.1201/9781003558804-16

etc. (Rajput et al. 2025). Specifically, rare earth elements (REEs), also called lanthanides (Ln), have emerged as potential pollutants released from various industries. REEs can be defined as elements having the same physicochemical properties and constituting less than 20% of the total elements present (Choudhary et al. 2023). Based on their geochemical properties, REEs can be categorized into light REEs, which include La, Ce, Pr, Nd, and Sm; medium REEs, like Eu, Gd, Tb, Dy, and Y; and heavy REEs, consisting of Ho, Er, Tm, Yb, and Lu (Yun et al. 2024). Due to their good metallic characteristics like low resistivity, ductility, and malleability, they are suitable for various applications such as renewable energy, ceramics, fertilizers, etc. (Massari and Ruberti 2013). Globally, there is an uneven distribution of REE generation, consumption, and processing, with the domination of China (70%), followed by the USA (14%) and Australia (4%) (Liu et al. 2023). India and Brazil each produce about 1%, with smaller contributions from Malaysia and Vietnam (Dev et al. 2020). Over-exploitation of these elements in various industries has made them emerging contaminants in the environment. Various technologies like solvent extraction, electrolytic reduction, and oxidation exist to address these contaminants, but they often face significant challenges, including excessive costs, high energy demands, environmental disruption, and the risk of generating secondary pollutants (Bautista 1992). Different technological advancements like the development of bionanomaterial adsorbents and biosorption using various microorganisms have been developed to remediate toxic contaminants in the environment. A synergy of nanotechnological methodologies incorporating bionanomaterials has been thoroughly examined for the remediation of metal contaminants (Chellasamy et al. 2022).

Nanotechnology is recognized as a sustainable, environmentally friendly, cost-effective, and cleaner approach that deals with the synthesis, properties, and application of nanosized particles (1–100 nm) for the effective removal of contaminants (Chandran et al. 2023). Nanotechnology combined with bioremediation has generated new advancements and avenues in the field of remediation. Engineered nanoparticles provide a more efficient and cost-effective approach when compared with traditional methods due to the high reactivity of nanomaterials and the possibility of onsite treatment (Mueller and Nowack 2010). This chapter deals with various methods of microplastic remediation along with lanthanides and resource recovery of lanthanides through smart sustainable nanoparticles (as shown in Figure 1).

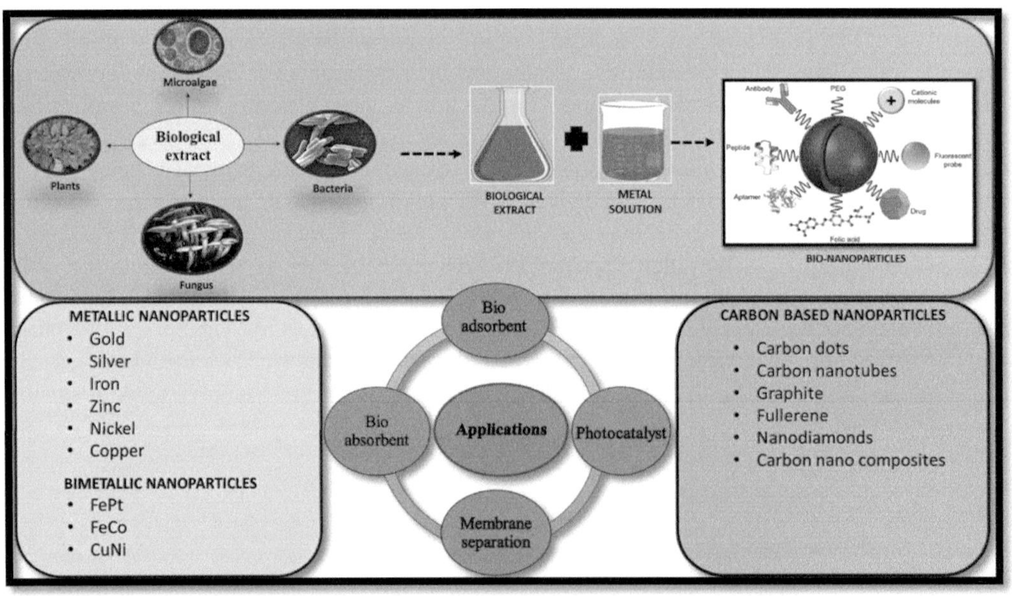

Figure 1: Overview of bio nanomaterial, including synthesis and type.

2. Microplastics—A Universal Threat

Microplastics are widespread and can be found in various environmental matrices, including marine waters, sediments, freshwater systems, terrestrial ecosystems, and even within living organisms. Plastic is a complex mixture of chemicals that are perilous for the environment and cause bio-toxicity in the ecosystem (Nguyen et al. 2023). An aquatic ecosystem is adversely affected by MPs due to their high accumulation rate and increasing toxicity to aquatic organisms (Ya et al. 2021). In 2018, around 114 aquatic species contained traces of MPs, leading to effects such as false satiation, reproductive challenges, enzyme inhibition, stunted growth, and oxidative stress in aquatic life (Auta et al. 2017). Large aquatic animals can consume MPs either directly from the environment or indirectly by ingesting other organisms that have eaten microplastics. MPs act as a carrier for several toxic chemicals like fire retardants and emollients, which are added to enhance the final properties of microplastics, ultimately causing a threat to the life of aquatic organisms (Guzzetti et al. 2018).

Many investigations have been conducted to study the effect of microplastic contamination on a variety of algae, like Cyanophyta, Chlorophyta, Bacillariophyta, and Euglenophyta. Previous studies on adsorption of microplastics by microalgae such as *Chlorella, Scenedesmus*, etc., showed reduced photosynthetic efficiency along with a diminished ability to cope with oxidative stress (Guzzetti et al. 2018). A study on toxicological effects showed that with an increased concentration of MPs, the amount of chlorophyll a, cell abundance, and size of microalgae decreased (Guo et al. 2024). A study of the dose-response relationship of MPs on *Chlorella pyrenoidosa* suggested that with an increase in MP concentration, there was a subsequent decrease in pigment content, like chl a, chl b, and carotenoids, which indicates harm caused to photosystem II due to oxidative stress triggered by MPs (Moreri and Yalei 2024). Studies on bioaccumulation and biomagnification of MPs within the food chain, starting from the lowest trophic level to the highest, emphasize the potential effects of MPs on human beings (Gao et al. 2024). A normal human being consumes approximately 4,000 microplastic particles annually through drinking water, 11,000 particles from marine fishes, and between 37 and 1,000 particles via edible sea salt each year (Wang et al. 2021). Exposure of humans to MPs presents significant health risks, as their small size leads to ingestion, intestinal absorption, biotransformation, and subsequent excretion, potentially contributing to various pathophysiological conditions (Roslan et al. 2024). Traces of microplastics were reported in human milk, the amniotic layer, and the liver. It was observed that humans ingest about a "credit card's worth" of MPs weekly (Chartres et al. 2024). Not only from aquatic systems, these particles can also enter humans via the respiratory tract by inhalation of contaminated air. Previous work reported that humans breathe around 272 particles of MPs every day from enclosed environments (Vianello et al. 2019). When inhaled, these particles cause pulmonotoxicity of the respiratory system, showing symptoms like coughing, sneezing, dyspnea, and hypoxemia (Wright and Kelly 2017). A study by Lee et al. (2023) suggests that nanosized plastic particles, particularly polystyrene (PS), can damage mitochondrial cells in the human pulmonary system and increase the risk of chronic obstructive pulmonary disease. So, there is an urgent need for the removal and remediation of MPs from nature. Several traditional techniques have been employed for the remediation of these particles, like membrane filtration, biodegradation, adsorption, photocatalytic disintegration, coagulation, and electrocoagulation. These technologies can work separately or together with other techniques for the effective removal of MPs. Previous studies show the efficiency of coagulation in the removal of microplastics from aqueous environments (Khan et al. 2023). However, there are some limitations of conventional techniques, such as inefficiency in the removal of smaller particles that pass through filters or remain in treated water, and the formation of secondary pollutants due to excessive use of chemicals for the elimination of MPs through coagulation and flocculation (Ajao 2024). Therefore, there is a need for environmentally sustainable methods of MP remediation.

Silicates	Halides	Arsenates & Sulfates	Carbonates	Oxides	Phosphates
Vicanite $(Ca,Ce,La)_{15}(Ti,Nb)(SiO_4)_6(Si_2O_7)_3O_4F_3$	Tysonite $(La,Ce)F_3$	Bayleyite $Mg_2(UO_2)(CO_3)_3 \cdot 18H_2O$	Ancylite $Sr(Ce,La)(CO_3)_2(OH) \cdot H_2O$	Euxenite $(Y,Ca,Ce,U,Th)(Nb,Ta,Ti)_2O_6$	Euxenite $(Y,Ca,Ce,U,Th)(Nb,Ta,Ti)_2O_6$
Steenstrupine $Na_{14}Ce_6Mn_2Fe_2(PO_4)_6(SiO_4)_6(O,OH,H_2O)_3$	Gagarinite NaCaREE(F,Cl)	Arsenoflorencite $(La,Ca)(Al,Fe)_3(AsO_4)_2(OH)_6$	Burbankite $(Na,Ca)_3(Sr,Ba,REE)(CO_3)_5$	Samarskite $(Y,Fe^{3+},U,Th,Ca)(Nb,Ta,Ti)_2O_8$	Samarskite $(Y,Fe^{3+},U,Th,Ca)(Nb,Ta,Ti)_2O_8$
Allanite $(Ce,La,Ca)_2(Al,Fe^{3+})_3(SiO_4)_3(OH)$	Yttrofluorite $(Ca,Y)F_2$	Goudeyite $(Ce,La,Nd,Y)AsO_4 \cdot 2H_2O$	Coalingite $Ba(Ca,Y,Ce)(CO_3)_2(OH)F \cdot 2H$	Polycrase $(Y,Ca,Ce,U,Th)(Ti,Nb,Ta)_2O_6$	Polycrase $(Y,Ca,Ce,U,Th)(Ti,Nb,Ta)_2O_6$
Eudialyte $Na_{15}Ca_6(Fe,Mn)_3Zr_3(Si,Nb)(Si_3O_9)_2(Si_9O_{27})_2(OH,Cl,H_2O)_3$	Fluocerite $(La,Ce)F_3$	Chemnitzite $YAsO_4$	Tengerite $(Y,Ho)_2(CO_3)_3 \cdot 2–3H_2O$	Aeschynite $(REE,Fe,Ca,Th)(Ti,Nb)_2O_6$	Aeschynite $(REE,Fe,Ca,Th)(Ti,Nb)_2O_6$
Hingganite $(Y,REE)_2BeSi_2O_8(OH)_2$		Gasparite $CeAsO_4$	Parisite $Ca(Ce,La)_2(CO_3)_3F_2$	Yttrocolumbite $(Y,U)(Nb,Ta)O_4$	Yttrocolumbite $(Y,U)(Nb,Ta)O_4$
Tuhualite $NaFe^{3+}Si_3O_8(OH)$		Agardite $(REE,Ca)Cu_6(AsO_4)_3(OH)_6 \cdot 3H_2O$	Carbocernaite $(Ca,Sr,Ce,La)(CO_3)_2$	Fergusonite $(Y,Ce,U,Th)(Nb,Ta)$	Fergusonite $(Y,Ce,U,Th)(Nb,Ta)$
Chevkinite $(Ce,La,Ca,Th)(Fe^{2+},Fe^{3+},Ti)_2Si_2O_{10}$		Retzian $Ca_2(AsO_4)(REE)OH$	Lanthanite $(La,Ce)_2(CO_3)_3 \cdot 8H_2O$	Yttrotantalite $YTaO_4$	Yttrotantalite $YTaO_4$
Lovozeroite $(Na,Ca,Ce)_2Zr(Si_2O_7)O(OH$			Bastnaesite $(Ce,La)(CO_3)F$	Betafite $(Ca,U,Ce)_2(Ti,Nb,Ta)_2O_7$	Betafite $(Ca,U,Ce)_2(Ti,Nb,Ta)_2O_7$
Thalenite $Y_3Si_3O_{10}(OH)$			Synchysite $Ca(Ce,La)(CO_3)_2F$		Mosandrite $Na_2Ca_4REE(SiO_3)_6(Si_3O_9)(OH,F)_2$
Gadolinite $Ce,La,Nd,Y)_2FeBe_2Si$			Lijimingite $Ca(Ce,La)(CO_3)_3$		Florencite $(La,Ce)Al_3(PO_4)_2(OH)_6$
Cenosite $CaCe(SiO_4)_2 \cdot 2H_2O$			Huanghoite $(Ba,REE)(CO_3)_2F$		Gorceixite $BaAl_3(PO_4)_2(OH)_5 \cdot H_2O$
Agrellite $NaCa_2Si_4O_{10}F$			Doylesite $BaY(CO_3)_2(OH) \cdot 2H_2O$		Belovite $(Sr,Na)(Ce,Nd)(PO_4)F$
Torendrikite $(Y,REE)_2Ti_2Si_2O_{10}(OH)_2$			Calcioancylite $(Ca,Ce)(CO_3)_2(OH) \cdot H_2O$		Crandallite $(REE,Ca,Na)Al_3(PO_4)_2(OH)_5 \cdot H_2O$
Zircon $ZrSiO_4$			Zhonghuacerite $Ba_2Ce(CO_3)_4F$		Frondelite $(Mn,Fe)_4(REE)(PO_4)_3(OH)_5 \cdot H_2O$
Telyushenkoite $Ca_2(La,Ce)_3(SiO_4)_3(OH)_3$			Lecoqite $Ca(Ce,Na)(CO_3)_3F$		Rhabdophane $(Ce,La,Nd)PO_4 \cdot H_2O$
Tuyukite $Ca_2Y_2(Si_2O_7)O(OH,F$			Kukisvumite $Ba_2(Ca,Y)(CO_3)_4 \cdot H_2O$		Vlasovite $Na_2ZrSi_4O_{11}$
Thorite $ThSiO_4$			Baiyuneboite NaBaCe2(CO3)4F		Monazite (Ce, La,Nd,Th)PO
			Bastnaesite $(La,Ce)(CO_3)F$		Tranquillityite $(Y,Ca,Ce)(Ti,Fe)_2SiO_5$
			Burbankite (Na,Ca)3(Sr,Ba,Ce) 3 (CO3)5		Xenotime -(Y) YPO_4
			Cebaite (Ba,Ca)4(Y,Ce)2(CO3)8·H2O		Churchite $YPO_4 \cdot 2H_2O$
			Barytolanthanite $(Ba(CeLa)_2(CO_3)F_2$		Brocktite $CePO_4$
			Ewaldite $Ba(Ce,La)_2(CO_3)F_2$		
			Gysinite Ba(Ce,La)2(CO3)F2		
			Hydroxybasenesite Ba(Ce,La)2(CO3)F2(OH)		
			Lokkaite Ba(Ce,La)2(CO3)F2		
			Kimuraite $(Y,Ce)(CO_3)F$		

Figure 2: List of Key Minerals of REEs/Lanthanides Figure 2(A) showing Silicates, Halides, Arsenate and Sulfates, Figure 2(B) representing Carbonates and Figure 2(C) showing Oxides and Phosphates Minerals of REEs (After Miyawaki and Nakai 1987, Bayliss and Levinson 1988, Burt 1989, Cesbron 1989, Leigh 1990, Clark 1984, 1993, Fleischer and Jones 1996, Gaines et al. 1997, Singh 2020).

3. Toxicity and Sources of Lanthanides

Rare Earth Elements (REEs- total 17 elements), including lanthanides, have chemical affinity and occur together, although categorized into two subgroups as LREE and HREE. Light elements are from La (57) to Eu (63) and abbreviated as LREE, and the heavy elements are from Gd (64) to Lu (71), along with Y, abbreviated as HREE. REEs have distinctive characteristics: electrical, catalytic, optical (luminescence), metallurgical (high density, high melting point), and nuclear properties. These properties make them appropriate for routine to advanced technological applications along with green and clean technology aspects. In modern times, demand for rare earth has been increasing predominantly in energy efficiency and also in green and clean technology. The dependence of green and clean technologies on the vulnerable supply of lanthanides classifies them as a widely recognized critical raw material. Conventional applications of lanthanides are limited to a few industries like coloring, glass polishing, ceramics, magnets, catalysts, and metallic alloys in metallurgy. Nowadays, their applications in various fields are on a broader scale in emerging technologies related to transport, information, environment, energy, defense, nuclear, and aerospace industries. The peak demand of REEs shortly is in automobiles (electronic and hybrid vehicles), mobiles, wind turbine generators, and fuel cell technology due to different components like light bulbs, glass, batteries, super magnets, superconductors, and plasma screen multi-level electronic components. Because of their critical role in the functionality of advanced technological devices, these elements are often referred to as "The Vitamins of Modern Industry."

REE-bearing minerals are bastnaesite, monazite, xenotime, and loparite. Monazite and bastnaesite comprise 70% and 75% combined rare earth oxides (REO), including around 2% and 0.05% yttrium oxide theoretically; however, in practice, they produce 55–65% and 60% REO. While xenotime can theoretically consist of 67% Y_2O_3, average xenotime concentrates contain about 25% Y_2O_3 (Hedrick 1985). Detailed global REE deposits are shown in Table 1 and Figure 2.

With rapid technological innovation and advancement, there has been a rapid increase in demand for REEs in the last two decades, leading to environmental contamination and occupational disorders. Ln is termed as technology-critical elements (TCE) due to their unique properties, which include ferromagnetism and high sensitivity, making them highly valuable for advanced-technology practices, which include hi-tech devices and equipment, revolutionized ammunition systems and platforms, green energy, and medical devices (Hissler et al. 2016, Culicov et al. 2021). Other important uses include the application of these elements in fertilizers to enrich soil quality for the growth and germination of seeds. Some studies reported that China applied 11,000 tons of REEs in 2011 to 6.5 mha of soil, which increased its fertility and hence became a major contaminant in the ecosystem (Egler et al. 2022). The following reasons are responsible for REE regulation that can be a new source of emerging contaminants: (1) absence of the highest permissible limit to protect human beings as well as ecological health makes REE regulations challenging; (2) exclusion of REEs in ecological and human health monitoring and regulation systems; (3) environmental concentrations of these elements are extremely low (ppm to ppb), which can be measured through highly sensitive advanced analytical equipment like ICP-MS, AAS, etc.; and (4) poor understanding of ecotoxicity in human and disease surveillance systems is also a major roadblock for the formulation of maximum permissible limits (Gwenzi et al. 2021). There are different paths through which REEs can enter the human body and accumulate in different organs and cells, thus posing harmful effects. Workers involved in different industries are exposed to significant amounts of rare earth elements through respiration, oral intake, and dermal contact (Brouziotis et al. 2022a). REEs enter the human body through various routes like hair sacs and sebaceous glands, and long-term exposure leads to intrauterine damage, causing harm to the body and growing fetus (Wang et al. 2024). A small accumulation of REEs in the brain cortex of humans causes subclinical damage. Significant amounts of La, Ce, and Gd were found in brain tumor patients compared to normal brain tissues (Brouziotis et al. 2022b). La causes a variety of biophysiochemical reactions in plant tissues, which change nutrient absorption, ROS, and GSH concentrations in *E. nuttallii* (Choudhary et al. 2023). It has

Table 1: Potential global REE deposits including Lanthanides and Actinides.

Locality	Principle Minerals	Total Resources REO(t)	Grade	References	HREE%
Bayan Obo, China	Bastnasite, Monazite,	10,000 X10^4	5.6	(Lawrence et al. 1990)	1.13%
Mountain Pass carbonatite in California, USA	Bastnasite	418.3X10^4	8.9	(Castor 2008)	0.49%
Mount Weld in Western Australia, Australia	Monazite	300X10^4	5.4	(Lottermoser 1994)	3.97
Kangankunde carbonatite, Malawi	Monazite, strantianite	32.45X10^4	5.58	(Wall and Mariano 1996, Holt 1965)	NA
Maoniuping, Sichuan Province, China	Bastnasite	317X10^4	2.95	(Xu et al. 2003, 2004, Kynicky et al. 2012)	11.1%
Ashram, North America	Bastnasite, monazite, xenotime	468.686	1.88	(Xu et al. 2003)	3.50%

even been investigated that Ln causes an increase in specific components that enhance plant growth and development. For instance, *Salvia miltiorrhiza* (bioindicator of REE), when cultivated in the presence of different concentrations of Pr, showed enhanced plant development and greater content of secondary metabolites (Fan et al. 2020).

4. Bionanomaterials: Sustainable Solutions for Remediation

Nanotechnology emerged as a potential technology that exploits the physicochemical properties of matter at the nanoscale. It covers the science of comprehending, developing, and using nanomaterials (NMs) in all branches of engineering, technology, and science. Various approaches, such as the top-down and bottom-up methods, are utilized in nanoparticle development. Among these, bionanomaterial-based synthesis stands out as an eco-friendly, cost-effective, and sustainable technique for producing nanoparticles. Bionanosynthesis involves the use of different micro- and macroscopic organisms like plant extracts, bacteria, fungi, viruses, and algae, which are considered nanofactories due to their ability to effectively synthesize nanocomposites in a cost-effective and environmentally friendly manner (Ahmed et al. 2022). Green synthesis is the most preferred process of nanomaterial generation due to less use of chemicals and toxic metals compared to physical and chemical synthesis (Ahmed et al. 2016). Microalgal biomass has been used for industrial, health, and pharmaceutical applications, along with biological treatments and toxicology studies, due to elevated content of nutrients like polysaccharides, peptides, vitamins, and essential minerals (Jain 2022, Dahoumane et al. 2016). Compared to larger plants, algae have high biomass productivity, shorter generation time, and an abundance of bioactive compounds and secondary metabolites, making them suitable for the development of metal- and metal oxide-assisted nanoparticles (Premarathna et al. 2024, Manivasagan and Kim 2015, Mukherjee et al. 2021). Different algal strains like *Chlorella, Sargassum,* and *Spirulina* have been used for the development of metallic nanoparticles (gold, silver, platinum) by extracellular and intracellular processes that involve bioactive compounds as reducing and stabilizing agents (Khanna et al. 2019a).

The distinctive features of bionanomaterials are their efficiency in environmental applications, especially in pollution cleanup. Bionanomaterials possess different properties like a large surface area compared to bulk materials, which help in the proper adsorption of pollutants and catalyze reactions, making them effective in removing heavy metals and organic pollutants (Kaur et al. 2024). This feature is particularly useful in adsorption-driven remediation techniques, allowing for the efficient removal of pollutants from polluted water and soil. Yang et al. (2024) showed the synthesis of cellulose nanomaterials for the successful degradation of dyes and reduction of chromium ions. Different modulations of nanomaterials are possible to enhance their characteristics. Due to the high area-to-volume ratio and the presence of different active sites, bionanomaterials allow greater contact with pollutants, leading to the reduction of contaminants (Roy et al. 2021). Some bionanomaterials exhibit the properties of photocatalysts, which enable the degradation of heavy metals. Previous studies discussed the synthesis of CdO–TiO_2 hybrid nanocomposites for the photocatalytic degradation efficiency of organic dyes, e.g. methyl orange and rhodamine (Dhanalekshmi et al. 2021). Similarly, chitosan-based nanostructures have demonstrated remarkable performance in adsorbing heavy metals, with their functional groups (e.g., amine and hydroxyl) playing a key role in binding metal ions selectively (Wan Ngah et al. 2011). Their stability under diverse environmental conditions and multifunctionality, ranging from adsorption and catalysis to antimicrobial activity, positions them as versatile tools for water, air, and soil remediation (Roy et al. 2021).

Intracellular and extracellular accumulation of heavy metals by microalgae helps in the synthesis of metal nanoparticles at varying physicochemical conditions like temperature, pH, and concentration, etc. These synthesized nanomaterials have low toxicity, are eco-friendly, highly efficient, and have low environmental impact (Agarwal et al. 2019, Sathishkumar et al. 2019). Metallic nanoparticles have a huge diversity with numerous applications, as shown in Table 2.

Nanoparticle synthesis involves the following series of steps: 1. Boiling the algal extract either in organic or inorganic solvent, 2. Preparation of ionic metallic compounds of molar concentration, and 3. Incubating and combining the algal extract solution with the ionic metallic compound solution; further stirring and time duration can be optimized according to the need (LewisOscar et al. 2016). The synthesis of NPs is directly proportional to the dosage of the metal ion precursor and the type of algal strain. The biomolecules involving polysaccharides, pigments, phenolics, flavonoids, and peptides are responsible for the reduction of metal ions. Amino groups or cysteine residues, along with sulfated polysaccharides, are useful in the capping and stabilizing of the nanoparticles (Shera and Banik 2021). The wide usage of algae for the synthesis of bio-based nanoparticles is due to their potential for donating their charge, which ultimately reduces the metal ion (Mukherjee et al. 2021). Bio-based reduction involves a three-step process, starting from the metal ion reduction, followed by nucleation (confirmed through the color changes in the solution). The last step of this process involves the final shaping of the synthesized nanoparticles. Temperature, pH, substrate amount, time, etc., are some of the important parameters responsible for the physical characteristics of synthesized nanoparticles (Chaudhary et al. 2020). The possible extracellular metallic nanoparticle production has been proposed to be attributed to polysaccharides, reducing carbohydrates, proteins, peptides, or other reducing factors present in the algal culture, precipitating reducing metal ions to nanoparticles (Gahlawat and Choudhury 2019). Biological resources can be used for nanoparticle synthesis either extracellularly or intracellularly.

4.1 Algal Intracellular Material-Based Synthesis of Nanoparticles

Intracellular synthesis is the process that occurs inside the cell. There is no pre-treatment of microalgae needed, as it relies on metabolic pathways like photosynthesis, respiration, and nitrogen fixation (Khanna et al. 2019b). This process generates reducing agents that aid in the reduction of ions within the plasma membrane, cytoplasm, and thylakoid (Sharma et al. 2016). Blue-green algae are known for the intracellular synthesis of metal nanoparticles. The mechanism behind metal nanoparticle synthesis involves the uptake and reduction of metal ions through NADPH-dependent reductase enzyme, generated via the electron transport system (ETS) either as a result of photosynthesis or through the oxidation reaction in the respiratory ETS at the chloroplast, cell membrane, and cytoplasm (Jeffryes et al. 2015). Stabilization and capping of nanoparticles are done by polysaccharides and other macromolecules (Dahoumane et al. 2014). An example of intracellular gold nanoparticle (Au-NP) synthesis can be observed in *Spirogyra submaxima*, where the bioconversion of Au^{3+} to Au was found to be highly efficient. Upon exposure to a gold solution, the biomass exhibited a distinct pinkish-purple coloration, indicating nanoparticle formation. The extraction of Au-NPs was facilitated using sodium citrate as a capping agent. Additionally, the

Table 2: List of metallic nanoparticles from microalgae, their type, size, and potential applications.

Algae	Nanoparticle	Size (nm)	Morphology	Application	Bio reduction agent	References
Coelastrella aeroterrestrica	Silver nanoparticles (Ag-NPs)	14.5	Hexagonal	Antimicrobial, Anticancer and Antioxidant Agent	proteins	(Hamida et al. 2022)
Spirulina platensis	Gold nanoparticles (Au-NPs)	22.18 ± 2	Triangular, spherical	Antimicrobial, Antioxidant, and Anticancer Agent	Protein and carboxylic acid	(Aamir et al. 2025)
Chlorococcum sp.	Iron nanoparticle (Fe- NPs)	20–50	Spherical	Remediation of toxic chromium (Cr^{+6}) from wastewater	Protein and carbohydrates	(Subramaniyam et al. 2015)
Synechocystis sp	Fe_2O_3 composites	0.2	Spherical	Removal of a range of heavy metal	OH, group and COOH group	(Shen et al. 2020)
Chlorella vulgaris *Microchaete* NCCU-342 *Galaxaura elongata* *Chlorella vulgaris* *Cystoseira myrica* *Chlorella vulgaris*	Gold nanoparticle (Au-NP) Silver nanoparticle (Ag-NP) Gold nanoparticle (Au-NP) Pd nanoparticle (Pd-NP) Zinc nanoparticle (Zn-NP) Ag-$CuFe_2O_4$	9–20 60–80 3.85–77.13 5–20 0.3–20 20nm	Spherical Spherical, polydisperse Triangular Spherical, crystalline Spherical Spherical	catalysis, antimicrobial properties Degradation of azo dye methyl red. Antibacterial activity Bio reduction Antioxidant, antidiabetic antibacterial	Biomolecules (amino, carboxylic, phosphate, thiol) Amino acid, pigments, phenolic compounds Fatty acid, polyphenols Polyester, polyamide Flavonoids, polysaccharides Amine group of protein	(Wei and Qian 2008) (Husain et al. 2019) (Abdel-Raouf et al. 2017) (Arsiya et al. 2017) (Diab et al. 2024) (Kahzad and Salehzadeh 2020)

presence of crystalline gold was confirmed through Bragg reflections observed in the purple-colored biomass, further supporting the intracellular synthesis of gold nanoparticles (Khanna et al. 2019a).

4.2 Algal Extracellular Material-Based Synthesis of Nanoparticles

Extracellular synthesis of nanoparticles using algae is considered an environmentally conscious approach, leveraging the secreted biomolecules from algal cells to mediate nanoparticle formation. Extracellular synthesis involves the reduction of metal ions attached to the algal surface by exudate bioactive compounds like peptides, antioxidants, pigments, enzymes, etc., as exhibited in Table 3.

The biomass was removed via centrifugation, followed by the addition of a metal precursor solution (Vijayan et al. 2014). This reduction process is facilitated by the biochemical activities of the metabolites, which donate electrons to the metal ions and lead to the formation of nanoparticles (Fayyad et al. 2024).

4.3 Harnessing Bio-Nanomaterials for Efficient Remediation of Heavy Metals

Nanoparticle/nanocomposites synthesis from different organisms like plants, algae, fungi, etc., enhances the microbes' properties by enhancing the microbial activities, which help in the elimination of pollutants. This technique is called nano-bioremediation (Malik et al. 2022). The basic concept of metal–microbe interaction has fueled the production of metal-mediated bio-nanoparticles (Sunanda et al. 2022). The advantage of bionanotechnology over conventional techniques involves less manpower, capital, and versatile and authentic techniques for the development and functionalization of bionanomaterials. These nanoparticles can act as nanocatalysts, nanofilters, or be exploited for other functions. Adenigba et al. (2020) synthesized highly stable polydisperse spherical gold nanoparticles using algal biomass of *Nannochloropsis* sp. and *Chlorella vulgaris* for their adsorption efficiency of Zn and Pb from the discharge of the pharma industry. Secondary metabolites present in algal biomass serve as reducing and capping agents for the nanomaterial. Adsorption efficiency shown by *Chlorella vulgaris* and *Nannochloropsis* sp.-based gold nanoparticles for the removal of zinc showed 66.83% and 60.32% removal efficiency, respectively. Additionally, for the removal of lead through *Chlorella vulgaris* and *Nannochloropsis* sp.-based nanoparticles, the removal

Table 3: Synthesis conditions of bio-nanoparticles and their manufacturing location.

Algae	NP	Location	Synthesis conditions	References
Gracilaria edulis	Ag	Extracellular	The sample maintained at 40^0 C with continuous shaking	(Chaudhary et al. 2020b)
Colpmenia sinusa	Ag	Extracellular	Continuous agitations in alkaline pH for 20 min at 70^0 C	(El-Rafie et al. 2013)
Chlorella vulgaris	Ag	Extracellular	Sample maintained at 40^0 C	(Xie et al. 2007)
Phormidium tenue	Au	Intracellular	Incubation at 27^0 C	(MubarakAli et al. 2012)
Tetraselmis kochinensis	Au	Intracellular	Sample maintained for 48h at room temperature on a 200 rpm shaker	(Senapati et al. 2012)
Coelastrella sp.	Au	Intracellular	Dark period for 1h centrifugation for 7min at 6000 rpm	(MubarakAli et al. 2013)
Chlorella sp.	ZnO	Extracellular	Incubation at 58^0 C for 1hr with rapid stirring at 150rpm	(Khalafi et al. 2019)
Anabaena cylindrical	CuO	Extracellular	Continuous stirring at 60^{0}C and dried overnight at 100^{0}C	(Bhattacharya et al. 2019)
Ulva lactuca	Carbon black	Extracellular	Hydrolysis, carbonization, and pyrolysis	(Elmaghraby et al. 2024)

efficiency was about 57.41% and 66.53%, respectively. This high adsorption capacity shown by algal nanoparticles is due to the presence of functional groups on them (Adenigba et al. 2020). Biosynthesis of silver nanoparticles was accomplished using an algal extract of isolated *Scenedesmus* sp. for the removal of the heavy metals Pb and Cd. Synthesis of Ag nanoparticles was confirmed by UV-visible adsorption peak at 411 nm, and maximum adsorption occurred at pH 8 with an adsorption dose of $0.5 gL^{-1}$ (Negi et al. 2021).

Khilji et al. (2022) utilized *Oedogonium* sp. for the biosynthesis of ZnO nanoparticles for the treatment of effluent released from the leather industry and heavy metal removal. These were subsequently used for the reduction of BOD, COD, TDS, and harmful heavy metals present in the eluent. The mechanism that enabled the removal of heavy metals involved adsorption, complexation, electrostatic attraction, hydrogen bonding, and precipitation facilitated by their large volume-to-surface area and functional groups from the algal extract.

Mantovani et al. (2023) used an environmentally friendly approach for the removal of heavy metals from the waste stream based on the synthesis of carbon-based iron nanoparticles (ME nFe) through a hydrothermal synthesis process. Different iron sources like iron (III) nitrate nonahydrate and ammonium iron (III) sulfate dodecahydrate were used for the development of nanoparticles. Jar tests, a common laboratory practice, included optimum conditions like variation in pH and dose of coagulant for wastewater treatment, and showed high removal efficiencies for copper (99.6%), zinc (97.8%), cadmium (96.4%), and nickel (80.3%) at a starting concentration of 1 mg/L. The main mechanism involved was a combination of adsorption and reduction, aided by the iron-carbon structure. Carbon-encapsulated iron nanoparticles were synthesized from de-oiled algal biomass for the removal of pharmaceutically active compounds, as reported by Mantovani et al. (2024) through a hydrothermal carbonization process. Synthesized nanoparticles showed high adsorption capacity for the removal of antibiotics, painkillers, and other compounds. Encapsulation with carbon enhanced surface area for efficient adsorption of the compounds. Nanoparticles retained efficiency for multiple rounds of utilization.

Previous studies demonstrated that the adsorption capacity of iron and manganese sulfide can be increased by the addition of carbon quantum dots synthesized from algal biomass for the efficient removal of Cr ions from waste. The research demonstrates that the incorporation of 1 wt% carbon quantum dots improved the Cd^{2+} adsorption capacity, reaching 481 mg/g at 25°C and 648.6 mg/g at 45°C, surpassing many previously reported other metal sulfides (Wang et al. 2023). Microalgae are also used as bioadsorbents for the recovery of rare earth elements from waste. In this regard, *Ulva* sp. biomass was utilized for the synthesis of bioadsorbent to recover lanthanum, neodymium, and dysprosium by the NaOH-assisted method. The recovery rate was as follows: 90.71%, 95.12%, and 93.83% (Manikandan et al. 2022). Synthesis of *G. sulphuraria* beads for the removal of La, Y, and Sm from ammonia-rich wastewater has also been attempted. These immobilized beads act as biosorbents due to the presence of different functional groups, which provide several active sites for adsorption (Sun et al. 2022). Algae can be used in a variety of ways for the bioremediation of heavy metals. They can be transformed into bioadsorbents or bio-nanomaterials, which promote environmentally friendly and sustainable approaches for the removal of toxic contaminants.

4.4 Bio-Nano Remediation of Microplastics

Various methods are employed for the removal of microplastics, including physical (filtration, adsorption, coagulation), chemical (photoactive catalysis), and biological (biodegradation) approaches. However, the removal efficiency of these particles remains between 40% and 90%, depending upon the size of MPs (Li et al. 2022). Bio-nanomaterials emerge as a promising tool for the removal and degradation of microplastics from different waste streams. Many studies have been done on the synthesis of bionanomaterials from different biological sources like plant extract, bacterial extract, and plant nanocomposites, for the elimination of microplastics. Many investigations on carbon-based nanoparticles like graphite nanotubes, carbon nanotubes, and flu dust have been done for the removal of MPs (Sajid et al. 2023). Carbon-based bioadsorbents prove to be efficient for the removal of MPs; in this regard, nanocellulose has gained immense attention. Properties of nanocellulose, like strong adsorption capacity and large surface area for the retention of microplastic, make it efficient for the removal of these particles (Burhani et al. 2024). Previous studies showed the synthesis of plant-based nanocellulose to capture colloidal micro- or nanoplastics from aquatic environments. The hygroscopic nature and large surface area enhance the entrapment of MPs on the surface of nanocellulose fibrils. The study also highlighted that a lower anionic charge increases the adsorption capacity of nanocellulose fibrils (Leppänen et al. 2022). Synthesis of Fe-mediated nanoparticles from *Anthemis pseudocotula* removed 92–93% of MPs from marine water (Abdullah et al. 2018). Carbon-based iron oxide nanoparticle synthesis through the microwave-assisted method from glucose and Fe_2O_3 solution acted as a good adsorbent and showed good efficiency in MPs removal (Elmaci 2020). Removal of polyethylene (PE) microplastic from water through the synthesis of Black Tea Magnetic nanoparticles has been previously demonstrated. The mechanism behind the removal involves the adhesion of the nanoparticle on the microplastic surface, which removes 90% of MPs from the water, and the nanoparticle can be recycled for the next round (Abbasi et al. 2024). Some studies also suggested the generation of a carbon nanoparticle-based water filtration system for the effective removal of microplastics and microorganisms present in water. This study serves as an innovative approach to the water purification system (Bano and Gupta 2023). Previously, Lu et al. (2023) developed a bacterial cellulose composite that acts as a flocculant for the elimination of polyethylene terephthalate (PET) micro- and nanoplastics, demonstrating 90% removal efficiency. The high removal efficiency is attributed to the extensive presence of amino groups in the composite, which act as adsorbents. Peydayesh et al. (2021) synthesized bionanomaterials from lysozyme amyloid fibrils, which act as a natural bio-flocculant for the remediation of MPs and metal contaminants with greater efficiency. Biosynthesis of ZnO nanoparticles from areca nut demonstrated multiple applications along with photocatalytic degradation of LDPE. In the presence of sunlight and ZnO-NP, LDPE shows 67% weight loss (Raghavendra et al. 2022). Synthesis of Cd-Mg ferrite nanocomposite decorated with titanium dioxide from the extract of *M. koenigii* was found to be effective in the elimination of plastic additives under variable lab conditions. The addition of 20 mg of composite dose degraded 2 mg/L of BPA (94%) and TBBPA (91%) and could be reused for up to 8 cycles without any damage. This shows the sustainability and eco-friendly nature of green synthesis for the removal and degradation of microplastics. Green synthesis of iron-doped ZnO nanoparticles from the leaves of *Hibiscus rosa-sinensis* for the degradation of low-density polyethylene under the irradiation of sunlight was attempted. Photo-degradation results highlighted 41.3% weight loss for 120 hr sunlight exposure as compared to undoped ZnO nanoparticles. Fe doping enhances optical density and charge carrier separation, which increases photocatalytic LDPE degradation, confirmed by the formation of a carbonyl group (Lam et al. 2021). Microalgal biomass also acts as an effective adsorbent for the biodegradation of MPs due to their ability to secrete extracellular polymeric compounds, which surround the surface of MPs and may result in their disintegration (Dar et al. 2025). From previous studies, it has been found that bionanomaterials have immense potential for the removal of microplastics and heavy metals from various wastes. Algae can be utilized as a promising source for the synthesis of nanomaterials, which will aid in the efficient removal of MPs and heavy metals without affecting the surrounding environment.

5. Conclusion

This chapter provides a holistic overview of the role of bio-nanomaterials in environmental pollutant remediation. Among distinct types of biologically synthesized nanoparticles, algal-based nanoparticles have emerged as the most promising and effective nanomaterials for the remediation of toxic pollutants. This chapter also widens the avenue for understanding the necessity of biocompatibility, biodegradability, and minimal ecological impact to develop such strategies. The role of algae-based nanomaterials highlights their potential specifically to remediate lanthanides and microplastics, emphasizing the sustainability, efficiency, and versatility of microalgae.

6. Acknowledgments

The author(s) are grateful for the motivation and guidance received from colleagues and mentors throughout the preparation of this book chapter. The author(s) also give special thanks to the UGC endorsed UGC-BSR SRG F. 30-528/2020 (BSR) and DST endorsed BT/PR45985/BCE/8/1777/2023 projects for the support. Author(s) also acknowledge the support received from the Department of Environmental Science, School of Earth Sciences, Central University of Rajasthan.

References

Aamir, M., Hassan, S., Khan, A. H., Ibrar, M., Sarwar, S., Mahmood, K. et al. (2025). Spirulina-mediated biosynthesis of gold nanoparticles: an interdisciplinary study on antimicrobial, antioxidant, and anticancer properties. *J. Sol-Gel Sci. Technol.* https://doi.org/10.1007/s10971-024-06649-6.

Abbasi, I. A., Aurangzeb, J., Ri, P. D., Soomro, A. F., and Tae, Y. I. (2024). Green Synthesis of Magnetic NanoparticlesUsing Black Tea Extract (BTE) to RemovePolyethylene (PE) Microplastics from Water. *Pol. J. Environ. Stud.*, https://doi.org/10.15244/pjoes/193133.

Abdel-Raouf, N., Al-Enazi, N. M., and Ibraheem, I. B. M. (2017). Green biosynthesis of gold nanoparticles using Galaxaura elongata and characterization of their antibacterial activity. *Arab. J. Chem.*, 10: S3029–S3039. https://doi.org/10.1016/j.arabjc.2013.11.044.

Abdullah, M. M. S., Atta, A. M., Allohedan, H. A., Alkhathlan, H. Z., Khan, M., and Ezzat, A. O. (2018). Green Synthesis of Hydrophobic Magnetite Nanoparticles Coated with Plant Extract and Their Application as Petroleum Oil Spill Collectors. *Nanomaterials*, 8: 855. https://doi.org/10.3390/nano8100855.

Adenigba, V., Omomowo, I., Oloke, J., Fatukasi, B., Odeniyi, M., and Adedayo, A. (2020). Evaluation of microalgal-based nanoparticles in the adsorption of heavy metals from wastewater. *IOP Conf. Ser. Mater. Sci. Eng.*, 805: 012030. https://doi.org/10.1088/1757-899X/805/1/012030.

Agarwal, P., Gupta, R., and Agarwal, N. (2019). Advances in Synthesis and Applications of Microalgal Nanoparticles for Wastewater Treatment. *J. Nanotechnol.*, 2019: 1–9. https://doi.org/10.1155/2019/7392713.

Ahmed, S., Saifullah, Ahmad, M., Swami, B. L., and Ikram, S. (2016). Green synthesis of silver nanoparticles using Azadirachta indica aqueous leaf extract. *J. Radiat. Res. Appl. Sci.*, 9: 1–7. https://doi.org/10.1016/j.jrras.2015.06.006.

Ahmed, S. F., Mofijur, M., Rafa, N., Chowdhury, A. T., Chowdhury, S., Nahrin, M. et al. (2022). Green approaches in synthesising nanomaterials for environmental nanobioremediation: Technological advancements, applications, benefits and challenges. *Environ. Res.*, 204: 111967. https://doi.org/10.1016/j.envres.2021.111967.

Ajao, A. (2024). Conventional and Advanced Treatment Technologies for Microplastics in Water Treatment Facilities. https://doi.org/10.20944/preprints202406.0075.v1

Arsiya, F., Sayadi, M. H., and Sobhani, S. (2017). Green synthesis of palladium nanoparticles using *Chlorella vulgaris. Mater. Lett.*, 186: 113–115. https://doi.org/10.1016/j.matlet.2016.09.101.

Auta, H. S., Emenike, C. U., and Fauziah, S. H. (2017). Distribution and importance of microplastics in the marine environment: A review of the sources, fate, effects, and potential solutions. *Environ. Int.*, 102: 165–176. https://doi.org/10.1016/j.envint.2017.02.013.

Bano, S., and Gupta, P. (2023). Advanced Carbon Nanoparticle-Based Filtration Systems for Water Disinfection and Microplastics Removal. *Migr. Lett.*, 20: 151–160. https://doi.org/10.59670/ml.v20iS13.6279.

Bautista, R.G. (1992). Advances in Rare Earths Separation and Purification. Miner. *Process. Extr. Metall. Rev.*, 8: 175–182. https://doi.org/10.1080/08827509208952685.

Bayliss, P., and Levinson, A. A. (1988). A system of nomenclature fro rare-earth mineral species. *American Mineralogist*, 73(3-4): 422–423.

Brouziotis, A. A., Giarra, A., Libralato, G., Pagano, G., Guida, M., and Trifuoggi, M. (2022a). Toxicity of rare earth elements: An overview on human health impact. *Front. Environ. Sci.*, 10: 948041. https://doi.org/10.3389/fenvs.2022.948041.

Brouziotis, A. A., Giarra, A., Libralato, G., Pagano, G., Guida, M., and Trifuoggi, M. (2022b). Toxicity of rare earth elements: An overview on human health impact. *Front. Environ. Sci.*, 10: 948041. https://doi.org/10.3389/fenvs.2022.948041.

Burhani, D., Voet, V. S. D., Folkersma, R., Maniar, D., and Loos, K. (2024). Potential ofNanocellulose for Microplastic removal: Perspective and challenges. *Tetrahedron Green Chem.*, 3: 100045. https://doi.org/10.1016/j.tgchem.2024.100045.

Burt, D. M. (1989). Compositional and phase relations among rare-earth element minerals. *Journal of Geochemical Exploration*, 32(1–3): 1–25. https://doi.org/10.1016/0375-6742(89)90001-0.

Cesbron, F. (1989). Mineralogy and geochemistry of rare earth elements. (pp. 45–70). *In*: Henderson, P. (Ed.). Rare Earth Element Geochemistry. Elsevier.

Chandran, R. R., Thomson, B. I., Natishah, A. J., Mary, J., and Nachiyar, V. (2023). Nanotechnology in Plastic Degradation. *Biosci. Biotechnol. Res. Asia*, 20: 53–68. https://doi.org/10.13005/bbra/3068.

Chartres, N., Cooper, C. B., Bland, G., Pelch, K. E., Gandhi, S. A., BakenRa, A. et al. (2024). Effects of Microplastic Exposure on Human Digestive, Reproductive, and Respiratory Health: A Rapid Systematic Review. *Environ. Sci. Technol.* acs.est.3c09524. https://doi.org/10.1021/acs.est.3c09524.

Chaudhary, R., Nawaz, K., Khan, A. K., Hano, C., Abbasi, B. H., and Anjum, S. (2020). An Overview of the Algae-Mediated Biosynthesis of Nanoparticles and Their Biomedical Applications. *Biomolecules*, 10: 1498. https://doi.org/10.3390/biom10111498.

Chellasamy, G., Kiriyanthan, R. M., Maharajan, T., Radha, A., and Yun, K. (2022). Remediation of microplastics using bionanomaterials: A review. *Environ. Res.*, 208: 112724. https://doi.org/10.1016/j.envres.2022.112724.

Choudhary, R., Parveen, S., and Patidar, S. K., (2023). Ecotoxicology of REEs in Aquatic Macrophytes and Prospect for Bioremediation of REEs. pp. 139–158. *In:* Kumar, S., Bauddh, K., Singh, R., Kumar, N., and Kumar, R. (Eds.). Aquatic Macrophytes: Ecology, Functions and Services. Springer Nature Singapore, Singapore, https://doi.org/10.1007/978-981-99-3822-3_7.

Clark, A. M. (1984). Mineralogy of the rare earth elements. pp. 33–54. *In*: Ribbe, P. H. (Ed.). Reviews in Mineralogy. Mineralogical Society of America.

Clark, A. M. (1993). Rare earth minerals and their properties. Mineralogical Magazine, 57(389): 297–312. https://doi.org/10.1180/minmag.1993.057.389.04.

Culicov, O. A., Trtić-Petrović, T., Balvanović, R., Petković, A., and Ražić, S. (2021). Spatial distribution of multielements including lanthanides in sediments of Iron Gate I Reservoir in the Danube River. *Environ. Sci. Pollut. Res.*, 28: 44877–44889. https://doi.org/10.1007/s11356-021-13752-6.

Dahoumane, S. A., Wijesekera, K., Filipe, C. D. M., and Brennan, J. D. (2014). Stoichiometrically controlled production of bimetallic Gold-Silver alloy colloids using micro-alga cultures. *J. Colloid Interface Sci.*, 416: 67–72. https://doi.org/10.1016/j.jcis.2013.10.048.

Dahoumane, Si Amar, Mourad, Mechouet, Alvarez, Francisco, Agathos, Spiros and Jeffryes, Clyton. (2016). Microalgae: An outstanding tool in nanotechnology. *Revista Bionatura*, 1(4). https://dio.org/10.21931/RB/206.01.04.7.

Dar, M. A., Palsania, P., Satya, S., Dashora, M., Bhat, O.A., Parveen, S. et al. (2025). Microplastic pollution: A global perspective in surface waters, microbial degradation, and corresponding mechanism. *Mar. Pollut. Bull.*, 210: 117344. https://doi.org/10.1016/j.marpolbul.2024.117344.

Dev, S., Sachan, A., Dehghani, F., Ghosh, T., Briggs, B. R., and Aggarwal, S. (2020). Mechanisms of biological recovery of rare-earth elements from industrial and electronic wastes: A review. *Chem. Eng. J.*, 397: 124596. https://doi.org/10.1016/j.cej.2020.124596.

Dhanalekshmi, K. I., Magesan, P., Umapathy, M. J., Zhang, X., Srinivasan, N., and Jayamoorthy, K. (2021). Enhanced photocatalytic and photodynamic activity of chitosan and garlic loaded CdO–TiO2 hybrid bionanomaterials. *Sci. Rep.*, 11: 20790. https://doi.org/10.1038/s41598-021-00242-5.

Diab, A. S., Alqhtani, H. A., Bin-Jumah, M., Rudayni, H. A., Elsayed, K. N. M., Al Zoubi, W. et al. (2024). Antioxidant and antidiabetic potential of Cystoseira myrica-mediated green metallic zinc nanoparticles using the algal extract: synthesis and characterization. *Mater. Res. Express*, 11: 115402. https://doi.org/10.1088/2053-1591/ad8e2f.

Egler, S. G., Niemeyer, J. C., Correia, F. V., and Saggioro, E. M. (2022). Effects of rare earth elements (REE) on terrestrial organisms: current status and future directions. *Ecotoxicology*, 31: 689–699. https://doi.org/10.1007/s10646-022-02542-6.

Elmaci, G. (2020). Microwave-assisted rapid synthesis of C@Fe3O4 composite for removal of microplastics from drinking water. *Adıyaman Univ. J. Sci.* https://doi.org/10.37094/adyujsci.739599.

Fan, Z., Zhang, K., Wang, F., Zhao, X., Bai, R., and Liu, B. (2020). Effects of Rare Earth Elements on Growth and Determination of Secondary Metabolites under In Vitro Conditions in *Salvia miltiorrhiza. HortScience,* 55: 310–316. https://doi.org/10.21273/HORTSCI14661-19.

Fayyad, R. J., Ali, A. N. M., Ali, T. R., and Faris, S. A. (2024). Extracellular Algae based Nanoparticles, Overview, Mechanisms and some Medical Applications. *South Asian Res. J. Biol. Appl. Biosci.*, 6: 114–119. https://doi.org/10.36346/sarjbab.2024.v06i04.003.

Fleischer, M., and Jones, R. L. (1996). Glossary of mineral species (7th ed.). Mineralogical Record Inc.

Gahlawat, G., and Choudhury, A. R. (2019). A review on the biosynthesis of metal and metal salt nanoparticles by microbes. *RSC Adv.* 9: 12944–12967. https://doi.org/10.1039/C8RA10483B.

Gaines, R. V., Skinner, H. C. W., Foord, E. E., Mason, B., and Rosenzweig, A. (1997). Dana's New Mineralogy (8th ed.). John Wiley & Sons.

Gao, S., Zhang, S., Feng, Z., Lu, J., Fu, G., and Yu, W. (2024). The bio–accumulation and –magnification of microplastics under predator–prey isotopic relationships. *J. Hazard. Mater.*, 480: 135896. https://doi.org/10.1016/j.jhazmat.2024.135896.

García Rellán, A., Vázquez Ares, D., Vázquez Brea, C., Francisco López, A., and Bello Bugallo, P. M. (2023). Sources, sinks and transformations of plastics in our oceans: Review, management strategies and modelling. *Sci. Total Environ.*, 854: 158745. https://doi.org/10.1016/j.scitotenv.2022.158745.

Guo, Z., Li, J., and Zhang, Z. (2024). Meta-analysis for systematic review of global micro/nano-plastics contamination versus various freshwater microalgae: Toxicological effect patterns, taxon-specific response, and potential eco-risks. *Water Res.*, 258: 121706. https://doi.org/10.1016/j.watres.2024.121706.

Guzzetti, E., Sureda, A., Tejada, S., and Faggio, C. (2018). Microplastic in marine organism: Environmental and toxicological effects. *Environ. Toxicol. Pharmacol.*, 64: 164–171. https://doi.org/10.1016/j.etap.2018.10.009.

Gwenzi, W., Mupatsi, N. M., Mtisi, M., and Mungazi, A. A. (2021). Sources and Health Risks of Rare Earth Elements in Waters, pp. 1–36. *In:* Inamuddin, Ahamed, M. I., and Lichtfouse, E. (Eds.). Water Pollution and Remediation:

Heavy Metals, Environmental Chemistry for a Sustainable World. Springer International Publishing, *Cham.* https://doi.org/10.1007/978-3-030-52421-0_1.

Hamida, R. S., Ali, M. A., Almohawes, Z. N., Alahdal, H., Momenah, M. A., and Bin-Meferij, M. M. (2022). Green Synthesis of Hexagonal Silver Nanoparticles Using a Novel Microalgae Coelastrella aeroterrestrica Strain BA_Chlo4 and Resulting Anticancer, Antibacterial, and Antioxidant Activities. *Pharmaceutics,* 14: 2002. https://doi.org/10.3390/pharmaceutics14102002.

Hissler, C., Stille, P., Iffly, J. F., Guignard, C., Chabaux, F., and Pfister, L. (2016). Origin and Dynamics of Rare Earth Elements during Flood Events in Contaminated River Basins: Sr–Nd–Pb Isotopic Evidence. *Environ. Sci. Technol.*, 50: 4624–4631. https://doi.org/10.1021/acs.est.5b03660.

Husain, S., Afreen, S., Hemlata, Yasin, D., Afzal, B., and Fatma, T. (2019). Cyanobacteria as a bioreactor for synthesis of silver nanoparticles-an effect of different reaction conditions on the size of nanoparticles and their dye decolorization ability. *J. Microbiol. Methods*, 162: 77–82. https://doi.org/10.1016/j.mimet.2019.05.011.

Jain, A. (2022). Algae-mediated synthesis of biogenic nanoparticles. *Adv. Nat. Sci. Nanosci. Nanotechnol.*, 13: 043001. https://doi.org/10.1088/2043-6262/ac996a.

Jeffryes, C., Agathos, S. N., and Rorrer, G. (2015). Biogenic nanomaterials from photosynthetic microorganisms. *Curr. Opin. Biotechnol.*, 33: 23–31. https://doi.org/10.1016/j.copbio.2014.10.005.

Kahzad, N., and Salehzadeh, A. (2020). Green Synthesis of CuFe2O4@Ag Nanocomposite Using the Chlorella vulgaris and Evaluation of its Effect on the Expression of norA Efflux Pump Gene Among *Staphylococcus aureus* Strains. *Biol. Trace Elem. Res.*, 198: 359–370. https://doi.org/10.1007/s12011-020-02055-5.

Kaur, S., Lal, M., and Sharma, P. (2024). Nanomaterials for Environmental Remediation, pp. 1–30. *in:* Sharma, P., and Khanna, V. (Eds.), Nanomaterials: An Approach Towards Environmental Remediation. *Bentham Science Publishers.* https://doi.org/10.2174/97898152236131240l0004.

Khan, M. T., Ahmad, M., Hossain, M. F., Nawab, A., Ahmad, I., Ahmad, K. et al. (2023). Microplastic removal by coagulation: a review of optimizing the reaction conditions and mechanisms. *Water Emerg. Contam. Nanoplastics,* 2. https://doi.org/10.20517/wecn.2023.39.

Khanna, P., Kaur, A., and Goyal, D. (2019a). Algae-based metallic nanoparticles: Synthesis, characterization and applications. J. Microbiol. Methods 163, 105656. https://doi.org/10.1016/j.mimet.2019.105656.

Khanna, P., Kaur, A., and Goyal, D. (2019b). Algae-based metallic nanoparticles: Synthesis, characterization and applications. J. Microbiol. Methods 163: 105656. https://doi.org/10.1016/j.mimet.2019.105656.

Khilji, S. A., Munir, N., Aziz, I., Anwar, B., Hasnain, M., Jakhar, A. M. et al. (2022). Application of Algal Nanotechnology for Leather Wastewater Treatment and Heavy Metal Removal Efficiency. Sustainability 14: 13940. https://doi.org/10.3390/su142113940.

Lam, S.-M., Sin, J.-C., Zeng, H., Lin, H., Li, H., Chai, Y.-Y. et al. (2021). Green synthesis of Fe-ZnO nanoparticles with improved sunlight photocatalytic performance for polyethylene film deterioration and bacterial inactivation. Mater. Sci. Semicond. Process. 123: 105574. https://doi.org/10.1016/j.mssp.2020.105574.

Leigh, R. J. (1990). Rare earth element mineral deposits: Geology and Mineral. *Economic Geology*, 85(3): 483–498.

Leppänen, I., Lappalainen, T., Lohtander, T., Jonkergouw, C., Arola, S., and Tammelin, T. (2022). Capturing colloidal nano- and microplastics with plant-based nanocellulose networks. *Nat. Commun.* 13: 1814. https://doi.org/10.1038/s41467-022-29446-7.

LewisOscar, F., Vismaya, S., Arunkumar, M., Thajuddin, N., Dhanasekaran, D., and Nithya, C. (2016). Algal Nanoparticles: Synthesis and Biotechnological Potentials. *In*: Thajuddin, N., Dhanasekaran, D. (Eds.). Algae - Organisms for Imminent Biotechnology. InTech. https://doi.org/10.5772/62909.

Li, W., Wu, C., Xiong, Z., Liang, C., Li, Z., Liu, B. et al. (2022). Self-driven magnetorobots for recyclable and scalable micro/nanoplastic removal from nonmarine waters. *Sci. Adv.*, 8: eade1731. https://doi.org/10.1126/sciadv.ade1731.

Li, Y., Tao, L., Wang, Q., Wang, F., Li, G., and Song, M. (2023). Potential Health Impact of Microplastics: A Review of Environmental Distribution, Human Exposure, and Toxic Effects. *Environ. Health*, 1: 249–257. https://doi.org/10.1021/envhealth.3c00052.

Liu, S.-L., Fan, H.-R., Liu, X., Meng, J., Butcher, A. R., Yann, L. et al. (2023). Global rare earth elements projects: New developments and supply chains. *Ore Geol. Rev.*, 157: 105428. https://doi.org/10.1016/j.oregeorev.2023.105428.

Lu, H., Sun, S., Sun, J., Peng, X., Li, N., Wajid Ullah, M. et al. (2023). Sustainable production of flocculant-containing bacterial cellulose composite for removal of PET nano-plastics. *Chem. Eng. J.*, 469: 143848. https://doi.org/10.1016/j.cej.2023.143848.

Malik, S., Kishore, S., Shah, M. P., and Kumar, S. A. (2022). A comprehensive review on nanobiotechnology for bioremediation of heavy metals from wastewater. *J. Basic Microbiol.*, 62: 361–375. https://doi.org/10.1002/jobm.202100555.

Manivasagan, P., and Kim, S. (2015). Biosynthesis of Nanoparticles Using Marine Algae: A Review, pp. 295–304. *In:* Kim, S., and Chojnacka, K. (Eds.). *Marine Algae Extracts*. Wiley, https://doi.org/10.1002/9783527679577.ch17.

Mantovani, M., Collina, E., Lasagni, M., Marazzi, F., and Mezzanotte, V. (2023). Production of microalgal-based carbon encapsulated iron nanoparticles (ME-nFe) to remove heavy metals in wastewater. *Environ. Sci. Pollut. Res.*, 30: 6730–6745. https://doi.org/10.1007/s11356-022-22506-x.

Mantovani, M., Collina, E., Passalacqua, E., Lasagni, M., and Mezzanotte, V. (2024). Microalgal-based carbon encapsulated iron nanoparticles for the removal of pharmaceutical compounds from wastewater. *J. Environ. Manage.*, 368: 122171. https://doi.org/10.1016/j.jenvman.2024.122171.

Massari, S., and Ruberti, M. (2013). Rare earth elements as critical raw materials: Focus on international markets and future strategies. *Resour. Policy,* 38: 36–43. https://doi.org/10.1016/j.resourpol.2012.07.001.

Miyawaki, R., and Nakai, I. (1987). Crystal chemistry of rare-earth minerals. *Mineralogy Journal*, 13(5): 283–297.

Moreri, G. B., and Yalei, Z. (2024). Impacts of Microplastics on Photosynthetic Efficiency and Pigment Composition in *Chlorella pyrenoidosa. Eur. J. Theor. Appl. Sci.*, 2: 551–564. https://doi.org/10.59324/ejtas.2024.2(3).42.

MubarakAli, D., Gopinath, V., Rameshbabu, N., and Thajuddin, N. (2012). Synthesis and Characterization of CdS nanoparticles using C-phycoerythrin from the marine cyanobacteria. Materials Letters, 74: 8–11. https://doi.org/10.1016/j.matlet.2012.01.026.

MubarakAli, D., Arunkumar, J., Nag, K. H., SheikSyedIshack, K. A., Baldev, E., Pandiaraj, D. et al. (2013). Gold nanoparticles from Pro and eukaryotic photosynthetic microorganisms—Comparative studies on synthesis and its application on biolabelling. *Colloids and Surfaces B: Biointerfaces*, 103: 166–173. https://doi.org/10.1016/j.colsurfb.2012.10.014.

Mueller, N.C., and Nowack, B. (2010). Nanoparticles for Remediation: Solving Big Problems with Little Particles. *Elements*, 6: 395–400. https://doi.org/10.2113/gselements.6.6.395.

Mukherjee, A., Sarkar, D., and Sasmal, S. (2021). A review of green synthesis of metal nanoparticles using algae. *Front. Microbiol.*, 12: 693899. https://doi.org/10.3389/fmicb.2021.693899.

Negi, S., Singh, V., and Rawat, J. (2021). Green Synthesis of Silver Nanoparticles Using Microalgal Extract and its Application in Metal Ion Removal from Aqueous Solution.

Nguyen, M.-K., Rakib, M. R. J., Lin, C., Hung, N.T.Q., Le, V.-G., Nguyen, H.-L. et al. (2023). A comprehensive review on ecological effects of microplastic pollution: An interaction with pollutants in the ecosystems and future perspectives. TrAC Trends Anal. *Chem.*, 168: 117294. https://doi.org/10.1016/j.trac.2023.117294.

Peydayesh, M., Suta, T., Usuelli, M., Handschin, S., Canelli, G., Bagnani, M. et al. (2021). Sustainable Removal of Microplastics and Natural Organic Matter from Water by Coagulation–Flocculation with Protein Amyloid Fibrils. *Environ. Sci. Technol.*, 55: 8848–8858. https://doi.org/10.1021/acs.est.1c01918.

Premarathna, K. S. D., Lau, S. Y., Chiong, T., Show, P.-L., Vithanage, M., and Lam, M. K. (2024). Greening up the fight against emerging contaminants: algae-based nanoparticles for water remediation. *Clean Technol. Environ. Policy*. https://doi.org/10.1007/s10098-024-02862-7.

Raghavendra, V. B., Shankar, S., Govindappa, M., Pugazhendhi, A., Sharma, M., and Nayaka, S. C. (2022). Green Synthesis of Zinc Oxide Nanoparticles (ZnO NPs) for Effective Degradation of Dye, Polyethylene and Antibacterial Performance in Waste Water Treatment. *J. Inorg. Organomet. Polym. Mater.*, 32: 614–630. https://doi.org/10.1007/s10904-021-02142-7.

Rahman, A., Sarkar, A., Yadav, O. P., Achari, G., and Slobodnik, J. (2021). Potential human health risks due to environmental exposure to nano- and microplastics and knowledge gaps: A scoping review. *Sci. Total Environ.*, 757: 143872. https://doi.org/10.1016/j.scitotenv.2020.143872.

Rajput, P., Benjwal, S., and Pandey, R. (2025). A Comprehensive Review on the Role of Bioremediation in Heavy Metal Contamination. *Nat. Environ. Pollut. Technol.*, 24: 235–245. https://doi.org/10.46488/NEPT.2024.v24iS1.017.

Roslan, N. S., Lee, Y. Y., Ibrahim, Y. S., Tuan Anuar, S., Yusof, K. M. K. K., Lai, L. A. et al. (2024). Detection of microplastics in human tissues and organs: A scoping review. *J. Glob. Health*, 14: 04179. https://doi.org/10.7189/jogh.14.04179.

Roy, A., Elzaki, A., Tirth, V., Kajoak, S., Osman, H., Algahtani, A. et al. (2021). Biological Synthesis of Nanocatalysts and Their Applications. *Catalysts*, 11: 1494. https://doi.org/10.3390/catal11121494.

Sajid, M., Ihsanullah, I., Tariq Khan, M., and Baig, N. (2023). Nanomaterials-based adsorbents for remediation of microplastics and nanoplastics in aqueous media: A review. *Sep. Purif. Technol.*, 305: 122453. https://doi.org/10.1016/j.seppur.2022.122453.

Sathishkumar, R. S., Sundaramanickam, A., Srinath, R., Ramesh, T., Saranya, K., Meena, M. et al. (2019). Green synthesis of silver nanoparticles by bloom forming marine microalgae Trichodesmium erythraeum and its applications in antioxidant, drug-resistant bacteria, and cytotoxicity activity. *J. Saudi Chem. Soc.*, 23: 1180–1191. https://doi.org/10.1016/j.jscs.2019.07.008.

Sharma, A., Sharma, S., Sharma, K., Chetri, S. P. K., Vashishtha, A., Singh, P. et al. (2016). Algae as crucial organisms in advancing nanotechnology: a systematic review. *J. Appl. Phycol.*, 28: 1759–1774. https://doi.org/10.1007/s10811-015-0715-1.

Shen, L., Wang, J., Li, Z., Fan, L., Chen, R., Wu, X. et al. (2020). A high-efficiency Fe2O3@Microalgae composite for heavy metal removal from aqueous solution. *J. Water Process Eng.*, 33: 101026. https://doi.org/10.1016/j.jwpe.2019.101026.

Shera, S. S., and Banik, R. M. (2021). Algal nanoparticles: synthesis and characterization. pp. 25–69. *In*: Thangadurai, D., Sangeetha, J., Prasad, R. (Eds.). Bioprospecting Algae for Nanosized Materials, Nanotechnology in the Life Sciences. Springer International Publishing, *Cham.*, https://doi.org/10.1007/978-3-030-81557-8_2.

Singh, Y. (2020). Rare earth elements: Occurence, mineralogy, and applications. *Journal of the Geological Society of India*, 95(6): 601–612. https://doi.org/10.1007/s12594.

Subramaniyam, V., Subashchandrabose, S. R., Thavamani, P., Megharaj, M., Chen, Z., and Naidu, R. (2015). *Chlorococcum* sp. MM11—a novel phyco-nanofactory for the synthesis of iron nanoparticles. *J. Appl. Phycol.*, 27: 1861–1869. https://doi.org/10.1007/s10811-014-0492-2.

Sun, Yabo, Lu, T., Pan, Y., Shi, M., Ding, D., Ma, Z., Liu, J. et al. (2022). Recovering rare earth elements via immobilized red algae from ammonium-rich wastewater. *Environ. Sci. Ecotechnology*, 12: 100204. https://doi.org/10.1016/j.ese.2022.100204.

Sunanda, Misra, M., and Ghosh Sachan, S. (2022). Nanobioremediation of heavy metals: Perspectives and challenges. *J. Basic Microbiol.*, 62: 428–443. https://doi.org/10.1002/jobm.202100384.

Tiwari, N., Santhiya, D., and Sharma, J. G. (2020). Microbial remediation of micro-nano plastics: Current knowledge and future trends. *Environ. Pollut.*, 265: 115044. https://doi.org/10.1016/j.envpol.2020.115044.

Vianello, A., Jensen, R. L., Liu, L., and Vollertsen, J. (2019). Simulating human exposure to indoor airborne microplastics using a Breathing Thermal Manikin. *Sci. Rep.*, 9: 8670. https://doi.org/10.1038/s41598-019-45054-w.

Vijayan, S. R., Santhiyagu, P., Singamuthu, M., Kumari Ahila, N., Jayaraman, R., and Ethiraj, K. (2014). Synthesis and characterization of silver and gold nanoparticles using aqueous extract of seaweed, *Turbinaria conoides*, and Their Antimicrofouling Activity. *Sci. World J.*, 2014: 1–10. https://doi.org/10.1155/2014/938272.

Wan Ngah, W. S., Teong, L. C., and Hanafiah, M. A. K. M. (2011). Adsorption of dyes and heavy metal ions by chitosan composites: A review. *Carbohydr. Polym.*, 83: 1446–1456. https://doi.org/10.1016/j.carbpol.2010.11.004.

Wang, C., Bi, L., Liu, J., Huang, B., Wang, F., Zhang, Y. et al. (2023). Microalgae-derived carbon quantum dots mediated formation of metal sulfide nano-adsorbents with exceptional cadmium removal performance. *J. Colloid Interface Sci.*, 629: 994–1002. https://doi.org/10.1016/j.jcis.2022.08.188.

Wang, C., Zhao, J., and Xing, B. (2021). Environmental source, fate, and toxicity of microplastics. *J. Hazard. Mater.*, 407: 124357. https://doi.org/10.1016/j.jhazmat.2020.124357.

Wang, W., Yang, Y., Wang, D., and Huang, L. (2024). Toxic Effects of Rare Earth Elements on Human Health: A Review. *Toxics*, 12: 317. https://doi.org/10.3390/toxics12050317.

Wei, D., and Qian, W. (2008). Facile synthesis of Ag and Au nanoparticles utilizing chitosan as a mediator agent. Colloids Surf. B Biointerfaces 62: 136–142. https://doi.org/10.1016/j.colsurfb.2007.09.030.

Wright, S. L., and Kelly, F. J. (2017). Plastic and Human Health: A Micro Issue? *Environ. Sci. Technol.*, 51: 6634–6647. https://doi.org/10.1021/acs.est.7b00423.

Ya, H., Jiang, B., Xing, Y., Zhang, T., Lv, M., and Wang, X. (2021). Recent advances on ecological effects of microplastics on soil environment. *Sci. Total Environ.*, 798: 149338. https://doi.org/10.1016/j.scitotenv.2021.149338.

Yang, Y., Li, X., Wan, C., Zhang, Z., Cao, W., Wang, G. et al. (2024). A comprehensive review of cellulose nanomaterials for adsorption of wastewater pollutants: focus on dye and heavy metal Cr adsorption and oil/water separation. *Collagen Leather*, 6: 35. https://doi.org/10.1186/s42825-024-00179-1.

Yun, J., Yang, Q., and Liu, G. (2024). Rare earth elements unintentionally released from global industrial activities. *J. Hazard. Mater.*, 480: 136146. https://doi.org/10.1016/j.jhazmat.2024.136146.

Zhang, K., Hamidian, A. H., Tubić, A., Zhang, Y., Fang, J. K. H., Wu, C. et al. (2021). Understanding plastic degradation and microplastic formation in the environment: *A review. Environ. Pollut.*, 274: 116554. https://doi.org/10.1016/j.envpol.2021.116554.

Index

V

W

Z

About the Editors

Megha Barot is an Assistant Professor (Research Cadre) in the Department of Environmental Science, Parul Institute of Applied Sciences, Parul University, India. She has over a decade of experience in environmental science, life science research, and clinical diagnostics. She has served as the National Coordinator of the Biotech Innovation Ignition School (BIIS) at SRISTI, in collaboration with DBT–BIRAC, Government of India, and as a Junior Scientist and NABL Internal Auditor at the Neuberg Centre for Genomic Medicine, Ahmedabad. A recipient of the Research Contribution Award from GCPC, she also attended a fully funded training program on Marine Environmental Health and Safety at Xiamen University, China. Dr. Barot has finalized seven academic books with CRC Press, Springer, and CABI, and has numerous research publications to her credit.

Azamal Husen served as a Full Professor and Head of the Department of Biology at the University of Gondar, Ethiopia, and currently serves as a Foreign Delegate at Wolaita Sodo University, Ethiopia. He is also an Adjunct Professor at Graphic Era University, India. Previously, he was a Full Professor at Sankalchand Patel University in India and a Visiting Faculty at the Forest Research Institute and Doon College of Agriculture and Forest in Dehradun, India. With 25 years of expertise in Nanobiotechnology, Plant Physiology, Environmental Botany, Medicinal Plants, and Clonal Propagation, he has led research funded by organizations such as the World Bank, the National Agricultural Technology Project, the Indian Council of Agricultural Research, the Indian Council of Forest Research Education, and the Japan Bank for International Cooperation. He has delivered numerous talks at conferences and workshops, led academic staff orientation courses, and developed BSc, MSc, and PhD curricula. He was honored with an award from the University of Gondar for his decade of service in teaching, research, and community work. Husen has published extensively and served on editorial boards of prestigious journals. Acknowledged as one of the "World's Top 2% Scientists" (2020 onwards) by Stanford University, he has also contributed as a Book Series Editor and Journal Guest Editor for several scientific publications in Plant Biology, Medicinal Plants, and Nanotechnology.

For Product Safety Concerns and Information please contact our EU representative GPSR@taylorandfrancis.com Taylor & Francis Verlag GmbH, Kaufingerstraße 24, 80331 München, Germany